W9-BPS-672

Differential Equations

Differential Equations

An Introduction with *Mathematica*®

Clay C. Ross
The University of the South

Springer-Verlag
New York Berlin Heidelberg London Paris
Tokyo Hong Kong Barcelona Budapest

Textbooks in Mathematical Sciences

Mathematica is a registered trademark of Wolfram Research, Inc.

Library of Congress Cataloging-in-Publication Data

Ross, Clay C.
Differential equations : an introduction with Mathematica / Clay C. Ross.
p. cm. — (Textbooks in mathematical sciences ; 1)
Includes bibliographical references and index.
ISBN 0-387-94301-3
1. Differential equations. 2. Mathematica (Computer file)
I. Title. II. Series.
QA371.R595 1995
515'.35'078—dc20 94-36401
CIP

Printed on acid-free paper.

Production managed by Laura Carlson; manufacturing supervised by Jacqui Ashri.
Composition by Integre Technical Publishing Company, Inc., Albuquerque, NM.
Printed and bound by R.R. Donnelley & Sons, Harrisonburg, VA.
Printed in the United States of America.

9 8 7 6 5 4 3 2 1

ISBN 0-387-94301-3 Springer-Verlag New York Berlin Heidelberg
ISBN 3-540-94301-3 Springer-Verlag Berlin Heidelberg New York

This book is dedicated

To my parents, Vera and Clay Ross,

- *for setting standards for me to live by;*
- *for inspiring me to become a teacher; and*

To my wife, Andrea,

- *for holding to the belief that this book should be completed; and*
- *for encouraging me, to ensure that it would.*

Preface

Goals and Emphasis of the Book

Over the next few years, mathematicians are going to find productive ways to incorporate computing power into the mathematics curriculum. There is no attempt here to avoid doing differential equations. The goal is to make some first explorations in the subject accessible to students who have had one year of calculus. Some of the sciences are now using the symbol-manipulative power of *Mathematica* to make more of their subject accessible. This book is one way of doing so for differential equations.

I believe that if a student's first exposure to a subject is pleasant and exciting, then that student will seek out ways to continue the study of the subject. The theory of differential equations permeates the discussion. Every topic is supported by a statement of the theory. But the primary thrust here is obtaining solutions, rather than proving theorems. There are other courses where proving theorems is central. The goals of this text are to establish a solid understanding of the notion of solution, and an appreciation for the confidence that the theory gives during a search for solutions. Later the student can have the same confidence while personally developing the theory.

When a study of the book has been completed, many important elementary concepts of differential equations will have been encountered. In addition, the use of *Mathematica* makes it possible to solve problems that are formidable without computational assistance. *Mathematica* is an integral part of the presentation, because in introductory differential equations courses it is too often true that simple tasks like finding the roots of a polynomial of relatively high degree—even when the roots are all rational—completely obscure the differential equations that are being studied. The complications encountered in the manual solution of a realistic problem of four first-order linear equations with constant coefficients can totally obscure the beauty and centrality of the theory. But having *Mathematica* available to carry out the complicated steps frees the student to think about what is happening, how the ideas work together, and what everything means.

The text contains many examples. Most are followed immediately by the same example done in *Mathematica*. The form of a *Mathematica* notebook is reproduced almost exactly so that the student knows what to expect when trying problems by him/herself. Having solutions by *Mathematica* included in the text also provides a sort of encyclopedia of working approaches to doing things in *Mathematica*. In addition, each of these examples exists as a real *Mathematica* notebook that can be executed, studied, printed out, or modified to do some other problem. Other *Mathematica* notebooks may be provided by the instructor. Occasionally a problem will request that new methods be tried, but by the time these occur, students should be able to write effective *Mathematica* code of their own.

Mathematica can carry the bulk of the computational burden, but this does not relieve the student of knowing whether or not what is being done is correct. For that reason, periodic checking of results is stressed. Often an independent manual calculation will keep a *Mathematica* calculation safely on course. *Mathematica*, itself, can and should do much of the checking, because as the problems get more complex, the calculations get more and more complicated. A calculation that is internally consistent stands a good chance of being correct when the concepts that are guiding the process are correct.

Since all of the problems except those that are of a theoretical nature can be solved and checked in *Mathematica*, very few of the exercises have answers supplied. As the student solves the problems in each section, they should save the notebooks to disk—where they can serve as an answer book and study guide if the solutions have been properly checked. A *Mathematica* package is a collection of functions that are designed to perform certain operations. Several notebooks depend heavily on a package that has been provided. More often, the functions required appear right in the notebook in which they are needed. This is especially true of functions that are primarily of interest in just one notebook. Most of the packages supplied undertake very complicated tasks, where the functions are genuinely intimidating, so the code does not appear in the text of study notebooks.

Topics Receiving Lesser Emphasis

The solutions of most differential equations cannot be described the way simple functions can be as combinations of elementary functions and operations. The solutions of such equations are often examined numerically. We indicate some ways to have *Mathematica* solve differential equations numerically. Also, properties of a solution are often deduced from careful examination of the differential equation itself, but a study of qualitative differential equations must wait for a more advanced course.

Some differential equations have solutions that are very hard to describe either analytically or numerically because the equations are sensitive to small changes in the initial values. Chaotic behavior is a topic of great current interest; we present some examples of such equations, but do not develop the concepts.

Materials Available to the Instructor and Student

A Solutions Manual and an Instructor's Manual are available from the publisher. The emphasis in the Instructor's Manual is on the effective incorporation of *Mathematica* into the course.

The notebooks and packages referred to in the text are all available by anonymous ftp at ftp.springer-ny.com or on the World Wide Web (WWW) at www.springer-ny.com. There is a README file at the ftp.springer-ny.com site that explains where to put the packages and the notebooks on your own system. The WWW pages give expanded explanations of instructions and descriptions of the various notebooks and packages. Instructors are encouraged to send the author additional notebooks and ideas for notebooks. Students often create exceptionally interesting notebooks; these can be submitted to the author for inclusion. Please contact the author directly for details.

Acknowledgments

I would like to thank Kim Evely Ly for reading several of the chapters in their early stages of development; Jeffrey K. Denny for critically reading the text and for reworking the *Mathematica* supplement; several science department colleagues for enduring questions and for responding so kindly; and Kirk Mathews for valuable suggestions.

Reviews were received from Professors Matthew Richey, Margie Hale, Stephen L. Clark, Stan Wagon, and William Sit. Dave Withoff contributed expert help on technical aspects of *Mathematica* programming. Special thanks go to each of these for their contributions to so many different aspects of the work. Of course, any errors that remain are solely the responsibility of the author.

Sewanee, 1994

Clay C. Ross
cross@sewanee.edu

Contents

1 About Differential Equations and Linear Algebra

1.0 Introduction

What Are Differential Equations? Who Uses Them?

The subject of differential equations is large, diverse, powerful, useful, and full of surprises. Differential equations can be studied on their own—just because they are intrinsically interesting. Or, they may be studied by a physicist, engineer, biologist, economist, physician, or political scientist because they can model (quantitatively explain) many physical or abstract systems. Just what is a differential equation? A differential equation having y as the dependent variable (unknown function) and x as the independent variable has the form

$$F\left(x, y, \frac{dy}{dx}, \dots, \frac{d^n y}{dx^n}\right) = 0$$

for some positive integer n. (If n is 0, the equation is an algebraic or transcendental equation, rather than a differential equation.) Here is the same idea in words:

Definition 1.1 *A* **differential equation** *is an equation that relates in a nontrivial manner an unknown function and one or more of the derivatives or differentials of that unknown function with respect to one or more independent variables.*

The phrase "in a nontrivial manner" is added because some equations that appear to satisfy the above definition are really identities. That is, they are always true, no matter what the unknown function might be. An example of such an equation is:

$$\sin^2\left(\frac{dy}{dx}\right) + \cos^2\left(\frac{dy}{dx}\right) = 1.$$

This equation is satisfied by every differentiable function of one variable. Another example is:

$$\left(\frac{dy}{dx} - y\right)^2 = \left(\frac{dy}{dx}\right)^2 - 2y\left(\frac{dy}{dx}\right) + y^2.$$

This is clearly just the binomial squaring rule in disguise: $(a+b)^2 = a^2+2ab+b^2$; it, too, is satisfied by every differentiable function of one variable. We want to avoid calling such identities differential equations.

One quick test to see that an equation is not merely an identity is to substitute some function such as $\sin(x)$ or e^x into the equation. If the result is ever false, then the equation is not an identity and is perhaps worthy of our study. For example, substitute $y = \sin(x)$ into $y' + y = 0$. The result is $\cos(x) + \sin(x) = 0$, and this is not identically true. (It is false when $x = \pi$, for instance.) If you have a complicated function and are unsure whether or not it is identically 0, you can use *Mathematica* to plot the function to see if it ever departs from 0. This does not constitute a proof, but it is evidence, and it suggests where to look if the function is not identically 0. A plot can be produced this way:

```
In[1]:=  Plot[Cos[x] + Sin[x], {x, 0, 2Pi}];
```

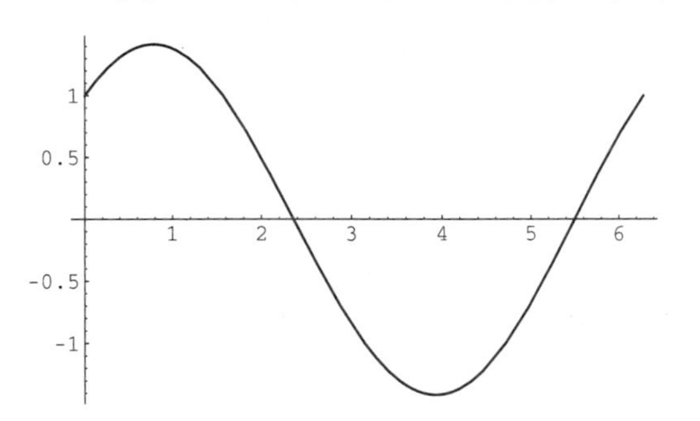

Another extreme that we would like to avoid is an equation that is never true for real functions, such as

$$\left(\frac{dy}{dx}\right)^2 + y^2 = -1.$$

No matter what the real differentiable function y is, the left-hand side of the equation is nonnegative and the right-hand side is negative—and this cannot happen. So the equations we want to study are those that can have some solutions, but not too many solutions. The meaning of this will become clear as we proceed. Unless stated otherwise, the solutions we seek will be real.

Classification of Differential Equations

Differential equations are classified in several different ways: **ordinary** or **partial**; **linear** or **nonlinear**. There are even special subclassifications: **homogeneous** or **nonhomogeneous**; **autonomous** or **nonautonomous**; **first-order**, **second-order**, ..., n**th order**. Most of these names for the various types have been inherited from other areas of mathematics, so there is some ambiguity in the meanings. But the context of any discussion will make clear what a given name means in that context. There are reasons for these classifications, the primary one being to enable discussions about differential equations to focus on the subject matter in a clear and unambiguous manner. Our attention will be on ordinary differential equations. Some will be linear, some nonlinear. Some will be first-order, some second-order, and some of higher order than second. What is the **order** of a differential equation?

Definition 1.2 *The* **order** *of a differential equation is the order of the highest derivative that appears (nontrivially) in the equation.*

At this early stage in our studies, we need only be able to distinguish ordinary from partial differential equations. This is easy: a differential equation is an **ordinary differential equation** if the only derivatives of the unknown function(s) are ordinary derivatives, and a differential equation is a **partial differential equation** if the only derivatives of the unknown function(s) are partial derivatives.

Example 1.1 Here are some ordinary differential equations:

$$\frac{dy}{dt} = 1 + y^2 \qquad \text{(first-order)} \qquad \text{[nonlinear]}$$

$$\frac{d^2y}{dx^2} + y = 3\cos(x) \qquad \text{(second-order)} \qquad \text{[linear, nonhomogeneous]}$$

$$\frac{d^3y}{dx^3} + 3\frac{d^2y}{dx^2} - 5y = 0 \qquad \text{(third-order)} \qquad \text{[linear, homogeneous]}$$

◇

Example 1.2 Here are some partial differential equations:

$$\frac{\partial u}{\partial x} = \frac{\partial u}{\partial y} \qquad \text{(first-order in } x \text{ and } y\text{)}$$

$$\frac{\partial u}{\partial t} = c^2\frac{\partial^2 u}{\partial x^2} \qquad \text{(first-order in } t\text{; second-order in } x\text{)}$$

$$\frac{\partial^2 u}{\partial x^2} + \frac{\partial^2 u}{\partial y^2} = 0 \qquad \text{(second-order in } x \text{ and } y\text{)}$$

$$\frac{\partial^2 u}{\partial x \partial y} = 3 \qquad \text{(second-order)}$$

◇

Solutions of Differential Equations

Definition 1.3 *To say that $y = g(x)$ is a* **solution** *of the differential equation*

$$F\left(x, y, \frac{dy}{dx}, \ldots, \frac{d^n y}{dx^n}\right) = 0$$

on an interval I means that

$$F(x, g(x), g'(x), \ldots, g^n(x)) = 0$$

for every choice of x in the interval I. In other words, a solution, when substituted into the differential equation, makes the equation identically true for x in I.

Example 1.3 The function $y = e^{-x}$ is a solution of the differential equation $y' + y = 0$, because $y' + y = -e^{-x} + e^{-x} = 0$ for all x. ◇

To have *Mathematica* verify this for you, conduct this dialog in an active *Mathematica* window:

```
In[2]:=  Clear[y,x]
         y[x_] = Exp[-x]

Out[2]=   -x
         E

In[3]:=  y'[x] + y[x] = 0

Out[3]=  True
```

The `True` that *Mathematica* returned indicates that $y'(x)+y(x) = 0$ (always), and hence we indeed have a solution. It is not necessary to `Clear` variables regularly, but if you get some unusual behavior, `Clear` the names involved, redefine them, and try the calculation again. *Mathematica* remembers definitions you may have forgotten, and these may interfere with a subsequent calculation.

Here are other examples of solutions of ordinary differential equations. They are from the notebook *Solutions of DE's*. You should execute ideas such as these yourself in *Mathematica.*

```
In[4]:=  y[x_] = c Exp[x^2]

Out[4]=      2
            x
         c E

In[5]:=  Simplify[y'[x]-2 x y[x] == 0]

Out[5]=  True

In[6]:=  Clear[y]
         y[t_] = c1 Sin[a t] + c2 Cos[a t]

Out[6]=  c2 Cos[a t] + c1 Sin[a t]
```

```
In[7]:=  Simplify[y''[t]+a^2 y[t]] == 0

Out[7]=  True
```

Direction Fields and Solutions

The solutions of the first-order differential equation $dy/dx = f(x, y)$ can be represented nicely by a picture. Given a point $P = (x, y)$, the differential equation tells what the slope of the tangent line to a solution is at the point P. If m is such a slope then the differential equation says that

$$m = \left.\frac{dy}{dx}\right|_P = f(P) = f(x, y).$$

The idea of a direction field is similar to that of a vector field, where $f(x, y)$, instead of giving a vector that is to be associated with (x, y), gives a slope that is to be associated with (x, y). If representatives of these slopes are indicated on a graph at enough points, some visual indication of the behavior of the solutions of the differential equation is suggested.

For example, in Figure 1.1 we have plotted some representative members of the direction field associated with the differential equation $dy/dx = (3/2) - 3y + e^{-3x/2}$. Then in Figure 1.2 some solutions of the differential equation are superimposed on the direction field. Notice how the direction field gives a sense of the behavior of the solutions. Solutions may be close together, but they do not cross. You may use the notebook *Direction Fields & Solutions* to produce similar pictures. These can help you understand the behavior of the solutions of any differential equation that has the form $dy/dx = f(x, y)$.

How Many Solutions Are There?

Once we understand that some differential equations have solutions, it is natural to ask several questions. How many solutions can a given differential equation

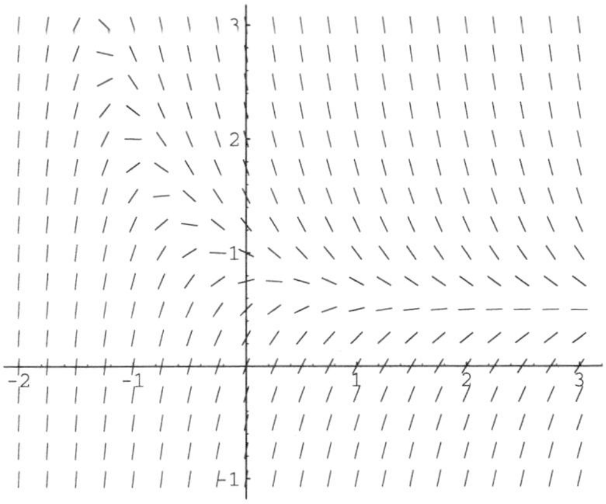

Figure 1.1: A portion of the direction field of $dy/dx = (3/2) - 3y + e^{-3x/2}$.

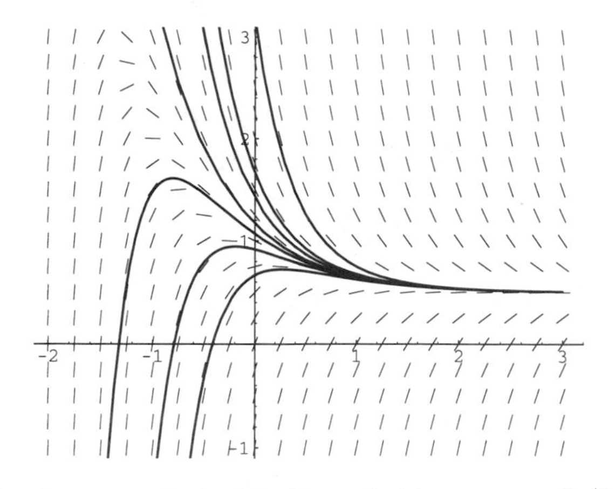

Figure 1.2: The direction field of $dy/dx = (3/2) - 3y + e^{-3x/2}$ and some solutions.

have? (In general there are many; they may be easy or extremely difficult to find.) When there are many solutions to choose from, is it possible to select one or more having certain properties? When, if ever, is there exactly one solution having the properties we want?

We will state and often prove theorems that will provide us with guidance as we seek answers to questions such as these. Some differential equations courses are structured so that you are asked to prove theorems yourself. In this text, what is required of you is not the ability to prove these theorems (though you are encouraged to prove them if you wish), but rather the ability to understand what the theorems mean, so that you can apply them and thereby profit from the work others have done on your behalf. Recall that Sir Isaac Newton[1] said "If it be that I have seen further than other men, it is because I have stood upon the shoulders of giants." It is upon the shoulders of Newton, himself a giant, and many others since, that we proceed to stand in the hopes of seeing further than we otherwise might do.

Here is the first such theorem. It is concerned with a differential equation that has an additional condition specified (an initial condition), having the form given in this equation:

$$\begin{cases} \dfrac{dy}{dx} = f(x, y) \\ y(x_0) = y_0. \end{cases} \tag{1.1}$$

[1]Sir Isaac Newton (1642–1727), British mathematician and natural philosopher. He, along with Leibniz, created both the differential and integral calculus. He proposed the fundamental laws of gravitation, was the first to adequately describe properties of light and color, and constructed the first reflecting telescope. In his later years, Newton was Warden of the Mint, where he reformed the coinage of the realm, was President of the Royal Society, and was a member of Parliament.

Theorem 1.1 (Existence and Uniqueness) *Suppose that the real-valued function $f(x, y)$ is defined and continuous on the rectangle $R = [a, b] \times [c, d]$ in the xy-plane, and that $(\partial/\partial y)f(x, y)$ exists and is continuous throughout R. Suppose further that (x_0, y_0) is an interior point of R. Then there is an open subinterval (a_1, b_1) of $[a, b]$ centered on x_0 and exactly one solution of the differential equation $dy/dx = f(x, y)$ that is defined on the subinterval (a_1, b_1) and passes through the point (x_0, y_0).*

This theorem tells us that a large class of differential equations have solutions, and that these solutions are particularly nice: not only do solutions exist, but if you specify a particular point through which you would like a solution to pass, then there is *exactly one solution* that passes through that point. Two concepts are central here: **existence** of a solution (there are solutions) and **uniqueness** of solutions (there is exactly one solution having the property we want). *Existence* says that there is at least one solution; *uniqueness* says that there is at most one solution. Together, they say that there is only one solution. This is important, because if you know that the problem you are solving has a unique solution, and you find a solution, then you need look no further: the solution you have is the only solution there is. Of greater importance to those who apply differential equations is the knowledge that if a process is governed by a differential equation having a unique solution, then if the process can be performed at all, there is only one way to perform it.

Consider the differential equation $dy/dx = \sin(y)$. Here $f(x, y) = \sin(y)$ has a continuous partial derivative with respect to y : $\cos(y)$. Given any point in the plane, this differential equation has a unique solution that passes through the point. However, the differential equation $dy/dx = y^{2/3}$ does not (necessarily) have a unique solution in the vicinity of any point where $y = 0$, because $(\partial/\partial y)(y^{2/3}) = (2/3)y^{-1/3}$ which is not continuous when $y = 0$.

How does one visualize the concept of uniqueness? In Figures 1.2, 3.4, and 3.5, you can see portions of various families of curves. It is not hard to imagine that each point of the plane lies on some solution. Furthermore, the solutions do not seem to cross one another. This is the idea of uniqueness: through each point there is only one solution. At any point where two solutions cross, we would not have uniqueness. Look at Figures 1.6 and 1.9 to see examples where this fails. We primarily study situations where solutions are unique.

Exercises 1.0

Determine whether or not these equations are differential equations. Classify the differential equations as being ordinary or partial. State the order of each differential equation.

1. $\left(\frac{dy}{dx} + 2y\right)^2 = \left(\frac{dy}{dx}\right)^2 + 4\frac{dy}{dx} + 4y^2$

2. $\left(\dfrac{dy}{dx}+2y\right)^2=\left(\dfrac{dy}{dx}\right)^2-4\dfrac{dy}{dx}+4y^2$

3. $\dfrac{d^3y}{dx^3}+4\dfrac{dy}{dx}-17y=x^4-\sin 5x$

4. $\dfrac{d^3y}{dx^3}+4\left(\dfrac{dy}{dx}\right)^3-17y^2=x^4-\sin 5x$

5. $\dfrac{\partial^2 u}{\partial x\partial y}=\dfrac{\partial u}{\partial y}$

6. $\dfrac{\partial^2 u}{\partial x\partial y}=\dfrac{\partial^2 u}{\partial y\partial x}$

7. $\dfrac{\partial u}{\partial x}=\dfrac{\partial u}{\partial y}$

8. $\left(\dfrac{\partial u}{\partial x}\right)^2=\left(\dfrac{\partial u}{\partial y}\right)^2$

There follow two columns of equations. In the first column is a differential equation; in the second column is a function or set of functions that is a solution of the differential equation. Verify that the given functions satisfy the corresponding equations. Do this manually and by *Mathematica*. Consider c_1, c_2, and λ to be arbitrary constants.

9. $\dfrac{d^2y}{dt^2}+36y=0$ — $y(t)=c_1\cos(6t)+c_2\sin(6t)$

10. $\dfrac{d^2y}{dt^2}+36y=72t+$ — $y(t)=c_1\cos(6t)+c_2\sin(6t)+2t+\dfrac{1}{36}$

11. $\dfrac{d^2y}{dx^2}-36y=0$ — $y(x)=c_1e^{6x}+c_2e^{-6x}$

12. $\dfrac{d^2y}{dx^2}-36y=18x+1$ — $y(x)=c_1e^{6x}+c_2e^{-6x}-\dfrac{x}{2}-\dfrac{1}{36}$

13. $\dfrac{dy}{dt}=1+y^2$ — $y(t)=\tan(t)$

14. $\dfrac{\partial u}{\partial x}=\dfrac{\partial u}{\partial y}$ — $u(x,y)=f(x+y)$; f arbitrary and differentiable

15. $2\dfrac{\partial u}{\partial x}=3\dfrac{\partial u}{\partial y}$ — $u(x,y)=f(3x+2y)$; f arbitrary and differentiable

16. $\dfrac{\partial u}{\partial t}=a^2\dfrac{\partial^2 u}{\partial x^2}$ — $u(x,t)=e^{-\lambda^2 t}\left(c_1\cos\left(\dfrac{\lambda x}{a}\right)+c_2\sin\left(\dfrac{\lambda x}{a}\right)\right)$

17. $\dfrac{\partial^2 u}{\partial x^2}+\dfrac{\partial^2 u}{\partial y^2}=0$ — $u(x,y)=e^{\lambda x}(c_1\cos(\lambda y)+c_2\sin(\lambda y))$

1.1 Some Linear Algebra

Linear Functions

This section will provide us with a first look at some aspects of linear algebra. One can study differential equations without knowing anything about linear algebra, but the study is so much more productive if you are aware of linear algebra as new ideas are introduced. There will still be plenty left over to study in a linear algebra course, but you will know some of the central ideas when we have finished. Topics from linear algebra will occur at several places in the chapters that follow.

We begin by introducing a working definition of the term *linear.*

Definition 1.4 *The function f is said to be* **linear** *provided that if u and v are in its domain, then $u + v$ is in the domain and*

$$f(u+v) = f(u) + f(v).$$

Furthermore, if c is a number, then cu is in the domain of f and

$$f(cu) = cf(u).$$

Using this definition, we can, for instance, show that the derivative function D defined by $Df(x) = (d/dx)f(x) = f'(x)$ is linear. (We use the sum and constant multiple rules from differential calculus.)

$$D(f(x)+g(x)) = \frac{d}{dx}(f(x)+g(x)) = f'(x)+g'(x) = Df(x)+Dg(x)$$

and

$$D(cf(x)) = \frac{d}{dx}(cf(x)) = c\frac{d}{dx}f(x) = cf'(x) = cDf(x).$$

Here $f(x)$ and $g(x)$ play the roles of u and v in the definition. In cases such as this, where a function is acting on a set of functions, we often refer to such a function as an **operator**. Thus we would refer to the derivative operator D.

There are other processes that are linear. Simple multiplication by the number 3 (or any other number) is such an example: Let $f(t) = 3t$. Then

$$f(u+v) = 3(u+v) = 3u+3v = f(u)+f(v)$$

and

$$f(ct) = 3(ct) = (3c)t = (c3)t = c(3t) = cf(t).$$

From algebra we needed the distributive law, $a(b+c) = ab+ac$, the commutative law, $ab = ba$, and the associative law, $(ab)c = a(bc)$. It is also clear that there was nothing special about the choice of 3 as the multiple. (Replace the 3 by k, throughout.)

It is easy to show that the function (operator) L defined by $L(y) = y' + 3y$ is linear. Let u and v be once-differentiable functions defined on a common domain.

Then, since sums and constant multiples of once-differentiable functions are once-differentiable,

$$\begin{aligned} L(u+v) &= (u+v)' + 3(u+v) \\ &= u' + v' + 3u + 3v \\ &= (u' + 3u) + (v' + 3v) \\ &= L(u) + L(v). \end{aligned}$$

When c is a constant,

$$\begin{aligned} L(cu) &= (cu)' + 3(cu) \\ &= cu' + 3cu \\ &= c(u' + 3u) \\ &= cL(u). \end{aligned}$$

We make two observations about the definition of linear as proposed above. Let V be the domain of f. Then

1. given any two objects in V, their sum had to be in V for the first part of the definition to hold, and
2. any multiple of an object in V must also be in V.

This means that not only does f have to behave properly on sums and multiples, but so does the domain of f. We will see this idea again in Section 4.2 when the concept of a **linear space** is defined.

Solving Linear Equations

Here are some properties of linear functions that we will use heavily in the chapters to come: Given a linear function L,

(a) The equation $L(u) = 0$ always has at least one solution. (Since $L(0) = 0$, 0 is such a solution.)
(b) If $L(u) = 0$ and $L(v) = 0$ then $L(u+v) = 0$, and for any constant c, $L(cu) = 0$. (If $L(u) = 0$ and $L(v) = 0$ then $L(u+v) = L(u) + L(v) = 0 + 0 = 0$, so $u+v$ is in the domain of L and is a solution, and $L(cu) = cL(u) = c0 = 0$, so cu is in the domain of L and is a solution.)
(c) If $L(p) = b$ and $L(q) = b$, then $L(p-q) = 0$. ($L(p-q) = L(p) - L(q) = b - b = 0$.)
(d) If $L(p) = b$ and $L(q) = b$, then there is a (unique) member u in the domain of L such that $p = q + u$. (Take $u = p - q$. Then $L(u) = L(p) - L(q) = b - b = 0$. If $p = q + u_1$, then $u_1 = p - q = u$.)
(e) If $L(q) = b$, $L(u) = 0$, and $p = q + u$, then $L(p) = b$. ($L(p) = L(q+u) = L(q) + L(u) = b + 0 = b$.)

It will be helpful to have names for the various objects that we encounter as we study linear problems.

Definition 1.5 *The set of all solutions of the linear equation $L(u) = 0$ is called the* **null space** *or* **kernel** *of L.*

A linear problem such as $L(y) = 0$, having right-hand side 0, is called **homogeneous**.

A linear problem such as $L(y) = b$, having nonzero right-hand side, is called **nonhomogeneous**.

Properties (d) and (e) will guide us as we begin solving nonhomogeneous linear (differential equations) problems. They are used this way:

Given the linear problem $L(y) = b$:

1. Find a typical member, u, of the kernel of L.
2. Find any one object, q, so that $L(q) = b$.
3. Then $y = q + u$ represents *every solution* of $L(y) = b$.

This process is a property of *linearity* not a property of differential equations, although properties of differential equations will dictate how we go about finding q and u. Any object that satisfies the nonhomogeneous equation, such as q above, is called a **particular solution**. A typical member of the set u, which represents the entire kernel of L, is often called a **complementary function** of L. A set of functions such as $y = q + u$ that represents every solution of $L(y) = b$ is called a **general solution** or **complete solution** of $L(y) = b$ because there are no other solutions.

Examples Using *Mathematica*: Solving Linear Algebraic Systems

A linear problem has either no solution, exactly one solution, or an infinite number of solutions. This can be illustrated by considering how two planes in 3-space can intersect. If the planes are parallel and different, then there is no point in common, and hence there is no solution to the two equations. Otherwise, the two planes intersect in a line or coincide. In either of these latter cases there are infinitely many solutions. If a third plane is also considered, then there can be exactly one solution (where the line of intersection of the first two intersects the third) or no solutions (from several interesting geometric arrangements) or an infinite number of solutions (where all planes coincide, or where all three share a line in common). It is of some interest to sketch each of the possibilities for the intersection, or lack of it, of three planes in ordinary 3-space.

The examples that follow illustrate analogous situations for two lines in the plane.

Here is a system that has a unique solution:

```
In[8]:=  Solve[{ 3x+2y == 4,
                 5x+3y == 2}]

Out[8]=  {{x -> -8, y -> 14}}
```

Next is a system that has infinitely many solutions. Note the warning that *Mathematica* issues. Indeed it does solve only for x in terms of y:

```
In[9]:=  Solve[{ 6x+2y == 4,
                 3x+ y == 2}]

Solve::svars:
   Warning: Equations may not give solutions for all "solve"
     variables.

Out[9]=
                2   y
         {{x -> - - -}}
                3   3
```

Capture the solution(s). The % means the results of the last calculation, and `[[1]]` is `Part[%,1]`, the inner quantity enclosed in braces. See the discussion in Appendix A2:

```
In[10]:= {x1,y1} = {x,y}/.%[[1]]

Out[10]= 2   y
        {- - -, y}
         3   3
```

Check the solution.

```
In[11]:= Simplify[{6x1+2y1 == 4,
           3x1+ y1 == 2}]

Out[11]= {True, True}
```

The two `True`'s indicate that both equations are satisfied.

Here is a system that has no solutions. Observe that *Mathematica* uses the notation "`{}`" for the empty set, signifying no solutions:

```
In[12]:= Solve[{ 6x+2y == 4,
           3x+ y == 1}]

Out[12]= {}
```

The expression $\{2/3 - y/3, y\}$ that appears as `Out[10]` can be rewritten $\{2/3, 0\} + y\{-1/3, 1\}$. The linear function is $L(x, y) = \{6x + 2y, 3x + y\}$ and the stated problem is $L(x, y) = \{4, 2\}$. From the definition of L we verify that $\{2/3, 0\}$ is a particular solution, and $y\{-1/3, 1\} = \{-y/3, y\}$ gets sent to 0.

Here is the verification: $L(2/3, 0) = \{6(2/3)+2(0), 3(2/3)+(0)\} = \{4, 2\}$. And $L(-y/3, y) = \{6(-y/3)+2(y), 3(-y/3)+(y)\} = \{2y-2y, -y+y\} = \{0, 0\}$. Thus our solution had the desired form: a typical member of the kernel plus a particular solution. The symbol y in the solution served as a simple constant multiplier: each choice of y gave a solution. The points on the solution are merely the points on the original curve; the form is that of a line expressed parametrically, rather than a form that you are used to. The line in parametric form can be written:

$$\begin{cases} x = 2/3 - y/3, \\ y = y. \end{cases}$$

A better form would be

$$\begin{cases} x = 2/3 - t/3, \\ y = t, \end{cases}$$

where t is used as the parameter, rather than y.

EXERCISES 1.1

1. Use *Mathematica* to solve each of these linear systems. Problem (a) would be entered as `Solve[{2x+3y == 5, 3x+4y == 7},{x,y}]`. Interpret each response. Each problem can be plotted to aid your interpretation. Plot problem (a) as `Plot[{(5-2x)/3, (7-3x)/4},{x, -2, 2}]`, for instance. Note that to do the plot, each equation was solved for y as a function of x and the resulting two functions plotted. If there is a unique solution, try to include it in the range over which you plot. The range just specified was $-2 \leq x \leq 2$.

   (a) $\begin{cases} 2x + 3y = 5 \\ 3x + 4y = 7 \end{cases}$

   (b) $\begin{cases} 5x - 3y = 5 \\ 3x + 4y = 3 \end{cases}$

   (c) $\begin{cases} x + 7y = 5 \\ 2x + 14y = 10 \end{cases}$

   (d) $\begin{cases} 2x + 3y = 5 \\ 4x + 6y = 70 \end{cases}$

   (e) $\begin{cases} 2x - 5y = -5 \\ x + 6y = 4 \end{cases}$

2. Verify that each of these functions is linear. Propose a reasonable domain and range for each of the functions.

   (a) $f(x, y) = (x + y, 2x - y)$.
   (b) $g(x, y, z) = (x - z, x, y, x + 2y)$.
   (c) $h(x, y) = (x - y, x, y, x + 2y)$.
   (d) $k(x, y, z, w) = (x - z + 2w, x + 2y)$.
   (e) $i(F) = \int_a^b F(x)\, dx$.
   (f) $p(G)(x) = \int_a^b G(x, t)\, dt$.

3. Sketch possible geometric configurations of three planes that result in

   (a) no solutions;
   (b) exactly one solution;
   (c) infinitely many solutions.

4. Each bracketed expression below is a line in the xy-plane expressed in parametric form. In each case eliminate t from the two equations to obtain an equation for the line containing the variables x and y, but not t. (Solve one equation for t and substitute into the other, or multiply each equation by an appropriate constant so that when the equations are added, t will drop out.) Sketch the lines that result.

   (a) $\begin{cases} x = 5 + 3t \\ y = 7 - t \end{cases}$

   (b) $\begin{cases} x = -2 + 5t \\ y = 3 - 2t \end{cases}$

   (c) $\begin{cases} x = 3t \\ y = 5 + t \end{cases}$

1.2 Differential Equations from Solutions

In the exercises following the introduction to this chapter, we verified that a function or set of functions satisfies a given differential equation. In Section 1.1 we looked at some ideas from linear algebra. Here we use ideas from linear algebra to find a (linear) differential equation of minimal order whose kernel contains a given set of functions.

Our technique is summarized as follows. Suppose that we are given a set $\{f_1(x), f_2(x), \ldots, f_n(x)\}$ of n functions having a common domain and each possessing at least n continuous derivatives. Let V be the set (linear space) of all linear combinations of $\{f_1(x), f_2(x), \ldots, f_n(x)\}$, that is, V is the set of all functions having the form

$$y(x) = c_1 f_1(x) + c_2 f_2(x) + \cdots + c_n f_n(x),$$

where the coefficients $c_1, c_2, \ldots, c_n$ are numbers. We want to find a (linear) differential equation of order n whose kernel is V. Here is one way to do this.

(a) Solve these n equations

$$\begin{aligned} y(x) &= c_1 f_1(x) + c_2 f_2(x) + \cdots + c_n f_n(x) \\ y'(x) &= c_1 f_1'(x) + c_2 f_2'(x) + \cdots + c_n f_n'(x) \\ &\vdots \\ y^{(n-1)}(x) &= c_1 f_1^{(n-1)}(x) + c_2 f_2^{(n-1)}(x) + \cdots + c_n f_n^{(n-1)}(x) \end{aligned}$$

for the coefficients $c_1, c_2, \ldots, c_n$ in terms of $y(x)$ and its derivatives and the functions $f_i(x)$ and their derivatives. This can be done uniquely if the functions $\{f_1(x), f_2(x), \ldots, f_n(x)\}$ are essentially different. (That is, they are linearly independent. Linear independence is discussed at length in Section 4.2. The examples and problems that follow all have the necessary conditions satisfied, so this procedure will work.)

(b) Having determined $c_1, c_2, \ldots, c_n$ uniquely from the above equations, substitute them into

$$y^{(n)}(x) = c_1 f_1^{(n)}(x) + c_2 f_2^{(n)}(x) + \cdots + c_n f_n^{(n)}(x).$$

The resulting equation involves $y(x)$, its derivatives, and the functions $f_i(x)$ and their derivatives, but none of the $c_1, c_2, \ldots, c_n$.

The equation that results from part (b) is the (linear) differential equation we seek. It is wise to check to see that the given functions do indeed satisfy the differential equation that was found.

We present a progression of examples, some very simple and some quite complicated. These examples illustrate the process, both manually and by *Mathematica.*

Example 1.4 (Simple antiderivatives)

1. Given the set of constant functions $y = c$, differentiate once to see that the differential equation is $y' = 0$. This is the differential equation since it involves y' and does not contain c.
2. Given a differentiable function $f(x)$ and the set of functions $y = f(x) + c$, differentiate once to see that the differential equation is $y' = f'(x)$. This is the differential equation for the same reason as before.
3. Given a differentiable function $f(t)$ and the set of functions $y = f(t) + c_1 t + c_2$, differentiate once to get $dy/dt = f'(t) + c_1$. Though it turns out to be unnecessary here, solve the pair of equations

$$y = f(t) + c_1 t + c_2$$
$$y' = f'(t) + c_1$$

for c_1 and c_2 to get

$$c_1 = y' - f'(t)$$
$$c_2 = y - f(t) - (y' - f'(t))t.$$

Substitute these into the second derivative $y'' = f''(t)$. This equation contains neither constant, so the differential equation we want is $y'' = f''(t)$. Note that if $f''(t) = -g$, with g being the acceleration due to gravity, then this differential equation describes the motion of an object falling freely due to the force of gravity. ◇

The constants do not go away so easily in most examples. Here is another way in which the constants appear to go away as if by magic.

Example 1.5 Find the (linear) differential equation of minimum order that is satisfied by the functions $y = c_1 \sin x + c_2 \cos x$.

Solution. Here $y' = c_1 \cos x - c_2 \sin x$ and $y'' = -c_1 \sin x - c_2 \cos x = -y$, so the differential equation is $y'' = -y$, or $y'' + y = 0$. You should solve the y and

y' equations for the coefficients c_1 and c_2 and substitute the values for c_1 and c_2 into y'' to see that the same equation results. ◇

One should not expect the coefficient constants to disappear. This example is more typical.

Example 1.6 Find the (linear) differential equation of minimum order that is satisfied by the functions $y = c_1 e^x + c_2 x$.

Solution. Solve the system

$$y = c_1 e^x + c_2 x$$
$$y' = c_1 e^x + c_2$$

for the coefficients c_1 and c_2 to get

$$c_1 = \frac{y - x\,y'}{e^x(1-x)}$$

and

$$c_2 = \frac{-y + y'}{(1-x)}.$$

Then substitute into $y'' = c_1 e^x$ to get the differential equation

$$y'' = c_1 e^x = \frac{y - x\,y'}{1-x}.$$

Simplify to get

$$(1-x)y'' + x\,y' - y = 0.$$

It is easy to verify that $y = c_1 e^x + c_2 x$ satisfies this differential equation. (Do so.) ◇

Here is one last example and its solution by *Mathematica.*

Example 1.7 Find the (linear) differential equation of minimum order that is satisfied by the functions $y = c_1 e^x + c_2 e^{-2x} + c_3 e^{3x}$.

Solution. Solve the system

$$y = c_1 e^x + c_2 e^{-2x} + c_3 e^{3x}$$
$$y' = c_1 e^x - 2c_2 e^{-2x} + 3c_3 e^{3x}$$
$$y'' = c_1 e^x + 4c_2 e^{-2x} + 9c_3 e^{3x}$$

for c_1, c_2, and c_3 to get

$$c_1 = \frac{6y + y' - y''}{6e^x}$$

$$c_2 = \frac{e^{2x}(3y - 4y' + y'')}{15}$$

$$c_3 = \frac{-2y + y' + y''}{10e^{3x}}.$$

Substitute into

$$y''' = c_1 e^x - 8c_2 e^{-2x} + 27c_3 e^{3x}$$

to get

$$\begin{aligned} y''' = & \frac{6y + y' - y''}{6e^x} e^x \\ & -8\frac{e^{2x}(3y - 4y' + y'')}{15} e^{-2x} \\ & +27\frac{-2y + y' + y''}{10e^{3x}} e^{3x} \\ = & -6y + 5y' + 2y''. \end{aligned}$$

The differential equation can be rewritten as

$$y''' - 2y'' - 5y' + 6y = 0,$$

and it is easy to verify manually that $y = c_1 e^x + c_2 e^{-2x} + c_3 e^{3x}$ is a solution. (Do so.) ◇

Example 1.7M Use *Mathematica* to find the (linear) differential equation of minimum order that is satisfied by the functions $y = c_1 e^x + c_2 e^{-2x} + c_3 e^{3x}$.

Solution.

```
In[13]:= expr = (y[x] == c1 Exp[x]+c2 Exp[-2x]+c3 Exp[3x])

Out[13]=
              c2          x         3 x
     y[x] == ---- + c1 E  + c3 E
              2 x
             E
```

These are the equations from which c1, c2, and c3 are to be eliminated.

```
In[14]:= equations = Table[D[expr,{x,i}],{i,0,3}]

Out[14]=
               c2          x         3 x
     {y[x] == ---- + c1 E  + c3 E     ,
               2 x
              E

               -2 c2          x           3 x
      y'[x] == ----- + c1 E  + 3 c3 E     ,
                2 x
               E
```

```
                 4 c2         x          3 x
    y''[x] ==   ----- + c1 E  + 9 c3 E     ,
                 2 x
                E

      (3)          -8 c2         x           3 x
    y    [x] ==   ------ + c1 E  + 27 c3 E     }
                   2 x
                  E
```

Solve for y''' [x] while eliminating c1, c2, and c3 to get an equation containing none of the constants c1, c2, and c3. When Solve is given a list of variables in its third parameter, it tries to give a solution that does not contain any of these variables. This is what happens on our example:

```
In[15]:= the3DerivRule =Simplify[
           Solve[equations, y'''[x], {c1,c2,c3}]]

Out[15]=     (3)
         {{y    [x] -> -6 y[x] + 5 y'[x] + 2 y''[x]}}
```

Name the differential equation so we can check our solutions. The substitution applies only to the right-hand side of the equation.

```
In[16]:= de[x_,y_] = (y'''[x] == (y'''[x]/.the3DerivRule[[1]]))

Out[16]=  (3)
         y    [x] == -6 y[x] + 5 y'[x] + 2 y''[x]

In[17]:= Clear[y]
         y[x_] = c1 Exp[x]+c2 Exp[-2x]+c3 Exp[3x]

Out[17]=  c2         x          3 x
         ---- + c1 E  + c3 E
          2 x
         E
```

Now substitute $y[x]$ into the differential equation

```
In[18]:= de[x,y]

Out[18]= -8 c2         x           3 x
         ------ + c1 E  + 27 c3 E      ==
          2 x
         E

              c2          x          3 x
         -6 (---- + c1 E  + c3 E     ) +
              2 x
             E

             -2 c2         x            3 x
          5 (------ + c1 E  + 3 c3 E      ) +
              2 x
             E

             4 c2         x            3 x
          2 (----- + c1 E  + 9 c3 E      )
              2 x
             E
```

It appears necessary to `Simplify` before we can tell whether or not the differential equation is satisfied. The `Option Trig->False` tells `Simplify` that it is not necessary to use trigonometric identities during simplification.

```
In[19]:= Simplify[%, Trig->False]

Out[19]= True
```

◇

EXERCISES 1.2

1. Manually verify that $y = c_1 e^x + c_2 x$ satisfies the differential equation $(1-x)y'' + xy' - y = 0$. (See Example 1.6.)

2. Manually verify that $y = c_1 e^x + c_2 e^{-2x} + c_3 e^{3x}$ satisfies the differential equation $y''' - 2y'' - 5y' + 6y = 0$. (See Example 1.7.)

3. Use the technique of Example 1.7M to find a second-order differential equation that is satisfied by the two-parameter family of functions $y = c_1 e^x + c_2 x + x^2 - 1$.

4. Use the technique of Example 1.7M to find a third-order differential equation that is satisfied by the three-parameter family of functions $y = c_1 e^x + c_2 e^{-2x} + c_3 e^{3x} + \sin x$.

5. Use the technique of Example 1.7M to find a third-order differential equation that is satisfied by the three-parameter family of functions $y = c_1 x e^x + c_2 e^{-2x} + c_3 e^{3x}$.

1.3 Numerical Methods

Only a very few differential equations can actually be solved—in the sense that we can write down an expression for a solution. This is especially true of nonlinear differential equations. In Chapter 2 we discuss many of the special cases where a solution can actually be obtained. If we need a solution to a differential equation, but are unable to obtain a closed-form expression for such a solution, how do we get useful information about the solution? We may just need a few points that lie on the solution, or we may need to know where our solution crosses some given curve, or we may wish to determine a maximum or a minimum on the solution. We need such information in the absence of a function to evaluate.

There are situations where we have to rely on approximating a solution, rather than obtaining a solution. Leonhard Euler has observed that the direction fields we saw in the introduction to this chapter can be exploited to give us useful information.

Suppose that the equation to solve is $dy/dx = f(x, y)$ and the initial point is $y(x_0) = y_0$. We want a solution over an interval $[a, b]$ with $x_0 = a$. Since, when h is small,

$$\frac{y(x+h) - y(x)}{h} \sim y'(x),$$

Euler reasoned that we should solve

$$\frac{y(x+h)-y(x)}{h} = f(x, y(x))$$

for

$$y(x+h) = y(x) + hf(x, y(x)).$$

Then, knowing the solution at $(x, y(x))$, estimate that the solution will pass through $(x+h, y(x)+hf(x, y(x)))$. This just says that a solution essentially follows its tangent line, whose direction is that of the direction field element at $(x, y(x))$. If h is small, then this guess, though probably wrong, is nevertheless reasonable. This guessing process is repeated enough times to estimate the solution over the entire interval $[a, b]$ by finding ordinates corresponding to $x_0 = a$, $x_1 = x_0 + h$, $x_2 = x_0 + 2h, \ldots, x_n = x_0 + nh = b$. The technique is called Euler's method. We denote $y(x_k)$ by y_k, and produce the data points $(x_0, y_0), (x_1, y_1), \ldots, (x_n, y_n)$ by the following rule:

Given x_0 and y_0, the coordinates of the initial point (x_0, y_0), and a small number h, calculate

$$\begin{cases} x_{k+1} = x_k + h \\ y_{k+1} = y_k + h\, f(x_k, y_k) \end{cases}, \qquad 0 \leq k \leq n-1.$$

It is worth noting that if $h < 0$ then $x_n < \cdots < x_1 < x_0$, so the points are formed from right to left.

Euler's method can easily be used with any reasonable spreadsheet. Many spreadsheets even permit plotting the results. Try doing so.

Here is an example that you can use for comparison purposes.

Example 1.8 Use Euler's method to approximate the solution to

$$\frac{dy}{dx} = x + 2y, \qquad y(1) = 1/2$$

over the interval from $x = 1$ to $x = 3$ in steps of $h = 0.1$.

Solution. For this problem $x_0 = 1$ and $y_0 = 1/2 = 0.5$. Since $h = 0.1$, from $x_n = x_0 + n\,h = 1 + (0.1)n = 3$ we find that $n = (3-1)/(0.1) = 20$. The process we need to iterate (repeat) is

$$\left.\begin{aligned} x_{k+1} &= x_k + 0.1, \\ y_{k+1} &= y_k + 0.1\, f(x_k, y_k) \\ &= y_k + 0.1\,(x_k + 2y_k) \\ &= 0.1\, x_k + 1.2\, y_k, \end{aligned}\right\} \quad 0 \leq k \leq 19.$$

We stop when $n = 19$ because the point (x_{20}, y_{20}) is produced at that step. This gives us 21 data points. The calculation actually proceeds as follows:

k	x_{k+1}	Euler Step y_{k+1}	Exact
—	1.0	0.5	0.5
0	$1.0 + 0.1 = 1.1$	$(0.1)(1.0) + (1.2)(0.5) = 0.7$	0.72675
1	$1.1 + 0.1 = 1.2$	$(0.1)(1.1) + (1.2)(0.7) = 0.95$	1.01478
2	$1.2 + 0.1 = 1.3$	$(0.1)(1.2) + (1.2)(0.95) = 1.26$	1.37765
3	$1.3 + 0.1 = 1.4$	$(0.1)(1.3) + (1.2)(1.26) = 1.642$	1.83193
4	$1.4 + 0.1 = 1.5$	$(0.1)(1.4) + (1.2)(1.642) = 2.1104$	2.39785
5	$1.5 + 0.1 = 1.6$	$(0.1)(1.5) + (1.2)(2.1104) = 2.68248$	3.10015
6	$1.6 + 0.1 = 1.7$	$(0.1)(1.6) + (1.2)(2.68248) = 3.37898$	3.96900
7	$1.7 + 0.1 = 1.8$	$(0.1)(1.7) + (1.2)(3.37898) = 4.22477$	5.04129
8	$1.8 + 0.1 = 1.9$	$(0.1)(1.8) + (1.2)(4.22477) = 5.24973$	6.36206
9	$1.9 + 0.1 = 2.0$	$(0.1)(1.9) + (1.2)(5.24973) = 6.48967$	7.98632
10	$2.0 + 0.1 = 2.1$	$(0.1)(2.0) + (1.2)(6.48967) = 7.98760$	9.98127
11	$2.1 + 0.1 = 2.2$	$(0.1)(2.1) + (1.2)(7.98760) = 9.79513$	12.4290
12	$2.2 + 0.1 = 2.3$	$(0.1)(2.2) + (1.2)(9.79513) = 11.9742$	15.4297
13	$2.3 + 0.1 = 2.4$	$(0.1)(2.3) + (1.2)(11.9742) = 14.5990$	19.1058
14	$2.4 + 0.1 = 2.5$	$(0.1)(2.4) + (1.2)(14.5990) = 17.7588$	23.6069
15	$2.5 + 0.1 = 2.6$	$(0.1)(2.5) + (1.2)(17.7588) = 21.5605$	29.1157
16	$2.6 + 0.1 = 2.7$	$(0.1)(2.6) + (1.2)(21.5605) = 26.1326$	35.8551
17	$2.7 + 0.1 = 2.8$	$(0.1)(2.7) + (1.2)(26.1326) = 31.6292$	44.0978
18	$2.8 + 0.1 = 2.9$	$(0.1)(2.8) + (1.2)(31.6292) = 38.2350$	54.1765
19	$2.9 + 0.1 = 3.0$	$(0.1)(2.9) + (1.2)(38.2350) = 46.1720$	66.4977

We have a table of numbers. What do they mean? For comparison purposes, the last column is calculated from the exact solution: $y(x) = -1/4 - x/2 + (5/4)e^{2x-2}$. This means that we didn't do very well. Our final y-values are off by more than 20, an approximately 31% error. We show how to perform these steps in *Mathematica* in the next example. Figure 1.3 is a picture of the results.

Example 1.9M Implement and use the standard Euler method for solving the problem of Example 1.8 in *Mathematica*.

Solution. After defining the Euler method for this problem we use `NestList`, which repeatedly applies a function to some initial data. This is one way to repeatedly apply the Euler method. Set the initial point by making a double assignment.

```
In[20]:= {x0, y0}={1, 1/2}

Out[20]=      1
         {1, - }
              2
```

```
In[21]:= f[x_,y_]=x+2y

Out[21]= x + 2 y
```

Define the value of h.

```
In[22]:= h=0.5

Out[22]= 0.5
```

Define the Euler method itself.

```
In[23]:= euler[{x_,y_}]={x+h,y+h*f[x,y]}

Out[23]= {0.5 + x, y + 0.5 (x + 2 y)}
```

Create the table of approximations using standard Euler.

```
In[24]:= e1t=NestList[euler,{x0,y0},20]

Out[24]=      1
         {{1, -}, {1.1, 0.7}, {1.2, 0.95}, {1.3, 1.26},
              2

           {1.4, 1.642}, {1.5, 2.1104}, {1.6, 2.68248},

           {1.7, 3.37898}, {1.8, 4.22477}, {1.9, 5.24973},

           {2., 6.48967}, {2.1, 7.9876}, {2.2, 9.79513},

           {2.3, 11.9742}, {2.4, 14.599}, {2.5, 17.7588},

           {2.6, 21.5605}, {2.7, 26.1326}, {2.8, 31.6292},

           {2.9, 38.235}, {3., 46.172}}
```

Plot the resulting table using `ListPlot`. Name the plot `p1p`.

```
In[25]:= p1p=ListPlot[e1t];
```

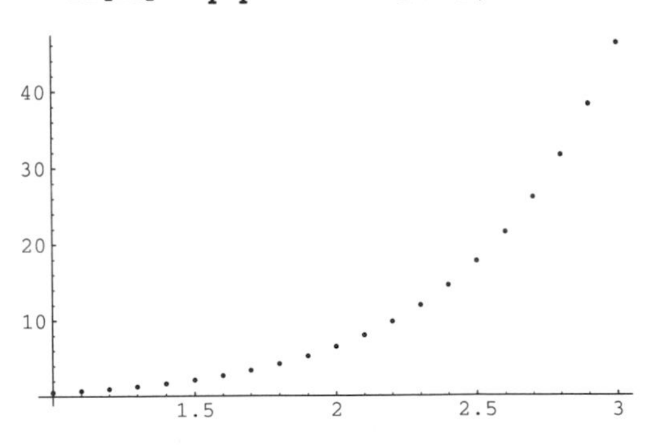

```
In[26]:= p1j=ListPlot[e1t, PlotJoined->True];
```

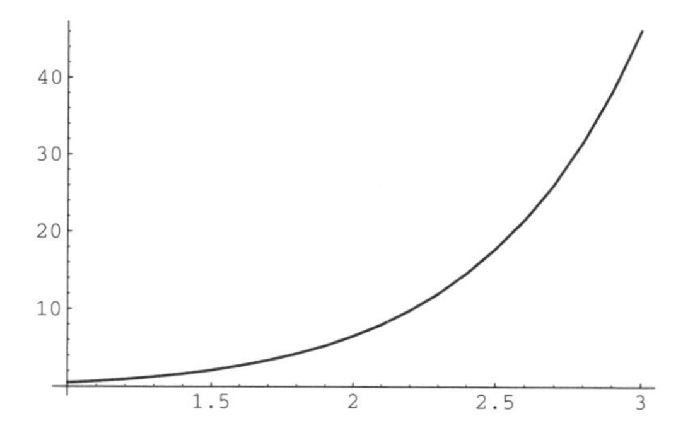

◇

The plot above by `ListPlot` of the points in table `e1t` produced a collection of dots. These are combined with a plot of the exact solution in Figure 1.3. The option `PlotJoined->True` in `ListPlot` connects the dots, as the plot above and Figures 1.4 and 1.5 illustrate.

At least we can say that our approximate solution tried to climb with the exact solution. But notice that if the exact solution curves upward, our approximation will be too small at each step because the tangent line at each point lies below the curve, thereby making our approximation get progressively worse. We have a systematic error here. It has manifested itself in the poor approximation that we found. The problem lies in the fact that once an error has taken us off of the actual solution, the slope is calculated incorrectly for the next point, and the errors may (and in this case do) get worse as we proceed. ◇

Modifying Euler's Method

Often Euler's method is not as bad as this example makes it appear, and the results can be improved by taking more steps with a smaller value of h. Can anything be done to eliminate the systematic error that we observed? To make the solution bend better, we can incorporate the second derivative of our solution

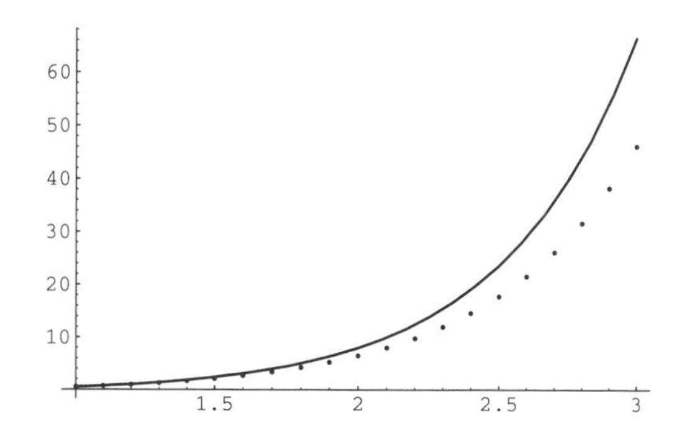

Figure 1.3: Euler (dots) versus exact solution (curve).

into Euler's method. But, how can the second derivative be found if we do not know the solution? The problem is not as great as it might seem—we know the derivative of our solution: $dy/dx = x + 2y$, and this enables us to implicitly differentiate $dy/dx = x + 2y$ to find that

$$\frac{d^2y}{dx^2} = \frac{d}{dx}(x+2y) = 1 + 2\frac{dy}{dx} = 1 + 2(x+2y).$$

We can take advantage of this once we recall that one form of Taylor's theorem says that

$$y(x+h) = y(x) + hy'(x) + \left(\frac{h^2}{2!}\right)y''(x) + \left(\frac{h^3}{3!}\right)y'''(x) + \cdots.$$

Euler's method used only the first two terms, $y(x)+hy'(x) = y(x)+hf(x,y(x))$. We can use three terms since we now know the second derivative $y''(x)$. This improves our results. The new process might be called Euler$_2$ since it uses the second derivative. We can actually calculate as many derivatives as we wish and make a much more accurate method. The Euler$_2$ method for the differential equation of Example 1.8, using $h = 0.1$, is

$$\begin{cases} x_0 = 1 \\ y_0 = 1/2 = 0.5. \end{cases}$$

$$\begin{cases} x_{k+1} = x_k + 0.1, \\ y_{k+1} = y_k + (0.1)\, y_k' + \dfrac{(0.1)^2}{2} y''_k \\ \quad = y_k + (0.1)\, (x_k + 2y_k) + \left(\dfrac{(0.1)^2}{2}\right)(1 + 2(x_k + 2y_k)) \\ \quad = 0.005 + (0.11)x_k + (1.22)y_k, \end{cases}$$

for $0 \leq k \leq 19$. For this differential equation, using the second derivative required very little extra work. This may not always be the case.

In general, the calculation of $y''(x)$ requires partial derivatives. For instance,

$$\begin{aligned} y''(x) &= \frac{d}{dx} f(x, y(x)) \\ &= f_x(x, y(x)) + f_y(x, y(x))y'(x) \\ &= f_x(x, y(x)) + f_y(x, y(x))f(x, y(x)). \end{aligned}$$

This derivative is easy to calculate in *Mathematica.*

```
In[27]:= f[x_,y_]=x+2y

Out[27]= x + 2 y

In[28]:= f2[x_,y_]=D[f[x,y],x]+D[f[x,y],y]f[x,y]

Out[28]= 1 + 2 (x + 2 y)
```

The third derivative is just as simple.

```
In[29]:= f3[x_,y_]=D[f2[x,y],x]+D[f2[x,y],y]f[x,y]

Out[29]= 2 + 4 (x + 2 y)
```

The results given are for the problem of Example 1.8. It is clear that being able to calculate these higher-order derivatives permits us to produce a Euler method that has any desired number of terms. In numerical analysis, one learns that using more terms really does improve the accuracy for normal problems. There the topic of accuracy of a solution is analyzed in thorough detail.

Here is the new Euler$_2$ method. Notice the new term that has been added.

```
In[30]:= euler2[{x_,y_}]={x+h,y+h*f[x,y]+(h^2/2)*f2[x,y]}

Out[30]= {0.5 + x, y + 0.5 (x + 2 y) + 0.125 (1 + 2 (x + 2 y))}
```

This is Euler$_2$ for our problem.

```
In[31]:= Expand[euler2[{x,y}]]

Out[31]= {0.1 + x, 0.005 + 0.11*x + 1.22*y}
```

Make a table of points using Euler$_2$, with x_0 and y_0 known from before.

```
In[32]:= e2t=NestList[euler2,{x0,y0},20]

Out[32]=       1
         {{1, -}, {1.1, 0.725}, {1.2, 1.0105}, {1.3, 1.36981},
               2

          {1.4, 1.81917}, {1.5, 2.37839}, {1.6, 3.07163},

          {1.7, 3.92839}, {1.8, 4.98463}, {1.9, 6.28425},

          {2., 7.88079}, {2.1, 9.83956}, {2.2, 12.2403},

          {2.3, 15.1801}, {2.4, 18.7778}, {2.5, 23.1779},

          {2.6, 28.557}, {2.7, 35.1305}, {2.8, 43.1612},

          {2.9, 52.9697}, {3., 64.9471}}
```

Again, `ListPlot` could produce a plot of this new approximation. Such a plot with points joined appears as a portion of Figure 1.4.

Runge-Kutta and NDSolve

There is a standard method, called the **Runge-Kutta method** for its creators, that effectively incorporates terms through the fourth derivative, requires no partial derivatives, and needs only four evaluations of the original function $f(x, y)$ to obtain the next point. Given that the point (x_k, y_k) is known, the next point (x_{k+1}, y_{k+1}) is calculated this way:

$$\begin{cases} K_1 = hf(x_k, y_k) \\ K_2 = hf(x_k + \frac{1}{2}h, y_k + \frac{1}{2}K_1) \\ K_3 = hf(x_k + \frac{1}{2}h, y_k + \frac{1}{2}K_2) \\ K_4 = hf(x_k + h, y_k + K_3) \\ x_{k+1} = x_k + h \\ y_{k+1} = y_k + \frac{1}{2}(K_1 + 2K_2 + 2K_3 + K_4) \end{cases}$$

The Runge-Kutta method is easy to program and is in wide use, even though there are much more sophisticated methods available. Runge-Kutta and the Euler method(s) are for use with first-order differential equations. There is a Runge-Kutta package available with *Mathematica.*

Here is how one might define and use the process just defined. The built-in **function** `Module` that is used is analogous to a Pascal function declaration. Definitions of h, x_0, y_0 and $f(x, y)$ are used globally; `K1, K2, K3, K4` are declared as local variables. The explicit use of `Return` was unnecessary, since *Mathematica* always returns the last expression that is evaluated inside the function.

```
In[33]:=  RK[{x_,y_}]:=Module[{K1,K2,K3,K4},
          K1=h*f[x,y];
          K2=h*f[x+h/2,y+K1/2];
          K3=h*f[x+h/2,y+K2/2];
          K4=h*f[x+h,y+K3];
          Return[{x+h,y+(1/6)(K1+2K2+2K3+K4)}]
          ]

In[34]:= RKt=NestList[RK,{x0,y0},20]

Out[34]=      1
         {{1, -}, {1.1, 0.72675}, {1.2, 1.01477}, {1.3, 1.37763},
              2

          {1.4, 1.8319}, {1.5, 2.39781}, {1.6, 3.10009},

          {1.7, 3.96892}, {1.8, 5.04118}, {1.9, 6.36191},

          {2., 7.98611}, {2.1, 9.98099}, {2.2, 12.4286},

          {2.3, 15.4292}, {2.4, 19.1052}, {2.5, 23.6061},

          {2.6, 29.1146}, {2.7, 35.8537}, {2.8, 44.0959},

          {2.9, 54.1741}, {3., 66.4946}}
```

For our purposes, when we need a numerical solution of a differential equation, we will rely on the built-in function `NDSolve`. Its use will be demonstrated on several occasions in the chapters that follow. `NDSolve` can be applied to higher-order differential equations as well as to first-order equations.

There are extensions of these methods that can be used when systems of differential equations must be solved. We see these in Chapter 8.

Table 1.1: Summary of results of several Euler methods.

k	x_k	Euler	Euler_2	Euler_3	Euler_4	Exact
0	1.0	0.5	0.5	0.5	0.5	0.5
1	1.1	0.7	0.725	0.726667	0.72675	0.726753
2	1.2	0.95	1.0105	1.01457	1.01477	1.01478
3	1.3	1.26	1.36981	1.37726	1.37763	1.37765
4	1.4	1.642	1.81917	1.83129	1.8319	1.83193
5	1.5	2.1104	2.37839	2.39689	2.39781	2.39785
6	1.6	2.68248	3.07163	3.09873	3.10009	3.10015
7	1.7	3.37898	3.92839	3.96698	3.96892	3.96900
8	1.8	4.22477	4.98463	5.03848	5.04118	5.04129
9	1.9	5.24973	6.28425	6.35819	6.36191	6.36206
10	2.0	6.48967	7.88079	7.98107	7.98611	7.98632
11	2.1	7.98760	9.83956	9.97422	9.98099	9.98127
12	2.2	9.79513	12.2403	12.4196	12.4286	12.4290
13	2.3	11.9742	15.1801	15.4172	15.4292	15.4297
14	2.4	14.5990	18.7778	19.0895	19.1052	19.1058
15	2.5	17.7588	23.1779	23.5855	23.6061	23.6069
16	2.6	21.5605	28.5570	29.0878	29.1146	29.1157
17	2.7	26.1326	35.1305	35.8189	35.8537	35.8551
18	2.8	31.6292	43.1612	44.0510	44.0959	44.0978
19	2.9	38.2350	52.9697	54.1162	54.1741	54.1765
20	3.0	46.1720	64.9471	66.4201	66.4946	66.4977

Table 1.1 summarizes the results of using our two Euler methods and Euler_3 and Euler_4 as well. These incorporate the third and fourth derivatives, respectively, into the process. You will observe that these latter higher-order methods become very accurate over the entire interval.

Figure 1.4 shows all of these results plotted on a single set of axes. Notice that the more accurate methods are indistinguishable from the exact solution at the resolution possible in this graph.

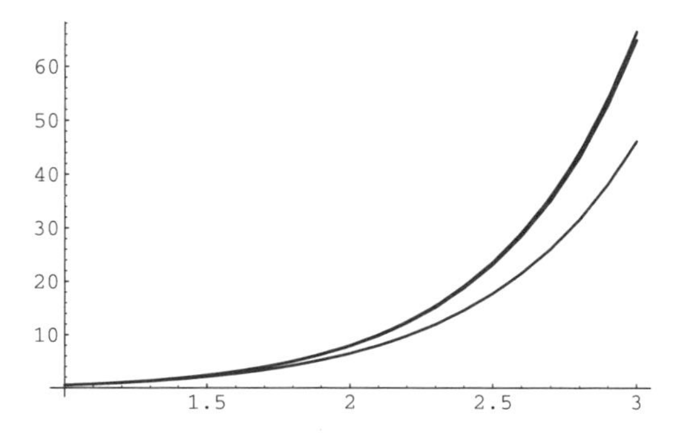

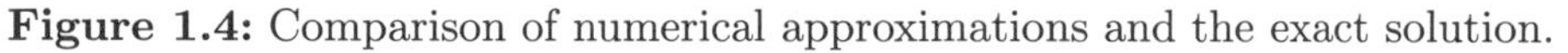

Figure 1.4: Comparison of numerical approximations and the exact solution.

Reducing the Step Size

We have discussed how improving the method can improve the accuracy for a fixed step size. Another common way to improve accuracy with a given method is to reduce the step size and take more steps to cross the desired interval. This is a valid approach. The primary negative aspect of reducing the step size is that it takes longer to cross an interval. This may or may not be important. One important consideration is that more steps with inaccurate information may cause the inaccuracies to compound into quite a large effect. This is a topic for extensive study in numerical analysis courses. Figure 1.5 demonstrates that reducing the step size, as well as improving the method, can reduce the error. Four plots appear. The enormous value of h is 0.5. This was chosen to amplify the effects for easier visualization. The four plots, from top to bottom, are the exact solution, the Euler$_2$ method that uses a quadratic polynomial, the standard Euler method in four steps of $h/4 = 0.125$, and the standard Euler method in a single step of size $h = 0.1$. Notice how using the standard Euler method in four steps allows the solution to bend at three interior points, and thus follow the correct solution more exactly. The Euler$_2$ method has a bend built in, but it, too, would give more accurate answers if it were applied more times using a reduced step size.

From time to time there will be an opportunity to discuss some important aspects of the numerical solution of differential equations.

EXERCISES 1.3

1. Evaluate `NestList[g,a,4]` to see what `NestList` does. Explain how this is applicable to iterative methods such as Euler's method or the Runge-Kutta method. What is "a" for Euler's method?

2. Consider the differential equation $dy/dx = 1 + y^2$ with $y(0) = 0$.

   (a) Use Euler's original method with $h = 0.1$ to estimate points on the solution of the stated problem. Find your solution on the interval $[0, 1.5]$.

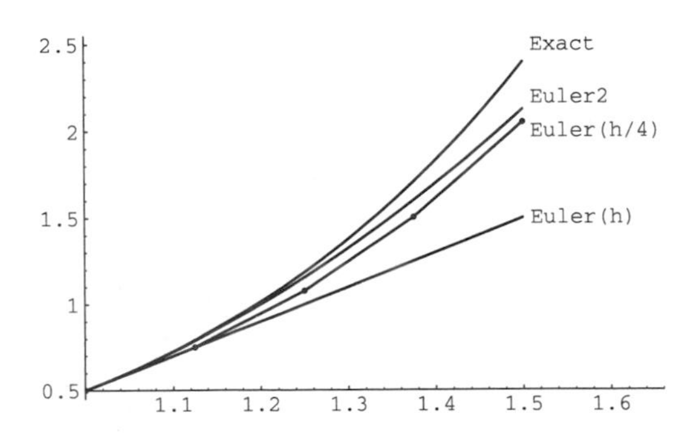

Figure 1.5: A closer look at Euler, Euler with reduced step, Euler$_2$, and the exact solution.

(b) Use `NDSolve` to find an approximate solution. Let:

```
{x0, y0}={0, 0}
h=0.1
s=NDSolve[{y'[x]==1+y[x]^ 2,y[x0]==y0},y[x],x]
```

Capture your solution using

```
w[x_]=y[x]/.First[s]
```

Then make a table of values of the solution function `w[x]`.

```
t=Table[{x0+k*h,w[x0+k*h]}{k,0,15}]}
```

(c) Compare these values to those that you calculated. If you used *Mathematica* to calculate the points from Euler's method, you can use the built-in function `ListPlot` to plot them. You can also `Plot` the function `w[x]`. The exact solution is $y(x) = \tan(x)$. You can compare both methods to this, if you like.

3. Incorporate the second derivative into Euler's method for the previous problem. Recalculate the estimated solution and compare with the previous results.
4. Incorporate the third derivative as well, calculate, and compare.
5. Use the Runge-Kutta method presented on the same problem. Compare results.

1.4 Uniqueness Considerations

Theorem 1.1, our existence and uniqueness theorem, says that existence and uniqueness are local properties of a differential equation. In this section, we examine a differential equation that fails to have a unique solution at any point through which a solution passes. In addition, no solutions pass through the half-plane where $y < 0$.

We seek a differential equation whose solutions are precisely of the form

$$y = (x - a)^2$$

where a is a real number. This one-parameter **family of curves** (Figure 1.6) consists of all horizontal translates of the parabola $y = x^2$. Observe that for each x, the corresponding point on any solution curve lies on or above the x-axis. This means that no solution will ever be negative. The differential equation of the family is found by taking a derivative:

$$y' = 2(x - a).$$

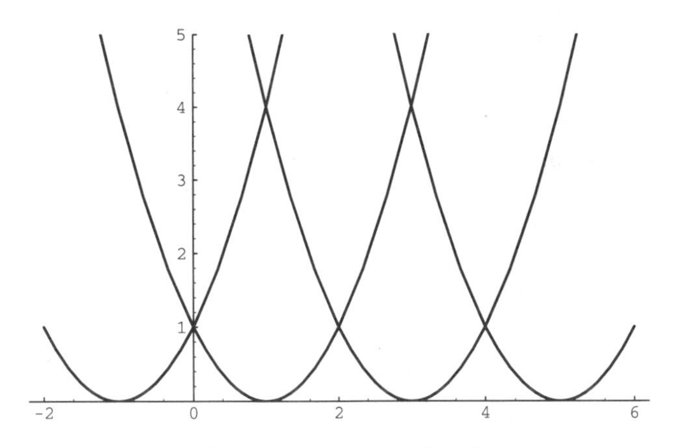

Figure 1.6: The one-parameter family of solutions.

Then from $(x-a) = y'/2$, one obtains

$$y = \left(\frac{y'}{2}\right)^2,$$

or the simpler equation $(y')^2 = 4y$ as a differential equation of the family.

It is easy to see that through each point (p, q) where $q \geq 0$ there pass exactly two members of the family of curves. To show this, suppose that $q > 0$. Then $q = (p-a)^2$ gives two choices: $a = p \pm \sqrt{q}$ for the parameter a. If $q = 0$, then from $0 = (p-a)^2$ one finds that $a = p$ is the only choice for the parameter. But $y = 0$ is another solution of the differential equation that passes through $(p, 0)$. This is the second solution that passes through $(p, 0)$. Note that the solution $y = 0$ of the differential equation was not a member of the family of solutions. Because it is somehow a different kind of solution, it is called a **singular solution**. This singular solution is tangent to each member of the family exactly once.

From the differential equation itself one sees that it is necessary that $y \geq 0$, since the left-hand side of the equation is a square. It also follows that at the point (p, q) if $q > 0$, then there are two choices, $y' = \pm\sqrt{q}$, for the slope of a solution curve at (p, q). But if $q = 0$, then it is required that $y' = 0$.

Most of the upper half-plane, except for the positive y-axis, is filled with solutions that pass through the origin: through each point (p, q) with $0 \leq q \leq p^2$ there is at least one solution that passes through the origin. We describe some of them.

If $p \geq 0$ take

$$y(x) = \begin{cases} 0, & x < p - \sqrt{q} \\ (x - p + \sqrt{q})^2, & x \geq p - \sqrt{q} \end{cases}.$$

See Figure 1.7.

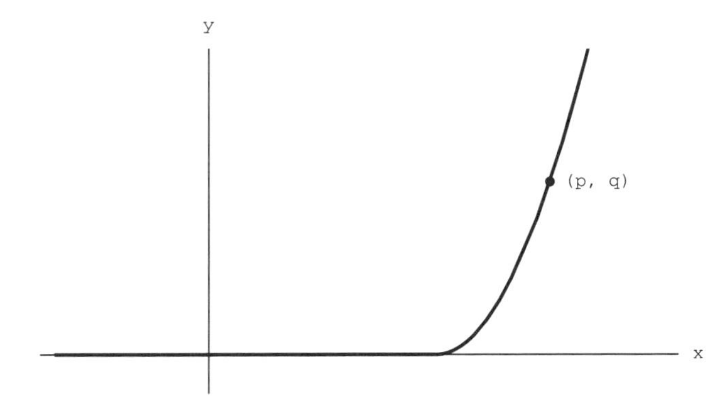

Figure 1.7: Solution passing through the origin, $p > 0$.

If $p \leq 0$ take

$$y(x) = \begin{cases} 0, & x > p + \sqrt{q} \\ (x - p - \sqrt{q})^2, & x \leq p + \sqrt{q} \end{cases}.$$

See Figure 1.8.

Let us write down the complete set of solutions that pass through the point $(2, 1)$. To aid us in our description, Figure 1.9 is a picture of the set we are attempting to describe.

The two curves in Figure 1.9 that cross at $(2, 1)$ are

$$y = (x - 1)^2$$

and

$$y = (x - 3)^2.$$

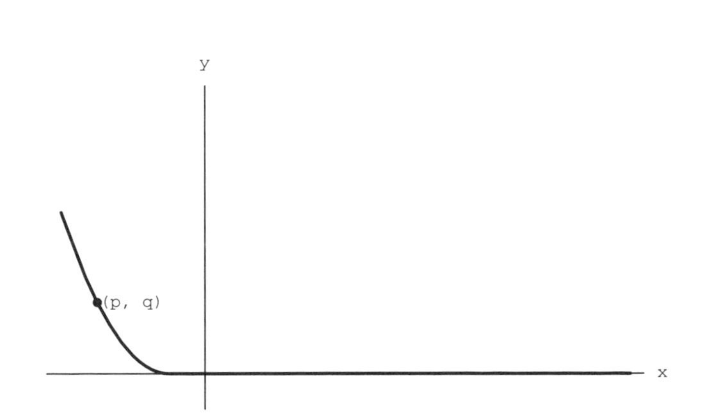

Figure 1.8: Solution passing through the origin, $p < 0$.

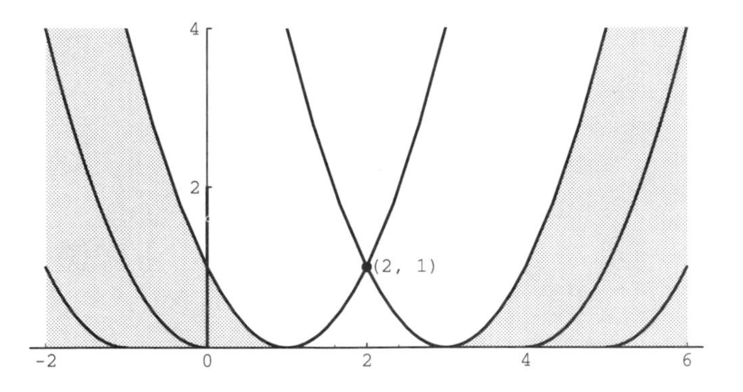

Figure 1.9: The set of solutions passing through the point $(2, 1)$.

The left gray area consists of all curves of the form

$$y(x) = \begin{cases} (x-p)^2, & x < p \\ 0, & p \le x \le 1 \\ (x-1)^2, & 1 < x \end{cases},$$

where $p < 1$. The right gray area consists of all curves of the form

$$y(x) = \begin{cases} (x-3)^2, & x < 3 \\ 0, & 3 \le x \le p \\ (x-p)^2, & p < x \end{cases},$$

where $p > 3$. Each of these curves has a "bathtub" shape, being a portion of the x-axis with half a parabola at either end. The two remaining solutions are

$$y(x) = \begin{cases} 0, & x < 1 \\ (x-1)^2, & x \ge 1 \end{cases},$$

and

$$y(x) = \begin{cases} (x-3)^2, & x < 3 \\ 0, & x \ge 3 \end{cases}.$$

These latter solutions have a "ski ramp" shape, with each consisting of a ray on the x-axis and half of a parabola. They look similar to Figures 1.8 and 1.9.

In summary, near the point $(2, 1)$ there are only two choices for the solution, but on any interval containing $x = 2$ that extends beyond $x = 1$ to the left or beyond $x = 3$ to the right there are infinitely many solutions of $(y')^2 = 4y$ that pass through the point $(2, 1)$. If we specify the sign of $y'(2)$ at the point $(2, 1)$ then near $(2, 1)$ the solution is unique, but not beyond $x = 1$ on the left if the slope $y'(2) > 0$ or beyond $x = 3$ on the right if the slope $y'(2) < 0$.

Theorem 1.1 warned us that there might be problems with uniqueness along the x-axis. We have $f(x, y) = \sqrt{y}$ or $f(x, y) = -\sqrt{y}$. In either case the partial derivative with respect to y is undefined, and hence not continuous, when $y = 0$. The theorem was unable to guarantee uniqueness where $y = 0$. With either $f(x, y) = \sqrt{y}$ or $f(x, y) = -\sqrt{y}$ we would have had uniqueness away from the

x-axis, but we had both since y' was squared. This gave us two solutions, one for $+$ and one for $-$, locally, off of the x-axis.

This example illustrates some of the things that can happen when a differential equation fails to have unique solutions.

EXERCISES 1.4

1. Repeat the ideas of this section for the family of cubics that are precisely of the form $y = (x-a)^3$. You may find it instructive to let *Mathematica* carry out the same sequence of operations that you do manually.
   (a) Find a differential equation for the family and show that $y \equiv 0$ is a solution. (*Mathematica* gives several differential equations, all of which are equivalent.)
   (b) Show that there are several kinds of solutions that involve part of one cubic, possibly part of $y = 0$, and then possibly part of another cubic.
   (c) Describe all of the kinds of solutions there are.
   (d) Find all of the solutions that pass through the point $(1, 2)$.

1.5 Differential Inclusions (Optional)

Rather than insist that $y(x)$ be a differentiable solution of a differential equation such as $dy/dx = f(x, y)$, suppose we merely ask that dy/dx be in some set S. We might write this as $dy/dx \in S$. This is an example of a **differential inclusion**.

Definition 1.6 *Let S be a set and I an interval of real numbers. An inclusion such as*

$$\frac{dy}{dx} \in S \tag{1.2}$$

is called a first-order **differential inclusion**, *because it asks that dy/dx be a member of a set, rather than giving an equation defining dy/dx. A continuous function $y(x)$ is called a* **solution** *of the differential inclusion* (1.2) *on I provided that $dy/dx \in S$ except possibly at a finite number of points of I at which dy/dx may fail to exist. If x_0 is in I and y_0 is a number, an* **initial value problem** *for the differential inclusion* (1.2) *asks that $y(x)$ satisfy*

$$\frac{dy}{dx} \in S \quad \text{and} \quad y(x_0) = y_0.$$

The set S can have parameters such as x or y or both.

A differential inclusion generally places fewer restrictions on a function that can be called a solution than does a differential equation. Since solutions of differential equations have to be differentiable everywhere, they are better behaved than some solutions of differential inclusions. Solutions of differential inclusions can have "corners" at points where they have no slope. Furthermore, if S has parameters x and y and there is only one member $f(x, y)$ in S for each permissible x and y, then the differential inclusion is really a differential equation: $dy/dx = f(x, y)$. All of this suggests that the requirement of differentiability everywhere for a solution to a differential equation is not necessary. This is true, but we leave the study of the implications of this remark to a later course in differential equations.

Let's look at an example of a differential inclusion. Let $S = \{-1, 1\}$ and consider $dy/dx \in S = \{-1, 1\}$. A solution $y(x)$ is continuous, and either $dy/dx = -1$, or $dy/dx = 1$ at each point where $y(x)$ has slope. On any finite interval, we only allow a finite number of points where $y(x)$ fails to have slope. What do our solution functions look like? In general, they consist of a broken line where each segment either has slope 1 or slope -1. Figure 1.10 gives a typical picture.

Of course a solution is permitted to be differentiable. Any function $y(x) = x + c$ or $y(x) = -x + c$, with c a constant, is a differentiable solution of $dy/dx \in \{-1, 1\}$. Suppose that we specify that each solution pass through the point (p, q). Then the two differentiable solutions that pass through (p, q) are the lines $y - q = +(x - p)$ and $y - q = -(x - p)$.

If a solution to $dy/dx \in \{-1, 1\}$ is not required to be differentiable everywhere, what are the solutions that pass through the point $(0, 2)$, for example? Figure 1.11 shows a picture that represents members of the set of solutions in the half-plane $x \geq 0$.

The particular solution that is drawn with thicker lines in Figure 1.11 is given by

$$y(x) = \begin{cases} -x + 2, & 0 \leq x \leq 1 \\ x, & 1 < x \leq 2 \\ -x + 4, & 2 < x \leq 4 \end{cases}.$$

Note that for $x \in (0, 1) \cup (1, 2) \cup (2, 4)$, either $dy/dx = 1$ or $dy/dx = -1$, so that $dy/dx \in \{-1, 1\}$.

Definition 1.7 *The functions $h(x)$ and $g(x)$ are called a* **maximal solution** *and a* **minimal solution**, *respectively, of the initial value problem $dy/dx \in S$, $y(x_0) = y_0$ on an interval I, if each is a solution of the initial value problem and*

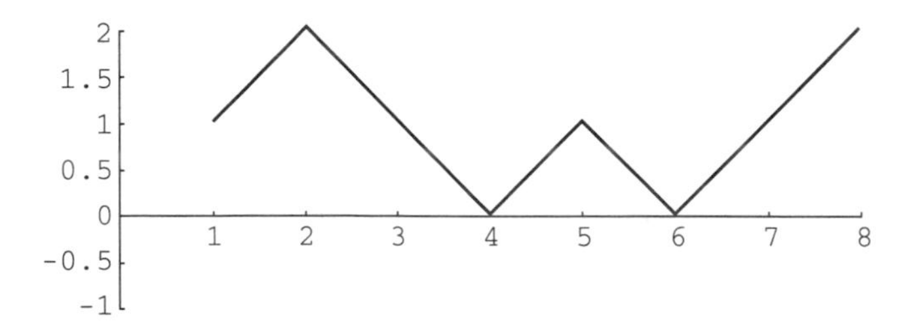

Figure 1.10: One solution of $dy/dx \in \{-1, 1\}$ over $1 \leq x \leq 8$.

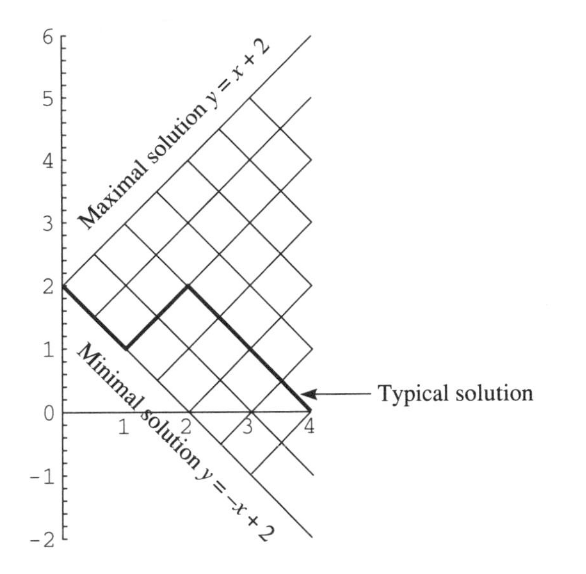

Figure 1.11: Solutions of $dy/dx \in \{-1, 1\}$, $y(0) = 2$.

for each x in I, whenever $y(x)$ is a solution of the initial value problem, then $g(x) \leq y(x) \leq h(x)$.

Note the apparent existence of a **maximal solution** and a **minimal solution** in Figure 1.11. Every solution must remain between these. That is, for $x \geq 0$, each solution $y(x)$ satisfies $-x+2 \leq y(x) \leq x+2$. (Why?) This is typical of differential inclusions. What are the maximal and minimal solutions for $x \leq 0$? Can you explain why they are different?

Sometimes the maximal and minimal solutions are the same over an interval. Then the solution is unique over any interval where this occurs. Look back at the differential equation $(y')^2 = 4y$ of the last section. Isn't that a differential inclusion: $dy/dx \in \{-2\sqrt{y}, 2\sqrt{y}\}$? What we changed in this section is that there may now be points at which the derivative does not exist. What happens to the solution of the example in Section 1.4 if the derivative of the solution can fail to exist at isolated points? Note that in Section 1.4, there is a maximal solution and a minimal solution for $x \geq 1$. The same is true for $x \leq 1$, but the maximal solution and minimal solution on the left are different from those on the right.

Exercises 1.5

1. Determine the maximal and minimal solutions for $x \geq 0$ to the differential inclusion $dy/dx \in \{-1, 1\}$, with initial condition $y(0) = 2$.

2. Find all differentiable solutions to the differential inclusion $dy/dx \in \{-x, x\}$. You will need to integrate the two equations $dy/dx = x$ and $dy/dx = -x$.

3. Given the differential inclusion $dy/dx \in \{-x, x\}$ of Problem 2 with initial condition $y(0) = 3$.
   (a) Depict the set of all solutions.
   (b) Determine the maximal and minimal solutions.
   (c) Find descriptions of those solutions that follow the maximal or minimal solution for a while, then branch off and continue onward as differentiable functions. That is, they have only one corner.
   (d) Describe the set of all points (p, q) with $p > 0$ through which at least one of the solutions passes.
   (e) Find the set of points (a, b) with $0 < a < p$ such that some solution to $dy/dx \in \{-x, x\}$ passes through each of the points $(0, 2)$, (a, b), and (p, q). How far can this process of finding intermediate points be continued?

PROJECT A. Examine the differential inclusion $dy/dx \in \{-2\sqrt{y}, 2\sqrt{y}\}$. (This is really the differential equation of Section 1.4 expressed as a differential inclusion.)

(a) All of the solutions given in Section 1.4 are still solutions, but now there can be corners. Write down typical solutions, including those that contain a segment of the x-axis. Use Section 1.4 as a guide.
(b) Suppose that the initial condition is given as before: $y(2) = 1$. What are the maximal and minimal solutions of this initial value problem? (Distinguish between $x \leq 2$ and $x \geq 2$.) Write down formulas for some solutions that may have corners, in terms of the maximal and minimal solutions. Be careful to state the domain of definition of each portion of each solution.
(c) How does Figure 1.9 of Section 1.4 change under the conditions of part (b)? How complete are your lists of solutions that you made in parts (a) and (b)?

PROJECT B. Consider the differential inclusion $dy/dx \in [-1, 1]$. That is, wherever a solution function has slope, that slope is not greater than 1 in absolute value.

(a) Show that some multiple of every function that has bounded slope is a solution.
(b) Show that $y = e^x$ is a solution over a restricted domain. (What is that domain?)
(c) Can you find functions that are not solutions over any interval?
(d) What are the maximal and minimal solutions that pass through the origin?
(e) What are the maximal and minimal solutions that pass through the point (p, q)?

2 First-Order Differential Equations

2.0 Introduction

First-order differential equations provide a rich example of differential equations of many forms, most of which we can solve easily in the formal sense, and many of which we can solve and actually get answers. From calculus, we need the rules of differentiation, both the formulas (for sums, products, quotients, chain rule, and so on) and the derivatives of the standard *Mathematica* functions (x^n, trigonometric functions, logarithms, exponentials, hyperbolic functions, and the like), and techniques of integration. We will likely see an example of most of the kinds of integrals that you ever attempted. If this sounds like bad news, the good news is that *Mathematica* can do these integrations for you. You will serve as the mastermind, and *Mathematica* will do the labor. Your responsibility is to ensure the correctness of the work that you are having *Mathematica* do, but *Mathematica* will do these correctness and consistency checks for you. You control what is being done; *Mathematica* does the hard work. It is important that you manually do some examples of problems of each type. The reason for this is that **if you have no idea how to do a problem yourself, then it is not too likely that you will know how to guide *Mathematica* through the solution process.**

Following each example or group of examples you will find sample code showing how to get *Mathematica* to do what was just done manually. If you work through the examples yourself, you will quickly begin to appreciate just how much of the computational burden *Mathematica* can assume.

2.1 First-Order Linear Differential Equations

First-Order Linear Differential Equations

These equations are among the simplest to solve—at least in theory. If you are good at calculating integrals, you will be good at this. If you have had trouble with integrals, at least you have the confidence that *Mathematica* can do these problems easily. If you formulate the problem correctly, *Mathematica* will give you the correct answer. Even when *Mathematica* cannot give you an answer in closed form (from `DSolve`), it can give you an answer that you can plot and from which you can get accurate numeric function values (from `NDSolve` in versions 2.0 and later).

The general first-order differential equation has the form:

$$a_1(x)y'(x) + a_0(x)y(x) = b(x). \tag{2.1a}$$

This can be rewritten in the simpler form

$$y'(x) + p(x)y(x) = q(x), \tag{2.1b}$$

with $p(x) = a_0(x)/a_1(x)$ and $q(x) = b(x)/a_1(x)$, provided that $a_1(x)$ is never 0 on any interval I over which we desire a solution. Assume that $a_1(x)$ and $a_0(x)$ are continuous on I. Initially we will also want $b(x)$ to be continuous on I, though some very interesting problems occur where $b(x)$, and hence $q(x)$, is discontinuous.

To draw your attention to the linear algebraic side of things, note that equations of this form consist of two parts,

1. a **function** L, defined by $L(y) = a_1(x)y' + a_0(x)y$ for y a differentiable function defined on I, and
2. a **function** b, which is the desired result when we evaluate L at y.

The functions $a_1(x)$ and $a_0(x)$ are the **coefficients**; they are functions of x alone (not y). They may be constant.

We deliberately separated from the definition of L any mention of the relationship of y to the variable x to emphasize the fact that L has as its domain a set of functions, of which y is a representative, and to say how L operates on each of these functions. $L(y)$ is also a function, but its domain is the interval I. A more complete definition of L says what $L(y)$ does to x. Here is this fuller definition:

$$L(y)(x) = a_1(x)y'(x) + a_0(x)y(x).$$

This notation says that $L(y)$ is a function, albeit with a complicated name, whose value at the number x is $L(y)(x)$. This is exactly like using $f(x)$ to denote the value of f at x. The operator L can be considered independently of $b(x)$, and indeed it is most productive to do so! We can look at $L(\sin x), L(e^x), L(x^2 - 3x)$ and so on. We could try many functions and not find one for which $L(y)(x) = b(x)$, but the point is: *if we substitute a function into L, we get back a function.* What we need is a simple way to decide which function to substitute into L so that the result is $b(x)$. That is what it means to **solve** the differential equation. It turns out that there are lots of functions that work, but that they are all related in a simple way, and our look at linear algebra in the last chapter showed us how to approach the problem.

If $b(x) \neq 0$, we have a **nonhomogeneous** problem. In the last chapter, we saw that to solve a nonhomogeneous linear problem completely, we should

1. completely solve the **homogeneous** problem,
2. find one solution of the **nonhomogeneous** problem and then
3. **add these** to obtain every solution.

We can use an observation from calculus to get ourselves started and then, having obtained the solutions, look back and analyze them.

A place to start is one of the simplest differential equations there is: Given the function $g(x)$, find $f(x)$ so that $(d/dx)f(x) = g(x)$. You recognize that the solution is

$$f(x) = \int g(x)\,dx + c.$$

There is linear algebra underlying even this simple problem:

- What is the linear function? The differentiation operator, (d/dx).
- What is its kernel? The set of all constant functions, represented by $u(x) = c$.
- What is $q(x)$, the function that (d/dx) sends into $g(x)$? That is merely $q(x) = \int g(x)\,dx$.
- The solution function $f(x)$ given above then can be thought of as following our prescription for solving linear problems: find the kernel $(= c)$ and something that satisfies the nonhomogeneous equation $(= \int g(x)\,dx)$ and then add.

We will call the "something that satisfies the nonhomogeneous equation" a **particular solution**. This will distinguish it from what is usually called the **general solution** or **complete solution** $(= \int g(x)\,dx + c)$, which is a description of all of the solutions to the problem.

We now proceed to find the general solution of the differential equation

$$y'(x) + p(x)y(x) = q(x). \tag{2.2}$$

We solve this by making an observation: (admittedly this is an unfair approach, but it is traditional)

$$\begin{aligned}\frac{d}{dx}\left(e^{\int p(x)\,dx}\,y(x)\right) &= e^{\int p(x)dx}\,y'(x) + p(x)e^{\int p(x)dx}\,y(x)\\ &= e^{\int p(x)dx}\,(y'(x) + p(x)y(x)).\end{aligned}$$

The second factor of this expression is the left-hand side of our differential equation, so if we multiply both sides of our differential equation through by $e^{\int p(x)\,dx}$, we have

$$\frac{d}{dx}\left(e^{\int p(x)\,dx}\,y(x)\right) = e^{\int p(x)\,dx}\,q(x).$$

We just solved an equation of this form in the paragraph above. In the present case, the unknown function is $e^{\int p(x)\,dx}\,y(x)$, for which we can solve to get

$$e^{\int p(x)\,dx}\,y(x) = \int e^{\int p(x)\,dx}\,q(x)\,dx + c.$$

From this we isolate our solution $y(x)$:

$$\begin{aligned} y(x) &= e^{-\int p(x)\,dx}\left(\int e^{\int p(x)\,dx}\,q(x)\,dx + c\right) \\ &= e^{-\int p(x)\,dx}\int e^{\int p(x)\,dx}\,q(x)\,dx + ce^{-\int p(x)\,dx}. \end{aligned} \tag{2.3}$$

If you look closely at equation 2.3, you will observe that the symbol x has three different meanings:

1. the argument of y,
2. the variable of integration in $e^{\int p(x)\,dx}$, and
3. the variable of integration in $\int e^{\int p(x)\,dx}\,q(x)\,dx$.

This has the potential for confusion, and we will eventually remedy this problem by using definite integrals.

It may be difficult to identify the two parts of the solution given in (2.3) that we expected to get. Here is a hint: the part that describes the kernel will always have arbitrary constants in it (here the part with c in it) and what remains is a particular solution. Let's check out these statements.

Given that $L(y) = y' + p(x)y$, what does L do to the two terms of our solution, equation 2.3?

(a) First,

$$\begin{aligned} L\left(ce^{-\int p(x)\,dx}\right) &= \frac{d}{dx}\left(ce^{-\int p(x)\,dx}\right) + p(x)\left(ce^{-\int p(x)\,dx}\right) \\ &= -p(x)\left(ce^{-\int p(x)\,dx}\right) + p(x)\left(ce^{-\int p(x)\,dx}\right) \\ &= 0. \end{aligned}$$

(b) Second,

$$\begin{aligned} &L\left(e^{-\int p(x)\,dx}\int e^{\int p(x)\,dx}q(x)\,dx\right) \\ &\quad= \frac{d}{dx}\left(e^{-\int p(x)\,dx}\int e^{\int p(x)\,dx}q(x)\,dx\right) \\ &\qquad+ p(x)\left(e^{-\int p(x)\,dx}\int e^{\int p(x)\,dx}q(x)\,dx\right) \end{aligned}$$

$$= -p(x)\left(e^{-\int p(x)\,dx}\int e^{\int p(x)\,dx}q(x)\,dx\right) + e^{-\int p(x)\,dx}e^{\int p(x)\,dx}q(x)$$
$$+\,p(x)\left(e^{-\int p(x)\,dx}\int e^{\int p(x)\,dx}q(x)\,dx\right)$$
$$= q(x).$$

Part (a) verifies that

$$u_c(x) = \left(ce^{-\int p(x)\,dx}\right)$$

is in the kernel of L, and part (b) verifies that

$$y_p(x) = \left(e^{-\int p(x)\,dx}\int e^{\int p(x)\,dx}\,q(x)\,dx\right)$$

is a particular solution of $L(y) = q(x)$. As the theory states, $y(x) = u_c(x)+y_p(x)$ is the complete solution of the first-order linear differential equation $L(y) = q(x)$. In the *Mathematica* notebook *Check Linear Theory* these two calculations are done by *Mathematica*.

We have left one thing to do. We know that the function $u_c(x) = ce^{-\int p(x)\,dx}$ is never 0 (so long as $p(x)$ is continuous and $c \neq 0$) and that $L(u\,c) = 0$, from (a) above. For any choice of c, $L(u) = 0$, but is every function $z(x)$ in the kernel of L representable as $z(x) = u_c(x)$ for some c? The answer is "Yes," as we now show: Suppose that $z(x)$ is in the kernel of L. Consider the function $w(x) = z(x)/v(x)$, where $v(x) = e^{-\int p(x)\,dx}$. Calculate

$$w'(x) = \frac{z'(x)v(x) - z(x)v'(x)}{(v(x))^2}$$
$$= \frac{-p(x)z(x)v(x) - z(x)(-p(x)v(x))}{(v(x))^2}$$
$$= 0$$

Since $w'(x) = 0$, $w(x) = c$. Thus $z(x) = v(x)w(x) = v(x)c = ce^{-\int p(x)\,dx}$, and hence is one of the functions represented by $u_c(x)$. This means that $u_c(x)$ represents the entire kernel of L.

Theorem 2.1 *Let $L(y) = y' + p(x)y$. Then*

(a) $u_c(x) = ce^{-\int p(x)\,dx}$ represents the entire kernel of L: $L(u_c)(x) = 0$ for every c, and every member of the kernel of L has this form.

(b) $y_p(x) = e^{-\int p(x)\,dx}\int e^{\int p(x)\,dx}\,q(x)dx$ is a particular solution of $L(y)(x) = q(x)$.

(c) The expression $y(x) = y_p(x)+u_c(x)$ represents all *of the solutions of $L(y)(x) = q(x)$: it is the general solution. Every solution of $L(y)(x) = q(x)$ has this form.*

The interaction between linear algebra and differential equations was this: linear algebra told us the form that the solution would take; differential equations invoked certain methods from calculus to enable us to actually find the solutions. This is the way it will continue to be throughout our study of linear differential equations: **Linear algebra will suggest the form of the solution, and differential equations will dictate the methods we must use to find the solution.**

Exercises 2.1

Part I. Solve these first-order linear differential equations manually and by *Mathematica*. Check your answers.

1. $\dfrac{dy}{dx} - 5y = 0.$

2. $\dfrac{dy}{dx} - 5y = e^{2x}.$

3. $\dfrac{dy}{dx} + (\cos x)y = 3\cos x.$

4. $\dfrac{dy}{dx} + (\tan x)y = \sin x.$

5. $\dfrac{dy}{dx} + 2xy = 0.$

6. $\dfrac{dy}{dx} + 2xy = 4x.$

7. $x\dfrac{dy}{dx} + 2y = 0.$

8. $x\dfrac{dy}{dx} + 2y = e^{x^2}.$

9. $x\dfrac{dy}{dx} + 2y = 3x.$

10. $(x+3)\dfrac{dy}{dx} + 2y = 0.$

11. $(x+3)\dfrac{dy}{dx} + 2y = (x+3)^4.$

12. $x\dfrac{dy}{dx} - 2y = 0.$

13. $x\dfrac{dy}{dx} - 2y = x^3.$

PART II. Solve and check these differential equations by *Mathematica*. [Look up the proper *Mathematica* representation for ln, tanh, sinh and cosh.]

14. $\dfrac{dy}{dx} - 5y = \cos x.$

15. $\dfrac{dy}{dx} + (\cos x)y = 3\sin x.$

16. $\dfrac{dy}{dx} + 2x\,y = 4x^2.$

17. $x\dfrac{dy}{dx} - 2y = x^9 \ln x.$

18. $x\dfrac{dy}{dx} - 2y = 3(\ln x)^2 - 2(\ln x)^3.$

19. $\dfrac{dy}{dx} + (\tanh x)y = \sinh x.$

20. $\dfrac{dy}{dx} + (\cosh x)y = 3\cosh x.$

2.2 Linear Equations by *Mathematica*

First-Order Linear Differential Equations by *Mathematica*

Having seen the theory of first-order linear differential equations and studied several examples, it is time to see how to get *Mathematica* to solve these problems for you. We will look at the same examples that were done manually in the last section. The notebook *Using DSolve* guides you through these same examples. You may read through that notebook rather than the text of this section. In that notebook, the explanations are fuller than they are here. This is basically a summary of that notebook.

Reminder, in *Mathematica* use == when defining an equation or for testing equality between two expressions. Use = for assignment. This distinction is important.

Example 2.1M Solve the differential equation $y' - 4y = 0$ in *Mathematica*.

Solution. Name the equation we are trying to solve and the variables x and y. It is easy to verify that $y(x) = c\exp(4x)$ is a solution from $y'(x) - 4y(x) = 4c\exp(4x) - 4c\exp(4x) = 0$. Here is how to find this solution using *Mathematica*.

```
In[1]:=  de[x_,y_] = (y'[x] - 4 y[x] == 0)

Out[1]=  -4 y[x] + y'[x] == 0
```

When DSolve is used, the result is given in the form of (one or more) rules. These rules do not define y[x], but say how y[x] should be defined.

```
In[2]:=  SolnRule = DSolve[de[x,y],y[x],x]

Out[2]=               4 x
         {{y[x] -> E      C[1]}}
```

We can capture our first solution and call it y1[x] this way: Substitute our rule into y[x], and use this expression to define y1[x], which is our solution.

```
In[3]:=  y1[x_] = Simplify[y[x]/.SolnRule[[1]]]

Out[3]=   4 x
         E    C[1]
```

The use of Simplify was not necessary here. But often it is useful, if not necessary, so it is a good habit to use it. The [[1]] extracts the contents of the outer list. {{a}}[[1]]=={a}. Otherwise y1[x] would look like y1[x] = {expr}, rather than y1[x] = expr. You will want to remember to extract the contents this way when capturing solutions. Alternatively, you can use First[expr] to accomplish the same thing as expr[[1]].

That y1[x] is a solution, no matter what value the parameter C[1] has, can be checked this easy way:

```
In[4]:=  Simplify[de[x,y1]]

Out[4]=  True
```

The symbol de[x,y] is an equation. When y1 is substituted for y, this equation is identically (not conditionally) True. Here, to remind you, is the definition of de[x, y]:

```
In[5]:=  de[x,y]

Out[5]=  -4 y[x] + y'[x] == 0
```

◇

Example 2.2M Solve the differential equation $xy' - 4y = x^7 e^x$ in *Mathematica*.

Solution. State the problem.

```
In[6]:=  de[x_,y_] = (x y'[x] - 4 y[x] == x^7 Exp[x])

Out[6]=                                x  7
         -4 y[x] + x y'[x] == E  x
```

Solve using DSolve.

```
In[7]:=  SolnRule = DSolve[de[x,y],y[x],x]

Out[7]=               x  4            2      4
         {{y[x] -> E  x  (2 - 2 x + x ) + x  C[1]}}
```

One can capture this solution, simplify its form, and call it y2[x] this way:

```
In[8]:=  y2[x_] = Simplify[y[x]/.SolnRule[[1]] ]

Out[8]=   4       x        x        x  2
         x  (2 E  - 2 E  x + E  x  + C[1])
```

There is no requirement to simplify, but why keep a solution that is unnecessarily complicated? That y2[x] is a solution can be checked this way:

```
In[9]:=  Simplify[de[x,y2]]

Out[9]=  True
```

Note: y2[x] can be obtained in one line by composing steps 7 and 8.

```
In[10]:= y2[x_] = Simplify[(y[x]/.DSolve[de[x,y],y[x],x][[1]]) ]

Out[10]=  4       x        x        x  2
         x  (2 E  - 2 E  x + E  x  + C[1])
```

◇

Example 2.3M Solve the differential equation $xy' + 4y = x^7 e^x$. (Note that the only change from the previous problem is a single sign: the "-" became "+". But observe how different the solution is!)

Solution. This time, for a change, let's not name our equation, but use it verbatim. Note how we lose generality.

```
In[11]:= SolnRule = DSolve[x y'[x] + 4 y[x] == x^7 Exp[x],y[x],x]

Out[11]=
                   x                                      2
         {{y[x] -> (E  (3628800 - 3628800 x + 1814400 x  -

                            3            4          5         6
                    604800 x  + 151200 x  - 30240 x  + 5040 x  -

                         7       8       9    10        4   C[1]
                    720 x  + 90 x  - 10 x  + x  )) / x  + ----}}
                                                               4
                                                              x
```

One can capture this solution, call it y3[x], this way:

```
In[12]:= y3[x_] = Simplify[y[x]/.SolnRule[[1]]]

Out[12]=   x                                    2             3
         (E  (3628800 - 3628800 x + 1814400 x  - 604800 x  +

                          4          5         6        7       8
                151200 x  - 30240 x  + 5040 x  - 720 x  + 90 x  -

                    9    10        4   C[1]
                10 x  + x  )) / x  + ----
                                        4
                                       x
```

That y3[x] is a solution can be checked this way:

```
In[13]:= Simplify[x y3'[x] + 4 y3[x] == x^7 Exp[x]]

Out[13]= True
```

Observe how much more typing is required when the equation is not named, and how much greater chance of error this introduces. Unlike the previous examples, we cannot just copy Input lines 11 or 13 for use elsewhere: these lines are special to this problem, and not general, as the others were. ◇

Exercises 2.2

Use DSolve on the problems in Exercises 2.1.

2.3 Exact Equations

Conservative Vector Fields

When you were studying multidimensional calculus, you certainly studied vector fields. A vector field on n-dimensional Euclidean space $\boldsymbol{R^n}$ is a function that assigns to each point of $\boldsymbol{R^n}$ a direction (vector) in $\boldsymbol{R^n}$. In symbols,

$$F(x_1, x_2, \ldots, x_n) \\ = (f_1(x_1, x_2, \ldots, x_n), f_2(x_1, x_2, \ldots, x_n), \ldots, f_n(x_1, x_2, \ldots, x_n)).$$

Examples are $F(x, y) = (y, x)$ on $\boldsymbol{R^2}$, and $F(x, y, z) = (yz, xz, xy)$ on $\boldsymbol{R^3}$.

Vector fields that are **conservative** are of particular importance and interest. (Both of the examples above are conservative.)

Definition 2.1 *Let U be a connected open set in $\boldsymbol{R^n}$ and F be a continuously differentiable vector field on U. A real function ϕ defined on U is called a* **potential function** *for F if the gradient of ϕ is F. The vector field F is called* **conservative** *if it has a potential function.*

One of the properties of conservative vector fields is that given a conservative vector field F, with potential function ϕ in a domain D, the integral of F along a piecewise continuous curve C from P to Q which lies completely in D can be calculated by: $\int_C F = \phi(Q) - \phi(P)$, and the value of the integral does not depend on the choice of the curve C. This integral looks like one of the forms for the fundamental theorem of calculus, and it does say that a potential function is an antiderivative of its associated vector field.

You also learned the theorem that if $f(x, y)$ is a function on $\boldsymbol{R^n}$ having continuous second partial derivatives, then

$$\frac{\partial^2 f}{\partial x \partial y} = \frac{\partial^2 f}{\partial y \partial x}.$$

A consequence of this is that so long as the derivatives involved are continuous,

$$\frac{\partial^{m+n} f}{\partial x^m \partial y^n} = \frac{\partial^{m+n} f}{\partial y^n \partial x^m}.$$

That is, when a given pair of mixed higher-order partial derivatives are of the same order in each variable, then they are equal. Thus the order of differentiation does not matter. This provides for an immense simplification in keeping track of higher-order derivatives. You may also have learned that if a vector field $(f(x,y), g(x,y))$ has the property that $\partial f/\partial y = \partial g/\partial x$ then the vector field is conservative. The obvious question is: *how do we find a potential function for the vector field, knowing that it is conservative?*

There is a process that is presented in most multidimensional calculus courses and most differential equations courses that will produce a potential function. The process is normally presented only for the case of two variables, and you are left to guess what it might be in higher dimensions.

A simple formula produces a potential function in one step (two integrations) for the two-dimensional case. Once you understand it, the generalization to higher dimensions is easy, and it is not at all difficult to remember. Here is the formula in two dimensions expressed as a theorem.

Theorem 2.2 *If the vector field $V(x,y) = (f(x,y), g(x,y))$ is conservative in the rectangle $R = [a,b] \times [c,d]$, and $P = (p_1, p_2)$ and $Q = (X, Y)$ are in R, then*

$$\phi(X,Y) = \int_{p_1}^{X} f(x, p_2)\, dx + \int_{p_2}^{Y} g(X, y)\, dy$$

is a **potential function** *for the vector field $V(x,y)$.*

Proof. Observe that the path of integration is along the horizontal line segment from P to the vertical line through Q and then along that vertical line segment to Q. To illustrate how to remember this formula, consider these parameterizations for the two lines, as illustrated in Figure 2.1:

HORIZONTAL: $x = t, \quad y = p_2, \quad p_1 \le t \le X,$

and

VERTICAL: $x = X, \quad y = s, \quad p_2 \le s \le Y.$

This is for the case when Q is to the right of and above P. Using this parameterization, the definition of ϕ is

$$\phi(X,Y) = \int_{p_1}^{X} f(t, p_2)\, dt + \int_{p_2}^{Y} g(X, t)\, dt.$$

The proof that ϕ is a potential function is just a matter of calculating the gradient to see that $\text{grad}(\phi(x,y)) = V(x,y)$. This is left as an exercise for you to do. □

There is a notebook, *Exact Differential Equations*, that generates potential functions in this manner and checks the results.

In three-space, the analog of "rectangle" is "box;" you need to get from the fixed corner P to the diagonally opposite "general" corner Q; and you do so

Figure 2.1: The path of integration to find a potential function: two variables.

by following a sequence of three connected edges from P to Q as Figure 2.2 indicates. You should parameterize the three parts of this curve.

Exact Differential Equations

Definition 2.2 *A differential equation in the form* $M(x,y)dx + N(x,y)$ $dy = 0$ *is called* **exact** *provided that the vector field* $V(x,y) = (M(x,y), N(x,y))$ *is conservative.*

To solve such an exact differential equation, we will find a potential function $\phi(x,y)$ and say that the solution is $\phi(x,y) = c$, where c is an arbitrary constant. Should we want a solution that passes through the point $P = (p_1, p_2)$, take $c = \phi(P)$, so that the solution is $\phi(x,y) = \phi(P)$. This defines the solution implicitly, rather than explicitly.

Example: Two Variables

Example 2.4 Show that the differential equation $ydx + xdy = 0$ is exact, and find the solution that passes through the point $(3,4)$.

Solution. Here $M(x,y) = y$ and $N(x,y) = x$. The equation is exact because

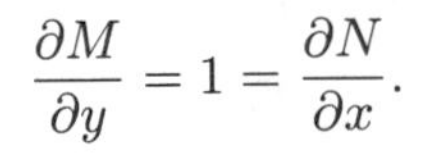

$$\frac{\partial M}{\partial y} = 1 = \frac{\partial N}{\partial x}.$$

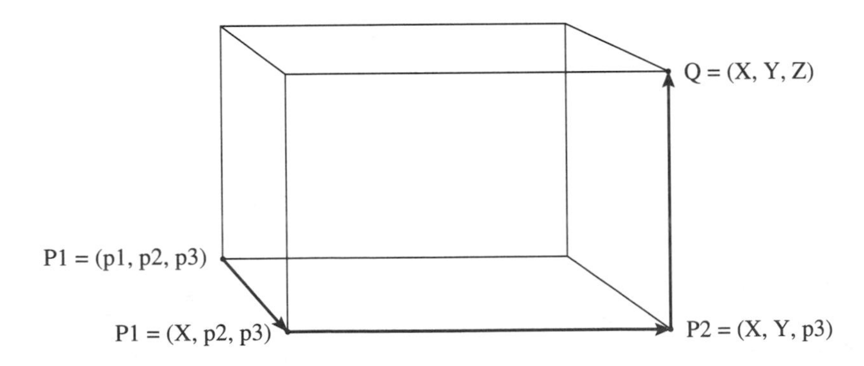

Figure 2.2: The integration path to find a 3-variable potential function.

From our theorem and the fact that we are given the fixed point $(3,4)$ from which to start the solution, we will want to express $\phi(x,y)$ this way:

$$\begin{aligned}\phi(X,Y) &= \int_3^X 4\,ds + \int_4^Y X\,dt \\ &= 4s\,\Big|_{s=3}^{X} + X\,t\,\Big|_{t=4}^{Y} \\ &= 4(X-3) + X(Y-4) \\ &= XY - 12\end{aligned}$$

so $\phi(X,Y) = XY - 12$. This checks in the differential equation and, in addition, $\phi(3,4) = 12 - 12 = 0$. ◇

Example 2.4M Here is Example 2.4, done in *Mathematica*.
Solution.

```
In[1]:=  MC[x_,y_] = y; (* use names MC and NC, because *)

In[2]:=  NC[x_,y_] = x; (* N, alone, is an illegal name *)

In[3]:=  {p1,p2} = {3,4};
```

Check for exactness.

```
In[4]:=  Simplify[D[MC[x,y],y] == D[NC[x,y],x]]

Out[4]=  True
```

Therefore the equation is exact. Solution using the method of the theorem:

```
In[5]:=  f[X_,Y_] = Integrate[MC[x,p2],{x,p1,X}]+
                    Integrate[NC[X, y],{y,p2,Y}]

Out[5]=  -12 + X Y
```

The solution we want is `f[x,y]==0`.

```
In[6]:=  f[X,Y] == 0

Out[6]=  -12 + X Y == 0
```

Define a gradient function

```
In[7]:=  grad[fcn_,vars_List] := Map[Function[t, D[fcn,t]],vars]
```

Check the solution using the function `grad` that we just defined.

```
In[8]:=  grad[f[x,y],{x, y}]

Out[8]=  {y, x}
```

◇

Example: Three Variables

Example 2.5 Show that the differential equation $y\,z\,dx + x\,z\,dy + x\,y\,dz = 0$ is exact and find the solution that passes through the point $(3,4,5)$.

Solution. Here $M_1(x,y,z) = y\,z$, $M_2(x,y,z) = x\,z$, and $M_3(x,y,z) = x\,y$. The equation is exact since

$$\frac{\partial M_1}{\partial y} = z = \frac{\partial M_2}{\partial x}, \quad \frac{\partial M_1}{\partial z} = y = \frac{\partial M_3}{\partial x}, \quad \text{and} \quad \frac{\partial M_2}{\partial z} = x = \frac{\partial M_3}{\partial y}.$$

The solution we desire is then

$$\begin{aligned}
\phi(X,Y,Z) &= \int_3^X (4)(5)\,dt_1 + \int_4^Y (X)(5)\,dt_2 + \int_5^Z (X)(Y)\,dt_3 \\
&= (20)t_1 \Big|_{t_1=3}^{X} + (5X)t_2 \Big|_{t_2=4}^{Y} + (XY)t_3 \Big|_{t_3=5}^{Z} \\
&= 20(X-3) + (5X)(Y-4) + (XY)(Z-5) \\
&= 20X - 60 + 5XY - 20X + XYZ - 5XY \\
&= XYZ - 60
\end{aligned}$$

So the result is $\phi(X,Y,Z) = X\,Y\,Z - 60 = 0$. This checks. The parameterizations used in defining the three connected lines were:

$$\begin{array}{lllll}
C_1: & x = t_1, & y = 4, & z = 5, & 3 \le t_1 \le X, \\
C_2: & x = X, & y = t_2, & z = 5, & 4 \le t_2 \le Y, \\
C_3: & x = X, & y = Y, & z = t_3, & 5 \le t_3 \le Z.
\end{array}$$

◇

Example 2.5M Here is Example 2.5 done in *Mathematica*.

Solution.

```
In[1]:=  M1[x_,y_,z_] = y*z; (* number the coefficients *)

In[2]:=  M2[x_,y_,z_] = x*z; (* to distinguish between  *)

In[3]:=  M3[x_,y_,z_] = x*y; (* them.                    *)

In[4]:=  {p1,p2,p3}    = {3,4,5};
```

Test for exactness:

```
In[5]:=  And[D[M1[x,y,z],y] == D[M2[x,y,z],x],
             D[M1[x,y,z],z] == D[M3[x,y,z],x],
             D[M2[x,y,z],z] == D[M3[x,y,z],y]]

Out[5]=  True
```

Therefore the equation is exact.

```
In[6]:=  f3[X_,Y_,Z_] = Integrate[M1[t1,p2,p3],{t1,p1,X}]+
                          Integrate[M2[X,t2 ,p3],{t2,p2,Y}]+
                          Integrate[M3[X, Y ,t3],{t3,p3,Z}]

Out[6]=  -60 + X Y Z
```

The solution we want is `f3[x, y, z]==0`:

```
In[7]:=  f3[X,Y,Z] == 0

Out[7]=  -60 + X Y Z == 0
```

Check that this is a solution. The grad function defined in Example 2.4M works for three variables.

```
In[8]:=  grad[f3[x,y,z],{x, y, z}]==
                {M1[x,y,z],M2[x,y,z],M3[x,y,z]}

Out[8]=  True
```

◇

Integrating Factors

Sometimes a differential equation $M(x,y)dx + N(x,y)dy = 0$ is not exact, but when the equation is multiplied by a function $m(x,y)$ the resulting equation

$$\mu(x,y)M(x,y)dx + \mu(x,y)N(x,y)dy = 0$$

is exact. Such a function $\mu(x,y)$ is called an integrating factor for the equation. If an equation $M(x,y)dx + N(x,y)dy = 0$ is not exact then, in general, the search for an integrating factor is not easy. Here is why: In order for $\mu(x,y)$ to be an integrating factor, we need to have $(\partial/\partial y)(\mu M) = (\partial/\partial x)(\mu N)$. When evaluated, this necessary equality becomes

$$\begin{aligned}\frac{\partial \mu(x,y)}{\partial y}M(x,y) + \mu(x,y)\frac{\partial M(x,y)}{\partial y}&\\ = \frac{\partial \mu(x,y)}{\partial x}N(x,y) + \mu(x,y)\frac{\partial N(x,y)}{\partial x}&,\end{aligned}$$

which is a partial, not ordinary, differential equation, and finding a solution can be a formidable task except in certain special cases when more is known about the form of $\mu(x,y)$.

Here are some of the special cases where it is possible to find an integrating factor without too much work:

SPECIAL CASE I. $\mu = \mu(x)$ is a function of x, alone. Then we need to solve:

$$\begin{aligned}\frac{\partial}{\partial y}(\mu(x)M(x,y)) &= \mu(x)\frac{\partial M(x,y)}{\partial y}\\ &= \frac{d\mu(x)}{dx}N(x,y) + \mu(x)\frac{\partial N(x,y)}{\partial x}\\ &= \frac{\partial}{\partial x}(\mu(x)N(x,y)).\end{aligned}$$

After a little rearranging, we see that we can solve

$$\frac{d\mu(x)}{dx}N(x,y)+\mu(x)\left(\frac{\partial N(x,y)}{\partial x}-\frac{\partial M(x,y)}{\partial y}\right)=0$$

as an *ordinary* differential equation if

$$\frac{1}{N(x,y)}\left(\frac{\partial N(x,y)}{\partial x}-\frac{\partial M(x,y)}{\partial y}\right)$$

contains only the variable x. Here $\mu(x)$ is easily obtained, because when this is true, the equation is merely $d\mu(x)/dx + P(x)\mu(x) = 0$, which is a homogeneous first-order linear ordinary differential equation. We studied these equations earlier and found that $\mu(x) = e^{-\int P(x)\,dx}$.

Example 2.6 Solve the differential equation $(3xy^2+y^2)dx+2\,x\,y\,dy = 0$, given that there is an integrating factor of the form $\mu = \mu(x)$.

Solution. The differential equation is not exact, since

$$\frac{\partial}{\partial y}M(x,y) = 6xy+2y \neq 2y = \frac{\partial}{\partial x}N(x,y)$$

except when either $x = 0$ or $y = 0$, which is not a rectangle in the plane. The condition that there exist an integrating factor of the desired form is

$$\frac{1}{N(x,y)}\left(\frac{\partial N(x,y)}{\partial x}-\frac{\partial M(x,y)}{\partial y}\right)=\frac{2y-(6xy+2y)}{2xy}=\frac{-6xy}{2xy}=-3,$$

which, being constant, qualifies as a function of x, alone. So it is feasible to search for an integrating factor that is a function of x alone. To do this, require that $(\partial/\partial y)(\mu M) = (\partial/\partial x)(\mu N)$. This means that $\mu(x)(6xy+2y) = 2\mu'(x)xy + 2\mu(x)y$. After subtracting $2\mu(x)y$ from both sides, we find that $2xy(\mu'(x) - 3\mu(x)) = 0$. Since $x = 0$ does not determine a function, we want $\mu'(x)-3\mu(x) = 0$, and $y \neq 0$, which determines $\mu(x) = e^{3x}$.

Multiply through the differential equation by $\mu(x) = e^{3x}$ to obtain

$$e^{3x}(3xy^2+y^2)dx + 2xe^{3x}ydy = 0,$$

which is exact, since

$$\frac{\partial M(x,y)}{\partial y} = e^{3x}(6xy+2y) = 6e^{3x}xy + 2e^{3x}y = \frac{\partial N(x,y)}{\partial x}.$$

The solution that passes through (p_1, p_2) is

$$\begin{aligned}\phi(X,Y) &= \int_{p_1}^{x} e^{3x}(3\,x\,p_2^2+p_2^2)dx + \int_{p_2}^{Y} 2\,X\,e^{3X}\,y\,dy\\ &= e^{3X}X\,Y^2 - (e^{3p_1})p_1p_2^2\\ &= 0.\end{aligned}$$

◇

SPECIAL CASE II. $\mu = \mu(y)$ is a function of y alone. The theory parallels that for the case where the integrating factor has the form $\mu = \mu(x)$. You are encouraged to develop the theory for yourself.

SPECIAL CASE III. $\mu(x,y) = x^p y^q$, a simple product of a power of x and a power of y. Sometimes, when the differential equation $M(x,y)dx + N(x,y)dy = 0$ has coefficients that are polynomials in x and y, it is possible to find an integrating factor of the form $\mu(x,y) = x^p y^q$ by equating the coefficients of like terms in the test for exactness.

Example 2.7 Consider the differential equation

$$(5x^2y - 6y^4)dx + (4x^3 - 14xy^3)dy = 0,$$

which is not exact. Show that the differential equation has an integrating factor of the form $\mu(x,y) = x^p y^q$ and solve the resulting exact equation.

Solution. The differential equation becomes

$$x^p y^q(5x^2y - 6y^4)dx + x^p y^q(4x^3 - 14xy^3)dy = 0,$$

which simplifies to

$$(5x^{p+2}y^{q+1} - 6x^p y^{q+4})dx + (4x^{p+3}y^q - 14x^{p+1}y^{q+3})dy = 0.$$

The test for exactness requires that $\partial M/\partial y = \partial N/\partial x$, which means that

$$5(q+1)x^{p+2}y^q - 6(q+4)x^p y^{q+3} = 4(p+3)x^{p+2}y^q - 14(p+1)x^p y^{q+3}.$$

Equating coefficients gives the simultaneous equations

$$5(q+1) = 4(p+3)$$

and

$$-6(q+4) = -14(p+1).$$

When simplified and solved, these equations yield $p = 2$ and $q = 3$.

This means that the integrating factor we want is $\mu(x,y) = x^2y^3$ and the equation we are to solve is

$$(5x^4y^4 - 6x^2y^7)dx + (4x^5y^3 - 14x^3y^6)dy = 0.$$

This is exact since

$$\frac{\partial M}{\partial y} = 20x^4y^3 - 42x^2y^6 = \frac{\partial N}{\partial x}.$$

Using the same technique as before, we see that the solution that passes through the point (p_1, p_2) can be expressed as

$$\begin{aligned}\phi(X,Y) &= \int_{p_1}^{X}(5x^4p_2^4 - 6x^2p_2^7)dx + \int_{p_2}^{Y}(4X^5y^3 - 14X^3y^6)dy \\ &= X^5Y^4 - 2X^3Y^7 - p_1^5p_2^4 + 2p_1^3p_2^7 \\ &= 0.\end{aligned}$$

Verify that this checks in the exact equation and the original. ◇

Exercises 2.3

Part I. Show that each of the differential equations below is exact. Solve each one manually if you can; otherwise by *Mathematica*. Check your answers.

1. $y\,dx + x\,dy = 0.$
2. $(1 + 6x^2 + y)\,dx + (-4 + x)\,dy = 0.$
3. $(3x - y + 3(x + 2y))\,dx + (-x + 2(3x - y) - 2y)\,dy = 0.$
4. $(\sin y^3)\,dx + 3xy^2(\cos y^3)\,dy = 0.$
5. $(y\cos x + \cos y)\,dx + (\sin x - x\sin y)\,dy = 0.$
6. $\dfrac{x}{(x^2+y^2)^{3/2}}\,dx + \dfrac{y}{(x^2+y^2)^{3/2}}\,dy = 0.$
7. $(1 + e^x + xe^x)y^3\,dx + 3x(1 + e^x)y^2\,dy = 0.$
8. $((e^x + e^y)y^3 + xe^xy^3)\,dx + (3x(e^x + e^y)y^2 + xe^yy^3)\,dy = 0.$
9. $(e^x + 6x^2 + \cos y)\,dx - (4 + x\sin y)\,dy = 0.$
10. $(2x^3 + xy + (x - 4y)(6x^2 + y))\,dx + (x(x - 4y) - 4(2x^3 + xy))\,dy = 0.$

Part II. Each of these differential equations has an integrating factor of the form $\mu = \mu(x)$. Find an integrating factor and solve the resulting differential equation manually if you can; otherwise solve by *Mathematica*. Check your answers.

11. $(4xy + 3e^xy^2 + xe^xy^2)\,dx + (x^2 + 2xe^xy)\,dy = 0.$
12. $(2y\cos x)\,dx + \sin x\,dy = 0.$
13. $3y\,dx + dy = 0.$

Part III. Integrating factors of the form $\mu = \mu(y)$.

14. Show that if $M(x,y)\,dx + N(x,y)\,dy = 0$ is not exact but that

$$\frac{1}{M(x,y)}\left(\frac{\partial M}{\partial y} - \frac{\partial N}{\partial x}\right)$$

is a function of y alone (does not contain x), then there is an integrating factor of the form $\mu = \mu(y)$. You will encounter the differential equation

$$\mu'(y) + \mu(y)\frac{1}{M(x,y)}\left(\frac{\partial M}{\partial y} - \frac{\partial N}{\partial x}\right) = 0.$$

Each of the differential equations 15–18 has an integrating factor of the form $\mu = \mu(y)$. Use the results of problem 14 to find an integrating factor. Solve the resulting differential equation manually if you can; otherwise solve by *Mathematica*. Check your answers.

15. $y\cos(xy)dx + (x\cos(xy) + \sin(xy))dy = 0.$

16. $(y + e^x y + xe^x y)dx + (3x + 3xe^x)dy = 0.$

17. $(y + ye^y)dx + (4x + 4xe^y + xye^y)dy = 0.$

18. $(1 + y)dx + (3x + 2xy)dy = 0.$

PART IV. Verify that each of these differential equations has an integrating factor of the form $\mu = \mu(x, y)$ that is given. Solve each equation manually if you can; otherwise by *Mathematica*. Check your answers.

19. $(y + xy)dx + (x + xy)dy = 0$; $\mu(x, y) = e^{x+y}$.

20. $$\begin{aligned}&[xy + y\arctan(x + y) + x^2 y\arctan(x + y) \\ &\quad + 2xy^2\arctan(x + y) + y^3\arctan(x + y)]\,dx \\ &\quad + [xy + x\arctan(x + y) + x^3\arctan(x + y) \\ &\quad + 2x^2 y\arctan(x + y) + xy^2\arctan(x + y)]\,dy = 0;\end{aligned}$$
$$\mu(x, y) = \frac{1}{1 + (x + y)^2}.$$

21. $$\frac{x}{x^2 + y^2}\,dx + \frac{y}{x^2 + y^2}\,dy = 0;$$
[exact, both before & after]
$$\mu(x, y) = \frac{1}{\sqrt{x^2 + y^2}}.$$

PART V. Verify that each of these differential equations has an integrating factor of the form $\mu = x^m y^n$. Find the quantities m and n by solving a system of two linear equations in m and n. Solve the resulting exact differential equation manually if you can; otherwise by *Mathematica*. Check your answers.

22. $(9y + 3ye^{2y})dx + (12x + 4xe^{2y} + 2xye^{2y})dy = 0.$

23. $$\begin{aligned}&(12y + xy^2\cos(xy) + 4y\sin(xy))\,dx \\ &\quad + (9x + x^2 y\cos(xy) + 3x\sin(xy))\,dy = 0.\end{aligned}$$

PART VI. Theory.

24. Prove that the function described in Theorem 2.2 is a potential function for the vector field $V(x, y)$, and that $\phi(P) = 0$. When trying to calculate $\partial\phi/\partial X$ you will need to use the fact that $\partial g/\partial X = \partial f/\partial y$. Then the integral you have will be easy to evaluate.

25. Given a conservative vector field $V(x, y, z)$ in 3-space, develop a single formula for finding a potential function for V in a manner analogous to that given in Theorem 2.2. You will integrate along a broken-line curve from the

fixed corner $P = (p1, p2, p3)$ to the diagonally opposite corner $Q = (X, Y, Z)$, by following three connected edges from P to Q. Note which variables are constant on a given edge. Two variables will be constant; you integrate the one that is not constant.

26. Prove that your 3-space formula is correct. The manipulations you need in your proof will suggest clearly to you why the definition of conservative is defined the way it is in higher dimensions.

27. State the definition of a conservative vector field for 3-space. There will be more relationships to satisfy since the behavior in each variable must be checked against every other variable.

28. Generalize the previous two problems to n-space. Everything generalizes nicely, but the notation gets to be somewhat awkward.

2.4 Variables Separable

Separable Differential Equations

Another class of differential equations that is easy to solve formally is those equations with variables separable.

Definition 2.3 *A first-order differential equation in two variables is said to have its* **variables separable** *provided that it is of one of these two forms:*

$$f_1(x)g_2(y)dx + f_2(x)g_1(y)dy = 0 \tag{2.4a}$$

or

$$dy/dx = F(x)G(y), \tag{2.4b}$$

or can be put into one of these forms.

Differential equations of the form (2.4a) have an obvious integrating factor $\mu(x, y) = 1/(f_2(x)g_2(y))$. In order to use this integrating factor, we require that $f_2(x)g_2(y) \neq 0$. That is, we seek solutions that are defined on intervals where $f_2(x) \neq 0$, and lie between horizontal lines where $g_2(y) = 0$. This means that you can expect your solution(s) to reside inside some rectangular box that is bounded left-to-right by consecutive places where $f_2(x) = 0$, (or by $\pm\infty$), and top-to-bottom by consecutive places where $g_2(y) = 0$, (or by $\pm\infty$). If either $f_2(x)$ or $g_2(y)$ is never 0, then in the appropriate direction these restrictions do not apply. The examples will help clarify these ideas.

Another possibility that may occur when integrating factors of this form are used is that you may either *gain* or *lose* solutions. That is, the solutions that the original equation has and the set of solutions that you produce may not agree.

This is a disturbing state of affairs and deserves careful attention. **You want all of the solutions of the original equation, and nothing that is not a solution.**

Returning to the solution of differential equations in the form 2.4a, multiply through by the integrating factor $\mu(x,y) = 1/(f_2(x)g_2(y))$ to get:

$$\frac{f_1(x)}{f_2(x)}\,dx + \frac{g_1(y)}{g_2(y)}\,dy = 0, \quad \text{or} \quad \frac{f_1(x)}{f_2(x)} + \frac{g_1(y)}{g_2(y)}\,\frac{dy}{dx} = 0$$

which may be integrated with respect to x to get the formal solution

$$F(x,y) = \int \frac{f_1(x)}{f_2(x)}\,dx + \int \frac{g_1(y)}{g_2(y)}\,dy = c. \tag{2.5a}$$

We have used the fact that $\int G(y)(dy/dx)\,dx = \int G(y)\,dy$ from the change-of-variables formula of calculus. Solutions produced this way are implicitly defined. That is, they state no preference whether x or y is independent and the other dependent. In general one has to appeal to the implicit function theorem to be certain that the solution formally defines y as a function of x or vice versa. Actually solving for y in terms of x is often impossible or unreasonably difficult. So we usually say that the implicit form $F(x,y) = c$ of the solution is the solution, even though the presence of an arbitrary constant on the right hand side means that what we have is actually an infinite set of solutions: the level curves of $F(x,y) = c$ for various values of c.

Once the form of $F(x,y) = c$ has been determined, one should check for gained or lost solutions. If y_0 is a number where $g_2(y_0) = 0$, then the constant function $y(x) = y_0$ is a solution that we have lost, and if x_0 is a number where $f_2(x) = 0$, then the vertical line $x(y) = x_0$ is a solution that we have lost. When $x = x_0$ or when $y = y_0$, our integrating factor $\mu(x,y) = 1/(f_2(x)g_2(y))$ is undefined. Solutions can be gained as well.

When the equation is given in the form $dy/dx = F(x)G(y)$, treat dy/dx as a quotient of differentials and write the equation as $dy/G(y) = F(x)dx$. Then integrate each term. This gives the implicit solution

$$H(x,y) = \int \frac{dy}{G(y)} - \int F(x)\,dx = c. \tag{2.5b}$$

Our implied integrating factor was $\mu(y) = 1/G(y)$, which means that the places where $G(y) = 0$ deserve special attention. These are the places where our integrating factor is undefined. If $G(y_0) = 0$ for some number y_0, then the constant function $y(x) = y_0$ is a solution of the original differential equation $dy/dx = F(x)G(y)$, but is not to be found among the implicitly defined solutions 2.5b. So we lost a solution and 2.5b does not represent all of the solutions of $dy/dx = F(x)G(y)$.

The notebook *Variables Separable* illustrates the solution of differential equations with variables separable.

Examples of Variables Separable Differential Equations

Example 2.8 Solve the differential equation

$$x(y^2+y-2)dx+(x-4)(1+5y)dy=0.$$

Solution. The equation is separable, with integrating factor

$$\mu(x,y)=\frac{1}{(x-4)(y^2+y-2)}.$$

Since $(y^2+y-2)=(y+2)(y-1)$, we have constant solutions $y=-2$ and $y=1$, which make $y^2+y-2=0$. In addition, the other solutions are defined on intervals where $(x-4)\neq 0$, that is, where $x\neq 4$. It can be argued that $x=4$ is a perfectly satisfactory (vertical) solution of the equation. So, armed with this information multiply through by $\mu(x,y)$ to get

$$\frac{x}{x-4}dx+\frac{1+5y}{y^2+y-2}dy=0.$$

Integrate to get

$$\int\frac{x}{x-4}dx+\int\frac{1+5y}{y^2+y-2}dy=c,$$

which evaluates to

$$x+4\ln|x-4|+2\ln|y-1|+3\ln|y+2|=c.$$

Observe that we need to avoid $x=4$, $y=1$, and $y=-2$, as was noted above, because any one of these will cause the natural logarithm to become undefined. However, moving the x to the other side, combining the logarithms, and exponentiating gives solutions in the form

$$(x-4)^4(y-1)^2(y+2)^3=Ae^{-x}, \tag{2.6}$$

where $A=\pm e^c$. In addition, when the solution is given in this form we can see that if $A=0$, we get back all three of our constant solutions: $x=4$, $y=1$, and $y=-2$. So we did not lose any of the solutions that had these special forms. If you attempt to solve 2.6 for y as a function of x, you will become convinced that the effort is not worthwhile. How do we check this solution? If we assume that the implicit solution 2.6 defines y as a function of x, then, upon taking the derivative

$$\frac{d}{dx}\left(e^x(x-4)^4(y-1)^2(y+2)^3\right)=\frac{d}{dx}A,$$

we arrive at

$$\begin{aligned}&4e^x(x-4)^3(y-1)^2(y+2)^3+e^x(x-4)^4(y-1)^2(y+2)^3\\&\quad+3e^x(x-4)^4(y-1)^2(y+2)^2\frac{dy}{dx}+2e^x(x-4)^4(y-1)(y+2)^3\frac{dy}{dx}=0,\end{aligned}$$

which factors into

$$e^x(x-4)^3(y-1)(y+2)^2\left(-2x+xy+xy^2-4\frac{dy}{dx}\right.$$

$$\left.+\,x\frac{dy}{dx}-20y\frac{dy}{dx}+5xy\frac{dy}{dx}\right)=0.$$

Treat dy/dx as a quotient of differentials and multiply through by dx, to get

$$e^x(x-4)^3(y-1)(y+2)^2(x(y-1)(y+2)dx+(x-4)(1+5y)dy)=0.$$

This expression is zero only when $x=4$, $y=1$, $y=-2$ or when

$$x(y-1)(y+2)dx+(x-4)(1+5y)dy=0.$$

This last expression is equivalent to the given differential equation. So the solution checks provided $x\neq 4$, $y\neq 1$, and $y\neq -2$. But these values also checked through other techniques. We have completely solved the problem. ◇

Example 2.8M The solution and checking processes in *Mathematica*.

Solution.

```
In[9]:=  Integrate[x/(x-4),x]+
         Integrate[(1+5*y)/((-1+y)*(2+y)),y] == c

Out[9]=  x + 4 Log[4 - x] + 2 Log[1 - y] + 3 Log[2 + y] == c
```

Steps to check the transformed solution are: define it

```
In[10]:= solution = E^x (x-4)^4(y-1)^2(y+2)^3

Out[10]=  x         4         2        3
         E  (-4 + x)  (-1 + y)  (2 + y)
```

Use the built-in function `Dt` to calculate the total derivative (in terms of `Dt[x]` and `Dt[y]`). Then change these to `dx` and `dy`.

```
In[11]:= factored = Factor[Dt[solution]]/.{Dt[x]->dx,Dt[y]->dy}

Out[11]=  x         3                    2
         E  (-4 + x)  (-1 + y) (2 + y)

           (-4 dy - 2 dx x + dy x - 20 dy y + dx x y + 5 dy x y +

                   2
             dx x y )
```

Our interest is in the fifth factor of this result. So we look at the fifth part of factored: `factored[[5]]`. Then `Collect` terms.

```
In[12]:= differentialExpression = Collect[factored[[5]],{dx,dy}]

Out[12]=                                                    2
         dy (-4 + x - 20 y + 5 x y) + dx (-2 x + x y + x y )
```

Get the coefficient of dx.

```
In[13]:= dxPart = Factor[Coefficient[differentialExpression,dx]]
Out[13]= x (-1 + y) (2 + y)
```

Get the coefficient of dy.

```
In[14]:= dyPart = Factor[Coefficient[differentialExpression,dy]]
Out[14]= (-4 + x) (1 + 5 y)
```

Put the equation back together.

```
In[15]:= dxPart*dx + dyPart*dy == 0
Out[15]= dx x (-1 + y) (2 + y) + dy (-4 + x) (1 + 5 y) == 0
```

You will recognize this last result as the original differential equation with the coefficient of dx factored. This demonstrates that our solution checks. ◇

Exercises 2.4

Part I. Solve these problems manually and by *Mathematica* using the technique of separation of variables. Check your answers. Be alert for both vertical and horizontal constant solutions. Explicitly list these.

1. $3(x-3)^2dx + 4(y+1)^3dy = 0.$
2. $x^2(y+1)dx + y^2(x-1)dy = 0.$
3. $\dfrac{dy}{dx} = \dfrac{5y}{x(y+2)}.$
4. $\dfrac{dy}{dx} = \dfrac{3x^2y}{1+x^3}.$
5. $(3+2y)dx + (4-x^2)dy = 0.$
6. $y^3dx - x^3dy = 0.$

Part II. Manually separate variables then solve by *Mathematica*. Check your answers. Explicitly list both vertical and horizontal constant solutions.

7. $x^3e^{3y}dx + e^{2x}(y-4)^3dy = 0.$
8. $(y-3)(y-2)dx + x(y+1)dy = 0.$
9. $(\ln x)4(\cot y)dx + xdy = 0.$
10. $(y-1)(y-2)(y-3)dx + (x+1)(x+2)(x+3)dy = 0.$
11. $(1+x^2+y^2+x^2y^2)^3dy = y^2dx.$ [Use `Factor`.]
12. $\dfrac{dy}{dx} = \dfrac{xy+2y-x-2}{xy-2x+4y-8}.$ [Use `Factor`.]

13. $x\dfrac{dy}{dx} = y^2 - 5y + 6.$

14. $(e^x + e^{-x})\dfrac{dy}{dx} = y^3.$

PART III. Two projects.

15. Consider Euler's quadratic differential equation

$$(1 - x^2)\left(\frac{dy}{dx}\right)^2 = (1 - y^2).$$

(a) Find the two constant solutions for y.
(b) Show that if the variables x and y are interchanged, the same equation results. What does this suggest about symmetry? What about constant "solutions" for x?
(c) Show that in any region of the xy-plane where $(1 - x^2)$ and $(1 - y^2)$ have opposite signs, Euler's equation has no real solution.
(d) Show that in any region where $(1 - x^2)$ and $(1 - y^2)$ are both positive or both negative, solution functions must satisfy one of the two differential equations

$$\frac{dy}{dx} = \pm\sqrt{\frac{1 - y^2}{1 - x^2}}.$$

(e) Solve these two differential equations by separation of variables to obtain these families of solutions:

$$\arcsin y \pm \arcsin x = C \text{ when } |x| < 1 \text{ and } |y| < 1 \quad \text{and}$$

$$\operatorname{arccosh} y \pm \operatorname{arccosh} x = C \text{ when } |x| > 1 \text{ and } |y| > 1.$$

The parameter C is arbitrary. Each equation constrains C.
(f) Make the substitution $x = \sin u$ and $y = \sin v$ in the equations $\arcsin y \pm \arcsin x = C$ and take the cosine of the results. Show that the resulting algebraic equations each represent a conic with 45° axes.
(g) Do the solutions of $\operatorname{arccosh} y \pm \operatorname{arccosh} x = C$ also represent conics?
(h) (optional) Treat Euler's equation as a differential inclusion (see below). How (if at all) are the solutions different?

16. (If you have not studied Section 1.4 on differential inclusions, do so in conjunction with working this problem.) Consider the differential equation

$$\left(\frac{dy}{dx}\right)^2 + y^2 = 1.$$

(a) Describe the portion of the plane in which solution can exist.
(b) Write the differential equation as a differential inclusion by solving for dy/dx.
(c) Find all constant solutions of the differential equation and the differential inclusion.

(d) Show that for every number c the function $y(x) = \sin(x+c)$ is a solution of the differential inclusion. Solve the differential equations given by the two membership conditions of the differential inclusion to see that this is all of the solutions that are not constant on any interval.

(e) Show that if $y_1(x)$ is a solution of the differential inclusion, then $|y_1(x)|$ is a solution that is differentiable except where $y_1(x) = 0$ or where $y_1(x)$ is not differentiable.

(f) Given a point (x_0, y_0) with $-1 < y_0 < 1$, sketch the maximal and minimal solutions that pass through (x_0, y_0).

(g) Describe "typical" solutions that have continuous first derivatives everywhere.

(h) Examine the solutions of the generalization

$$\left(\frac{dy}{dx}\right)^2 + a^2y^2 = a^2r^2,$$

where a and r are positive constants. Discuss the ways a and r produce differing properties in the solution functions. Examine the limiting behavior as the parameter a approaches 0 or ∞. What are solutions like for very large a?

2.5 Homogeneous Nonlinear Differential Equations

Homogeneous Nonlinear Differential Equations

It is quite likely that in multidimensional calculus you studied functions $f(x, y)$ that were called **homogeneous**. The name suggests the fact that x and y and various combinations of them occur in the definition of f in "the same way."

Definition 2.4 *A function $f(x, y)$ is called* **homogeneous of degree n** *if*

$$f(tx, ty) = t^n f(x, y)$$

for all $t > 0$.

Examples of homogeneous functions of various degrees are:

FUNCTION	DEGREE
$x + 3y$	1
$\sqrt{x^2 - 5xy + y^2}$	1
$x - 5\sqrt{xy} + 3y$	1
$\dfrac{x^3 - 5xy^2 + y^3}{x - y}$	2
$x^2 - 5xy + y^2$	2
x/y	0
$3 + \dfrac{x}{y} + \dfrac{x^2}{y^2}$	0

Definition 2.5 *A differential equation of the form* $M(x, y)dx + N(x, y) dy = 0$ *is called* **homogeneous of degree n** *if each of the coefficients* $M(x, y)$ *and* $N(x, y)$ *is homogeneous of the same degree* n*. That is* $M(tx, ty) = t^n M(x, y)$ *and* $N(tx, ty) = t^n N(x, y)$*.*

Theorem 2.3 *If* $M(x, y)dx + N(x, y)dy = 0$ *is homogeneous of some degree then the substitution* $y = vx$ *will reduce the equation to one with variables separable. The substitution* $x = uy$ *will do the same.*

Proof. We use the substitution $y = vx$. Since we want to be able to recover $v = y/x$, we want to have $x \neq 0$. The product rule for differentials says $d(vx) = vdx + xdv$. The proof proceeds as follows. Suppose the degree of homogeneity is n.

$$\begin{aligned}
M(x, y)dx &+ N(x, y)dy \\
&= M(x, vx)dx + N(x, vx)(x\,dv + v\,dx) \\
&= x^n M(1, v)dx + x^n N(1, v)(x\,dv + v\,dx) \\
&= x^n (M(1, v) + vN(1, v))dx + x^{n+1} N(1, v)dv = 0,
\end{aligned}$$

which is separable into

$$\frac{dx}{x} + \frac{N(1, v)}{M(1, v) + vN(1, v)} dv = 0.$$

Integrating term by term gives a solution having the form

$$\ln |x| + G(v) = c.$$

Since $v = y/x$, the final (implicit) solution of the original equation has the form

$$\ln |x| + G\left(\frac{y}{x}\right) = c.$$

Observe that we still need to have $x \neq 0$. The proof in the case of the substitution $x = uy$ is similar, except that here we need $y \neq 0$. This proof is left as an exercise. □

The choice of whether to substitute $y = vx$ or $x = uy$ is determined mainly by whichever one results in the easiest integrations to perform. The notebook *Homogeneous Fcns & Eqns* actually performs both substitutions and successfully completes the integrations both ways. This notebook contains a function that does all of the steps for you. It is worthwhile to do several of these problems manually.

The solutions of homogeneous nonlinear differential equations are typically defined for (x, y) inside one of the four quadrants, but not necessarily at any point on either the x- or the y-axes. Sometimes, however, $x = 0$ or $y = 0$ is a constant solution of the equation.

Example 2.9 Show that the differential equation $(x^2 + y^2)dx + xydy = 0$ is homogeneous of degree 2 and find its solutions.

Solution. Both $M(x,y) = x^2+y^2$ and $N(x,y) = xy$ are homogeneous of order 2 since

$$M(tx,ty) = t^2x^2 + t^2y^2 = t^2(x^2+y^2)$$

and

$$N(tx,ty) = (tx)(ty) = t^2xy = t^2N(x,y).$$

We make the substitution $y = vx$. Then $dy = vdx + xdv$, and the differential equation becomes

$$\begin{aligned}(x^2+y^2)dx &+ xydy \\ &= (x^2+v^2x^2)dx + x(vx)(vdx+xdv) \\ &= x^2(1+v^2)dx + vx^2(v\,dx + x\,dv) \\ &= x^2(1+2v^2)dx + v\,x^3\,dv \\ &= x^3\left(\frac{dx}{x} + \frac{v}{1+2v^2}dv\right) = 0.\end{aligned}$$

This results in the equations $x = 0$ and $dx/x + v/(1+2v^2)dv = 0$, which have as solutions $x = 0$ (which *is* a solution of the original equation) and

$$\ln|x| + \frac{1}{4}\ln(1+2v^2) = c.$$

Substitute $v = y/x$ into this to get a family of solutions to the original equation:

$$\ln|x| + \frac{1}{4}\ln\left(1+2\left(\frac{y}{x}\right)^2\right) = c.$$

You should verify that this is a solution of the original differential equation for $x \neq 0$. ◇

Example 2.9M Solve example 2.9 by *Mathematica.* (This is done in the notebook *Homogeneous Fcns & Eqns.*)

Solution.

```
In[16]:= de[x_,y_] = (x^2+y^2)*dx+x*y*dy == 0

Out[16]=
                       2    2
         dy x y + dx (x  + y ) == 0
```

Assume y=v*x and make the substitution:

```
In[17]:= de[x,v*x]/.{dy->dv*x+v*dx}

Out[17]=    2                         2    2  2
         v x  (dx v + dv x) + dx (x  + v  x ) == 0
```

Divide both sides by x^2 (specific to this problem) and cancel common factors. The function `Map` has to be used to cause the operation to be applied to both sides of the equation.

```
In[18]:= Map[Cancel,Map[Function[t,t/x^2],%]]

Out[18]=                2
         dx + 2 dx v  + dv v x == 0
```

`Collect` terms to put into standard form.

```
In[19]:= Map[Function[u,Collect[u,{dx,dv}]],%]

Out[19]=                2
         dx (1 + 2 v ) + dv v x == 0
```

This is our separable equation, ready for solving. (In the notebook *Homogeneous Fcns & Eqns*, the solution from this point forward is left as an exercise.) ◇

Exercises 2.5

Part I. Determine whether the function is homogeneous. If so, state the degree of homogeneity.

1. $x^2 + 5xy - y^2$.
2. $\sqrt{x+y}$.
3. $x \sin\left(\frac{y}{x}\right)$.
4. $(x+1)(y+1) - (xy+1)$.
5. $\dfrac{y}{x^2 + 3\sqrt{x^4+y^4}}$.
6. $\dfrac{\ln x^2}{\ln y^2}$.

Part II. Solve these problems manually and by *Mathematica*. Show that each is homogeneous. State the degree of homogeneity and make the appropriate substitution $y = vx$ or $x = vy$ to convert the problem into one that has variables separable. Check your answers. Be alert for solutions that have some special form. Explicitly list these solutions having special form.

7. $(x-y)dx + x\,dy = 0$.
8. $(y^2 + x^2)dx + x^2 dy = 0$.
9. $2x^3 y\,dx + (x^4 + y^4)dy = 0$.

10. $(x^2 - 3y^2)dx + 2xy\,dy = 0.$

11. $(y + \sqrt{x^2 + y^2}\,)dx - x\,dy = 0.$

12. $(\sqrt{x+y} + \sqrt{x-y}\,)dx + (\sqrt{x-y} - \sqrt{x+y}\,)dy = 0.$

PART III. Manually convert these homogeneous nonlinear differential equations into equations with variables separated. Solve the resulting separable equations by *Mathematica*. Check your answers. When *Mathematica* cannot do one of the integrals, indicate a formal solution.

13. $(x^4 + y^4)dx - 2x^3ydy = 0.$

14. $xdx + (y - 2x)dy = 0.$

15. $\dfrac{dy}{dx} = \dfrac{x+3y}{3x+y}.$

16. $\dfrac{dy}{dx} = \dfrac{y}{x}\ln\dfrac{y}{x}.$

17. $\dfrac{dy}{dx} = \dfrac{x^2 - y^2}{x^2 + y^2}.$

18. $\dfrac{dy}{dx} = \sin\left(\dfrac{y}{x}\right).$

PART IV. Theory.

19. Prove that the substitution $x = uy$ also converts a homogeneous nonlinear differential equation $M(x,y)dx + N(x,y)dy = 0$ into a separable differential equation.

20. If $M(x,y)dx + N(x,y)dy = 0$ is homogeneous of some degree, show that the substitution $x = r\cos\theta$, $y = r\sin\theta$ reduces this equation to a separable equation in the variables r and θ.

21. Suppose that the equation $M(x,y)dx + N(x,y)dy = 0$ is homogeneous of degree n. Show that the substitution $x = k\zeta$, $y = k\eta$ gives the same equation back with the variables x and y changed to ζ and η. Here k is a positive constant.

22. If $f(x,y)$ is homogeneous of degree n, show that

$$x\frac{\partial f}{\partial x} + y\frac{\partial f}{\partial y} = nf.$$

This is called Euler's relation. An analogous relation holds for functions of more than two variables that are homogeneous of degree n.

23. Show that the homogeneous nonlinear differential equation $M(x,y)dx + N(x,y)dy = 0$ can be written in the alternative forms

$$\frac{dy}{dx} = G\left(\frac{y}{x}\right) \quad \text{and} \quad \frac{dx}{dy} = F\left(\frac{x}{y}\right).$$

24. Complete the solution process started in example 2.9**M**. Use *Mathematica*.

2.6 Bernoulli and Riccati Differential Equations (Optional)

Bernoulli[1] and Riccati[2] differential equations are special nonlinear equations that are seen from time to time in theory and in practice.

Bernoulli Differential Equations

A differential equation of the form

$$y' + P(x)y = Q(x)y^n, \quad \text{where} \quad n \neq 0 \quad \text{or} \quad 1, \tag{2.7}$$

is called a **Bernoulli differential equation**. In the cases where $n = 0$ or $n = 1$, the equation is actually linear and should be solved by those methods. But when $n \neq 0$ or 1, to solve an equation of this form, let $w = y^{1-n}$. Then $w' = (1-n)y^{-n}y'$, and $y' = w'/(1-n)y^{-n}$. So the differential equation becomes

$$\frac{w'}{(1-n)y^{-n}} + P(x)y = Q(x)y^n,$$

which, after multiplying through by y^{-n}, becomes

$$\frac{1}{1-n}w' + P(x)y^{1-n} = Q(x),$$

or,

$$w' + (1-n)P(x)w = (1-n)Q(x).$$

This is a linear differential equation from which we find w. Obtain y from $y = w^{1/(1-n)}$. Since the solution w involves one arbitrary constant, so does y. Thus the first-order nonlinear differential equation 2.7 has a one-parameter family of solutions, as it should.

Example 2.10 Solve the Bernoulli differential equation

$$y' - y = \frac{e^x}{y}.$$

Solution. This is a Bernoulli equation with $n = -1$, so we substitute $w = y^{1-(-1)} = y^2$. Then $y' = w'/(2y)$, so, $w'/(2y) - y = (e^x)/y$. The equation and these derivatives dictate that $y \neq 0$. Thus, after multiplying through by $(2y)$,

---

[1] James Bernoulli (1654–1705), the Swiss mathematician who first studied differential equations of this form.

[2] Jacopo Francesco Riccati (1676–1754) was the son of an Italian nobleman who worked for many years on several areas of differential equations. He corresponded at length with mathematicians around Europe, including Leibniz. His sons Vincenzo and Giordano were also mathematicians of note. Vincenzo studied the hyperbolic functions thoroughly years before Lambert, who is popularly credited with introducing them into mathematics.

which is nonzero, we get the equation

$$w' - 2y^2 = 2e^x,$$

or

$$w' - 2w = 2e^x.$$

Solve this linear equation to find $w = -2e^x + ce^{2x}$. Then, since $y = \pm\sqrt{w}$, either

$$y = \sqrt{w} = \sqrt{-2e^x + ce^{2x}}$$

or

$$y = -\sqrt{w} = -\sqrt{-2e^x + ce^{2x}}.$$

Observe that these expressions require that $-2e^x + ce^{2x} \geq 0$. So $-2 + ce^x \geq 0$. From this one sees that $c > 0$, and, when this is so, the domain is the set of all $x \geq \ln(2/c)$. When $c < 0$, the solution is imaginary. ◇

We can use *Mathematica* to look at some selected solution curves. Note how the domains of the curves differ, as was predicted.

```
In[20]:= curves = Flatten[Table[{Sqrt[-2E^x+c E^(2x)],
                  -Sqrt[-2E^x+c E^(2x)]},{c,1/4,3,3/4}]]

Out[20]=
                           2 x                        2 x
                   x      E                   x      E
         {Sqrt[-2 E  + ----], -Sqrt[-2 E  + ----],
                             4                          4

                  x    2 x                 x    2 x
          Sqrt[-2 E  + E   ], -Sqrt[-2 E  + E   ],

                          2 x                        2 x
                  x    7 E                  x    7 E
          Sqrt[-2 E  + ------], -Sqrt[-2 E  + ------],
                           4                           4

                          2 x                        2 x
                  x    5 E                  x    5 E
          Sqrt[-2 E  + ------], -Sqrt[-2 E  + ------]}
                           2                           2
```

```
In[21]:= Plot[Evaluate[curves],{x,-1,3},PlotRange->All];
```

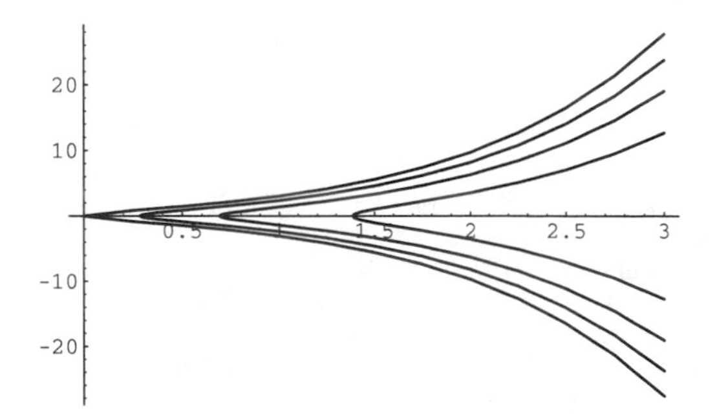

Example 2.10M Solve the same Bernoulli equation $y' - y = (e^x)/y$ in *Mathematica*.

Solution. (Note that except for minor changes, like putting all nonzero parts on the left-hand side the steps followed are exactly those of Example 2.10 above.)

Define the equation.

```
In[22]:= BernoulliEquation[x_,y_]:=D[y[x],x]-y[x]==E^x y[x]^(-1)
```

Verify the equation. (Display it.)

```
In[23]:= BernoulliEquation[x,y]

Out[23]=
                            x
                           E
         -y[x] + y'[x] == ----
                          y[x]
```

Determine the power of y[x]. (This allows the process to be done in a general way.)

```
In[24]:= BNum = -1

Out[24]= -1
```

Declare the required substitution.

```
In[25]:= w[x_] == y[x]^(1-BNum)

Out[25]=                2
         w[x_] == y[x]
```

Make the substitution.

```
In[26]:= yp = Solve[w'[x] == D[y[x]^(1-BNum),x], y'[x]]

Out[26]=             w'[x]
         {{y'[x] -> ------}}
                    2 y[x]
```

Put everything on the left, then divide through by y[x]^BNum term by term.

```
In[27]:= LeftSide = Map[Function[t, t/y[x]^BNum],
                        BernoulliEquation[x,y]/.yp[[1]]/.
                   (a_ == b_)->a-b]

Out[27]=   x        2   w'[x]
         -E  - y[x]  + -----
                         2
```

Obtain the operator (may have a nonhomogeneous part).

```
In[28]:= L[x_,w_] = LeftSide/. y[x]^(1-BNum)->w[x]

Out[28]=    x                w'[x]
         -E   - w[x] +  -----
                          2
```

Solve the linear differential equation.

```
In[29]:= SolnW = DSolve[L[x,w] == 0,w[x], x]

Out[29]=
                         x     2 x
         {{w[x] -> -2 E  + E     C[1]}}
```

Simplify the solution to the Bernoulli equation.

```
In[30]:= s[x_] = Simplify[(w[x]/. SolnW[[1]])^(1/(1-BNum))]

Out[30]=      x          x
         Sqrt[E  (-2 + E  C[1])]
```

Check the solution.

```
In[31]:= Simplify[BernoulliEquation[x,s]]

Out[31]= True
```

◇

Example 2.11 Solve the Bernoulli differential equation $y' + (2/3)xy = xy^4$.

Solution. Note first that $y = 0$ is a constant solution. This Bernoulli equation has $n = 4$, so for $y \neq 0$ we substitute $w = y^{1-4} = y^{-3}$, which produces the linear equation in w: $w' - 2xw = -3x$. This has solution $w = 3/2 + ce^{x^2}$, so

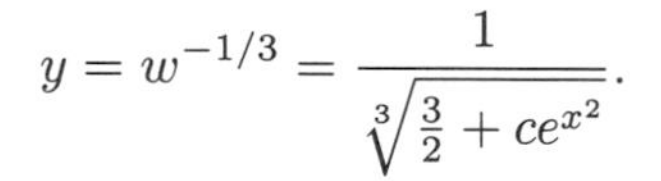

$$y = w^{-1/3} = \frac{1}{\sqrt[3]{\frac{3}{2} + ce^{x^2}}}.$$

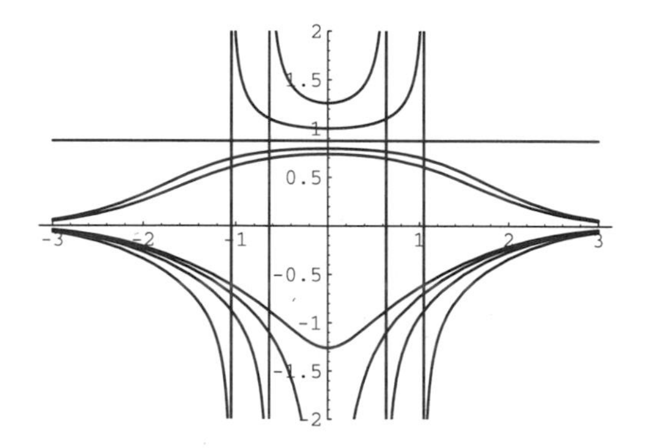

Figure 2.3: Some of the solutions produced in Example 2.11.

Several observations about the character of this solution are in order.

1. If $c = 0$, then y is constant with value $\sqrt[3]{2/3}$.
2. If $c > 0$, y is defined for all x, and is always positive.
3. If $-3/2 \le c < 0$, then y has a vertical asymptote when $e^{x^2} = -3/2c$. That is, when
$$x = \pm\sqrt{\ln\left(\frac{-3}{2c}\right)}.$$
4. If $c \le -3/2$, then y is defined for all x and is always negative.
5. If $c \ne 0$, $y \to 0$ (one of the constant solutions) as $x \to \pm\infty$.
6. When $c > 0$, but very near 0, y is near the constant solution $\sqrt[3]{2/3}$ for x near 0. ◇

Riccati Differential Equations

A differential equation that can be written in the form
$$y' = p(x) + q(x)y + r(x)y^2 \tag{2.8}$$
is called a **Riccati** differential equation.

Any Riccati differential equation where $r(x)$ is differentiable can be transformed into a homogeneous second order linear differential equation by the transformation $y = -w'/r(x)w$. It follows that
$$y' = -\frac{w''r(x)w - w'(r'(x)w + r(x)w)}{(r(x)w)^2},$$
and after some simplification,
$$r(x)w'' - (r'(x) + q(x)r^2(x))w' + (r^2(x)p(x))w = 0.$$
At this point, we merely observe that this is a fact. Later we will occasionally be able to solve a Riccati differential equation by finding one solution of such a second-order differential equation.

Any second-order linear differential equation $a_2(x)w'' + a_1(x)w' + a_0(x)w = 0$, where $a_2(x) \ne 0$ can be transformed into a Riccati equation by the transformation $w = e^{-\int u}$. Indeed, $w' = -ue^{-\int u}$, and $w'' = -u'e^{-\int u} + u^2e^{-\int u}$, so substitution yields
$$\begin{aligned}
0 &= a_2(x)w'' + a_1(x)w' + a_0(x)w \\
&= a_2(x)(-u'e^{-\int u} + u^2e^{-\int u}) + a_1(x)(-ue^{-\int u}) + a_0(x)e^{-\int u} \\
&= \left(a_2(x)(-u' + u^2) + a_1(x)(-u) + a_0(x)\right)e^{-\int u} \\
&= a_2(x)\left(-u' + u^2 + \frac{a_1(x)}{a_2(x)}(-u) + \frac{a_0(x)}{a_2(x)}\right)e^{-\int u},
\end{aligned}$$

which implies that

$$-u' + u^2 + \frac{a_1(x)}{a_2(x)}(-u) + \frac{a_0(x)}{a_2(x)} = 0.$$

This can be rewritten as

$$u' = p(x) + q(x)u + r(x)u^2,$$

where $p(x) = a_0(x)/a_2(x)$, $q(x) = -a_1(x)/a_2(x)$, and $r(x) = 1$. This completes the transformation of the second-order differential equation to one of Riccati type.

The complete solution of equations of Riccati type proceeds in three steps: find one solution; use that solution to further transform the equation into a first-order linear differential equation, whose complete solution is formally easy to obtain. Then recover all of the solutions from these solutions. We will find the complete solution to be of the form

$$y(x) = \frac{cf_1(x) + f_2(x)}{cf_3(x) + f_4(x)}.$$

It is left as an exercise to show that if $f_1(x), f_2(x), f_3(x)$, and $f_4(x)$ are differentiable functions then the function $u(x) = (cf_1(x) + f_2(x))/(cf_3(x) + f_4(x))$ satisfies a differential equation of Riccati type that is independent of the number c.

Continue the solution process by assuming that a solution $y_1(x)$ of the Riccati differential equation 2.8 has been found. Let $y = y_1(x) + 1/v(x)$, where $v(x) \neq 0$. Then we have these calculations

$$y' = p(x) + q(x)y + r(x)y^2$$

$$y_1(x) - \frac{v'(x)}{v^2(x)} = p(x) + q(x)\left(y_1(x) + \frac{1}{v(x)}\right) + r(x)\left(y_1(x) + \frac{1}{v(x)}\right)^2$$

$$y_1(x) - \frac{v'(x)}{v^2(x)} = p(x) + q(x)y_1(x) + \frac{q(x)}{v(x)}$$

$$+r(x)\left(y_1^2(x) + \frac{2y_1(x)}{v(x)} + \frac{1}{v^2(x)}\right)^2$$

This becomes

$$-\frac{v'(x)}{v^2(x)} = \frac{q(x)}{v(x)} + \frac{2r(x)y_1(x)}{v(x)} + \frac{r(x)}{v^2(x)},$$

since $y_1'(x) = p(x) + q(x)y_1(x) + r(x)y_1^2(x)$. Multiply through by $-v^2(x)$ to get the first-order linear differential equation

$$v'(x) + (q(x) + 2r(x)y_1(x))v(x) = -r(x).$$

This has a solution of the form $v(x) = v_1(x)c + v^2(x)$ that we can substitute back into

$$y = y_1(x) + \frac{1}{v(x)} = y_1(x) + \frac{1}{v_1(x)c + v_2(x)}.$$

This is a one parameter family of solutions that simplifies to the form that was desired.

Example 2.12 Solve the Riccati differential equation $u' = (1-2x^2)+xu+u^2$.
Solution. We observe that $u_1(x) = x$ is a solution. Let

$$u(x) = u_1(x) + \frac{1}{v(x)} = x + \frac{1}{v(x)}.$$

Then $u' = 1-(v'/v^2)$, and $1-(v/v^2) = 1-2x^2+x(x+(1/v(x))+(x+(1/v(x))^2$. This reduces to the linear differential equation $v' + 3xv = -1$, which has as solution

$$v = \exp\left(\frac{-3x^2}{2}\right) c - \exp\left(\frac{-3x^2}{2}\right)\left(\int \exp\left(\frac{3x^2}{2}\right) dx\right).$$

The one parameter family of solutions is obtained from $u = x + 1/v$ to be

$$u = x + \frac{1}{\exp\left(\frac{-3x^2}{2}\right) c - \exp\left(\frac{-3x^2}{2}\right)\left(\int \exp\left(\frac{3x^2}{2}\right) dx\right)}.$$

This solution is used in the solution of an exercise in Chapter 4. ◇

Example 2.12M Solve example 2.12 by *Mathematica.*
Solution.
The Riccati operator `R[x,u]`:

```
In[32]:= R[x_,u_] = (1 - 2 x^2)  + x u[x] + u[x]^2 - u'[x]

Out[32]=          2                   2
         1 - 2 x  + x u[x] + u[x]  - u'[x]
```

A solution by inspection.

```
In[33]:= u1[x_] = x

Out[33]= x
```

Check the solution.

```
In[34]:= R[x,u1]

Out[34]= 0
```

A second solution can be obtained by the further transformation `u[x]` = `u1[x]+1/v[x]`. The equation for `v[x]` is linear and can be solved completely.

```
In[35]:= Expand[v[x]^2 R[x,Function[t,t+1/v[t]]]]

Out[35]= 1 + 3 x v[x] + v'[x]
```

```
In[36]:= DSolve[% == 0,v[x],x]

Out[36]=
                                    Pi                3
                               Sqrt[--] Erfi[Sqrt[-] x]
                 C[1]                6                2
{{v[x] -> ------------- - ---------------------------------}}
                 2                     2
            (3 x )/2              (3 x )/2
           E                     E
```

This is a family of solutions. The parameter is `C[1]`. (Try it without the `Simplify`.)

```
In[37]:= u2[x_] = Simplify[(u1[x]+1/v[x])/.%[[1]]]

Out[37]=
                       2
                  (3 x )/2
                 E
x + -------------------------------------
                   Pi                3
    C[1] - Sqrt[--] Erfi[Sqrt[-] x]
                   6                 2
```

Check the family of solutions in the differential equation

```
In[38]:= Simplify[R[x,u2]]==0

Out[38]= True
```

◇

The function **Erfi** is a form of the error function that is encountered in statistics.

Exercises 2.6

Solve these Bernoulli differential equations manually and by *Mathematica* using Example 2.10M as a guide:

1. $\dfrac{dy}{dx} = 4y + xy^3.$

2. $y^5 \dfrac{dy}{dx} = 1 - 5y^6.$

3. $\dfrac{dy}{dx} = \dfrac{1}{x^5 + xy}$ [Hint: Consider dx/dy.]

4. $\dfrac{dy}{dx} = y(xy^3 - 1).$

5. $4(1 + x^2)\dfrac{dy}{dx} = 2xy(y^4 - 1).$

6. $\dfrac{dy}{dx} - 2y = e^x y^2.$

Solve these Riccati differential equations manually and by *Mathematica*, using Example 2.12M as a guide. In each case a solution has been supplied.

7. $\dfrac{dy}{dx} = x^2 - 2 + 2xy + y^2; \quad y_1 = 1 - x.$

8. $\dfrac{dy}{dx} = x - 1 + y - xy^2; \quad y_1$ is constant.

9. $(1-x^2)\dfrac{dy}{dx} = 1 - y^2; \quad y_1 = 1.$

10. $(1-x^2)\dfrac{dy}{dx} = 1 - y^2; \quad y_1 = x.$

11. $x(x^2-1)\dfrac{dy}{dx} + x^2 - (x^2-1)y = y^2; \quad y_1 = x^n$ for some n.

12. $\dfrac{dy}{dx} = -2 - y + y^2; \quad y_1$ is constant.

13. Show that if $f_1(x), f_2(x), f_3(x)$, and $f_4(x)$ are differentiable functions then the function

$$u(x) = \frac{cf_1(x) + f_2(x)}{cf_3(x) + f_4(x)}$$

satisfies a differential equation of Riccati type.

2.7 Clairaut Differential Equations (Optional)

Suppose that $y = F(x)$ is a real function. At the point $(x_0, F(x_0))$ an equation for the tangent line is

$$y - F(x_0) = F'(x_0)(x - x_0). \tag{2.9}$$

This is a one-parameter family with parameter x_0. We can find a differential equation for this family this way. If $y' = F'(x_0)$, and F' has an inverse g near x_0, then $x_0 = g(y')$ and we can rewrite the given equation for the tangent line as

$$y = xF'(x_0) - x_0F'(x_0) + F(x_0)$$

or

$$y = xy' + f(y'), \tag{2.10}$$

where we have used the assumption that $x_0 = g(y')$, and have defined

$$f(y') = -g(y')F'(g(y')) + F(g(y')).$$

Equation 2.10 is a differential equation for the original family of straight lines. It has the form of a **Clairaut**[3] differential equation. *We expect the solutions to be straight lines*, since this is supposed to be the differential equation for a family of straight lines.

Differential equations of the Clairaut type are solved as follows: differentiate

$$y = xy' + f(y')$$

to get

$$y' = y' + xy'' + f'(y')y''.$$

This simplifies to the product

$$(y'')(x + f'(y')) = 0.$$

Any function that satisfies this equation is a solution of $y = xy' + f(y')$. There are two factors. Either of them could be zero. So we get two kinds of solutions.

1. Suppose that $y'' = 0$. This means that $y' = c$, a constant, giving solutions to $y = xy' + f(y')$ of the form $y = xc + f(c)$. This is a one-parameter family of straight lines.
2. Suppose that $x + f'(y') = 0$. We pair this equation with $y = xy' + f(y')$ and let $y' = t$, to get this parametric representation for a plane curve:
$$\begin{cases} x = -f(t) \\ y = f(t) - tf'(t) \end{cases} \tag{2.11}$$
This curve satisfies both $x + f'(y') = 0$ and $y = xy' + f(y')$ at each of its points. In general the curve is not a straight line, but is tangent to each line of the family $y = xc + f(c)$. Such a solution is called a **singular solution** because it has a different form from the members in the family of straight lines. Its relationship to these members causes it to be referred to as an **envelope** of the family. The reason for this name is made clear by an example.

Example 2.13 Solve the differential equation $y = xy' + (1/2)(y')^2$.

Solution. This is a Clairaut equation with $f(y') = (1/2)(y')^2$. A one-parameter family of straight line solutions is $y = cx + (1/2)c^2$. The third output of Example 2.13**M** below shows some representative members of this family. It is apparent that the lines are tangent to some curve. From part (2) of our solution process, we find that this curve can be defined parametrically by

$$x = -f(t) = -t,$$
$$y = f(t) - tf(t) = \tfrac{1}{2}t^2 - t(t) = -\tfrac{1}{2}t^2.$$

We can eliminate the parameter t between these two equations and get an equation for the singular solution to be $y = -(1/2)x^2$. You should verify that this

[3] Alexis Claude Clairaut (1713–1765) French mathematician. He studied privately with his father. At age 19, below the legal age, his work on quadratic curves got him admitted to the French Academy of Science. His work on lunar orbits was well-respected, he calculated the perihelion of Halley's comet, and was among the first to observe the existence of singular solutions to differential equations.

satisfies the original differential equation and is tangent to each of the members of the family of straight lines. ◇

Example 2.13M In *Mathematica* Example 2.13 can be thought about this way:
Solution. This is the one-parameter family

```
In[39]:= y[x_] = x c +(1/2)c^2

Out[39]=  2
         c
         -- + c x
         2
```

Here are some representative lines.

```
In[40]:= samples = Table[y[x],{c,-3,3,1/2}]

Out[40]=  9          25    5 x             9   3 x 1         1    x    1   x
         {- - 3 x,-- - ---,2 - 2 x,- - ---,- - x,- - -,0 - + -,
          2         8     2               8    2  2          8    2    8   2

           1        9   3 x              25   5 x   9
           - + x,  - + ---,  2 + 2 x,  -- + ---,  - + 3 x}
           2        8    2               8    2    2
```

Plot this set of lines on a common axis.

```
In[41]:= lines = Plot[Evaluate[samples],{x,-3,3},PlotRange->{-5,5}];
```

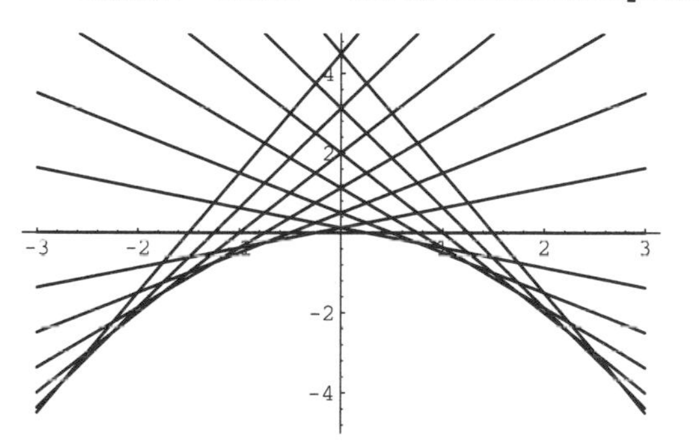

The plot certainly suggests that there is some curve (an envelope) to which each of the straight lines is tangent at one point. This curve is given by the parametric representation described above.

Note that the graph of the envelope is almost totally obscured by the tangent lines that are representatives of the one-parameter family. In this case, we can eliminate t from these two equations and find an equation for y in terms of x.

```
In[42]:= f[t_]=1/2t^2

Out[42]=  2
         t
         --
         2
```

```
In[43]:= Eliminate[{x == -f'[t], y == f[t] - t*f'[t]}, t]

Out[43]=  2
         x  == -2 y
```

This is the downward-turning parabola we saw above. ◇

Exercises 2.7

Solve these Clairaut equations. In each case, find a singular solution.

1. $y = xy' + 1 - \ln y'$.
2. $xy' - y = e^{y'}$.
3. $y = xy' - \tan y'$.
4. $y - xy' = 1 + 4(y')^2$.
5. Let $z = \sin x$ in $y = y' \tan x - (y')^2 \sec^2 x$.
6. $y = xy' + 2 + (y')^3$.

3 Applications of First-Order Equations

3.0 Introduction

In this chapter we will look at several applications of first-order differential equations. There are many of these that could be studied, but we will concentrate on those that can be described by linear differential equations or by separable differential equations. There are applications whose differential equations are first-order and which fall into each of the classifications that we saw in the last chapter. These two kinds of applications are chosen because the applications are interesting and useful, and the differential equations can usually be solved with a minimum of difficulty, especially since we have *Mathematica* to do the rather complicated integrations that may be necessary.

Among the applications are families of curves where the members cross each other perpendicularly. These are called orthogonal families. We will consider pairs of families that are mutually orthogonal.

In addition, we will consider applications from biology, chemistry, economics, and physics. Some of these are growth and decay, radiocarbon dating, simple cooling problems and chemical mixtures. These problems are all linear. We will also consider problems of air resistance, restricted population growth, and bimolecular reactions. These problems are nonlinear.

Each of these problems can suggest ideas for careful study. Many questions arise from the solutions we find and the meaning to be extracted from them. Most of the problems are really only approximately correct, and there is a great deal of room for further study of more exact formulations of the problems. These studies might well occupy your time for several years to come. There is room for experts in most of these applications areas.

3.1 Orthogonal Trajectories

We have seen one-parameter families of functions having the form $f(x, y, c) = 0$, where c is the parameter. The family is usually defined for c in some interval of the reals (maybe all of the reals). Often such a family has associated with it another family $g(x, y, d) = 0$ such that if any member of the first family crosses a member of the second family, the tangent lines to the respective curves are orthogonal, that is, they are perpendicular. Recall that two lines not parallel to the axes that have slopes m_1 and m_2 meet perpendicularly if the two slopes satisfy $m_1 m_2 = -1$. We can use this property of orthogonal slopes to find orthogonal families for some interesting families of curves.

In physics, an example of orthogonal families is the equipotential curves around one pole of a magnetic dipole or electrostatic monopole. They are orthogonal to the lines of force acting on external objects. The direction of the force (inward or outward) on an external object depends on the relative polarity of the external object. In meteorology, the isobars (lines of equal barometric pressure) on a weather map are orthogonal to the pressure gradient curves. In meteorology one also plots isotherms, curves of equal temperature. Heat flows orthogonally to the isotherms (from hot to cold). In true applications in spaces of three dimensions, the level objects are surfaces, rather than curves, and the differential equations are really partial differential equations. We have to look at the two-dimensional special cases, because we are studying ordinary differential equations.

The procedure for obtaining an orthogonal family is:

1. Find a differential equation whose solutions contain the members of the given family, by eliminating c from the two equations
$$f(x, y, c) = 0 \quad \text{and} \quad \frac{\partial f}{\partial x} + \frac{\partial f}{\partial y}\frac{dy}{dx} = 0.$$
2. Use the orthogonality relationship to obtain a differential equation of the orthogonal family.
3. Solve this latter equation to find the members of the orthogonal family.

Example 3.1 Given the family of all straight lines through the origin. Find a description of the orthogonal family.

Solution. The given family may be expressed as $y = cx$, for c real. Note that this formulation does not include the y-axis, which is one of these lines. Rewrite $y = cx$ as $x = y/c$ for $c \neq 0$. If you take the limit as $c \to \infty$, you get the missing line, $x = 0$. It is worth taking note that sometimes we can express missing objects as limits of objects we are able to express with formulas of a particular kind.

To get the differential equation of the given family, differentiate $y = cx$ to get $y' = c$. From these two equations we eliminate c to get $y = y'x$, or $y' = y/x$, $x \neq 0$. Rewrite this as $dy/dx = y/x$ and apply the orthogonality condition, to get

$$\frac{-1}{dy/dx} = -\frac{dx}{dy} = \frac{y}{x}$$

as the differential equation of the orthogonal family. This equation is separable into $ydy = -xdx$, and has as solutions $y^2/2 = -x^2/2 + d$, or $x^2 + y^2 = r^2$, by writing $r^2 = 2d$. This is a family of circles concentric with the origin. A picture of both families will make the geometric situation clear. See Figure 3.1. ◇

Example 3.2 Given the family of curves $xy = c$. Find the orthogonal family.

Solution. The given family is the two axes ($c = 0$) plus all of the hyperbolas in quadrants 1 and 3 ($c > 0$) or quadrants 2 and 4 ($c < 0$) having the axes as asymptotes. These are all rectangular hyperbolas, since their asymptotes are perpendicular. The differential equation of the family is $xy' + y = 0$, or $dy/dx = -y/x$. The orthogonality condition says that $-1/(dy/dx) = -dx/dy = -y/x$. Separating variables gives the differential equation $xdx = ydy$, which has as solution $x^2 = y^2 + d$. When $d = 0$, this gives $y = \pm x$. When $d = a^2 > 0$ the solution is $x^2/a^2 - y^2/a^2 = 1$, a family of hyperbolas opening left and right each of which has as asymptotes the lines $y = \pm x$. When $d = -a^2 < 0$ the solution is $y^2/a^2 - x^2/a^2 = 1$, a family of hyperbolas opening up and down each of which also has as asymptotes the lines $y = \pm x$. Thus, the orthogonal family of a set of rectangular hyperbolas (plus two lines) is a family of rectangular hyperbolas (plus two lines). See Figure 3.2. ◇

Example 3.3 The differential equation

$$\frac{dy}{dx} = \frac{\cos x}{e^y}$$

defines its family of solutions. Find the orthogonal family and plot both families.

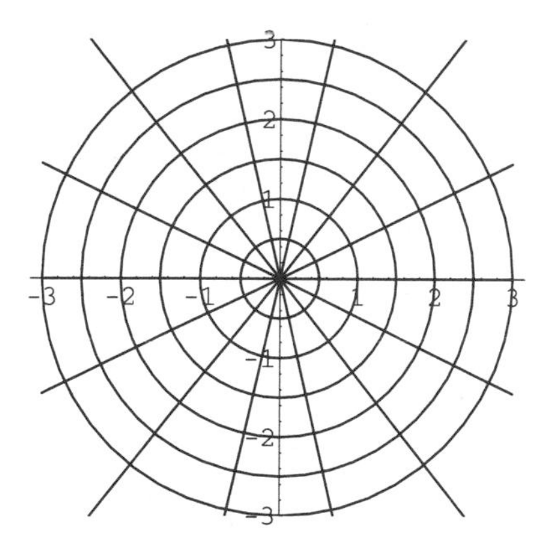

Figure 3.1: The orthogonal families $y = cx$ and $x^2 + y^2 = r^2$.

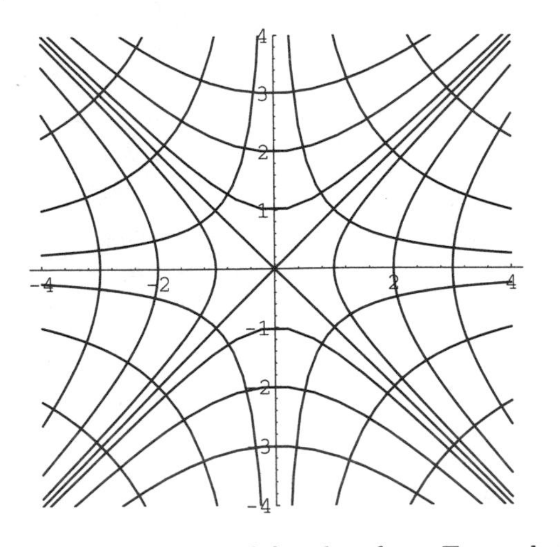

Figure 3.2: Orthogonal families from Example 3.2.

Solution. The equation has its variables separable. Separating gives $e^y dy = \cos x dx$, and hence $e^y = \sin x + c_1$ as the family of solutions. The orthogonal family has as its differential equation

$$\frac{dy}{dx} = \frac{-e^y}{\cos x}$$

which is also separable and has among its solutions the family

$$e^{-y} = \ln|\sec x + \tan x| + c_2$$

and the vertical lines $x = \pm\pi/2, \pm 3\pi/2$, etc. These families can be considered to be the level curves of $f(x, y) = e^y - \sin x$ and $g(x, y) = e^{-y} - \ln|\sec x + \tan x|$, respectively. The families can each be solved for y as a function of x and plotted. Combining the two plots yields Figure 3.3. ◇

In Figure 3.3, the solutions of the original family are horizontal at $x = \pm\pi/2$. This means that the members of the orthogonal family are very steep near $x = \pm\pi/2$. The vertical lines $x = \pm\pi/2$ are in the orthogonal family.

Exercises 3.1

1. Show that the family of circles with center on the y-axis that pass through the origin has as orthogonal family all of the circles with center on the x-axis that pass through the origin.

2. Show that the family of circles with center on the x-axis that pass through the origin has as orthogonal family all of the circles with center on the y-axis that pass through the origin.

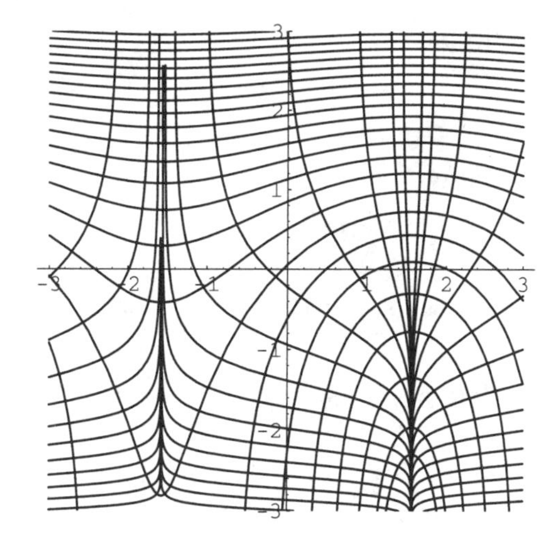

Figure 3.3: Some members of the orthogonal families of Example 3.3.

3. Show that the family of circles with center on the line $y = x$ that pass through the origin has as orthogonal family all of the circles with center on the line $y = -x$ that pass through the origin. [*Hint:* Think about the two previous problems.]

4. Consider

$$\frac{x^2}{a^2} + \frac{y^2}{a^2 - c^2} = 1,$$

for fixed c and a as parameter, $a \neq 0$ and $a \neq \pm c$. This is a family of confocal conics. (Each curve is a conic having foci at $(0, c)$ and $(0, -c)$). Show that this family is self-orthogonal. That means that the differential equation of the orthogonal family is the same as the differential equation of the original family. As a secondary question, what is the geometric, not algebraic, significance of the requirement that $a \neq 0$ and $a \neq \pm c$?

5. Find a family orthogonal to the family

$$y = \frac{2ce^{2x}}{1 + ce^{2x}}.$$

6. Find the orthogonal trajectories of

$$y = \frac{cx}{2 + x}.$$

7. Find the orthogonal trajectories of the family of solutions of the separable differential equation

$$\frac{dy}{dx} = f(x)g(y).$$

8. Find the family of solutions of

$$\frac{dy}{dx} = \frac{\sec x}{4y}.$$

9. Find the family of solutions of

$$\frac{dy}{dx} = \frac{\tan x}{8y}.$$

Find the orthogonal trajectories of the family of solutions.

3.2 Linear Applications

This section is devoted to applications of first-order linear differential equations in the biological sciences, the physical sciences, and the social sciences. Each of these applications is closely related to the notions of exponential growth and decay.

Growth and Decay

It is interesting to note that a wide range of problems such as the growth of bacteria, fungi, rabbits, cities, and continuous compound interest, and the decay of radioactive substances all can be modeled by a single kind of differential equation, namely,

$$\frac{dy}{dt} = ry. \tag{3.1}$$

The loss of temperature in warm-body cooling satisfies a closely related equation

$$\frac{dy}{dt} = r(c - y),$$

where c is a constant.

If $r > 0$, equation 3.1 describes a growth phenomenon (See Figure 3.4), and if $r < 0$ a decay phenomenon (See Figure 3.5). Whatever the value of r, $y = 0$ is a constant solution and every solution has the form $y(t) = Ce^{rt}$.

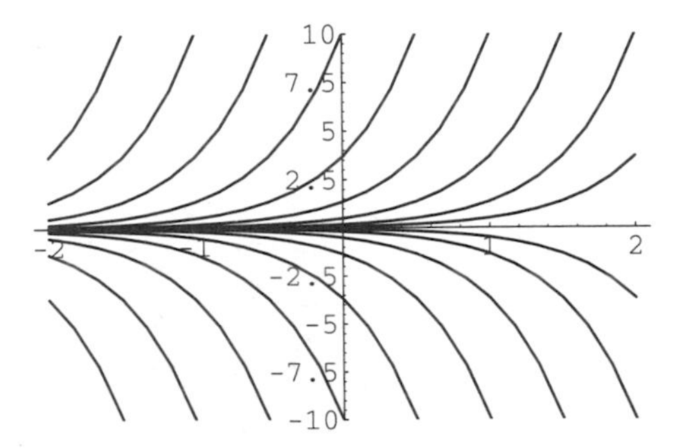

Figure 3.4: Exponential growth.

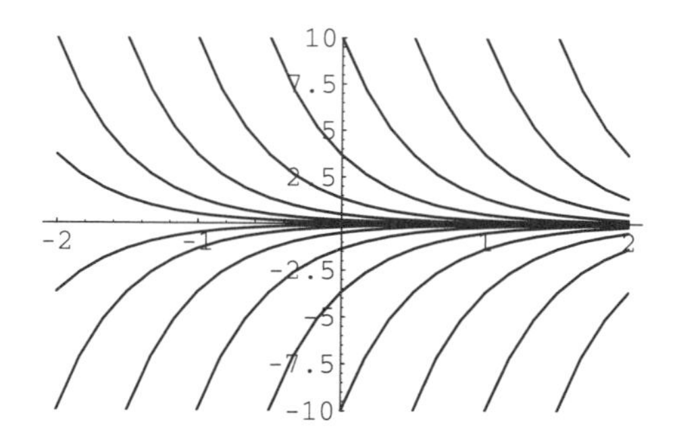

Figure 3.5: Exponential decay.

There are certain key phrases that identify a process as satisfying a differential equation of the form $y' = ry$. Such phrases as "**rate of change** is proportional to the amount present," or "the relative rate of change (of a quantity) is constant" are key to the proper statement of the problem. The **relative rate of change** is $(dy/dt)/y$. Sometimes r is given in absolute terms and sometimes as a percent. Be careful to express r in absolute terms. The word "initially" usually indicates a condition at time $t = 0$.

Applications in the Biological Sciences

Example 3.4 Suppose that the rate of growth of a population of organisms is 5% of the number present, t being measured in days. If there are 10,000 individuals present initially, how many are present in 10 days? When will the initial population have doubled?

Solution. Let $p(t)$ denote the number present at time t. Then $dp/dt = 0.05p$ and $p(0) = 10{,}000$, and the solution has the form $p(t) = Ce^{0.05t}$. From the given initial condition we find that $p(0) = C = 10{,}000$. Therefore there are $p(t) = 10{,}000e^{0.05t}$ individuals present t days later. Hence there are $10{,}000e^{0.5} = 16,487.2$, which we round to 16,487, individuals present at the end of day 10.

In order to find when the population has doubled, we need to solve the equation $p(T) = 10{,}000e^{0.05T} = 2(10{,}000) = 20{,}000$ for T. This means $e^{0.05T} = 2$, which means that $T = (\ln 2)/0.05 = 13.8629$ days. (From the size of the population after 10 days, we knew that doubling had to occur shortly after 10 days.) ◇

A common variation on this problem is the following:

Example 3.5 The population of a city was 20,000 according to the 1980 census and 25,000 according to the 1990 census. If the rate of growth of population is proportional to the population and is constant,

(a) what was the population in 1960? and
(b) when will the population be 40,000?

Solution. The statement of the problem suggests that $dP/dt = rP$, where $P(t)$ is the population at time t (in years), so $P(t) = Ce^{rt}$. We are given that

$$P(1980) = Ce^{1980r} = 20{,}000$$

and

$$P(1990) = Ce^{1990r} = 25{,}000.$$

Divide these to find that $e^{10r} = 25{,}000/20{,}000 = 5/4$, so that $10r = \ln(5/4)$, and $r = (1/10)\ln(5/4) \approx 0.0223144$. From $Ce^{(1980/10)\ln(5/4)} = 20{,}000$, we find that

$$C = 20{,}000e^{(-1980/10)\ln(5/4)} \approx 1.29672(10^{-15}).$$

This means that

$$P(t) = 20{,}000e^{((t-1980)/10)\ln(5/4)}.$$

To answer the question of part (a),

$$\begin{aligned} P(1960) &= 20{,}000e^{((1960-1980)/10)\ln(5/4))} \\ &= 20{,}000e^{-2\ln(5/4)} \\ &= 12{,}800. \end{aligned}$$

To answer the question of part (b), we need to determine t (remember that the question asked *when*) such that

$$20{,}000e^{((t-1980)/10)\ln(5/4)} = 40{,}000.$$

This means that

$$e^{((t-1980)/10)\ln(5/4)} = 2,$$

so

$$\frac{t-1980}{10}\ln(5/4) = \ln 2$$

and

$$\begin{aligned} t &= 1980 + 10\frac{\ln 2}{\ln(5/4)} \\ &= 2011.06 \text{ years}, \end{aligned}$$

or sometime in early 2011. The time of doubling of the 1980 population was therefore 31 years. ◇

The Spread of an Epidemic

In 1959, H. Muench introduced an elementary model for the progress of an epidemic. His model is

$$\frac{dy}{dt} = r(1-y),$$

where r is a constant rate. Here "1" denotes the entire population, and y denotes the portion of the population that has come down with the disease. Initially some small fraction $y(0)$ has the disease. From then on, the course of the outbreak is described completely by the solution

$$y(t) = 1 - (1 - y(0))e^{-rt},$$

where $0 \leq y \leq 1$, and t is in days. This model says that the disease remains present in the population forever. However, when everyone in the population has had the disease, we would say that the differential equation no longer applies, and the epidemic has passed. It is perhaps clearer to state the problem with a base population of size n that is large (tens of thousands or millions). Then the equation reads

$$\frac{dy}{dt} = r(n - y),$$

and $y(0)$ is a small positive integer. The epidemic is over when $t \geq T$ implies $y(t) \geq n-1$. This makes the duration of the epidemic finite. The solution in this case is $y(t) = n - (n - y(0))e^{-rt}$, $0 \leq t \leq T$.

Adding Tap Water to a Fish Bowl

Anyone who keeps or is interested in keeping freshwater fish in a bowl or aquarium should want to know how to renew the water that is lost naturally by evaporation. It is important to do more than just maintain the water level. If water is merely added, the concentration of sodium (as ordinary table salt, NaCl) that is present in almost every commercial or private water supply will increase until deadly levels can occur. We compare this scenario with that of systematically removing an extra quantity of water whenever fresh water is added. (As it has been stated, the problem is naturally a difference equation, with water additions being discrete events. We consider the fresh water to be (slowly) added continuously. When water is removed, whether by evaporation or manually, we assume that it is continuously being removed at a constant rate. This allows us to study the problem as a differential equation.)

In the first scenario, fresh (tap) water exactly replaces the water lost through evaporation. Suppose that r is the rate of addition of water, that e is the rate of evaporation, $x(t)$ is the amount of salt present at time t, with x_0 being present initially. Say that the concentration of NaCl in the fresh water is ε. Then

$$\frac{dV}{dt} = r - e = 0, \quad V(0) = V_0.$$

This means that $V(t) = V_0$ for $t > 0$. Also

$$\frac{dx}{dt} = r\varepsilon - 0 = r\varepsilon > 0,$$

and

$$x(0) = x_0 \geq 0.$$

Thus for $t > 0$, the amount of salt, $x(t) = (r\varepsilon)t + x_0$. This function increases without bound, and hence will cause the concentration $x(t)/V_0$ of salt in the fishbowl to exceed the survivable level for the fish living in the fishbowl. Of course the described process must effectively halt at some time because the amount of salt in the fishbowl cannot exceed the volume of the fishbowl (See Figure 3.6).

In the second scenario, more water than that which evaporates is removed (and disposed of) at the rate δ and fresh water exactly replaces the water being lost through evaporation and deliberate removal. Under these circumstances,

$$\frac{dV}{dt} = r - \delta - e = 0, \quad \text{since } r = \delta + e,$$

so that again

$$V(t) = V_0 \text{ for } t > 0.$$

Under these circumstances, the amount of salt in the fishbowl has a different differential equation,

$$\frac{dx}{dt} = r\varepsilon - \delta\left(\frac{x}{V_0}\right),$$

but it satisfies the same initial condition $x(0) = x_0 \geq 0$. We can already see potential benefits in that now it is possible for $dx/dt < 0$ and therefore the amount of salt in solution can decrease. The differential equation has a constant, or equilibrium, solution $x(t) = r\varepsilon V_0/\delta$ for which $x'(t) \equiv 0$ (that may or may not satisfy the initial condition), and the solution

$$x(t) = \frac{r\varepsilon V_0}{\delta} + \left(x_0 - \frac{r\varepsilon V_0}{\delta}\right) e^{-\frac{\delta}{V_0}t}.$$

This solution says that the limiting concentration

$$x_c = \frac{\frac{r\varepsilon V_0}{\delta}}{V_0} = \frac{r\varepsilon}{\delta} = \left(\frac{r}{r-e}\right)\varepsilon > \varepsilon.$$

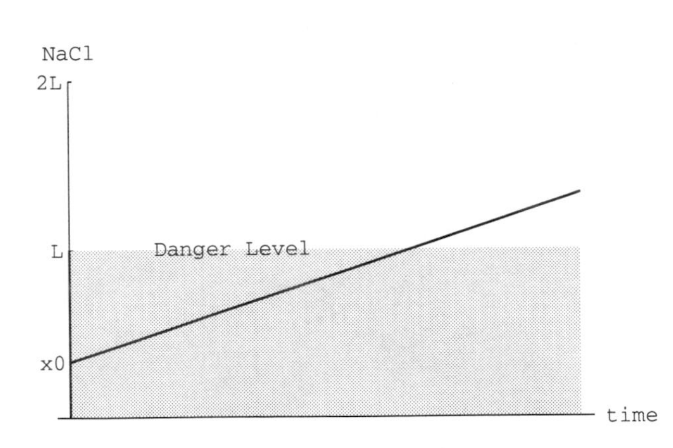

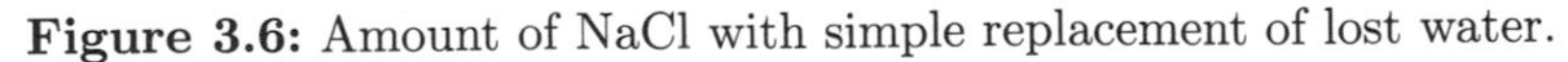

Figure 3.6: Amount of NaCl with simple replacement of lost water.

However, x_c can be large if r/δ is large. (That is, if the amount of water disposed of is only a small fraction of the fresh water being added.) If, for instance, extra water is disposed of at the same rate that evaporation takes place, then $\delta = e, r = 2e$ and $x_c = (2e/e)\varepsilon = 2\varepsilon$, which should be an acceptable concentration, because presumably the available tap water has an acceptably low concentration of salt. For a saltwater aquarium this procedure is inappropriate because it can cause the concentration of sodium to decrease to levels that may be dangerous to saltwater organisms that expect high concentrations of sodium. See Figure 3.7.

The limiting value of $x(t)$ as $t \to \infty$ is the equilibrium solution noted above. This means that no matter what the initial amount x_0 of salt, the solution ultimately approaches the equilibrium solution. There is more discussion of equilibrium solutions in Section 8.7.

It might interest some of you to ask a representative of your local water supplier about the concentration of sodium in your water supply. Then you could use real data to model the situations described here. In our small town in Tennessee, the concentration of sodium is 0.9 mg per liter. Nationally, the concentration ranges from 0.5 mg per liter to 25 mg per liter. When the concentration exceeds 20 mg per liter some patients with heart disease may have to take special precautions under the guidance of a physician.

Of course fishbowls are not the only places where sodium can concentrate if water is able to evaporate but sodium cannot be readily removed: just consider the Dead Sea and the Great Salt Lake! The Great Lakes have a substantial concentration of sodium that is not substantially increasing because the Great Lakes are constantly emptying to the sea by way of the St. Lawrence River. The oceans themselves are the ultimate example of extreme sodium concentration produced when only evaporation is able to take place.

Applications in the Physical Sciences

Cooling

Newton's law of cooling states that the rate that a warm body cools is proportional to the difference between its temperature, T, and the temperature, T_0, of

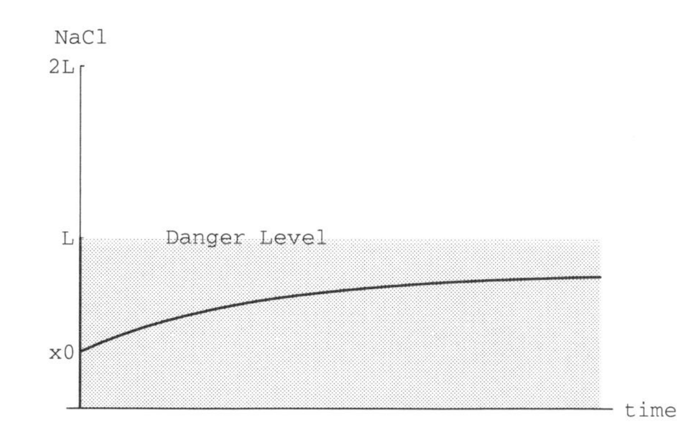

Figure 3.7: Amount of NaCl with systematic water removal.

the surroundings. As an equation, this says

$$\frac{dT}{dt} = k(T_0 - T). \tag{3.2}$$

This is a linear differential equation whose solution is obvious when the equation is written as

$$\frac{d(T - T_0)}{dt} = \frac{dT}{dt} = -k(T - T_0),$$

so that

$$T - T_0 = ce^{-kt},$$

or

$$T = T_0 + ce^{-kt}.$$

The constant c is the initial temperature differential $T(0) - T_0$. So the solution is

$$T(t) = T_0 + (T(0) - T_0)e^{-kt},$$

which decays to T_0 as $t \to \infty$. The parameter k is related to several factors such as the volume, surface area, mass, and the heat capacity of the body that is cooling. In many situations where this cooling law is used, values for k are tabulated.

Example 3.6 Suppose that a cake is removed from a 325 degree (Fahrenheit) oven into a 75 degree kitchen. If the cake cools to 180 degrees in 10 minutes, how long afterward will its temperature have fallen to 80 degrees?

Solution. From the discussion above, $T(t) = 75 + (325 - 75)e^{-kt} = 75 + 250e{-}kt$. From $T(10) = 75 + 250e^{-10k} = 180$, we find that $e^{-10k} = (180 - 75)/250 = 105/250 = 21/50$, so that $-10k = \ln(21/50)$, or

$$k = -(1/10)\ln(21/50) \approx 0.086750.$$

This means that

$$T(t) = 75 + 250e^{\frac{1}{10}\ln\left(\frac{21}{50}\right)t}.$$

We need to solve

$$75 + 250e^{\frac{1}{10}\ln\left(\frac{21}{50}\right)t} = 80$$

for t. This requires that

$$\frac{1}{10}\ln\left(\frac{21}{50}\right)t = \ln(5/250) = \ln(1/50),$$

so $t = 10(\ln(1/50)/\ln(21/50)) = 45.1$ minutes. This is the time since removal from the oven, so the time since the temperature was 180 degrees is $45.1 - 10 =$

35.1 minutes for the temperature to drop the remaining 100 degrees. It is of interest to note that the temperature at the end of one hour of cooling is 76.4 degrees. Most people would say that the cake had cooled. See Figure 3.8. ◇

Those of you who thought carefully about the problem will realize that the heat that leaves the cake has to go into the environment, thereby raising the temperature of the environment. So our assumption about the temperature of the kitchen "being" 75 degrees might have been slightly incorrect. There are two ways to think about this. First, that there is another differential equation that is associated with the one we defined, but which accounts for the heat that the cake loses. This is absolutely correct. On the other hand, the heat capacity of the kitchen environment is certainly much greater than the heat capacity of the cake, so the temperature in the kitchen will rise only negligibly, and can safely be called constant. This, too, is correct. There are delicate situations where it is vitally important to keep track of everything, and there are situations where a high degree of accuracy is wasteful. What is being suggested here is that there is a sort of continuity about differential equations: One can say when two differential equations are near one another. Under the appropriate assumptions, the solutions of two differential equations that are near one another are also near one another.

This model of cooling does not apply to cooling near a phase transition point, such as where water becomes ice or where steam becomes water. Other laws apply in such situations. It does apply on intervals that are not too near such points. Nor does this model apply when the temperature changes over too great a range. There are nonlinear differential equations that apply in more extreme situations such as these, but they all are approximated very closely by Newton's law of cooling when the temperature range is small.

A second application of Newton's law of cooling that you may have heard about for years on your local television news is the procedure used by coroners and medical examiners to determine the time of death of a body. The coroner

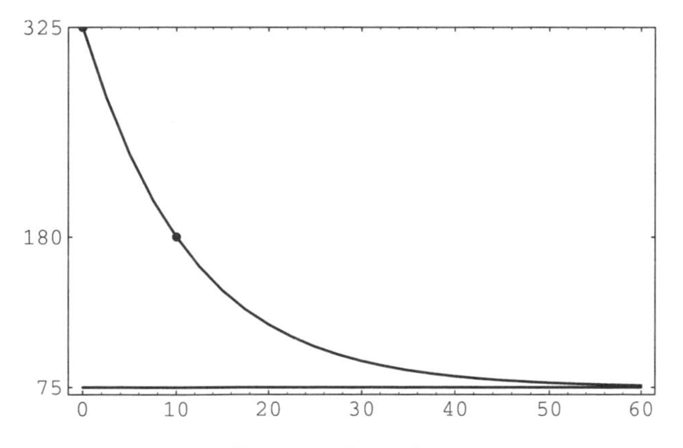

Figure 3.8: Cooling of a cake, Example 3.6.

measures the temperature of the body. Then, using this information the coroner can estimate the time of death. It is of some interest that coroners in Tennessee use the approximate rule $H = (99 - T)/1.5$ to estimate hours since death from measured (rectal) temperature. The "more precise" rule in use is that body temperature falls three degrees Fahrenheit per hour for the first three hours and one degree per hour thereafter until "ambient temperature is reached." You will recognize that this is merely a working approximation to the shape of a decreasing exponential. It implicitly assumes that the body is indoors, because a body outdoors in cold weather would certainly lose heat faster than one inside a building that is at a normal residential temperature.

We all know that "cooling" can go the other way. The temperature of a cold glass of water rises to the temperature of its surroundings, and Newton's law of cooling can be used to estimate the times and temperatures involved.

Mixtures

There are many types of problems that can be classified as mixture problems. Typically, a container holds a solution of a substance that is initially at some concentration. A new source of the substance is coming in at a different concentration. Optionally, the solution in the container is being drawn off. Assuming that the solution in the container is being kept well-stirred at all times, how does the concentration of the substance in the container vary over time? In later chapters we will consider a network of interconnected containers and ask about the concentrations in each of them. This will require several related differential equations, one for each container.

Two principles that must be understood in order to be able to set up problems of this type are: Rate of change of volume = Incoming change of volume – Outgoing rate of change of volume,

$$\frac{dV}{dt} = \frac{dV_{\text{in}}}{dt} - \frac{dV_{\text{out}}}{dt}; \tag{3.3}$$

and Change of mass = Incoming mass – Outgoing mass,

$$\Delta c = \Delta c_{\text{in}} - \Delta c_{\text{out}},$$

so that

$$\frac{\Delta c}{\Delta t} = \frac{\Delta c_{\text{in}}}{\Delta t} - \frac{\Delta c_{\text{out}}}{\Delta t},$$

and hence

$$\frac{dc}{dt} = \frac{dc_{\text{in}}}{dt} - \frac{dc_{\text{out}}}{dt}, \tag{3.4}$$

where c is the mass of the subject in question. This relationship also holds for weight.

Example 3.7 A 1000-gallon container initially contains 50 pounds of salt. A brine mixture of 1/4 pounds of salt per gallon is entering the container at 6 gallons per minute. The well-mixed contents of the container are being discharged from the container at the rate of 6 gallons per minute. Express the amount of salt in the container as a function of time. What is the limiting concentration of salt in the container?

Solution. Let $x(t)$ denote the amount of salt in the container at time t. First note that the volume of the mixture in the container is unchanging since $dV/dt = 6 - 6 = 0$. Salt is coming into the container at the rate of

$$\frac{dx_{\text{in}}}{dt} = (6\text{gal}/\min)(1/4 \text{ lb/gal}) = (3/2)\text{lb}/\min.$$

Also, since the concentration of salt in the container

$$\frac{dx_{\text{in}}}{dt} = \frac{\text{amount in the container}}{\text{volume of the container}} = \frac{x}{1000} \text{ lb/gal},$$

it follows that salt is leaving the container at the rate of

$$\frac{dx_{\text{out}}}{dt} = (6\text{gal}/\min)(x/1000 \text{ lb/gal}) = \frac{6x}{1000}\text{lb}/\min.$$

This means that the rate of change of salt in the container is

$$\frac{dx}{dt} = \frac{dx_{\text{in}}}{dt} - \frac{dx_{\text{out}}}{dt} = \frac{3}{2} - \frac{6x}{1000} = \frac{3}{2} - \frac{3x}{500},$$

and the initial condition is $x(0) = 50$. This is a linear differential equation. You should verify that the solution is

$$x(t) = 250 - 200e^{-3t/500}.$$

See Figure 3.9. ◇

This expresses the amount of salt in the container at time t as was requested, and it is easy to see that the limiting amount of salt in the container is 250 pounds, which results in a limiting concentration of $250/1000 = 1/4$ pound per

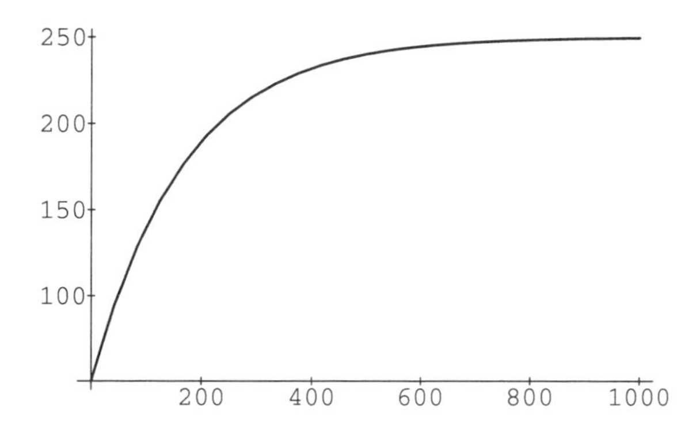

Figure 3.9: The amount of salt, constant volume.

gallon of salt in the container. Is it an accident that the limiting concentration is the same as the entering concentration? See the exercises.

Example 3.8 Suppose that the rate of discharge of the solution in the previous example is reduced to 5 gallons per minute, and the container can hold a maximum of 2000 gallons. Now express the amount of salt in the container as a function of time. When does the container begin to overflow? What is the concentration of salt in the container when the container begins to overflow?

Solution. Here $dV/dt = 6 - 5 = 1$ gallons per minute, so that $V(t) = (1000 + t)$ gallons after t minutes. This means that the container will overflow when $1000 + t = 2000$, or $t = 1000$. We need to account for the fact that the volume is not constant when we consider the concentration of salt. The rate of change of salt in the container is

$$\frac{dx}{dt} = \frac{dx_{\text{in}}}{dt} - \frac{dx_{\text{out}}}{dt} = \frac{3}{2} - \frac{5x}{1000 + t}.$$

This is still a linear differential equation, but it no longer has constant coefficients. Using the methods we studied before, the solution is found to be

$$x(t) = 250 + \frac{t}{4} - \frac{2(10^{17})}{(1000 + t)^5}.$$

This has a much different character than the solution with constant volume. In fact it is equivalent to a **rational function**: the quotient of two polynomials.

Notice in Figure 3.10 that the amount of salt is increasing without apparent limit (until the container overflows). But the concentration of the salt at $t = 1000$ when the container begins to overflow is $x(1000)/V(1000)) = 493.75/2000 = 0.246875$, which is near 0.250, the concentration of the incoming stream. It is also interesting to note that if the process is not stopped, but the container is permitted to continue overflowing, then the differential equation of the process becomes

$$\frac{dx}{dt} = \frac{dx_{\text{in}}}{dt} - \frac{dx_{\text{out}}}{dt} = \frac{3}{2} - \frac{6x}{2000}, \quad t \geq 1000,$$

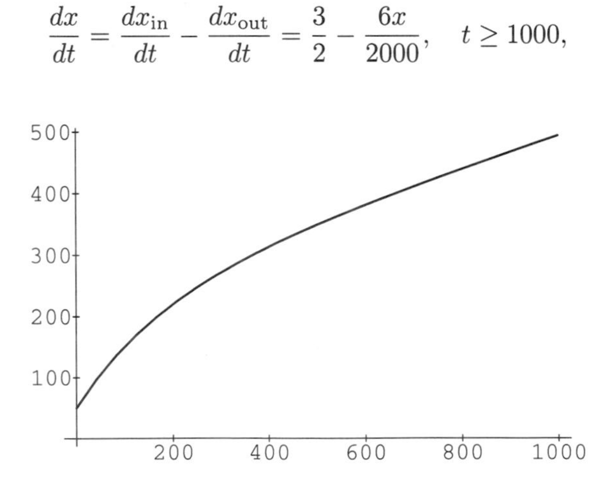

Figure 3.10: The amount of salt when volume is increasing.

with initial condition $x(1000) = 493.75$, the amount of salt in the container when overflow first begins. This overflow equation has solution

$$x(t) = 500 - \frac{25e^{3(1000-t)/1000}}{4},$$

from which can be easily seen that the limiting concentration in the overflowing container is 1/4 pound/gallon, the salt concentration in the entering stream. ◇

Example 3.9 Solve and check example 3.9 by *Mathematica*.
Solution. We let *Mathematica* solve the system.

```
In[1]:=  DSolve[{x'[t]==3/2-5 x[t]/(1000+t),x[0]==50},x[t],t]

Out[1]=               50000000000000000
         {{x[t] -> ------------------ +
                                5
                       (1000 + t)

                                                      2
            (6000000000000000 t + 15000000000000 t  +

                           3              4         5    6
              20000000000 t  + 15000000 t  + 6000 t  + t ) /

                          5
            (4 (1000 + t) )}}
```

Capture the solution. Write it expanded by partial fractions using `Apart`.

```
In[2]:=  Soln[t_]=Apart[x[t]/.First[%]]

Out[2]=         t   200000000000000000
         250 + - - ------------------
               4               5
                      (1000 + t)
```

It is interesting to try to have *Mathematica* check this answer, because the form of the derivative is very different from the form of the right hand side, and *Mathematica* considers both sides to be in a simple form. But the two forms do not compare well without further transformation. Here is what happens:

```
In[3]:=  Soln'[t]==3/2-5 Soln[t]/(1000+t)

Out[3]=  1   1000000000000000000
         - + ------------------- ==
         4             6
               (1000 + t)

                       t   200000000000000000
               5 (250 + - - ------------------)
                       4               5
            3                  (1000 + t)
            - - ------------------------------
            2             1000 + t
```

Even Simplifying does not help.

```
In[4]:=  Simplify[Soln'[t]==3/2-5 Soln[t]/(1000+t)]

Out[4]=  1      1000000000000000000
         - + ─────────────────────── ==
         4                  6
                 (1000 + t)

                          t    200000000000000000
               5 (250 + - - ──────────────────)
                          4                5
           3                      (1000 + t)
           - - ─────────────────────────────────
           2               1000 + t
```

We can force both sides to have the same form (a rational function) by using the Function Together on each side separately. Then comparison is simple:

```
In[5]:=  Together[Soln'[t]]==Together[3/2-5 Soln[t]/(1000+t)]

Out[5]=  True
```

The initial condition checks easily.

```
In[6]:=  Soln[0]==50

Out[6]=  True
```

◇

Applications in the Social Sciences

Compound Interest

An amount P_0 of money earns interest at an annual rate of i percent compounded n times per year at times $t_k = k/n$, $k = 1, 2, \ldots$. Let P_k denote the amount present at time t_k. Then

$$P_1 = P_0 + \frac{i}{100n} P_0.$$

$$P_2 = P_1 + \frac{i}{100n} P_1,$$

and in general

$$P_{k+1} = P_k + \frac{i}{100n} P_k, \quad k = 0, 2, \ldots.$$

This says that

$$\frac{P_{k+1} - P_k}{\frac{1}{n}} = \frac{i}{100} P_k.$$

If n is permitted to increase without bound, we have the **law of continuously compounded interest**:

$$\frac{dP}{dt} = \frac{i}{100}P, \quad P(0) = P_0.$$

This has as solution $P(t) = P_0 e^{it/100}$. Note also that continuously compounded interest obeys the same law as that of unrestricted growth.

Radioactive Decay

Radioactive substances decay at a rate proportional to the amount present. That is, $dy/dt = -ry$. If an amount y_0 is present initially, then for any time $t > 0$, $y(t) = y_0 e^{-rt}$. A common term that occurs in discussions of radioactive phenomena is **half-life**. A substance has half-life $t_{1/2}$ if $y(t + t_{1/2}) = (1/2)y(t)$. The value of $t_{1/2}$ is determined completely by r and is independent of y_0 and t. We find $t_{1/2}$ from the calculation

$$y(t + t_{1/2}) = y_0 e^{-r(t+t_{1/2})} = (1/2)y_0 e^{-rt} = (1/2)y(t).$$

Simplify to get

$$y_0 e^{-rt} e^{-rt_{1/2}} = \frac{1}{2} y_0 e^{-rt},$$

or

$$e^{-rt_{1/2}} = \frac{1}{2}.$$

This simplifies to $t_{1/2} = (1/r)\ln 2$. The smaller r is, the larger $t_{1/2}$ is. Values of $t_{1/2}$ for various substances range from small fractions of a second to many billions of years.

Radiocarbon Dating

Around 1950 Willard Libby, a chemist, proposed a method for estimating the age of an object made from materials that had been alive. Such materials as wood, or natural fibers that occur in fabrics were well-suited for his methods. The idea is that the proportion of the isotope ^{14}C of carbon, called carbon-14, which is part of all naturally occurring carbon in the environment has been essentially constant for thousands of years and hence has been constant in living organisms, since all of them require carbon for life, and thus assimilate ^{14}C and ^{12}C with carbon. When an organism dies, the ^{14}C it contains decreases through the process of radioactive decay, but the ^{12}C does not, and hence the ratio of ^{14}C to ^{12}C decreases. The idea proposed by Libby, which won him a Nobel prize for chemistry in 1960, was that one could measure the ratio of ^{14}C to ^{12}C in an object, compare that with the (constant) ratio in the environment, and hence determine how long since the object died. This would enable the determination of the approximate date when the object died. This is obviously of great interest to archaeologists, biblical scholars, and other historians. This section is concerned with the nature of the errors in such a technique. The technological methodology of those who actually do such measurement is very sophisticated. This discussion is not.

We earlier mentioned the differential equation $dy/dt = -ry$ which models radioactive decay. The idea of a half-life, the time it takes for half of a substance to decay, is reasonably clear. What isn't clear is exactly how to accurately measure half-lives. Human understanding of radioactive decay and our measurement of decay phenomena spans only slightly more than one human lifetime. This is not a long enough span of time to be able to accurately estimate very long half-lives. It is true that two points on a decreasing exponential determine the exponential, and hence the half-life, but each such measurement contains some error, however small. These errors mean that we cannot know long half-lives accurately until we have been able to obtain two measurements that are very far apart. Scientists have made attempts at this by measuring the ^{12}C ratio of ancient materials whose dates have been corroborated independently by historians. For our purposes, assume that the half-life of ^{14}C is about 5600 ± 30 years. Libby used the figure 5568 years, and 5730 ± 40 is used in other circumstances. Let's find out what effect this indeterminacy has on our attempts to date objects using ^{14}C. It is standard to date objects as being so many years **B.P.** (before present), present being set at 1950, the date of Libby's method.

The techniques for radioactive decay we saw before say that if the true half-life is $5600 - 30 = 5570$, then ^{14}C decays like $C(t) = Ae^{-((\ln 2)/5570)\,t}$. If the true half-life is 5600, then ^{14}C decays like $C(t) = Ae^{-((\ln 2)/5600)\,t}$. And if the true half-life is $5600 + 30 = 5630$, then ^{14}C decays like $C(t) = Ae^{((-\ln 2)/5630)\,t}$. These three curves decay to the right of the point $(0, A)$. They are actually quite close together, with the 5570-curve below the 5600-curve below the 5630-curve. For the calculation below, we take $A = 1$, meaning we start with 1 unit of the substance whose age is to be determined. As time increases to the right, that is, as the actual age of the sample increases, the ratio of ^{14}C present in the sample should lie between the three curves. What effect does this have on the estimated age? In addition, measurement of the ratio of ^{14}C present in a given sample is subject to some measurement error, though these errors are of a statistical nature, involving the counting of actual particle decomposition. What is the effect of such measurement errors on the results we obtain? Here are some examples for objects that date throughout the entire gamut of time to which this method is applicable: 50,000 B.P. to the present.

These examples assume a fixed measurement error of 1%. This is an entirely fictitious estimate. We assume a ^{14}C to ^{12}C fraction, f, is present and produce 9 estimated ages, depending on measurements of $0.99\ f$, $1.00\ f$, and $1.01\ f$. The ages are estimated along each of the three decay curves from the last paragraph. This assumes that if a measurement error of 1% is made, then $0.99f \leq$ measured ratio $\leq 1.01f$. We therefore look at the age predictions at these measurement extremes and at the assumed fraction f. This gives us a 3×3 table where each row entry is an age estimated by a given measurement level and each column entry is an age estimated on one of the possible decay curves.

Table 3.1: Ratio present: $f = 0.00205226$

	$h = 5570$	$h = 5600$	$h = 5630$
$1.01\ f$	49652	49919	50187
$1.00\ f$	49732	50000	50268
$0.99\ f$	49813	50081	50350

Sample from Around 50,000 Years B.P.

Example 3.10 Ratio assumed present: f = 0.00205226 of the original. Table 3.1 summarizes the results.

If our measured result is too large, it falsely says that our sample is too recent. If our measured result is too small, it falsely says that the sample is too old. It seems clear that this sample is approximately 50,000 years old, because all of the estimated ages are near 50,000. But it might be as old as 50,350 years or as recent as 49,652 years. This indicates an uncertainty of 698 years. But along each row, at each level of measurement error, the error is about 536 years due to the uncertainty about the half-life of ^{14}C, and down each column, that is, along each decay curve, the uncertainty is about 162 years. This indicates that for samples whose age is in the range of 50,000 years B.P., most of the uncertainty is due to the uncertainty in the half-life of ^{14}C, because the deviation along each row is greater than the deviation down each column. See Figure 3.11. ◇

Sample from Around 1950 Years B.P., or 1 A.D.

For a sample dating from around 1 A.D., the data table is essentially: Ratio present: $f = 0.7856$ of the original. See Table 3.2.

Thus the total uncertainty is about 82 years, with about 20 years due to uncertainty in the half-life of ^{14}C and about 62 years due to measurement errors. See Figure 3.12.

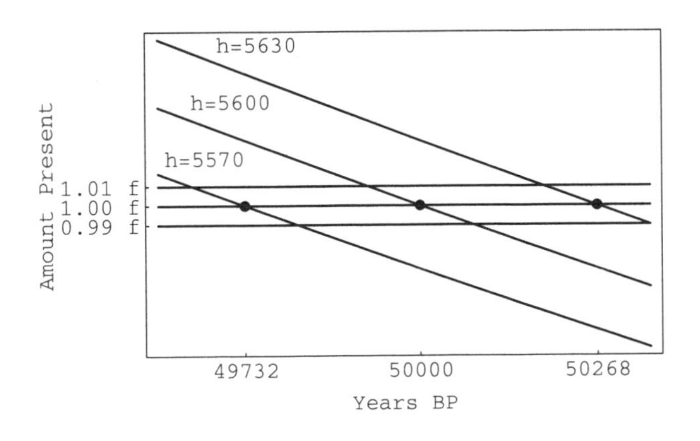

Figure 3.11: Radiocarbon dating uncertainty around 50,000 B.P.

Table 3.2: Ratio present: $f = 0.7856$

	$h = 5570$	$h = 5600$	$h = 5630$
1.01 f	1860	1870	1880
1.00 f	1940	1950	1960
0.99 f	2020	2031	2042

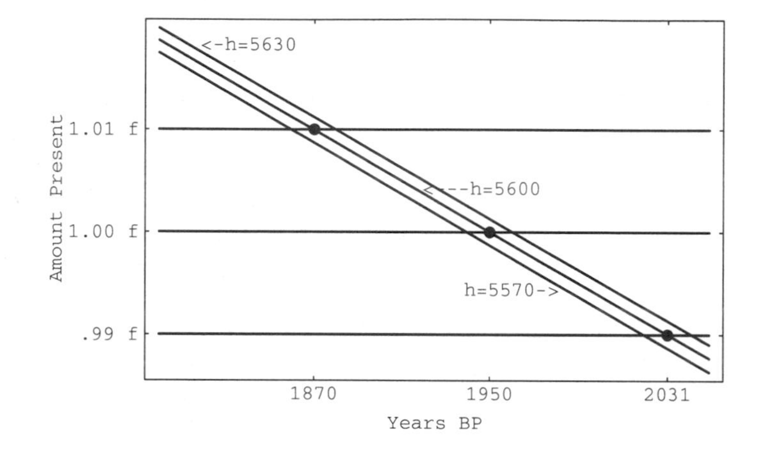

Figure 3.12: Radiocarbon uncertainty near 1950 B.P.

The Shroud of Turin

Example 3.11 The question of the authenticity of the shroud of Turin has fascinated people for centuries. It is claimed that the shroud is the burial shroud of Jesus Christ, which would date it from about 30 A.D., or 1920 B.P. The shroud is kept in the Royal Chapel of the Turin (Italy) Cathedral in a specially designed shrine. During the years 1986–1988, a major scientific investigation was undertaken under the auspices of the British Museum, with the approval of the Roman Catholic Church which owns the shroud. Improvements in radiocarbon dating methodology (using small gas-counters and accelerator-mass-spectrometry methods) had reached the point where reliable dating could be done from a very small sample. A sample measuring 10 mm×70 mm, and weighing approximately 150 mg was cut from the hem in the lower left of the shroud. This sample was divided into three equivalent parts and independently analyzed in three different laboratories, with these results [*Nature* 337:611–615]:

Table 3.3: Radiocarbon dating of the Shroud of Turin

Laboratory	Years B.P.	Range (B.P.)	Dates (A.D.)
Arizona	646 ± 31	615–677	1273–1335
Oxford	750 ± 30	720–780	1170–1230
Zurich	676 ± 24	652–700	1250–1298

Table 3.4: Shroud of Turin: ratio $f = 0.91825$.

	$h = 5600$	Date
1.01 f	609	1341 AD
1.00 f	689	1261 AD
0.99 f	770	1180 AD

Using our estimates the same way as before, except only along the nominal decay curve, and taking the nominal age of the shroud as given in the report to be 689 B.P., then the ratio of ^{14}C present would be $f = 0.91825$ of the original. See Table 3.4.

This analysis gives strong evidence that the shroud is not authentic, but is a medieval forgery of impressive quality.

You can see from Figure 3.13, the shroud of Turin graph, that the dominant error is from measurement error and not from uncertainty in the half-life. Our estimates, as depicted, essentially cover the range of dates returned from the laboratory measurements. (It is interesting that the apparent agreement is so close, considering that our assumed measurement error of 1% was a guess not based on any supporting data.) ◇

Supply and Demand

A simple model of the effects of **supply** and **demand** on the **price** of a commodity is

$$\frac{dp}{dt} = k(D - S),$$

where p is the price, S is the supply, D is the demand, and k is a positive constant. One sees immediately that if demand exceeds supply then $D - S > 0$

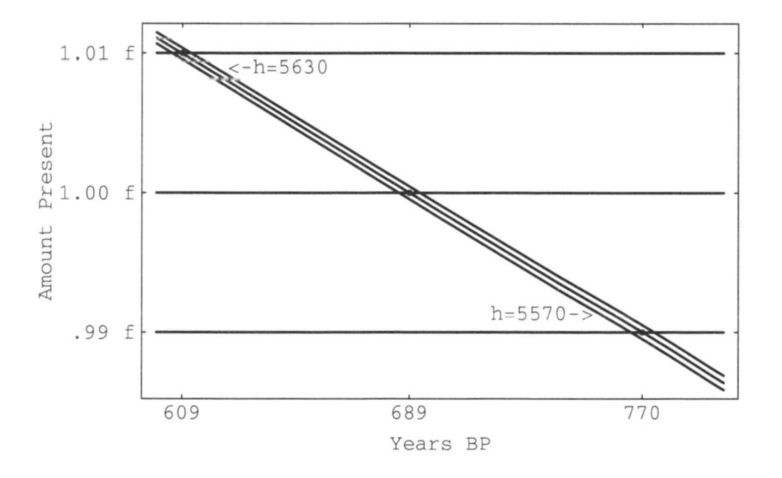

Figure 3.13: Estimated uncertainty for dating the shroud of Turin (years B.P.)

and $dp/dt > 0$, so the price rises. On the other hand, if supply exceeds demand, then $D-S < 0$ and $dp/dt < 0$ and the price falls. If supply equals demand, then $dp/dt = 0$ and price does not change.

One speaks of **seasonal** fluctuation in either supply or demand of a commodity. Saying the fluctuation is seasonal implies regular repetition that is usually modeled as being periodic. For instance one might assume $S = C_1 - C_2\cos(\alpha t)$. It is often the case that demand can reasonably be said to be a decreasing function of price so that $D = A - Bp$. This expresses "decreasing" as $dD/dt = -B < 0$. Furthermore, it says that if the commodity were free there would still be only the limited demand A, but if the price were to rise to A/B, then there would be no demand.

Example 3.12 Suppose that we have the simple supply and demand model $dp/dt = k(D - S)$, with $D = A - Bp$ and $S = C_1 - C_2\cos(\alpha t)$. How does price vary as a function of time?

Solution. We have

$$\frac{dp}{dt} = k(A - Bp - C_1 + C_2\cos(\alpha\, t))$$

or

$$\frac{dp}{dt} + (kB)p = k(A - C_1) + kC_2\cos(\alpha t)$$

This is a nonhomogeneous linear differential equation having solution

$$p(t) = Ke^{-kBt} + \frac{A - C_1}{B} + kC_2 e^{-kBt}\int e^{kBt}\cos(\alpha\, t)\,dt.$$

In Chapter 5 we will learn how to write the last term so that the solution has the form

$$\begin{aligned} p(t) &= Ke^{-kBt} + \frac{A - C_1}{B} + kC_2\sqrt{\alpha^2 + B^2k^2}\sin(\alpha t + \theta), \\ &= p_K(t) + p_\alpha(t), \end{aligned}$$

where

$$\begin{aligned} p_K(t) &= Ke^{-kBt}, \\ p_\alpha(t) &= \frac{A - C_1}{B} + kC_2\sqrt{\alpha^2 + B^2k^2}\sin(\alpha t + \theta), \end{aligned}$$

and $q = \arctan(\alpha/Bk)$. If the initial price is $p(0) = p_0$, then

$$K = p_0 - \frac{A - C_1}{B} - kC_2\sqrt{\alpha^2 + B^2k^2}\sin(\theta).$$

It is important to note that

$$\lim_{t\to\infty} p_K(t) = \lim_{t\to\infty} Ke^{-kBt} = 0,$$

so that ultimately the price is not dependent on the initial price and it fluctuates about the stable price $(A-C_1)/B$ with a maximum deviation of $kC_2\sqrt{\alpha^2+B^2k^2}$, and does so at the same rate that supply fluctuates. However, the fluctuation of price is "out of phase" with the supply by an amount q. For this reason q is called a **phase angle** in some applications. See Figure 3.14. ◇

The relationship between price and demand is something like the following: Another supply and demand model is

$$\frac{dS}{dt} = k(D - S),$$

which says that the rate of change of supply is proportional to the difference between demand and supply. In this model, if demand exceeds supply, then supply will decrease, whereas if supply exceeds demand, the supply will decrease. Notice that this model says that the rate of change of supply is not explicitly dependent on price.

Example 3.13 If the seasonal demand for a commodity has the form

$$D = C_1 - C_2\cos(\alpha t),$$

how does the supply vary?

Solution. Here the model becomes

$$\frac{dS}{dt} = k(C_1 - C_2\cos(\alpha t) - S),$$

or

$$\frac{dS}{dt} + kS = kC_1 - kC_2\cos(\alpha t).$$

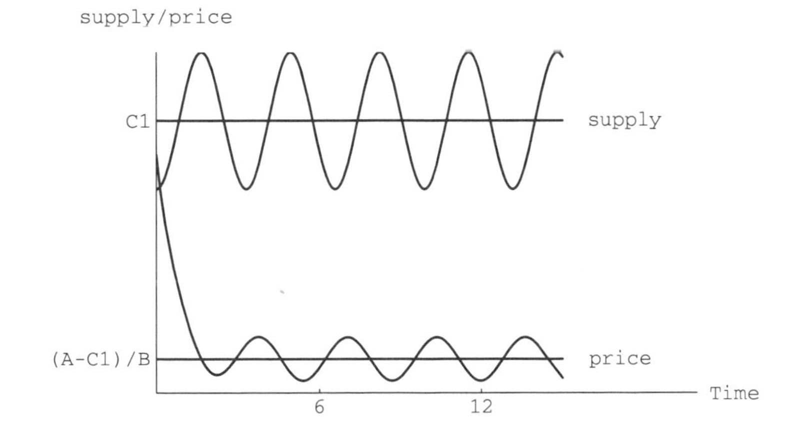

Figure 3.14: Effect of fluctuating supply on price.

This is the same form as the equation of Example 3.12 and the solution is

$$\begin{aligned} S(t) &= Ke^{-kt} + C_1 - kC_2\sqrt{\alpha^2 + k^2}\sin(\alpha t + \theta_1), \\ &= S_K(t) + S_\alpha(t), \end{aligned}$$

where

$$\begin{aligned} S_K(t) &= Ke^{-kt}, \\ S_\alpha(t) &= C_1 - kC_2\sqrt{\alpha^2 + k^2}\sin(\alpha t + \theta_1), \end{aligned}$$

and

$$\theta_1 = \arctan\left(\frac{\alpha}{k}\right).$$

Again K is determined by the initial supply, but since

$$\lim_{t\to\infty} S_K(t) = \lim_{t\to\infty} Ke^{-Kt} = 0,$$

the supply is ultimately independent of the initial supply. The fluctuation rate of the supply is the same as that of demand, but once again is out of phase by an amount q_1. Note also that the long-term fluctuations of supply are centered on C_1, which is precisely where demand is centered. ◇

Figure 3.15 is a typical picture of the relationship of supply to demand.

Exercises 3.2

Problems from the Biological Sciences

1. (*The Doomsday Problem*) Assume that the population of the world doubles every 75 years. It was 4.6 billion in 1982, according to the 1986 World Almanac. Under the hypothesis that the rate of growth is proportional to the

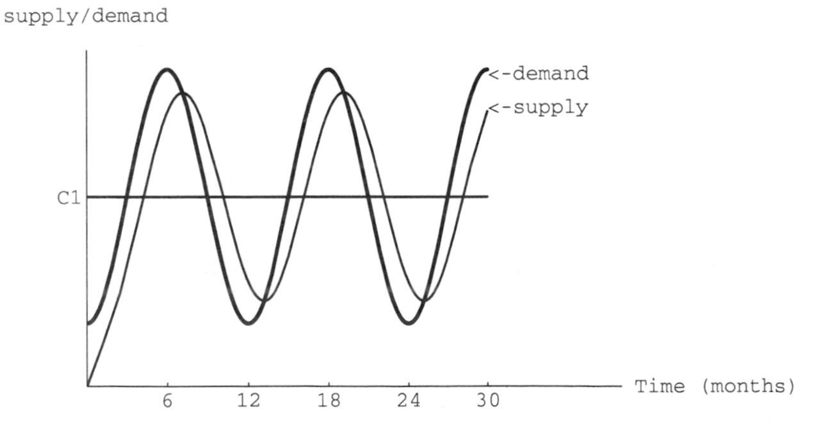

Figure 3.15: Effect of fluctuating demand on supply

population present, when will there be one person for each of the $1.5(10^{14})$ square meters of land on the earth? [Motivated by Shenk, p. 394, problem 15.]

2. Suppose that the rate of growth of a population of organisms is 3% of the number present, t being measured in days. If there are 5000 individuals present initially, how many are present in 10 days? When will the initial population have doubled?

3. Suppose that the rate of growth of a population of organisms is 3% of the number present, t being measured in days. If there are 10,000 individuals present in 10 days, how many were present initially? When will the initial population have doubled?

4. Suppose that the rate of growth of a population of organisms is 3% of the number present, t being measured in days. If there are 10,000 individuals present in 10 days and 15,000 individuals present in 15 days, when will the initial population have doubled?

5. At a university with 1000 students after a mid-semester break one student returns with type X flu. Let $y(t)$ denote the number of students who have contracted the disease by time t (in days). If the disease propagates throughout the population according to the rule $dy/dt = 0.05(1000 - y)$, how long does it take before 75% of the students have had the disease?

6. At a university with 1000 students after a mid-semester break one student returns with type X flu. If $y(t)$ is as in problem 5 and the disease propagates throughout the population according to the rule $dy/dt = 0.10(1000 - y)$, how many days does it take before 75% of the students have had the disease?

7. If the concentration of NaCl in the tap water of a community is 0.05 mg per liter, and water in a 5 gallon (18.93 liter) aquarium is kept at constant volume by adding tap water at the rate of 0.25 liter per week, after how many weeks will the concentration of NaCl in the aquarium be 0.20 mg per liter?

8. If the concentration of NaCl in the tap water of a community is 0.05 mg per liter, and water in a 5 gallon (18.93 liter) aquarium is kept at constant volume by discarding 0.01 liter per week and then adding tap water at the rate of $(0.01 + 0.25 = 0.26)$ liter per week, after how many weeks will the concentration of NaCl in the aquarium be 0.20 mg per liter? Compare with the previous problem.

Problems from the Physical Sciences

9. Repeat problem 8 with r as the concentration of NaCl in the entering solution and see if the limiting concentration is r. Does the final concentration of NaCl depend on the initial concentration of NaCl in the container? Set the initial amount of salt to x_0, and solve again.

10. An automobile engine is turned off and allowed to cool. Assume that Newton's law of cooling holds. Suppose that the internal temperature of the engine is 230°F when it is turned off and is 190°F after one hour. What is the constant of proportionality if the ambient temperature of the surroundings is 75°F? How long does it take for the engine to cool to 100°F?

11. In problem 10, if the ambient temperature of the surroundings is 95°F? How long does it take for the engine to cool to 100°F? Assume the same constant of proportionality.

12. In problem 10, if the ambient temperature of the surroundings is 35°F? How long does it take for the engine to cool to 100°F? Assume the same constant of proportionality.

13. In problem 10, if the ambient temperature of the surroundings is 0°F? How long does it take for the engine to cool to 100°F? Assume the same constant of proportionality.

14. A 100 gallon container initially contains 5 pounds of salt. A brine mixture of $1/4$ pound of salt per gallon is entering the container at 6 gallons per minute. The well-mixed contents of the container are being discharged from the container at the rate of 6 gallons per minute. Express the amount of salt in the container as a function of time. What is the limiting concentration of salt in the container?

15. A 100 gallon container initially contains 5 pounds of salt. A brine mixture of $1/4$ pounds of salt per gallon is entering the container at 6 gallons per minute. The well-mixed contents of the container are being discharged from the container at the rate of 7 gallons per minute. Express the amount of salt in the container as a function of time during the interval before the container empties.

PROBLEMS FROM THE SOCIAL SCIENCES

16. How long does it take an amount of money to double when it is invested at $i\%$ per year compounded continuously? Compare with the **rule of 72** which says that the number of years to doubling is approximately $72/i$.

17. If A invests P_0 dollars in an account at i_0 compounded continuously and B invests P_1 dollars in an account at i_1 compounded continuously, $P_1 < P_0$ and $i_0 < i_1$, when do the two accounts contain equal amounts? One year later, how much more is in the account owned by B?

18. Solve problem 17 with $P_0 = \$1000, P_1 = \$500, i_0\% = 6\%$ and $i_1\% = 8\%$.

19. We wish to determine the age of an artifact made of formerly living substances by Libby's method of radiocarbon dating. It is found that the ratio of ^{14}C present is 0.689817 the ratio in a similar living substance. Estimate a range of dates B.P. for the age of the sample using the half-life of ^{14}C as 5600 ± 30 years.

3.3 Nonlinear Applications

Nonlinear differential equations occur widely and are very interesting. It is sometimes quite a challenge to explain why some behavior that the solutions of a problem exhibit is not behavior that would be expected of the physical system that the equation purports to model. Often the defect can be remedied at the expense of clarity and ease of understanding. This can be an expensive trade-off: simplicity and clarity with possible fallacious behavior, versus obscurity and accuracy of behavior. Here are several applications of nonlinear differential equations to the biological, the physical, and the social sciences. Most of these equations have variables separable, but others are examples of special classes of differential equations that can be studied productively.

Applications in the Biological Sciences

The Logistic Equation

The nonlinear differential equation $dy/dt = y(b - ay)$ is called the logistic equation. This name is from the Greek word $\lambda o \gamma \iota \sigma \tau \iota \kappa o \varsigma$ (logistikos), which means "skilled in calculating." Presumably this is because this equation works so well in a wide range of applications. Here the topic is populations where the growth is restricted, rather than unrestricted as it was in Section 3.2. But differential equations similar to the logistic equation, such as the equation

$$\frac{dy}{dt} = (d - cy)(b - ay), \tag{3.5}$$

which is clearly a generalization of the logistic equation, govern such other processes as bimolecular chemical reactions and the spread of flu epidemics. Both of these equations are a special type of separable differential equation where $dy/dt = F(y)$. Notice that the independent variable t is not present on the right hand side. Such equations are called **autonomous**, and their solutions behave in special ways. For instance, if $y(t)$ is a solution, then so is $y(t + k)$ for any k. (Any horizontal translation of a solution is also a solution. This is left as an exercise.)

Autonomous differential equations are solved in the standard way for separable equations to obtain

$$G(y) = \int \frac{1}{F(y)}\, dy = t + c,$$

which we want to solve for y. Sometimes G has an identifiable inverse so that we can explicitly find $y(t) = G^{-1}(t + c)$.

In the last section we saw that the differential equation for unrestricted growth was $dy/dt = ry$. However, it is not reasonable to think that any support system can sustain unrestricted growth for any extended period of time. Early in a process, the growth may appear to be unrestricted, but since the earth and every habitat in it is finite in extent, restrictions have to appear. A closer look at the equation for unrestricted growth indicates that we have successfully accounted for deaths in our population in this way: Suppose that the death rate is d per unit population and the birth rate is b per unit population. Then the rate of population change is $r = b - d$. This says that the effective growth rate is the difference between the birth rate and the death rate. If the birth rate and death rate are equal, $dp/dt = 0$, and the population is stable. If the birth rate is numerically greater than the death rate, the population is increasing. If the death rate exceeds the birth rate, the population is in decline. But all of this assumes that any possible growth is unrestricted if $b > d$. How does one account for the fact that natural resources, be they an agar medium in a Petri dish, or a tropical rain forest, or the Pacific Ocean, are finite?

The standard technique for accounting for restricted growth is to assume that the death rate is not a constant, but depends on the size of the population. That is we take as our equation of growth, $dy/dt = by - (ay)y = by - ay^2 = y(b - ay)$. Here b is the birthrate, but the death rate is (ay), which depends on y. Strange as it may seem at first glance, this equation has served very well at predicting the size of populations living in restricted environments. Note that if y is small, the equation is essentially $dy/dt = by$, which is the equation of unrestricted growth. But as y increases, no matter how small a is, so long as it is nonzero, the term (ay) begins to have an effect, and the rate of population growth begins to slow. We assume an initial population of y_0.

The two constant solutions, $y = 0$ and $y = b/a$ have significance. If $y = 0$ the population has become **extinct**. No further growth of this population is possible. The other constant value $y = b/a$ is called the **carrying capacity** of the environment. The carrying capacity of the environment is the maximum population that the environment can sustain. If the population reaches this level then it cannot increase further because the death rate becomes larger than the birth rate. We will solve the logistic equation and observe how that equation models the phenomenon of restricted growth. It is of critical importance for a population (such as humanity) to know the value of a. Part of the problem of estimating a is that there are competitors for all natural resources, and the equation as formulated does not take this explicitly into account. All of the effects of competition are incorporated into the one coefficient a, so a is more than merely an observable "death rate."

Since the portion of the curve in which we have special interest (because it describes most populations living in restricted environments) lies between $y = 0$ and $y = b/a$, this interval will be implicitly used in our solution. For $0 < y < b/a$, separate variables to get

$$\int \frac{dy}{y(b - ay)} = \int dt + c_0.$$

Then integrate to get

$$\frac{1}{b} \ln y - \frac{1}{b} \ln(b - ay) = t + c_0.$$

When $t = 0,\ y(0) = y_0$, so

$$c_0 = \frac{1}{b}\ln\left(\frac{y_0}{b - ay_0}\right).$$

Multiply through by b, combine logarithms, and exponentiate to get

$$\frac{y}{b - ay} = \frac{y_0}{b - ay_0}e^{bt}.$$

Solving for y gives

$$y(t) = \frac{by_0}{ay_0 + (b - ay_0)e^{-bt}}.$$

Notice that $y(t) \to b/a$ as $t \to \infty$, and that $y(t) \to 0$ as $t \to -\infty$. See Figure 3.16.

A quick calculation reveals that $y'' = (b-2ay)(y)(b-ay)$, so that y is concave up between 0 and $b/(2a)$, and y is concave down between $b/(2a)$ and b/a. When $y = b/(2a)$, which is half-way between extinction and the carrying capacity, $y'' = 0$, so the population curve y has its maximum slope there. If earth's population is following a logistic curve, then we are not yet at one-half of the carrying capacity of the earth, because the rate of increase of population is still increasing. If we were past half-way, then the rate of increase would be decreasing. Of course the flaw in this reasoning is that we are probably not following a logistic curve, and we may be in greater trouble than this simple analysis would dictate. There certainly are regions of the earth that have essentially reached the carrying capacity of those regions to support their own populations.

Observe the disturbing fact that for most of its early history a population is small. It then experiences a short period of *very rapid growth* that increases the size of the population to near its possible maximum, and then the size of the population becomes nearly constant at this very large size (the carrying capacity). The population of the earth has dramatically increased during this century, so it would appear that mankind is in the short period of rapid growth.

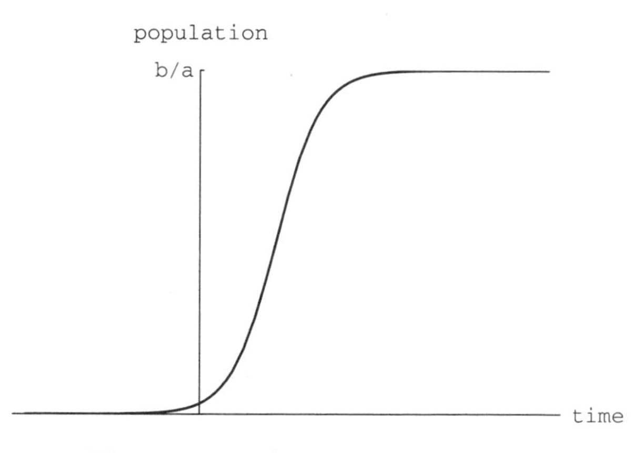

Figure 3.16: A typical logistic curve.

When will this rapid growth stop? Since the carrying capacity of the earth is unknown, no one knows. Several world models (Jay Forrester and others) have suggested that the global standard of living is likely to erode markedly as the population of the earth approaches its maximum.

Applications in the Physical Sciences

Air Resistance

The form of air resistance can be either linear (when the object is moving relatively slowly) or quadratic (when high speed is involved).

Example 3.14 (Quadratic air resistance). A 1600-pound box is dropped from a stationary helicopter. The magnitude of the force of air resistance on it is $(1/25)v^2$ pounds when its velocity is v feet per second. Find its velocity as a function of time. What is its terminal velocity? How high was the helicopter in order for the box to reach 90% of its terminal velocity before striking the ground?

Solution. We suppose that the box moves vertically on an axis with positive direction pointing down and units measured in feet. Measure time t in seconds starting with $t = 0$ when the box begins its fall, and let $g = 32\text{ft}/\text{sec}^2$ denote the acceleration due to gravity. The downward velocity is v feet per second, and $v(0) = 0$. Since the box is falling, air resistance is directed upward, and is given by $-(1/25)v^2$ pounds when its velocity is v. The downward force on the box is due to gravity: the weight of the box, 1600 pounds. This makes the mass of the box $m = w/g = 1600/32 = 50$ slugs. The crate's downward acceleration is $a = dv/dt$, so, from Newton's law $F = ma$, we find that the acceleration on the box is $50dv/dt = 1600 - (1/25)v^2$. Multiply through by 25 to get

$$1250\frac{dv}{dt} = 40{,}000 - v^2 = 200^2 - v^2.$$

The constant solutions are $v = \pm 200$ feet per second. Neither satisfies the initial conditions, so the solution we seek is not constant. In fact, since $v(0) = 0$, it starts between the two constant solutions. This makes $dv/dt > 0$, so $v(t)$ is increasing, and is bounded above by 200. Separating variables leads to

$$\int \frac{dv}{200^2 - v^2} = \int \frac{1}{1250}dt + C,$$

which, when integrated by partial fractions or *Mathematica* yields

$$\frac{1}{400}\ln\left|\frac{200 + v}{200 - v}\right| = \frac{1}{1250}t + C.$$

The initial condition $v(0) = 0$ (implied by the word 'stationary') implies that $1/400\ln(1) = 0 + C$, and hence that $C = 0$. Since $(200 + v)/(200 - v) > 0$ when $v = 0$, take

$$\left|\frac{200 + v}{200 - v}\right| = \frac{200 + v}{200 - v}$$

and solve

$$\frac{1}{400}\ln\left(\frac{200+v}{200-v}\right)=\frac{1}{1250}t$$

for v, obtaining

$$\begin{aligned} v &= 200\left(\frac{e^{\frac{1}{1250}t}-1}{(e^{\frac{1}{1250}t}+1}\right) \\ &= 200\left(\frac{e^{0.32t}-1}{e^{0.32t}+1}\right) \\ &= 200\left(\frac{1-e^{-0.32t}}{1+e^{-0.32t}}\right). \end{aligned}$$

From this latter expression we find the terminal velocity to be 200 feet/second, since $\lim_{t\to\infty} e^{-0.32t}=0$. The time until the box reaches 90% of terminal velocity is

$$t=\frac{1250}{400}\ln\left(\frac{200+180}{200-180}\right)=\frac{1250}{400}\ln\left(\frac{380}{20}\right)=(3.125)(2.9444)=9.20\,\text{sec}.$$

The height of the helicopter would then be

$$\text{height}=\int_0^{9.2}200\left(\frac{e^{0.32t}-1}{e^{0.32t}+1}\right)dt=1037.71\text{ feet}$$

in order for the box to reach 180 feet per second = 90% of the 200 feet per second terminal velocity.

The *Mathematica* cell that was used to find the height was:

```
In[7]:= NIntegrate[200(Exp[0.32t]-1)/(Exp[0.32t]+1),{t,0,9.2}]

Out[7]= 1037.71
```

◇

Bimolecular Reactions

Suppose that m molecules of chemical A and n molecules of chemical B react in a water solution to form one molecule of chemical C. Let $x(t)$ be the concentration (in grams per liter) of A and $y(t)$ be the concentration (same units) of B in the solution. Let $z(t)$ be the concentration of chemical C at time t. We suppose that initially $x(0)=a, y(0)=b$, and $z(0)=0$. What is the concentration $z(t)$ of C as a function of t?

We begin by deriving a differential equation for the rate of change of z. Since each molecule of C requires m molecules of A, $x(t)$ decreases m times as fast as $z(t)$ increases:

$$\frac{dx(t)}{dt}=-m\frac{dz(t)}{dt},$$

so

$$\frac{dx}{dt} + m\frac{dz}{dt} = 0.$$

This means that

$$x(t) + mz(t) = \text{constant} = a, \quad \text{or} \quad x(t) = a - mz(t).$$

By similar reasoning,

$$y(t) + nz(t) = \text{constant} = b, \quad \text{or} \quad y(t) = b - nz(t).$$

Under some circumstances (no catalysts present, for instance), because each unit of C requires m units of A and n units of B, it is reasonable to assume that

$$\frac{dz}{dt} = kx^m y^n.$$

This equation expresses the need for this relationship among the molecules within the solution. In this case, we have the equation

$$\frac{dz}{dt} = k(a - mz)^m (b - nz)^n \tag{3.6}$$

as the differential equation of the reaction. The exponents m and n are called the **orders** of the reaction with respect to the concentrations of A and B, and dz/dt is called the **rate of the reaction**.

Example 3.15 Suppose that one molecule of A reacts with one molecule of B to produce one molecule of C. Then with x, a, y, b, and z, defined as above, we obtain as the differential equation of the reaction,

$$\frac{dz}{dt} = k(a - z)(b - z).$$

This equation is separable, and

$$\int \frac{dz}{(a - z)(b - z)} = \int k\, dt + c_1.$$

We want the solution where $a - z \geq 0$ and $b - z \geq 0$.

Solution. If $a \neq b$, we integrate and apply the initial condition $z(0) = 0$, to obtain

$$\frac{1}{a - b}\ln(a - z) + \frac{1}{b - a}\ln(b - z) = kt + \frac{1}{a - b}\ln\frac{a}{b}.$$

Multiply through by $a - b$, collect the logarithms and exponentiate to find

$$\frac{a - z}{b - z} = \frac{a}{b}e^{k(a-b)t}.$$

Then solve this for z to get

$$z(t) = ab\left(\frac{e^{k(a-b)t} - 1}{ae^{k(a-b)t} - b}\right).$$

Having z, we can get $x(t) = a - z(t)$ and $y(t) = b - z(t)$.

Notice that if $a > b$, then $z \to b, x \to a - b$, and $y \to 0$, and if $a < b$, then $z \to a, x \to 0$, and $y \to b - a$. (What do these say about the physical setting?)

If $a = b$, then the differential equation is

$$\frac{dz}{dt} = k(a - z)^2,$$

which separates into

$$\int \frac{dz}{(a - z)^2} = \int k\,dt + c_1.$$

Upon integrating and substituting the initial condition $z(0) = 0$, we find that

$$\frac{1}{a - z} = kt + \frac{1}{a}.$$

Solving for z yields

$$z(t) = a\left(1 - \frac{1}{akt + 1}\right),$$

from which we can get $x(t)$ and $y(t)$ as before. Note also that since $x(t) = a - z(t)$, as $t \to \infty, z \to a, x \to 0$, and $y \to 0$. ◇

Applications in the Social Sciences

The Propagation of a Single Action in a Population

In some circumstances people perform a specific action because it is warranted and others because they have observed the action being taken. We assume there is an external stimulus that may precipitate an action in some individuals, and other individuals who then imitate that action. For example, some people may mow their lawn just because it needs mowing and others who do so because they see lawns being mowed. In time it becomes obvious to those who have not mown that the time has come to act or be identified as a laggard. Eventually all of the lawns get mowed. Another example might be yawning in a class where a boring lecture is the continuing external stimulus. Some yawn because they are bored. Some yawn not because they are bored, but because they see others yawning. Eventually everyone has yawned. An even more compelling example is the act of turning on headlights at sundown. It is clear that there is an ever more pressing reason to act as night approaches, but the mere presence of vehicles with lights on may cause some to respond where the darkness itself did not.

If the population is sufficiently large it has been suggested by Rapoport (1952) that the proportion $y(t)$ of the population who have acted can be described by the first-order nonlinear differential equation

$$\frac{dy}{dt} = (1-y)(x(t)+by), \tag{3.7}$$

where $x(t)$ is the nonnegative **external stimulus** and b is the positive **coefficient of imitation**. Note that there is a constant solution $y = 1$, which says that once the entire population has taken the action, the rate of change becomes 0. If you multiply out the right-hand side of equation 3.7 you get

$$\frac{dy}{dt} = x(t) + (b - x(t))y - by^2. \tag{3.8}$$

This is a **Riccati** differential equation and we know one solution, $y = 1$. The theory of Riccati differential equations from Section 2.6 suggests that to get a one parameter family of solutions let $y = 1 + 1/v$. This produces these calculations: $dy/dt = (-1/v^2)dv/dt$ and

$$\begin{aligned}\frac{-1}{v^2}\frac{dv}{dt} &= x(t) + (b - x(t))\left(1+\frac{1}{v}\right) - b\left(1+\frac{1}{v}\right)^2 \\ &= x(t) + b + \frac{b}{v} - x(t) - \frac{x(t)}{v} - b - \frac{2b}{v} - \frac{b}{v^2} \\ &= -\frac{x(t)+b}{v} - \frac{b}{v^2}.\end{aligned}$$

Now it must be true that $v \neq 0$, so we can multiply through by $-v^2$ to get

$$\frac{dv}{dt} = (b + x(t))v + b$$

which is a first-order linear differential equation and is easily solved (formally). One then gets a one parameter family of solutions y from $y = 1 + 1/v$. Even for simple functions $x(t)$ these solutions are complicated. For this reason, you might consider using `NDSolve` to obtain numerical solutions, rather than using `DSolve` to get formal solutions.

Even in the special case when $x(t) = at$, for positive constant a, the integrations are complicated. In situations such as this, we could have *Mathematica* do:

```
In[8]:=  NDSolve[{y'[t] == (1-y[t])(0.50t+y[t]/10), y[0] == .01},
                                          y[t],{t,0,5}]

Out[8]=  {{y[t] -> InterpolatingFunction[{0., 5.}, <>][t]}}

In[9]:=  p1 = Plot[Evaluate[y[t]/.%[[1]]],{t,0,5}];
```

Three plots such as this for $b = 0.1, x(t) = at$, and $a = 0.25$, 0.33, and 0.50 can be combined to compare the effects of varying the constant multiplier a to produce a picture such as Figure 3.17.

We saw curves of this general shape when we considered the logistic equation. It is worth noting that when $x(t)$ is small, equation 3.7 is very near

$$\frac{dy}{dt} = by(1-y)$$

which is a form of the logistic equation. Had $x(t)$ been constant, the equation would have been

$$\frac{dy}{dt} = (1 - y)(a + by),$$

which is similar to the equations of bimolecular reactions, so we might expect the shape of the solution curve to be similar to that of a solution of the bimolecular equation. The presence of $x(t)$ on the right-hand side causes the shape of the solution curves to change as $x(t)$ is changed. This was reflected in the three curves depicted in Figure 3.17.

EXERCISES 3.3

Mathematica may have trouble with some of these nonlinear problems. It is sometimes to your advantage to do some manual conversions on a problem before letting *Mathematica* do work for you. As a general rule, since most of these problems are solvable by separating variables, evaluate constants of integration before attempting to further solve for the function you desire. This can usually be accomplished by using definite integrals.

1. Show that if $y(t)$ is a solution of the autonomous differential equation $dy/dt = F(y)$, then $z(t) = y(t + a)$ is a solution for any fixed number a.

PART I. Problems from the Biological Sciences

2. Suppose that a population whose growth is following a logistic equation $dy/dt = y(b - ay)$ is presently at $1/4$ of the carrying capacity of the environment in which it lives. How long until the greatest rate of growth occurs?

3. Answer problem 2 in case $b = 0.1$ and $a = 0.0001$.

4. Suppose that a population whose growth is otherwise following a logistic equation $dy/dt = y(b - ay)$ is being harvested continuously. Harvesting

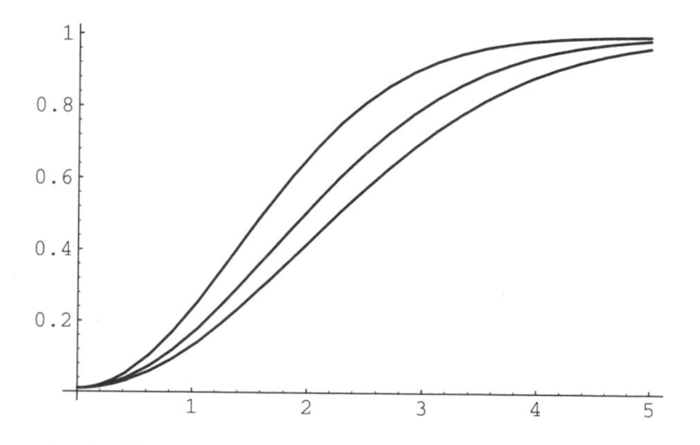

Figure 3.17: Propagation of a single action; three choices for a.

artificially increases the death rate. One model of this situation that was proposed by the biologist M. B. Schafer is $dy/dt = y(b - ay) - \varepsilon y$, where ε is constant.

(a) What is the effect of this harvest on the maximum size of the population? [There is a new artificial carrying capacity.] Consider several cases.
(b) What relationship between the population when harvesting begins and ε must exist in order for the population to actually decrease? The value of ε depends on the population at the time harvesting begins.

5. In problem 4 if the differential equation of the population is $dy/dt = y(0.1 - 0.0001y) - \varepsilon y$, what rate of harvest ε (depending on the population when harvesting begins) will result immediately in a constant population? What is the new behavior of the population if the rate of harvest is then reduced by half?

6. If the differential equation of the population is $dy/dt = y(0.1 - 0.0001y)$, and is essentially constant before harvesting begins, what rate of harvest ε will result in an equilibrium population that is half of the original size? What happens to the population if that rate of harvest is reduced by half?

7. Modify the model of exercise 4 so that $dy/dt = y(b - ay) - c$ where c is the (constant) rate of harvest. Harvesting is done at a fixed rate which is independent of the size of the population. In exercise 4, the rate of harvest decreased as the population decreased. That is not true here.

(a) What is the effect of c on the possible equilibrium positions the population can have? Consider several cases.
(b) Among these possibilities, how can c be chosen so that the population continues to increase. For an ocean resource is it reasonable to assume that the population is near the carrying capacity prior to beginning harvesting?
(c) What values of c will cause the the population to actually decrease toward extinction as the only possible equilibrium position?

8. Gompertz curves result from this modification of the logistic equation: $dy/dt = y(b - a \ln y)$. Find the constant solutions. Solve the equation by separating variables. Discuss the behavior of the solution curves as a function of the signs of a and b. [Gompertz curves have application in population biology; in actuarial studies; and in sales predictions for commercial products.]

PART II. Problems from the Physical Sciences

9. An object falls downward with velocity changing according to the rule

$$\frac{dv}{dt} = 400 - \frac{v^2}{16},$$

with downward being positive and units being feet per second.

(a) What is the only physically realizable constant solution v_∞?
(b) Show that if the object has initial velocity 0, then $\lim_{t\to\infty} v(t) = v_\infty$.
(c) Find the time of fall to the ground if the object was initially at rest at 2000 feet.
(d) What is the effect of an initial velocity of $v_0 = v_\infty + 100$ on the time of fall from 2000 feet?

10. Two chemicals A and B combine to produce a product X according to the differential equation $dx/dt = k(250 - x)(40 - x)$. Assume that $x(0) = 0$. Find k if $x(10) = 30$. Find an expression for $x(t)$ for $t \geq 0$, and find $\lim_{t\to\infty} x(t)$.

11. When three chemicals react to form a single product, one has a third-order chemical reaction. The mass of the product X may obey a differential equation of the form

$$\frac{dX}{dt} = k(\alpha - X)(\beta - X)(\gamma - X),$$

where α, β, and γ are positive.

(a) Solve the equation assuming that α, β, and γ are all different.
(b) Solve the equation assuming that $\alpha = \beta$, and γ is different.
(c) Solve the equation assuming that α, β, and γ are all the same.

12. When three chemicals react to form a single product, the mass of the product X more generally obeys a differential equation of the form

$$\frac{dX}{dt} = k(\alpha - X)^{m_1}(\beta - X)^{m_2}(\gamma - X)^{m_3},$$

where α, β, and γ are positive and m_1, m_2, and m_3 are positive integers. Explore the nature of the solutions for various choices of m_1, m_2, and m_3.

13. Newton's law of cooling is not the only law of cooling for radiative cooling. According to **Stefan's law** of radiation, the rate of change of temperature of a body at temperature T K is

$$\frac{dT}{dt} = k(T^4 - T_0^4),$$

where T_0 is the absolute temperature of the surroundings.

(a) Solve Stefan's equation

$$\frac{dT}{dt} = k(T^4 - T_0^4),$$

(b) Recall that $T^4 - T_0^4 = (T - T_0)(T + T_0)(T^2 + T_0^2)$. When $T - T_0$ is small and T_0 is large, $(T + T_0)(T^2 + T_0^2) \approx k_1$ a large constant and $T^4 - T_0^4 \approx k_1(T - T_0)$. Therefore $dT/dt = k(T^4 - T_0^4) \approx kk_1(T - T_0) = \kappa(T - T_0)$, which is Newton's law of cooling.

(c) Look at the series representation of $T^4 - T_0^4$ centered on T_0. What is k_1 when $T_0 = 300$ K?

(d) Investigate whether or not the solution of Stefan's equation is near the solution of Newton's equation when $T - T_0$ is small and T_0 is large. Use the same initial condition for both equations. Look at the series representation of both solutions centered on T_0.

PART III. Problems from the Social Sciences

14. Complete the solution of the differential equation 3.8.

15. A new electronic consumer product is introduced into the marketplace with initial sales of 10,000 and projected total sales $s(t)$ at time t governed by the differential equation

$$\frac{ds}{dt} = 10s(10{,}000{,}000 - s).$$

What is the expected saturation level (carrying capacity)? How long will it take for the market to be 95% saturated? Plot your solution curve.

16. At sundown, seeing someone with lights on suggests that you should turn on your lights. So does the onset of darkness. Suppose that initially no one has turned on their lights, that the external stimulus due to the onset of darkness is $x(t) = 2t$, and that the coefficient of imitation is 0.25. State and solve the appropriate differential equation of the form of equation 3.8.

17. The price of a commodity is proportional to the excess of demand over supply as presented in Section 3.2. Suppose that demand is inversely proportional to price and supply is directly proportional to price. Give a differential equation that governs price in this situation. Solve the equation. Plot your solution in the special case where the constants of proportionality are: for price, 1; for demand, 200; and for supply, 0.1. Use a reasonable initial condition.

4 Higher-Order Linear Differential Equations

4.0 Introduction

Linear differential equations of higher order have useful and interesting applications, just as first-order differential equations do. We study linear differential equations of higher order in this chapter. The word *linear* in the chapter title should suggest that techniques for solving linear equations will be important. What is somewhat unexpected is that we have to appeal to the theory of solving polynomial equations in one variable. Though the solution technique for first-order equations gave us a complete solution in essentially one step, this is not the case here. For the first time we have to solve the homogeneous and nonhomogeneous equations separately and by different methods.

When solving the homogeneous equation, we need to find the roots of polynomials. The notion of **linearly independent solutions** becomes centrally important for the first time. The **Wronskian determinant** is introduced to test for linear independence. It is the determinant of the **Wronskian matrix** upon which the technique of **variation of parameters** is based. Variation of parameters and the method of **undetermined coefficients** are used to find a **particular solution** to nonhomogeneous problems, once the homogeneous problem has been completely solved.

This chapter leans heavily on linear algebra for its theory. To help you to understand the differential equations better, more linear algebra is introduced. You are shown how to use *Mathematica* to perform many of the steps that are simple in theory but very hard in practice. The idea behind the use of *Mathematica* is to free you from computational burdens so that you can concentrate on the meaning of our activities, rather than on the many details that arise as we seek solutions.

In this chapter we once again see all of the elementary functions that you have studied: polynomials, the natural exponential, the sine and cosine functions, and the hyperbolic functions. The theory of linear differential equations with constant coefficients is built on these functions. The natural logarithm plays an important role in Chapter 7 where differential equations with variable coefficients are studied. Matrices become important to us in this chapter, and remain so throughout the rest of the text.

The *Mathematica* function `NDSolve` can be used to obtain accurate numerical solutions to the differential equations studied in this chapter, but this accuracy extends only to the values of the solution function, not the derivatives of the solution. This is because `NDSolve` approximates the solution by a **cubic spline**, a sequence of cubic polynomials that are defined over short intervals and adjacent polynomials agree in function value and slope at the endpoint that is common to their intervals of definition. This serves very well for accurately determining points along the solution curve, but these cubics do not accurately convey slope information.

If you need slope information, then you need techniques from Chapter 8 for expressing a higher order differential equation as a system of equations that explicitly define the derivatives you need. In that context, `NDSolve` does give accurate values for the derivatives that the differential system explicitly defines. Be alert to what you are asking a numerical method to do, to be sure that it is capable of providing what you seek.

4.1 The Fundamental Theorem

In order to see why we can expect to find solutions for differential equations, we state the fundamental theorem, which says that there are solutions and gives conditions under which there is only one solution. Of course, knowing that there are solutions is not at all the same thing as being able to actually find a solution. But there is a large class of important differential equations, the linear differential equations with constant coefficients, where we can actually write down a formula for the solutions. But even in cases such as these, when it is clear what has to be done to obtain a solution, the theory is much easier than the practice. We will find that *Mathematica* eases the computational burden immensely, but that there are places where the theory says something exists and another piece of theory says that we cannot necessarily write out an actual solution. In these cases we can often get approximate solutions that are very close to the theoretically exact solutions we seek. But when we are calculating with approximate objects an immediate question is: just how good is the approximation? Questions such as these are covered in courses in numerical analysis.

The Fundamental Theorem

Theorem 4.1 (Existence and Uniqueness) *Given an open subset U of $(n+1)$-space $\boldsymbol{R}^{n+1}$ and a point $P = (x_0, a_0, a_1, a_2, \ldots, a_{n-1})$ of U. Suppose that the real-valued function f is defined and continuous on U and there is a* *positive number M so that if $(x, u_1, u_2, \ldots, u_n)$ and $(x, v_1, v_2, \ldots, v_n)$ are in U then*

$$|f(x, u_1, u_2, \ldots, u_n) - f(x, v_1, v_2, \ldots, v_n)| \\ \leq M(|u_1 - v_1| + |u_2 - v_2| + \cdots + |u_n - v_n|).$$

Then there is exactly one solution of the nth-order initial value problem

$$y^{(n)}(x) = f(x, y(x), y(x), ..., y^{(n-1)}(x)) \tag{4.1a}$$

$$y(x_0) = a_0, y'(x_0) = a_1, \ldots, y^{(n-1)}(x_0) = a_{n-1} \tag{4.1b}$$

that is defined for x in a neighborhood of x_0.

Equations 4.1a and 4.1b constitute an initial value problem because the information needed to select one particular solution is given at a single point. Note that the initial conditions 4.1b merely define the value of $f(P)$. This is enough to determine the solution (near x_0). We say that Theorem 4.1 is an existence and uniqueness theorem because it says that there is a solution (**existence**) and that this is the only solution (**uniqueness**).

The constant M whose existence must be demonstrated for the theorem to hold, is an idea due to Lipschitz.[1] For the first-order case, if $\partial f/\partial y$ exists, then the constant M can be taken to be any upper bound on $\partial f/\partial y$, which can be seen from the inequality

$$\left|\frac{f(x,u) - f(x,v)}{u - v}\right| = \left|\frac{\partial f}{\partial y}(x, c)\right| \leq M.$$

In the nth-order case, M can be taken to be any bound

$$\text{Max}\left\{\left|\frac{\partial f}{\partial u_1}\right|, \ldots, \left|\frac{\partial f}{\partial u_n}\right|\right\} \leq M.$$

As the examples below indicate, a bound can be found through algebraic manipulation.

Example 4.1 The differential equation $y' = 1 + y^2$ has a unique solution that passes through the point $(0, 0)$. Here $f(x, u) = 1 + u^2$ and $f(x, v) = 1 + v^2$. Thus $|f(x, u) - f(x, v)| = |u^2 - v^2| = |u + v||u - v| \leq 2|u - v|$ in the rectangle $U : |x| < 1$, $|y| < 1$. So we can take $M = 2$, and Theorem 4.1 says that there is exactly one solution that passes through $(0, 0)$. You may check that this solution is given by $y(x) = \tan x$. ◇

The theorem guarantees a solution only in a neighborhood of the initial point. The next example demonstrates that this is the case when f is a polynomial, which is nearly as simple as a nonlinear function can get.

---

[1] Rudolf Lipschitz (1832-1903), German mathematician. Professor at Bonn for many years.

Example 4.2 Show that the differential equation $y' = -2xy^2$ has exactly one solution that passes through the point $(2, 1/2)$.

Solution. Choose U to be the rectangle $|x-2| < 1$, $|y-1/2| < 1/2$. Then for (x, u) and (x, v) in U, we have that $|f(x,u) - f(x,v)| = |-2xu^2 + 2xv^2| \leq 2|x||u+v||u-v| \leq 12|u-v|$, because if $|x-2| < 1$, then $1 < x < 3$, and if $|y-1/2| < 1/2$ then $0 < y < 1$. Hence $2|x||u+v| \leq 2(3)(1+1) = 12$. Theorem 4.1, with $M = 12$, says that there is a unique solution that passes through the point $(2, 1/2)$.

We find the solution. Separating variables gives

$$-\frac{y}{y^2} = 2x.$$

Integrating both sides gives

$$\frac{1}{y} = x^2 + c.$$

At the point $(2, 1/2)$, we find c from $2 = 2^2 + c$, or $c = -2$. Thus $y(x) = 1/(x^2 - 2)$. Note that $y(x)$ is defined on an interval containing $x = 2$ only for $x \geq \sqrt{2}$, and hence only for the subinterval $\sqrt{2} < x < 3$ of the original interval $|x-2| < 1$ that is in U. This illustrates why Theorem 4.1 only guarantees a solution on a subinterval of the original interval. ◇

Even very simple examples of equation 4.1a can be exceedingly difficult to solve. But if we specify additional properties that f should have, then we can begin to say something about the nature of any solutions that the equation may have.

Definition 4.1 *Let C be a set of real valued functions that are defined on an open interval $a < x < b$ of the reals and which have n continuous derivatives for $a < x < b$. A* **differential operator** *is a function F whose domain is C, such that for some continuous function g, if u is in C, then $F(u)(x) = g(x, u(x), u'(x), \ldots, u^{(n)}(x))$. In case*

$$F(u)(x) = a_n(x)u^{(n)}(x) + a_{n-1}(x)u^{(n-1)}(x) + \cdots + a_0(x)u(x),$$

where $a_n(x) \neq 0$, and the functions $a_i(x)$ are continuous for $a < x < b$, then F is called an **nth-order linear differential operator**. *Further, if each of the coefficient functions $a_i(x)$ is constant, then F is an* **nth-order linear differential operator with constant coefficients**.

Example 4.3 The linear differential operator $A(y) = y'' + xy$, so denoted because the homogeneous differential equation $y'' + xy = 0$ is known as Airy's differential equation, does not have constant coefficients, whereas $L(y) = y'' + 4y$ is a second-order linear differential operator with constant coefficients. ◇

A linear differential operator, as a function, is linear. That is, if u and v are in the domain of F, $F(u+v) = F(u) + F(v)$ and $F(cu) = cF(u)$ for any number c.

In the linear case the fundamental existence and uniqueness theorem goes this way.

Theorem 4.2 *Let L_n be an nth-order linear differential operator and q be a function defined and continuous for $a < x < b$, over which $a_n(x) \neq 0$. Then the initial value problem consisting of the nth-order linear differential equation*

$$L_n(y)(x) = a_n(x)y^{(n)}(x) + a_{n-1}(x)y^{(n-1)}(x) + \cdots + a_0(x)y(x) = q(x), \tag{4.2a}$$

and for any $a < x_0 < b$ the initial conditions

$$y(x_0) = a_0, y'(x_0) = a_1, \ldots, y^{(n-1)}(x_0) = a_{n-1} \tag{4.2b}$$

has a unique solution in a neighborhood of x_0.

Proof. Theorem 4.2 follows from Theorem 4.1 by taking

$$F(x, y(x), y'(x), \ldots, y^{(n-1)}(x)) = (1/a_n(x))[q(x) - a_{n-1}(x)y^{(n-1)}(x) - \cdots - a_0(x)y(x)].$$

□

Example 4.4 The linear differential equation $(x-1)y'' - xy' + y = 0$ has the solution $y = x$, which satisfies the initial conditions $y(0) = 0$, $y'(0) = 1$, and the solution $y = e^x$, which satisfies the initial conditions $y(0) = y'(0) = 1$. Both solutions successfully cross $x = 1$, where the coefficient of y'' is 0. ◇

Theorem 4.3 *Suppose that $y_p(x)$ is a solution of the nth-order linear differential equation $L_n(y)(x) = q(x)$ on some interval I, with $a_n(x) \neq 0$ on I. If $z(x)$ is also a solution of $L_n(y)(x) = q(x)$ on the same interval I then there is a function $u(x)$ that is a solution of $L_n(y)(x) = 0$ such that $z(x) = u(x) + y_p(x)$ for x in I.*

Proof. The proof of this theorem follows from this calculation: Both y_p and z are known solutions. Let $u = z - y_p$. Then $L_n(u)(x) = L_n(z - y_p)(x) = L_n(z)(x) - L_n(y_p)(x) = q(x) - q(x) = 0$. Solving for z gives $z = u + y_p$, which is what was to be proved. □

Theorem 4.3 is just an extension of the similar result we saw in chapter 1. It tells us the form that solutions of nonhomogeneous linear differential equations will take. The idea presented in Theorem 4.3 is sometimes called the **principle of superposition**. The next theorem tells us more about the form of solutions of homogeneous linear differential equations.

Theorem 4.4 *Given the nth-order homogeneous linear differential equation $L_n(y)(x) = 0$ defined on an interval $a < x < b$ over which $a_n(x) \neq 0$. There exist n functions $y_1, y_2, \ldots, y_n$ that are solutions of $L_n(y)(x) = 0$ such that if z is any solution of $L_n(y)(x) = 0$ for $a < x < b$, then there are numbers $c_1, c_2, \ldots, c_n$ having the property that*

$$z(x) = c_1y_1(x) + c_2y_2(x) + \cdots + c_ny_n(x)$$

for $a < x < b$.

Proof. This theorem is proved by actually exhibiting a collection of functions that works. Let $a < x_0 < b$. Define $y_i^{(k)}(x)$ so that $y_i^{(k)}(x_0) = \delta_{k+1}^i$, where δ_j^i is 0 when $i \neq j$ and 1 when $i = j$. [The symbol δ_j^i is called the **Kronecker[2] delta**.] With this definition,

$$\begin{array}{llll} y_1(x_0) = 1, & y_1'(x_0) = 0, & \ldots, & y_1^{(n-1)}(x_0) = 0, \\ y_2(x_0) = 0, & y_2'(x_0) = 1, & \ldots, & y_2^{(n-1)}(x_0) = 0, \\ \vdots & \vdots & \vdots & \vdots \\ y_n(x_0) = 0, & y_n'(x_0) = 0, & \ldots, & y_n^{(n-1)}(x_0) = 1. \end{array}$$

A direct calculation shows that if

$$w(x) = z(x_0)y_1(x) + z'(x_0)y_2(x) + \cdots + z^{(n-1)}(x_0)y_n(x),$$

then w is a solution of the differential equation and

$$w(x_0) = z(x_0), w'(x_0) = z'(x_0), \ldots, w^{(n-1)}(x_0) = z^{(n-1)}(x_0).$$

The function w satisfies the same initial conditions that z satisfies. Since there is only one solution that satisfies these initial conditions, $z(x) = w(x)$ for all x, $a < x < b$. □

Example 4.5 The functions $y_1(x) = e^x$ and $y_2(x) = e^{-x}$ are solutions of the differential equation $y'' - y = 0$. They can serve as the functions mentioned in Theorem 4.4. For instance, observe that $y(x) = \cosh x$ is also a solution of $y'' - y = 0 :\ y'(x) = \sinh x$ and $y''(x) = \cosh x$, so that $y''(x) - y(x) = \cosh x - \cosh x = 0$. To find the constants, let

$$\cosh x = c_1 e^x + c_2 e^{-x}.$$

Then when $x = 0$, $\cosh 0 = 1 = c_1 + c_2$. Take a derivative to get

$$\sinh x = c_1 e^x - c_2 e^{-x}.$$

Then take $x = 0$ to get $\sinh 0 = 0 = c_1 - c_2$. Solve the simultaneous equations.

$$\begin{cases} 1 = c_1 + c_2 \\ 0 = c_1 - c_2 \end{cases}$$

to find that the constants are $c_1 = c_2 = 1/2$. Then we recover the definition

$$\cosh x = \frac{e^x + e^{-x}}{2}.$$

Similarly, $y(x) = \sinh x$ is a solution of $y'' - y = 0$, and we recover the definition

$$\sinh x = \frac{e^x - e^{-x}}{2}$$

from $\sinh x = c_1 e^x - c_2 e^{-x}$ and $c_1 = c_2 = 1/2$. ◇

[2] Leopold Kronecker (1823–1891), German mathematician.

The results of Theorems 4.3 and 4.4 combine to give Theorem 4.5, which will direct our thoughts for much of the rest of the chapter.

Theorem 4.5 *Suppose that $y_p(x)$ is a solution of the nonhomogeneous linear differential equation $L_n(y)(x) = f(x)$ on $a < x < b$ over which $a_n(x) \leq 0$. Then there are n functions $y_1, y_2, \ldots, y_n$ that are solutions of $L_n(y)(x) = 0$ on $a < x < b$ with this property:*

If $y(x)$ is a solution of $L_n(y)(x) = f(x)$ then there are n numbers $c_1, c_2, \ldots, c_n$ such that for $a < x < b$,

$$y(x) = c_1 y_1(x) + c_2 y_2(x) + \cdots + c_n y_n(x) + y_p(x). \tag{4.3}$$

The functions $y_1, y_2, \ldots, y_n$ that this theorem requires can be the functions constructed during the proof of Theorem 4.4, though there are other sets of functions that will work.

Example 4.6 The family of functions $y = c_1 \cos x + c_2 \sin x + x^2 - 2$ is a solution of the differential equation $y'' + y = x^2$ for every choice of c_1 and c_2 and every real number x. (You should check this.) We will use *Mathematica* to check solutions of equations like this, because the equations we will consider and the functions that are their solutions can be so complicated that checking may be a daunting task. ◇

Example 4.6M Check the solution given in example 4.6. Then, in the family of solutions of Example 4.6, find numbers c_1 and c_2 so that $y(0) = 3$ and $y'(0) = -2$.

Solution.

```
In[1]:=  y[x_] = c1 Cos[x] + c2 Sin[x] + x^2 - 2

Out[1]=         2
         -2 + x  + c1 Cos[x] + c2 Sin[x]
```

Check:

```
In[2]:=  y''[x]+y[x] == x^2

Out[2]=  True
```

Impose the initial conditions to get two equations.

```
In[3]:=  eqns = {y[0] == 3, y'[0] == -2}

Out[3]=  {-2 + c1 == 3, c2 == -2}
```

Solve these equations.

```
In[4]:=  c1c2 = Solve[eqns,{c1,c2}]

Out[4]=  {{c1 -> 5, c2 -> -2}}
```

Capture the solution.

```
In[5]:=  soln[x_] = y[x]/.c1c2[[1]]

Out[5]=          2
         -2 + x   + 5 Cos[x] - 2 Sin[x]
```

Check the solution in the differential equation and the initial conditions. The infix operator "&&" means "And."

```
In[6]:=  soln''[x]+soln[x] == x^2 && soln[0] ==3 && soln'[0] == -2

Out[6]=  True
```

◇

Definition 4.2 *A collection of functions having the property indicated in Theorem 4.5 is called a* **fundamental set of solutions** *of* $L_n(y)(x) = 0$. *The function*

$$y_c(x) = c_1y_1(x) + c_2y_2(x) + \cdots + c_ny_n(x)$$

is also called the **complementary function** *of* $L_n(y)(x) = 0$.

The set $\{e^x, e^{-x}\}$ is a fundamental set of solutions for the equation $y'' - y = 0$ discussed in Example 4.4. So is the set $\{\cosh x, \sinh x\}$.

We need a way of determining whether or not a given collection of solutions of a homogeneous linear differential equation is fundamental. The necessary tool is the determinant function introduced by Wronski.[3]

Definition 4.3 *The Wronskian* $W(y_1, y_2, \ldots, y_n)$ *of a set of functions having* $n-1$ *continuous derivatives on some interval* I *is the determinant*

$$W(y_1, y_2, ..., y_n) = \begin{vmatrix} y_1 & y_2 & \cdots & y_n \\ y_1' & y_2' & \cdots & y_n' \\ \vdots & \vdots & \ddots & \vdots \\ y_1^{(n-1)} & y_2^{(n-1)} & \cdots & y_n^{(n-1)} \end{vmatrix} \tag{4.4}$$

We will use the notation $W[y_1(x), y_2(x), \ldots, y_n(x)]$ to mean the same thing as $W(y_1, y_2, \ldots, y_n)(x)$.

Example 4.7 The Wronskian of the pair $\{\cos x, \sin x\}$ is

$$W[\cos x, \sin x] = \begin{vmatrix} \cos x & \sin x \\ -\sin x & \cos x \end{vmatrix} = \cos^2 x + \sin^2 x = 1.$$

◇

The Wronskian is important to us for this reason:

---

[3]Hoëné Wronski (1788–1853), Polish mathematician. He spent most of his professional life working in France.

Theorem 4.6 *If the functions $y_1, y_2, \ldots, y_n$ are solutions of $L_n(y)(x) = 0$ on $a < x < b$ over which $a_n(x) \neq 0$, then $W[y_1(x), y_2(x), \ldots, y_n(x)]$ is either zero for every x, $a < x < b$, or is never zero for any x, $a < x < b$.*

Proof. (For two functions.) The proof in the case of two functions is easy. It suggests how the general proof goes, however.

Suppose that $p_1(x)$ and $p_0(x)$ are continuous for $a < x < b$, and let y_1 and y_2 be solutions of $L_2(y) = y'' + p_1(x)y' + p_0(x)y = 0$. Then

$$W(y_1, y_2) = \begin{vmatrix} y_1 & y_2 \\ y_1' & y_2' \end{vmatrix} = y_1 y_2' - y_1' y_2,$$

and

$$\begin{aligned}(W(y_1, y_2))' &= y_1 y_2'' - y_1'' y_2 \\ &= y_1(-p_1(x)y_2' - p_0(x)y_2) - (-p_1(x)y_1' - p_0(x)y_1)y_2 \\ &= -p_1(x)(y_1 y_2' - y_1' y_2) \\ &= -p_1(x)W(y_1, y_2).\end{aligned}$$

This first-order differential equation has as solution

$$W(y_1, y_2)(x) = c e^{\int -p_1(x)\, dx},$$

which is either identically zero, if $c = 0$, or never zero. This equation is called **Abel's identity** after Niels Henrik Abel.[4] □

We leave the complete proof of Theorem 4.6 to the exercises.

Example 4.7 illustrates that $W[\cos x, \sin x] \neq 0$ for any x.

Example 4.8 Here is a Wronskian that is identically zero.

$$\begin{aligned}W\left[\cos\left(x - \frac{\pi}{2}\right), \sin x\right] &= \begin{vmatrix} \cos\left(x - \frac{\pi}{2}\right) & \sin x \\ -\sin\left(x - \frac{\pi}{2}\right) & \cos x \end{vmatrix} \\ &= \cos\left(x - \frac{\pi}{2}\right)\cos x + \sin\left(x - \frac{\pi}{2}\right)\sin x \\ &= \cos\left(x - \frac{\pi}{2} - x\right) \\ &= \cos\left(-\frac{\pi}{2}\right) \\ &= 0.\end{aligned}$$

◇

In the next section we will define terms to use to describe the situations that occurred in the last two examples.

---

[4]Niels Henrik Abel (1802–1829). The premier Norwegian mathematician. A child genius, as a youth, he showed that the general quintic (fifth-degree polynomial) was unsolvable by radicals. His work on higher-order equations foreshadowed the work of Galois.

EXERCISES 4.1

PART I. Pairs of equations appear below. The first is a differential equation; in the second a function or set of functions that is a solution of the differential equation. Verify that the given functions satisfy the corresponding equations. Do this manually and by *Mathematica*. Consider c_i to be an arbitrary constant.

1. $y'' - 4y' + 3y = 0;$
   $y(x) = c_1 e^{3x} + c_2 e^{x}$

2. $y'' - 3y' = 0;$
   $y(x) = c_1 e^{3x} + c_2$

3. $y'' + 4y' + 3 = 0;$
   $y(x) = c_1 e^{-3x} + c_2 e^{-x}$

4. $y'' - 2y = 0;$
   $y(x) = c_1 e^{-\sqrt{2}x} + c_2 {-\sqrt{2}x}$

5. $y'' - 2y' + y = 0;$
   $y(x) = c_1 e^{x} + c_2 x e^{x}$

6. $y'' + 2y' + y = 0;$
   $y(x) = c_1 e^{-x} \cos \sqrt{3}x + c^2 e^{-x} \sin \sqrt{3}x$

7. $y'' - 4y' + 13y = 0;$
   $y(x) = c_1 e^{2x} \cos 3x + c_2 e^{2x} \sin 3x$

8. $y'' - 2y' + 2y = 0;$
   $y(x) = c_1 e^{x} \cos x + c_2 e^{x} \sin x$

9. $y(4) - 8y''' + 74y'' - 232y' + 841y = 0;$
   $y(x) = c_1 e^{2x} \cos 5x + c_2 e^{2x} \sin 5x + c_3 x e^{2x} \cos 5x + c_4 x e^{2x} \sin 5x$

10. $y''' - 3y'' + 4y' - 12y = e2x - 5 \sin 3x;$
    $y(x) = c_1 e^{3x} + c_2 \sin 2x + c_3 \cos 2x - (1/8)e^{2x} - (1/6) \sin 3x - (1/6) \cos 3x$

PART II. For each solution given in problems 1–10 above, evaluate the Wronskian determinant of the indicated basis for the kernel of the differential operator.

PART III. Prove Theorem 4.4. The proof requires several properties of determinants. Here is an outline of the proof for $n = 3$. You should extend it to the general case. Let y_1, y_2, and y_3 be solutions of the third-order homogeneous linear differential equation $L(y) = y''' + a_2(x)y'' + a_1(x)y' + a_0(x)y = 0$. Let

$$W(y_1, y_2, y_3) = \begin{vmatrix} y_1 & y_2 & y_3 \\ y_1' & y_2' & y_3' \\ y_1'' & y_2'' & y_3'' \end{vmatrix}.$$

Then

$$\begin{aligned}
(W(y_1, y_2, y_3))' &= \begin{vmatrix} y_1' & y_2' & y_3' \\ y_1' & y_2' & y_3' \\ y_1'' & y_2'' & y_3'' \end{vmatrix} + \begin{vmatrix} y_1 & y_2 & y_3 \\ y_1'' & y_2'' & y_3'' \\ y_1'' & y_2'' & y_3'' \end{vmatrix} + \begin{vmatrix} y_1 & y_2 & y_3 \\ y_1' & y_2' & y_3' \\ y_1''' & y_2''' & y_3''' \end{vmatrix} \\
&= \begin{vmatrix} y_1 & y_2 & y_3 \\ y_1' & y_2' & y_3' \\ y_1''' & y_2''' & y_3''' \end{vmatrix},
\end{aligned}$$

since the first two determinants are 0 (they have two identical rows). Note the pattern for the derivative of a determinant: it is the sum of n determinants where each successive row is differentiated. Continuing,

$$\begin{aligned}
(W(y_1, y_2, y_3))' &= \begin{vmatrix} y_1 & y_2 & y_3 \\ y_1' & y_2' & y_3' \\ y_1''' & y_2''' & y_3''' \end{vmatrix} \\
&= \begin{vmatrix} y_1 & y_2 & y_3 \\ y_1' & y_2' & y_3' \\ g(y_1) & g(y_2) & g(y_3) \end{vmatrix} \\
&= \begin{vmatrix} y_1 & y_2 & y_3 \\ y_1' & y_2' & y_3' \\ -a_2(x)y_1'' & -a_2(x)y_2'' & -a_2(x)y_3'' \end{vmatrix} \\
&= -a_2(x) \begin{vmatrix} y_1 & y_2 & y_3 \\ y_1' & y_2' & y_3' \\ y_1'' & y_2'' & y_3'' \end{vmatrix} \\
&= -a_2(x) W(y_1, y_2, y_3),
\end{aligned}$$

where $g(y) = -(a_2(x)y'' + a_1(x)y' + a_0(x)y)$. The simplification in the second step results by adding $a_0(x)$ times row one and $a_1(x)$ times row two to row three. These manipulations leave the value of the determinant unchanged. The resulting differential equation

$$(W(y_1, y_2, y_3))' = -a_2(x) W(y_1, y_2, y_3)$$

has as solution

$$W(y_1, y_2, y_3)(x) = c e^{\int -p_1(x)\, dx} = W(y_1, y_2, y_3)(x_0) \exp\left(\int_{x_0}^{x} -p_1(t)\, dt \right)$$

and hence is identically zero if and only if $W(y_1, y_2, y_3)(x_0) = 0$. Since x_0 can be any point, $W(y_1, y_2, y_3)$ is either identically zero or never zero.

4.2 More Linear Algebra

Because our examination of higher-order differential equations will concentrate on linear equations, we need to broaden our horizons. What we must do is computationally intensive except in the simplest of examples, so we will quickly turn to *Mathematica* to perform the necessary calculations. Once again, in order to be sure that what you have asked *Mathematica* to do is correct, you must be able to understand and do manual calculations.

Linear Spaces

In Section 1.1, we learned about linear functions, and saw the need for a fuller discussion of properties of their domains. We now undertake that discussion.

Definition (Partial). *A* **linear space** V *is a set that satisfies these properties:*

1. If u and v are in V, then $u + v$ is in V, and
2. If u is in V and c is a number, then cu is in V.

This definition is incomplete. It covers the essentials, but there are some (very important) technical matters that have been omitted. The omitted parts of the definition are there to insure that things that happen to objects in linear spaces obey nice, obvious rules. We have to state clearly all of the nice properties because there are strange sets where things do not happen the way we want. So we state exactly what we need in order to avoid trouble later. Here is the full definition:

Definition 4.4 *A* **linear space** V *is a set that satisfies these properties: If* u, v, *and* w *are in* V, *and* r *and* s *are numbers, then*

1. $u + v$ *is in* V, *and addition has these properties:* [Additive closure]
   (1a) $u + v = v + u$; [Commutative law]
   (1b) $u + (v + w) = (u + v) + w$; [Associative law]
   (1c) *there is a member* 0 *in* V *such that* $u + 0 = u$ *for every* u *in* V; [Zero]
   (1d) $u + (-u) = 0$ [Additive inverse]
2. ru *is in* V, *and multiples have these properties:* [Multiplicative closure]
   (2a) $r(su) = (rs)u$; [Associative law]
   (2b) $1u = u$ [1 is the number 'one'.]; [Unit]
   (2c) $r(u + v) = ru + rv$; [Distributive law]
   (2d) $(r + s)u = ru + su$. [Distributive law]

You can see that sums and multiples are the principal ideas. The other parts of the definition guarantee that things work correctly. For our purposes, the set V will usually be a set of functions having some specified properties.

The standard notation $f : V \to W$ is used to say that V is the **domain** of f and W is the **range** of f. The statement that f **maps** V to (or into) W conveys the same idea. The set $f(V) = \{y \in W \mid y = f(u) \text{ for some } u \in V\}$ is the **image** of V under f. The set $f(V)$ is a subset of W. It is also a linear space as this theorem states:

Theorem 4.7 *Suppose that the linear function f maps the linear space V to the linear space W and $U = f(V)$ is the image of V under f. Then U is a linear space.*

Proof. Let u and v be in U. Then $u = f(x)$ and $v = f(y)$ for some x and y in V. So $u + v = f(x) + f(y) = f(x + y) = f(z_1)$, where $z_1 = x + y$ is in V. Thus $u + v$ is in U. Similarly, $cu = cf(x) = f(cx) = f(z_2)$, where $z_2 = cx$ is in V. Thus cu is in U. □

The idea of a linear subset of a linear space is an important concept.

Definition 4.5 *Suppose that W is a subset of the linear space V. Then W is said to be a* **subspace** *of V if and only if W is a linear space.*

Linear spaces are also called **vector spaces** or **linear vector spaces**. These are commonly accepted terms.

The following theorem states two properties of linear functions that are easy to prove and are very useful.

Theorem 4.8 *Suppose that V is a linear space and f and g are linear functions defined on V such that $f(u) + g(u)$ is defined for each u in V. Then*

1. *the function $F = f + g$, such that $F(u) = f(u) + g(u)$ for each u in V is a linear function, and if r is a number, then*
2. *the function $G = rf$ defined for each u in V by $G(u) = rf(u)$ is a linear function.*

Proof. (Note how the assumptions that f and g be linear are used throughout, as are the properties of linear space.)

$$\begin{aligned}
1.\ F(u+v) &= f(u+v) + g(u+v) && \text{[Definition of } F\text{]}\\
&= (f(u) + f(v)) + (g(u) + g(v)) && [f, g \text{ are linear}]\\
&= (f(u) + g(u)) + (f(v) + g(v)) && \text{[Commutative, associative laws]}\\
&= F(u) + F(v) && \text{[Definition of } F\text{].}\\
F(cu) &= f(cu) + g(cu) && \text{[Definition of } F\text{]}\\
&= c\,f(u) + c\,g(u) && [f, g \text{ are linear}]\\
&= c\,(f(u) + g(u)) && \text{[Distributive law]}\\
&= c\,F(u) && \text{[Definition of F].}\\
2.\ G(u+v) &= r\,f(u+v) && \text{[Definition of G]}\\
&= r\,(f(u) + f(v)) && [f \text{ is linear}]\\
&= r\,f(u) + r\,f(v) && \text{[Distributive law]}\\
&= G(u) + G(v) && \text{[Definition of } G\text{].}\\
G(cu) &= r\,f(cu) && \text{[Definition of G]}\\
&= r(c\,f(u)) && [f \text{ is linear}]\\
&= (r\,c)f(u) && \text{[Associative law]}\\
&= (cr)f(u) && \text{[Commutative law]}\\
&= c(r\,f(u)) && \text{[Associative law]}\\
&= c\,G(u) && \text{[Definition of } G\text{].}
\end{aligned}$$

□

This theorem says that sums and multiples of linear functions are again linear. What happens if we perform one linear function and then perform a linear function on the result? The technical term is **composition**. Is the composition of two linear functions linear? We state the answer as a theorem.

Theorem 4.9 *The composition of two linear functions is linear. In other words, if F is linear, G is linear, and the domain of F contains the range of G, then the function L defined for each x in the domain of G by $L(x) = F(G(x))$ is linear.*

Proof. If u and v are in the domain of G and c is a number, then $L(u+v) = F(G(u+v)) = F(G(u)+G(v)) = F(G(u)) + F(G(v)) = L(u) + L(v)$, and $L(cu) = F(G(cu)) + F(cG(u)) = cF(G(u)) = cL(u)$. □

An immediate consequence of the fact that the composition of linear functions is linear is that the second derivative function is linear. (Why?) By induction, one learns that derivatives of any order are linear. (If the kth derivative is linear, why is the $(k+1)$st derivative linear?) Also by induction, any finite sum of linear functions is linear. (See the exercises.)

Span and Basis of a Linear Space

You are familiar with the fact that you can express any point (a,b) in the plane $\boldsymbol{R}^2$ this way: $(a,b) = a(1,0) + b(0,1)$. This means that we can express every member of $\boldsymbol{R}^2$ by using exactly two vectors: $(1,0)$ and $(0,1)$. Furthermore, if $a(1,0) + b(0,1) = (0,0)$, then $a = b = 0$.

We can do the same thing with the two vectors $(1,1)$ and $(1,-1)$:

$$(a,b) = \frac{a+b}{2}(1,1) + \frac{a-b}{2}(1,-1),$$

and if $c(1,1) + d(1,-1) = (0,0)$, then $c = d = 0$. The point (a,b) has been expressed as a linear combination (see Definition 4.6 below) of the members of two different sets.

Here is another familiar example in an unfamiliar setting: Given the differential equation $L(y) = (d^3y/dx^3) = 0$, we can see that $y_1 = 1, y_2 = x$, and $y_3 = x^2$ are each solutions of the differential equation. Furthermore, for any choice of numbers a, b, and c,

$$y(x) = a\,x^2 + b\,x + c = a\,y_3(x) + b\,y_2(x) + c\,y_1(x)$$

is also a solution of the differential equation. It is also true that if $ax^2+bx+c = 0$ for every number x, then $a = b = c = 0$. These properties of polynomials and of this differential equation look a lot like the properties we saw for the plane, $\boldsymbol{R}^2$. And they are.

We can multiply any nonzero member of a linear space by every real number and get new members in the space, so linear spaces that have one nonzero member have an infinite number of members. Most linear spaces we encounter can be completely described by simple "arithmetic" on a finite number of elements. Other than the fact that we have to be somewhat careful about how we choose these finite sets of elements, their form will be simple, and—at least in theory—the elements we want will be easy to find. We just need some definitions to see what is happening.

Definition 4.6 *Given a set $\{y_1, y_2, \ldots, y_m\}$ of vectors in a linear space V, and a set $\{c_1, c_2, \ldots, c_m\}$ of numbers, the vector $w = c_1 y_1 + c_2 y_2 + \cdots + c_m y_m$ of V is called a* **linear combination** *of the members of the set $\{y_1, y_2, \ldots, y_m\}$. We say that c_k is the* **coefficient** *of y_k, and the set $\{c_1, c_2, \ldots, c_m\}$ is the* **set of coefficients** *of the linear combination.*

The set of all linear combinations of a set of vectors has a name.

Definition 4.7 *Given a set $A = \{y_1, y_2, \ldots, y_m\}$ of vectors in a linear space V. The set of all linear combinations of members of A is called the* **linear span**, *or simply the* **span** *of the set A.*

It is worth noting that this set is a linear space.

Theorem 4.10 *The span of a set A is a linear space.*

Proof. This is proved easily by combining the definition of span and the definition of linear space and doing a little calculating. The proof of this is left as an exercise. □

The plane $\boldsymbol{R}^2$ is the span of $\{(1,0),(0,1)\}$, and the set of all polynomials of degree 2 or less is the span of $\{1, x, x^2\}$. Both of these sets did their spanning in an especially nice way.

Here is another example of the idea of a spanning set. Consider the set $\{1, x, x^2, x^3\}$. What is the span of this set? Typical members of the span are: $1-x$, x^2+3x-4, x^3, x^3-2, and so on. The most general member is ax^3+bx^2+cx+d, which you will recognize as a polynomial of degree 3 (or less if $a=0$ or $a=b=0$ or $a=b=c=0$ or even $a=b=c=d=0$). Notice that the sum of any two of these functions (the polynomials in x of degree three or less) is a polynomial in x of degree three or less. The same is true of a constant multiple of any one.

Here the members of our linear space are functions. So a 'vector' need not be a geometric "point," but may even be a function. The members of the vector spaces we will see as we study differential equations will often be more than simply a function. They will sometimes be vectors of functions or even matrices of functions. (See the discussion of matrices below.)

Definition 4.8 *Given a subset $A = \{y_1, y_2, \ldots, y_m\}$ of vectors that spans a linear space V. Suppose that the zero linear combination $c_1y_1+c_2y_2+\cdots+c_my_m = 0$ can be produced only when $c_1 = c_2 = \cdots = c_m = 0$. Then A is said to be* **linearly independent**.

A linearly independent spanning set for a linear space V is called a **basis** *for V.*

If it is possible to express $c_1y_1 + c_2y_2 + \cdots + c_my_m = 0$ *with not all of the* c_i *being 0, then A is said to be* **linearly dependent**.

We showed that the set $\{(1,0),(0,1)\}$ is a basis for $\boldsymbol{R}^2$, as is the set $\{(1,1),(1,-1)\}$, and the set $\{1,x,x^2,x^3\}$ is a basis for the polynomials of degree 3 or less. We are especially interested in sets that are a basis for the spaces in which we have interest, because the members of a basis are genuinely and essentially different from one another. They are exactly enough to describe the set of interest: fewer would not do the job; more would be redundant. When we solve homogeneous linear differential equations, we will always seek a basis for the set of solutions. When we looked at kernels before, the "complete description of the kernel" that we sought can be given in terms of a basis.

One property of the linear spaces we will use is this: every basis of a space has the same number of vectors in it. This means that if each of us solves a given linear problem and has found a basis for the kernel, then we each have found the same number of vectors. We may have *different* vectors in our bases, but each basis has the same number of vectors. This idea is worth stating as a theorem.

Theorem 4.11 *If any basis for a vector space V is finite, then every basis is finite, and any two bases have the same number of elements.*

Definition 4.9 *The number of vectors in a basis of a vector space V is called the* **dimension** *of V.*

Theorem 4.11 tells us that the idea of dimension is well defined. If we say that a space is five-dimensional, then any basis for this space will have five vectors in it. Since there are always many choices for a basis, we will often find that one basis reveals more about a problem than another. We will also find how to convert one basis into another, so that we can start with a basis that is easy to find and then convert it into the one that reveals the properties that we wish to exploit.

Matrices, Matrix Multiplication, Transposes, and Inverses

Usually by the time students have reached a course in differential equations, they know what matrices are and how to perform the basic operations on them. We will briefly cover these ideas.

Properties of Matrices

After each definition or property is an example in *Mathematica*.

A **matrix** is a rectangular array of objects. We will have **matrices** (the plural of matrix) whose entries are numbers, some whose entries are real- or complex-valued functions, and some whose entries are operators.

```
In[1]:=  A = {{2,3,-1},{0,2,5}}

Out[1]=  {{2, 3, -1}, {0, 2, 5}}

In[2]:=  M = {{Cos[x], Sin[x]},{-Sin[x],Cos[x]}}

Out[2]=  {{Cos[x], Sin[x]}, {-Sin[x], Cos[x]}}
```

The **shape** of a matrix is given in the form $m \times n$ where m is the number of **rows** in the matrix and n is the number of **columns** in the matrix. The rows of a matrix are horizontal and the columns are vertical. The entry in row i and column j of the matrix $\boldsymbol{A}$ is denoted $\boldsymbol{A}_{ij}$ and is called the i, j^{th} entry of $\boldsymbol{A}$.

```
In[3]:=  Dimensions[A]

Out[3]=  {2, 3}

In[4]:=  A[[1,2]]

Out[4]=  3
```

Square matrices have the same number of rows as columns. A square matrix has a main **diagonal**, those entries having the same row and column number. $\boldsymbol{A}_{ii}$ for instance is the ith main diagonal entry in the matrix $\boldsymbol{A}$.

```
In[5]:=  Dimensions[M]

Out[5]=  {2, 2}
```

For example from *In[1]*;

$$A = \begin{pmatrix} 2 & 3 & -1 \\ 0 & 2 & 5 \end{pmatrix}$$

is a 2×3 matrix having two rows and three columns, the first row being $(2, 3, -1)$ and second row $(0, 2, 5)$. The columns are

$$\begin{pmatrix} 2 \\ 0 \end{pmatrix}, \begin{pmatrix} 3 \\ 2 \end{pmatrix}, \text{ and } \begin{pmatrix} -1 \\ 5 \end{pmatrix}.$$

`A[[1]]` denotes the first row of $\boldsymbol{A}$. In general, `A[[k]]` denotes the kth row of $\boldsymbol{A}$.

```
In[6]:=  A[[1]]                    (* The first row of A *)

Out[6]=  {2, 3, -1}
```

Matrix equality is defined only for matrices of the same shape. The equality $\boldsymbol{A} = \boldsymbol{B}$ asserts that the matrices $\boldsymbol{A}$ and $\boldsymbol{B}$ have the same shape and that corresponding elements are equal. This is shorthand for a set of equations.

```
In[7]:=  B = {{2,3,-1},{0,-3,5}}

Out[7]=  {{2, 3, -1}, {0, -3, 5}}
```

Note that *Mathematica* uses == for equality.

```
In[8]:=  A == B

Out[8]=  False
```

Matrix addition is defined only for matrices of the same shape. The sum $\boldsymbol{A} + \boldsymbol{B}$ is a matrix whose entries are the sum of corresponding elements of $\boldsymbol{A}$ and $\boldsymbol{B}$. Matrix addition is commutative and associative.

```
In[9]:=  A+B

Out[9]=  {{4, 6, -2}, {0, -1, 10}}

In[10]:= A-B

Out[10]= {{0, 0, 0}, {0, 5, 0}}
```

Two special matrices are the $m \times n$ **zero matrix**, $\mathbf{0}_{mn}$, each entry of which is zero, and the $n \times n$ **identity matrix** $\boldsymbol{I}_n$, often written $\boldsymbol{I}$ when the context is clear. The entries of $\boldsymbol{I}$ are zero except down the main diagonal where each entry is 1. If $\boldsymbol{A}$ is an $m \times n$ matrix, then $\boldsymbol{A} + \mathbf{0} = \boldsymbol{A}$ and $\mathbf{I}_m \boldsymbol{A} = \boldsymbol{A}\boldsymbol{I}_n = \boldsymbol{A}$.

```
In[11]:= Z23 = Table[0, {i,2}, {j,3}]

Out[11]= {{0, 0, 0}, {0, 0, 0}}

In[12]:= A+Z23 == A

Out[12]= True
```

Define a 3×3 identity matrix.

```
In[13]:= IdentityMatrix[3]//MatrixForm

Out[13]= 1  0  0
         0  1  0
         0  0  1
```

Multiplication by a scalar is defined. If c is a scalar and $\boldsymbol{A}$ is a matrix, $c\boldsymbol{A}$ is the matrix of the same shape as $\boldsymbol{A}$ whose entries are c times the corresponding entries of $\boldsymbol{A}$.

```
In[14]:= c A

Out[14]= {{2 c, 3 c, -c}, {0, 2 c, 5 c}}
```

Vectors are single-row or single-column matrices. So vectors are either row-matrices or column-matrices. A vector is called an **n-vector** if it has n components.

```
In[15]:= v = {x,y,z}

Out[15]= {x, y, z}

In[16]:= w = {a,b,c}

Out[16]= {a, b, c}
```

A **dot product** or **scalar product** is defined for any two n-vectors. If $\boldsymbol{u} = (u_1, u_2, \ldots, u_n)$ and $\boldsymbol{v} = (v_1, v_2, \ldots, v_n)$ are n-vectors then $\boldsymbol{u} \cdot \boldsymbol{v} = u_1v_1 + u_2v_2 + \cdots + u_nv_n$. Dot is commutative: $\boldsymbol{u} \cdot \boldsymbol{v} = \boldsymbol{v} \cdot \boldsymbol{u}$. If c is a number then $c(\boldsymbol{u} \cdot \boldsymbol{v}) = (c\boldsymbol{u}) \cdot \boldsymbol{v} = \boldsymbol{u} \cdot (c\boldsymbol{v})$. When dotting, we do not distinguish whether a vector is a row-vector or a column-vector.

```
In[17]:= v.w

Out[17]= a x + b y + c z
```

One can **partition** an $m \times n$ matrix to emphasize and name its rows

$$\boldsymbol{A} = \begin{pmatrix} \boldsymbol{A}_{1\bullet} \\ \hline \boldsymbol{A}_{2\bullet} \\ \hline \vdots \\ \hline \boldsymbol{A}_{m\bullet} \end{pmatrix}$$

or its columns $\boldsymbol{A} = (\boldsymbol{A}_{\bullet 1} | \boldsymbol{A}_{\bullet 2} | \ldots | \boldsymbol{A}_{\bullet n})$. The notation $\boldsymbol{A}_{i\bullet}$ refers to row i of $\boldsymbol{A}$ and $\boldsymbol{A}_{\bullet j}$ refers to column j of $\boldsymbol{A}$.

```
In[18]:= A//MatrixForm

Out[18]= 2   3   -1
         0   2   5
```

A **matrix product** AB is defined for matrices $\boldsymbol{A}$ and $\boldsymbol{B}$ having shapes $m \times p$ and $p \times n$, respectively. Notice that the number of columns of $\boldsymbol{A}$ is the same as the number of rows of $\boldsymbol{B}$. Then if $\boldsymbol{C} = \boldsymbol{AB}$, the entry in row i and column j of $\boldsymbol{C}$ is defined to be $\boldsymbol{C}_{ij} = \boldsymbol{A}_{i\bullet} \cdot \boldsymbol{B}_{\bullet j}$, where $\boldsymbol{A}_{i\bullet}$ is row i of $\boldsymbol{A}$ and $\boldsymbol{B}_{\bullet j}$ is column j of $\boldsymbol{B}$. Thus there are mn dot products to perform to calculate the product of matrices $\boldsymbol{A}$ and $\boldsymbol{B}$. The product of $\boldsymbol{A}$ with a row m-vector on the left and with a column p-vector on the right are also defined.

```
In[19]:= B = {{2,1},{-1,1},{3,-4}}

Out[19]= {{2, 1}, {-1, 1}, {3, -4}}

In[20]:= A.B

Out[20]= {{-2, 9}, {13, -18}}

In[21]:= B.A

Out[21]= {{4, 8, 3}, {-2, -1, 6}, {6, 1, -23}}
```

```
In[22]:= A.v

Out[22]= {2 x + 3 y - z, 2 y + 5 z}

In[23]:= {c1,c2}.A

Out[23]= {2 c1, 3 c1 + 2 c2, -c1 + 5 c2}
```

We want to use matrices to denote systems of linear equations easily. For instance, the system of two equations in three unknowns

$$\begin{cases} 2x + 3y - z = 7 \\ -x + 5y + 4z = -3 \end{cases}$$

can be represented by a single equation $\boldsymbol{Av} = \boldsymbol{B}$, where

$$\boldsymbol{A} = \begin{pmatrix} 2 & 3 & -1 \\ -1 & 5 & 4 \end{pmatrix}, \quad \boldsymbol{v} = \begin{pmatrix} x \\ y \end{pmatrix}, \quad \text{and} \quad \boldsymbol{B} = \begin{pmatrix} 7 \\ -3 \end{pmatrix}.$$

```
In[24]:= B = {7,-3}

Out[24]= {7, -3}

In[25]:= A.v == B

Out[25]= {2 x + 3 y - z, 2 y + 5 z} == {7, -3}

In[26]:= LogicalExpand[%]

Out[26]= 2 x + 3 y - z == 7 && 2 y + 5 z == -3
```

A square $n \times n$ matrix $\boldsymbol{A}$ is **invertible** if and only if the **determinant** of $\boldsymbol{A}$, denoted $\det(\boldsymbol{A})$ is nonzero. In this case $\boldsymbol{A}$ is said to have an inverse. The matrix $\boldsymbol{B}$ is the inverse of the matrix $\boldsymbol{A}$ provided that $\boldsymbol{AB} = \boldsymbol{BA} = \boldsymbol{I_n}$, where $\boldsymbol{I_n}$ is the $n \times n$ identity matrix. Such a matrix $\boldsymbol{B}$ is unique and is denoted $\boldsymbol{A}^{-1}$. If the matrix $\boldsymbol{A}$ is invertible, then a system of equations $\boldsymbol{AX} = \boldsymbol{C}$ has the unique solution $\boldsymbol{X} = \boldsymbol{A}^{-1}\boldsymbol{C}$. If $\boldsymbol{C}$ has more than one column, then so does $\boldsymbol{X}$.

```
In[27]:= A = {{5,3},{3,2}}

Out[27]= {{5, 3}, {3, 2}}

In[28]:= Det[A]

Out[28]= 1

In[29]:= Inverse[A]

Out[29]= {{2, -3}, {-3, 5}}

In[30]:= A = {{5,3},{10,6}}

Out[30]= {{5, 3}, {10, 6}}

In[31]:= Det[A]

Out[31]= 0
```

Since `Det[A]==0`, the matrix A has no inverse. Such a matrix is called **singular**. Note the message that *Mathematica* prints before returning the expression unevaluated.

```
In[32]:= Inverse[A]

Inverse::sing: Matrix {{5, 3}, {10, 6}} is singular.

Out[32]= Inverse[{{5, 3}, {10, 6}}]
```

Mathematica has functions for finding a basis for the kernel of a matrix and for solving a matrix system of equations. Here again is an invertible matrix, A.

```
In[33]:= A = {{5,3},{3,2}}

Out[33]= {{5, 3}, {3, 2}}
```

The kernel (null space) of an invertible matrix contains no nonzero vectors.

```
In[34]:= A = {{5,3},{3,2}}

Out[34]= {{5, 3}, {3, 2}}

In[35]:= NullSpace[A]

Out[35]= {}
```

The kernel of a singular matrix contains nonzero vectors.

```
In[36]:= A = {{5,-3,4},{10,-6,1},{0,0,-3}}

Out[36]= {{5, -3, 4}, {10, -6, 1}, {0, 0, -3}}

In[37]:= ns = NullSpace[A]

Out[37]= {{3, 5, 0}}
```

Solve a nonhomogeneous system of linear algebraic equations.

```
In[38]:= b = {17, 27, -3}

Out[38]= {17, 27, -3}

In[39]:= Solve[A.{x,y,z} == b]

Out[39]=
                13   3 y
         {{x -> -- + ---, z -> 1}}
                5     5
```

This is what the solution vector s looks like.

```
In[40]:= s = {x,y,z}/.%[[1]]

Out[40]=  13   3 y
         {— + ———, y, 1}
          5    5
```

Check that s is a solution.

```
In[41]:= Simplify[A.s == b]

Out[41]= True
```

Mathematica gives you a function to find only a particular solution.

```
In[42]:= p = LinearSolve[A,b]

Out[42]=  13
         {—, 0, 1}
          5
```

Construct a solution z that has one arbitrary constant (c).

```
In[43]:= z = ns[[1]]*c+p          (* ns comes from In[37] *)

Out[43]=  13
         {— + 3 c, 5 c, 1}
          5
```

Check that z is a solution.

```
In[44]:= Simplify[A.z == b]

Out[44]= True
```

To say that the matrix $\boldsymbol{B}$ is the **transpose** of the matrix $\boldsymbol{A}$ means that the rows of $\boldsymbol{B}$ are the columns of $\boldsymbol{A}$ in the same order. That is, the entries in row i of $\boldsymbol{B}$ are the entries in column i of $\boldsymbol{A}$. The operation of transpose is denoted $\boldsymbol{B} = \boldsymbol{A}^T$. For example, if

$$\boldsymbol{A} = \begin{pmatrix} a_1 & a_2 & a_3 \\ c_1 & c_2 & c_3 \end{pmatrix}, \quad \text{then} \quad \boldsymbol{A}^T = \boldsymbol{B} = \begin{pmatrix} a_1 & c_1 \\ a_2 & c_2 \\ a_3 & c_3 \end{pmatrix}.$$

Note that for any matrix $\boldsymbol{A}$, $\boldsymbol{A}^{TT} = \boldsymbol{A}$.

```
In[45]:= Transpose[{{a1,a2,a3},{c1,c2,c3}}]

Out[45]= {{a1,c1},{a2,c2},{a3,c3}}

In[46]:= Transpose[{{a1,a2,a3},{c1,c2,c3}}]//MatrixForm

Out[46]= a1   c1
         a2   c2
         a3   c3
```

Exercises 4.2

Part I. Linear spaces.

1. Show that each of these sets is a linear space:

   (a) The set of all continuous functions defined on the interval $[a, b]$.
   (b) The set of all differentiable functions defined on the interval $[a, b]$.
   (c) The set of all differentiable functions defined on the interval $[a, b]$ such that if f is in the set, then $f(a) = 0$.
   (d) The set of all differentiable functions defined on the interval $[a, b]$ such that if f is in the set, then $f(a) = f(b) = 0$. Notice that each succeeding set of functions is a subset of the preceding sets, but is still a linear space.

2. Using properties of sums, multiples and composition of linear operators, explain why each of these operators is linear:

   (a) $L_1(y) = y' + 3y$;
   (b) $L_2(y) = 5y' + 2y$;
   (c) $L_3(y) = y'' + 4y' + y$;
   (d) $L_4(y) = y'' - 5y$;
   (e) $L_5(y) = y^{(5)} - 7y^{(4)} + 3y'' - 2y$.

3. Show by induction that derivatives of any order are linear.

4. Show by induction that any finite sum of linear functions is linear.

Part II. Perform the following operations on the given pairs $\boldsymbol{A}, \boldsymbol{B}$ of matrices.

$$\boldsymbol{AB}, \boldsymbol{BA}, \boldsymbol{A}+\boldsymbol{B}, \boldsymbol{B}-\boldsymbol{A}.$$

Do this manually and by *Mathematica*. If the indicated operation is undefined, say so.

5. $\boldsymbol{A} = \begin{pmatrix} 1 & 2 \\ -2 & 3 \end{pmatrix}$; $\quad \boldsymbol{B} = \begin{pmatrix} 3 & 2 \\ 0 & 0 \end{pmatrix}$.

6. $\boldsymbol{A} = \begin{pmatrix} 1 & 0 \\ 0 & 1 \end{pmatrix}$; $\quad \boldsymbol{B} = \begin{pmatrix} -2 & 0 \\ 5 & 7 \end{pmatrix}$.

7. $\boldsymbol{A} = \begin{pmatrix} 1 & 3 & 2 \\ 2 & 3 & 2 \\ 2 & 3 & 1 \end{pmatrix}$; $\quad \boldsymbol{B} = \begin{pmatrix} 0 & 2 & 0 \\ 1 & 0 & 0 \\ 2 & 3 & 1 \end{pmatrix}$.

8. $\boldsymbol{A} = \begin{pmatrix} 1 & 2 & 3 & 4 \\ 3 & 2 & -1 & 0 \\ 0 & 0 & 0 & 1 \end{pmatrix}$; $\quad \boldsymbol{B} = \begin{pmatrix} 2 & 1 & -9 & 0 \\ 0 & 0 & 0 & 4 \\ 2 & 0 & 0 & 1 \end{pmatrix}$.

9. $\boldsymbol{A} = \begin{pmatrix} 1 & 3 & 2 \\ 2 & 3 & 2 \\ 2 & 3 & 1 \end{pmatrix}$; $\quad \boldsymbol{B} = \begin{pmatrix} 0 & 2 \\ 1 & 5 \\ -3 & 2 \end{pmatrix}$.

PART III. The transpose of a matrix. For each of the pairs of matrices given in problems 5 through 9, perform the operations that are defined: (Do some manually; all by *Mathematica*.) State the cause whenever one of the operations is undefined.

$$\boldsymbol{A}^T, \boldsymbol{B}^T, \boldsymbol{A}^T\boldsymbol{B}^T, \boldsymbol{B}^T\boldsymbol{A}^T, (\boldsymbol{AB})^T, (\boldsymbol{BA})^T, \boldsymbol{A}+\boldsymbol{B}, \boldsymbol{A}^T+\boldsymbol{B}^T, \boldsymbol{A}^{TT}$$

Point out identical results. These are problems 5b–9b.

PART IV. The inverse of a matrix. For each of these matrices, find the inverse by *Mathematica*. In each case verify that the product of the inverse with the matrix on both sides is an identity matrix.

10. $\boldsymbol{A} = \begin{pmatrix} 1 & 2 \\ -2 & 3 \end{pmatrix}$.

11. $\boldsymbol{B} = \begin{pmatrix} \cos x & \sin x \\ -\sin x & \cos x \end{pmatrix}$.

12. $\boldsymbol{C} = \begin{pmatrix} 1 & 3 & 2 \\ 2 & 3 & 2 \\ 2 & 3 & 1 \end{pmatrix}$.

PART V. The inverse of a product. For each of these pairs of matrices, find the inverse of both matrices by *Mathematica*. In each case verify that the product of the inverses is the inverse of the product taken in the reverse order. That is, if $\boldsymbol{A}$ and $\boldsymbol{B}$ are invertible and $\boldsymbol{AB}$ is defined, then $\boldsymbol{AB}$ is invertible and $(\boldsymbol{AB})^{-1} = \boldsymbol{B}^{-1}\boldsymbol{A}^{-1}$.

13. $\boldsymbol{A} = \begin{pmatrix} \cos x & \sin x \\ -\sin x & \cos x \end{pmatrix}$, $\quad \boldsymbol{B} = \begin{pmatrix} \cos t & \sin t \\ -\sin t & \cos t \end{pmatrix}$.

14. $\boldsymbol{A} = \begin{pmatrix} 1 & 3 & 2 \\ 2 & 3 & 2 \\ 2 & 3 & 1 \end{pmatrix}$; $\quad \boldsymbol{B} = \begin{pmatrix} 0 & 2 & 0 \\ 1 & 0 & 0 \\ 2 & 3 & 1 \end{pmatrix}$.

15. $\boldsymbol{A} = \begin{pmatrix} 1 & 2 & 3 & 4 \\ 3 & 2 & -1 & 0 \\ 0 & 0 & 0 & 1 \\ 2 & 1 & 3 & 7 \end{pmatrix}$; $\quad \boldsymbol{B} = \begin{pmatrix} 2 & 1 & -9 & 0 \\ 0 & 0 & 0 & 4 \\ 2 & 0 & 0 & 1 \\ 3 & 0 & 1 & 1 \end{pmatrix}$.

16. $\boldsymbol{A} = \begin{pmatrix} 1 & -3 & 2 \\ -2 & 3 & 2 \\ 2 & 3 & -1 \end{pmatrix}$; $\quad \boldsymbol{B} = \begin{pmatrix} 0 & 2 & 0 \\ 1 & 0 & -3 \\ 2 & 3 & -1 \end{pmatrix}$.

PART VI. Theory.

17. Prove Theorem 4.10 by showing that if $u = \sum c_i y_i$ and $v = \sum d_i y_i$ are in V, then so are $u + v$ and ru.

18. Show that $\{(1,0,0), (0,1,0), (0,0,1)\}$ is a basis for $\boldsymbol{R}^3$.

19. Show that $\{\boldsymbol{e}_1 = (1, 0, \ldots, 0), \boldsymbol{e}_2 = (0, 1, \ldots, 0), \ldots, \boldsymbol{e}_n = (0, 0, \ldots, 1)\}$ is a basis for $\boldsymbol{R}^n$. The vectors have n components, $n-1$ of which are 0, and one is 1. The only 1 in the vector $\boldsymbol{e}_k$ is in the kth component of $\boldsymbol{e}_k$.

20. Show that $\{1, x, x^2, x^3\}$ is a basis for the set of polynomials of degree 3 or less.

21. Show that $\{1, x, x^2, \ldots, x^n\}$ is a basis for the set of polynomials of degree n or less.

4.3 Homogeneous Second-Order Linear Constant Coefficients

Because we can completely describe the set of all solutions to a homogeneous second-order linear differential equation with constant coefficients, we proceed to do it. Everything to come in the next section that is concerned with higher-order linear differential equations with constant coefficients will have been anticipated in this brief section.

It would be useful to have a reliable check for the linear independence of a finite set of functions. The Wronskian does this for us. Recall that for two functions, y_1 and y_2, the Wronskian determinant is simply

$$W(y_1, y_2)(x) = \begin{vmatrix} y_1(x) & y_2(x) \\ y_1'(x) & y_2'(x) \end{vmatrix} = y_1(x)y_2'(x) - y_1'(x)y_2(x)$$

Our interest in the Wronskian determinant is this theorem (second-order case).

Theorem 4.12 *Given a pair $\{y_1, y_2\}$ of functions having a continuous derivative on an interval I. If the functions are not linearly independent on I, then $W(y_1, y_2)(x) = 0$ for every x in I.*

If the functions $\{y_1, y_2\}$ are both solutions of the same linear homogeneous differential equation, then $W(y_1, y_2) \neq 0$ on I if and only if the pair of functions $\{y_1, y_2\}$ is linearly independent.

Example 4.9 The functions $\cos x$ and $\sin x$, two solutions of the linear differential equation $y'' + y = 0$, are linearly independent. Their Wronskian is

$$W[\cos x, \sin x] = \begin{vmatrix} \cos x & \sin x \\ -\sin x & \cos x \end{vmatrix} = \cos^2 x + \sin^2 x = 1 \neq 0.$$

On the other hand, x and $5x$, which are linearly dependent since $5(x) + (-1)(5x) = 0$, has the Wronskian

$$W[x, 5x] = \begin{vmatrix} x & 5x \\ 1 & 5 \end{vmatrix} = 5x - 5x = 0. \qquad \diamond$$

We showed in Example 4.8 that the functions $\cos(x + \pi/2)$ and $\sin x$ are linearly dependent.

We will often need to calculate such Wronskian determinants. When the determinants involved are 3×3 or larger, this is a real chore. There is a collection of rules for manipulating determinants that can simplify the work, but, done manually, such a calculation is prone to errors. *Mathematica* does these calculations easily, and we will rely heavily on it. In *Mathematica* Example 4.9 is the following.

Example 4.9M Use *Mathematica* to show the linearly independence of the sine and cosine functions and the linear dependence of the functions $y_1 = x$ and $y_2 = 5x$.

Solution.

```
In[47]:= y = {Cos[x],Sin[x]}

Out[47]= {Cos[x], Sin[x]}

In[48]:= w = {y,D[y,x]}

Out[48]= {{Cos[x], Sin[x]}, {-Sin[x], Cos[x]}}

In[49]:= Det[w]

                 2          2
Out[49]= Cos[x]  + Sin[x]

In[50]:= y = {x, 5x}

Out[50]= {x, 5x}

In[51]:= w = {y,D[y,x]}

Out[51]= {{x, 5x}, {1, 5}}

In[52]:= Det[w]

Out[52]= 0
```

The fact that the determinant is 0 means that these two functions are linearly dependent. ◇

The Dimension of the Kernel of a Second-Order Homogeneous Linear Differential Equation

Theorem 4.13 *The dimension of the kernel of a second-order linear differential operator is two. That is, suppose that I is an interval and the functions $a_1(x)$ and $a_0(x)$ are continuous on I, then a homogeneous linear differential equation of the form*

$$L_2(y) = \frac{d^2y}{dx^2} + a_1(x)\frac{dy}{dx} + a_0(x)y = 0$$

has two linearly independent solutions that are a basis for the set of all solutions.

Proof. We proved this theorem in Section 4.1. □

Theorem 4.13 tells us how large a task we have when we attempt to solve a homogeneous linear differential equation. It also gives us a measure of how much work remains as we proceed to find solutions to a problem. Though linear algebra may tell us how many vectors (in our case functions) we can expect in a basis for the kernel of a linear differential operator, it does not say how to find these basis vectors. This we learn in differential equations. But you may be surprised at how much of the theory leads us back to the algebra you learned in high school, as well as back to calculus. Actually, there is a very good fit between algebra and calculus. The present discipline "differential equations" responds well to the application of methods we already know.

Second-Order Linear Differential Equations with Constant Coefficients

Second-order linear differential equations give us a good look at what we can expect as we examine higher-order equations. For the moment we will concentrate our studies on linear differential equations with constant coefficients, because of the significance of their applications and the simplicity of the theory of their solution.

Let us define the second-order differential operator that we will study to be

$$L_2(y) = a\frac{d^2y}{dx^2} + b\frac{dy}{dx} + cy, \tag{4.5}$$

where a, b, and c are numbers and $a \neq 0$. The first thing to notice is that if we let $y = f(x)$, then $L_2(f(x))$ is a function. For instance, using the fact that $(d^k/dx^k)e^{rx} = r^k e^{rx}$ several times, observe that

$$\begin{aligned} L_2(e^x) &= a(e^x) + b(e^x) + c(e^x) &&= (a+b+c)e^x, \\ L_2(e^{2x}) &= a(4e^{2x}) + b(2e^{2x}) + c(e^{2x}) &&= (4a+2b+c)e^{2x}, \\ L_2(e^{-x}) &= a(e^{-x}) + b(-e^{-x}) + c(e^{-x}) &&= (a-b+c)e^{-x}; \end{aligned}$$

and $y = x$ and $y = x^2$ produce

$$L_2(x) = a(0) + b(1) + c(x) = b + cx,$$

and

$$L_2(x^2) = a(2) + b(2x) + c(x^2) = 2a + 2bx + cx^2.$$

We can take advantage of the way L_2 acts on exponentials of the form e^{rx} to see how to get solutions of the equation $L_2(y) = 0$. In order for

$$L_2(e^{rx}) = (ar^2 + br + c)e^{rx} = 0,$$

we must have $p(r) = ar^2 + br + c = 0$ since e^{rx} is never 0. This means that we need only choose r to be a root of the polynomial equation

$$p(r) = ar^2 + br + c = 0$$

to have a solution $y(x) = e^{rx}$ of the differential equation $L_2(y) = 0$. The polynomial equation $p(r) = 0$ is called the **characteristic equation** of the differential equation $L_2(y) = 0$, and the polynomial $p(r)$ is the **characteristic polynomial**. If the characteristic equation has distinct roots r_1 and r_2, we immediately have two linearly independent solutions, $y_1(x) = e^{r_1 x}$ and $y_2(x) = e^{r_2 x}$, of $L_2(y) = 0$.

Example 4.10 Solve the differential equation $L_2(y) = y'' - 5y' + 6y = 0$ by making the substitution $y = e^{rx}$.

Solution. When we do this, $y' = re^{rx}$ and $y'' = r^2 e^{rx}$, so that $L_2(e^{rx}) = r^2 e^{rx} - 5re^{rx} + 6e^{rx} = (r^2 - 5r + 6)e^{rx} = p(r)e^{rx}$. $L_2(e^{rx}) = 0$ only when $p(r) = r^2 - 5r + 6 = (r-2)(r-3) = 0$.

This means that $r_1 = 2$ and $r_2 = 3$ are the roots we seek. Thus

$$y_1(x) = e^{r_1 x} = e^{2x}$$

and

$$y_2(x) = e^{r_2 x} = e^{3x}$$

are two solutions. That they are solutions is verified by

$$L_2(e^{2x}) = 4e^{2x} - 5(2e^{2x}) + 6(e^{2x}) = (4 - 10 + 6)e^{2x} = 0,$$

and

$$L_2(e^{3x}) = 9e^{3x} - 5(3e^{3x}) + 6(e^{3x}) = (9 - 15 + 6)e^{3x} = 0.$$

These two solutions are also linearly independent, because their Wronskian is

$$W[e^{2x}, e^{3x}] = \begin{vmatrix} e^{2x} & e^{3x} \\ 2e^{2x} & 3e^{3x} \end{vmatrix} = e^{5x}(3-2) = e^{5x} \neq 0.$$

Since $\{e^{2x}, e^{3x}\}$ is a basis for the kernel of the differential equation $L_2(y) = 0$, any member of the kernel can be described by $y = c_1 e^{2x} + c_2 e^{3x}$. ◇

Example 4.10M These solutions can be found by *Mathematica* in two ways:

1. by working through the theory, and
2. by using DSolve.

Demonstrate these two ways.

Solution.

Define the operator:

```
In[53]:= L[x_,y_] = y''[x]-5y'[x]+6y[x]

Out[53]= 6 y[x] - 5 y'[x] + y''[x]
```

Get the characteristic polynomial by finding the coefficient of Exp[r x] after substituting Exp[r x] into the operator L.

```
In[54]:= CharacteristicPoly[r_] = Coefficient[L[x,Exp[r #]&],Exp[r x]]

Out[54]=          2
         6 - 5 r + r
```

Get the roots of the characteristic equation. (Note the use of the ReplaceAll operator '/.'.)

```
In[55]:= roots = r/.Solve[CharacteristicPoly[r] == 0, r]

Out[55]= {3, 2}
```

Turn these roots into functions.

```
In[56]:= Solns = Map[Exp[# x]&, roots]

Out[56]=   3 x    2 x
         {E   , E   }
```

Form the complete kernel. Here Dot is used. You saw it in multidimensional calculus.

```
In[57]:= AllSolns[x_] = Solns.{c1,c2}

Out[57]=  3 x          2 x
         E     c1 + E     c2
```

Check that this set of solutions works:

```
In[58]:= Simplify[L[x,AllSolns] == 0]

Out[58]= True
```

Here is the second way: the solution using the built-in function DSolve.

```
In[59]:= DSolve[y''[x]-5y'[x]+6y[x] == 0, y[x], x]

Out[59]=
                      2 x            x
         {{y[x] -> E     (C[1] + E  C[2])}}
```

◇

Notice that the first steps that *Mathematica* was requested to execute were exactly the steps that we executed manually in solving the same problem. Whenever such a second-order problem has a characteristic equation with distinct roots, the steps we first executed suffice to solve that problem. As was seen in the last step above, *Mathematica* can solve the differential equation. This is powerful, but, for now, we want to be able to check our understanding of each step of our work, so the intermediate steps are worthwhile.

Review of the Solution of Quadratic Equations

In the second-order case with real coefficients the characteristic equation is a quadratic, and since there are three different situations that can occur when solving such equations, we consider the effect on differential equations of each of these three cases. The cases are:

1. two distinct roots,
2. one double root (repeated roots), and
3. complex conjugate roots.

Since $L_2(y) = a(d^2y/dx^2) + b(dy/dx) + cy$ has as characteristic equation the general quadratic equation $ar^2 + br + c = 0$, which admits each of these three cases, we need to do some algebra to remind ourselves about how to recognize each case. Solving $ar^2+br+c = 0$ by completing the square leads to the quadratic formula, $r = (-b \pm \sqrt{b^2 - 4ac})/(2a)$. The quantity $q = b^2 - 4ac$ determines which case we have, and for this reason is called the **discriminant**. The discriminant, being a number, must be either positive or zero or negative.

These three cases correspond to the classification of roots mentioned above. To see why, look at $r = (-b \pm \sqrt{q})/(2a)$.

1. If $q > 0$, $\sqrt{q}$ is real and nonzero, so $-b \pm \sqrt{q}$ denotes two different numbers: $-b + \sqrt{q}$, and $-b - \sqrt{q}$. After dividing these two numbers by $2a$, we have two different roots $r_1 = (-b + \sqrt{q})/(2a)$. and $r_2 = (-b - \sqrt{q})/(2a)$.
2. When $q = 0, -b \pm \sqrt{q} = -b \pm 0 = \{-b, -b\}$. This is the double root case, where $r_1 = r_2 = -b/(2a)$.
3. When $q < 0$, $\sqrt{q}$ is a pure imaginary, which we can denote by $i\sqrt{|q|}$. The two roots are $r = (-b \pm i\sqrt{|q|})/(2a)$ which is more clearly seen to represent a pair of conjugate complex numbers when written $r_1 = -b/(2a) + (i\sqrt{|q|})/(2a)$, $r_2 = -b/(2a) - (i\sqrt{|q|})/(2a)$. We see that the two roots have the same real part $\mathrm{Re}(r_1) = \mathrm{Re}(r_2) = -b/(2a)$. The imaginary parts differ only in sign: $\mathrm{Im}(r_1) = \sqrt{|q|}/(2a)$, and $\mathrm{Im}(r_2) = -\sqrt{|q|}/(2a)$.

We are now left with the problem of deciding what kinds of solutions correspond to these three cases. In the two cases where the real or complex roots are different, we get linearly independent solutions, even though we do not yet know the form of the solution in the complex case.

Discriminant > 0: Distinct Real Roots ($r_1 \neq r_2$)

$L(e^{r_1x}) = p(r_1)e^{r_1x} = (0)e^{r_1x} = 0$, and $L(e^{r_2x}) = p(r_2)e^{r_2x} = (0)e^{r_2x} = 0$. The Wronskian determinant gives

$$W[e^{r_1x}, e^{r_2x}] = \begin{vmatrix} e^{r_1x} & e^{r_2x} \\ r_1e^{r_1x} & r_2e^{r_2x} \end{vmatrix} = (r_2 - r_1)e^{(r_1+r_2)x} \neq 0$$

since $r_1 \neq r_2$, and no exponential is zero. The two linearly independent solutions are $y_1(x) = e^{r_1x}$ and $y_2(x) = e^{r_2x}$.

Discriminant $= 0$: Repeated Real Roots ($r = r_1 = r_2$)

Suppose that the discriminant is zero. Then the characteristic equation $p(r) = 0$ has roots $r_1 = r_2 = -b/(2a)$. It is clear that $y_1(x) = e^{r_1x}$ and $y_2(x) = e^{r_1x}$ are not different solutions. The way to find a second solution involves a fact about polynomials.

Theorem 4.14 *Given a polynomial $p(x)$.*

1. *Then $p(x)$ has a root r provided that $p(x) = (x-r)q_0(x)$.*
2. *Also, $p(x)$ and its derivative $p'(x)$ have a common root r provided that $p(x) = (x-r)^2 q_1(x)$. That is, $p(x)$ has r as (at least) a double root.*
3. *Further, $p(x), p'(x)$ and $p''(x)$ have a common root provided that $p(x) = (x-r)^3 q_2(x)$. [Similarly for higher multiplicities.]*

Proof. The proof involves expanding $p(x)$ in a Taylor series about r and interpreting why some of the leading terms are missing. [The value $p(r)$, and the appropriate derivatives $p^{(k)}(r)$ are zero.] You have as an exercise the construction of a proof from this hint. □

This theorem and the interchange of order of differentiation property for partial derivatives tells us how to get a second linearly independent solution when the characteristic equation has a double root. Recall that $L_2(e^{rx}) = p(r)e^{rx}$, with the derivatives taken with respect to x. Since there was a parameter r present, we actually should have denoted the derivatives as partial derivatives. But we thought of x as the independent variable and r as essentially a constant whose value(s) we would determine. Let's upgrade the status of r. Calculate

$$\frac{\partial}{\partial r} L_2(e^{rx}) = \frac{\partial}{\partial r}(p(r)e^{rx}) = p'(r)e^{rx} + p(r)xe^{rx}.$$

Since r is a double root of $p(x) = 0$, both $p(r) = 0$ and $p'(r) = 0$. So, by interchanging the order of differentiation, we find that

$$\frac{\partial}{\partial r} L_2(e^{rx}) = L_2\left(\frac{\partial}{\partial r} e^{rx}\right) = L_2(xe^{rx}) = 0.$$

This says that xe^{rx} is a solution since $L_2(xe^{rx}) = 0$. In order to understand this, we examine a differential operator whose characteristic equation has a double root.

For r to be a double root of the quadratic $p(x) = 0$, the characteristic polynomial must be a constant multiple of $p(x) = (x-r)^2 = x^2 - 2rx + r^2$. In this case the corresponding differential operator is $L_2(y) = y'' - 2ry' + r^2 y$. We need to verify that $y_1(x) = e^{rx}$ and $y_2(x) = xe^{rx}$ are both solutions of the differential equation $L_2(y) = y'' - 2ry' + r_2 y = 0$ and that they are linearly independent. $L_2(y_1(x)) = L_2(e^{rx}) = (r^2e^{rx}) - 2r(re^{rx}) + r^2(e^{rx}) = (r^2 - 2r^2 + r^2)e^{rx} = 0$, so $y_1(x)$ is a solution. For $y_2(x)$, we have $y_2'(x) = (d/dx)(xe^{rx}) = e^{rx} + rxe^{rx}$, and $y_2''(x) = 2re^{rx} + r^2xe^{rx}$, so that

$$\begin{aligned} L_2(y_2(x)) &= 2re^{rx} + r^2xe^{rx} - 2r(e^{rx} + rxe^{rx}) + r^2xe^{rx} \\ &= (2r + r^2x - 2r - 2r^2x + r^2x)e^{rx} \\ &= ((2r - 2r) + (r^2 - 2r^2 + r^2)x)e^{rx} \end{aligned}$$

$$
\begin{aligned}
&= (p'(r) + p(r)x)e^{rx} \\
&= (0 + 0x)e^{rx} \\
&= 0e^{rx} \\
&= 0.
\end{aligned}
$$

It remains to show that y_1 and y_2 are linearly independent. The Wronskian determinant

$$
W[y_1(x), y_2(x)] = \begin{vmatrix} e^{rx} & xe^{rx} \\ re^{rx} & e^{rx} + rxe^{rx} \end{vmatrix} = e^{2rx} \neq 0,
$$

so y_1 and y_2 are linearly independent. This idea looks ahead to what we will have to do to find missing solutions when we encounter the repeated-root situation in higher-dimensional settings.

The one remaining case is when our discriminant is negative, so that the two roots of the characteristic equation are complex conjugates.

Complex Numbers and Functions of Complex Numbers

When studying complex variables you find that most elementary manipulations with complex numbers behave just like they would with real numbers. For instance, addition is commutative and associative, as is multiplication, there are additive and multiplicative inverses (except for 0 which has no multiplicative inverse), $0z = 0$ for any z, and if $z_1 z_2 = 0$, then either $z_1 = 0$ or $z_2 = 0$. The differentiation rules for real functions all carry over to complex functions, and the formulas for the derivatives of familiar functions, such as $(d/dz)z^2 = 2z$, still look the same. The rules for derivative of a sum, product, and quotient, all look and work the same, as does the chain rule.

The formula that we need most is called **Euler's formula**[5] and it says that if q is real, then $e^{iq} = \cos q + i \sin q$. Thus $\mathrm{Re}(e^{iq}) = \cos q$, and $\mathrm{Im}(e^{iq}) = \sin q$. The functions Re and Im are the real part and imaginary part of their argument, respectively. Using them, $\mathrm{Re}(a + bi) = a$ and $\mathrm{Im}(a + bi) = b$. Since the length or modulus of the complex number $z = a + bi$ is defined to be $|z| = \sqrt{a^2 + b^2}$,

$$
\left|e^{iq}\right| = \sqrt{\cos^2 q + \sin^2 q} = 1, \text{when } q \text{ is real.}
$$

This means that all of the values of e^{iq} for real q lie on the unit circle in the complex plane. Furthermore, since the usual law of exponents $e^{z_1+z_2} = e^{z_1}e^{z_2}$ holds for complex z_1 and z_2, we have that $e^{a+bi} = e^a e^{bi} = e^a(\cos b + i \sin b)$. This is the key to representing the solutions corresponding to complex roots.

[5]Leonhard Euler (1707–1783), Switzerland's greatest scientist, and one of the greatest mathematicians who has ever lived. A prolific author, he averaged around 800 published pages per year of his life. He founded or expanded many of the present themes of mathematics. His name is attached to theorems from all branches of modern mathematics.

Discriminant < 0: Conjugate Complex Roots ($r_1, r_2 = a \pm bi$)

If $r_1 = a + bi$ is a complex root of the characteristic equation, the corresponding solution is

$$y(x) = e^{r_1 x} = e^{(a+bi)x} = e^{ax}(\cos bx + i \sin bx) = e^{ax}\cos bx + ie^{ax}\sin bx.$$

The reason this is so important to us is contained in Theorem 4.15.

Theorem 4.15 *If L is a real-valued linear function and the elements in the domain of L are all real-valued, then the domain of L can be enlarged to include elements that are complex-valued as follows: If u and v are in the domain of L (so they are real-valued), and $y = u + iv$, then $L(y) = L(u) + iL(v)$ and if $L(y) = 0$, then $L(u) = L(v) = 0$.*

Proof. The theorem follows from the calculation $L(y) = L(u + iv) = L(u) + iL(v) = 0$. Since the representation $0 = 0 + i0$ is unique and both $L(u)$ and $L(v)$ are real, equate coefficients to see that $L(u) = L(v) = 0$. □

The equations $L(u) = 0$ and $L(v) = 0$ are equations involving real-valued, not complex-valued objects. We use this discovery this way: If the characteristic equation corresponding to the differential equation $L(y) = 0$ has the pair $r_1, r_2 = \alpha \pm i\beta$ of conjugate complex roots, then even though $e^{r_1 x}$ and $e^{r_2 x}$ are complex-valued solutions, $y_1(x) = e^{\alpha x}\cos\beta x$ and $y_2(x) = e^{\alpha x}\sin\beta x$ are two linearly independent real solutions that span the same set of real-valued solutions that the complex-valued functions $e^{r_1 x}$ and $e^{r_2 x}$ span. We get the same real pair from r_2 as from r_1.

That the two real functions y_1 and y_2 are solutions follows from Theorem 4.15. Here is the proof that they are linearly independent.

$$\begin{aligned}
&W[c^{\alpha x}\cos\beta x, c^{\alpha x}\sin\beta x] \\
&= \begin{vmatrix} e^{\alpha x}\cos\beta x & e^{\alpha x}\sin\beta x \\ e^{\alpha x}\alpha\cos\beta x - e^{\alpha x}\beta\sin\beta x & e^{\alpha x}\beta\cos\beta x + e^{\alpha x}\alpha\sin\beta x \end{vmatrix} \\
&= e^{2\alpha x}\beta(\cos^2\beta x + \sin^2\beta x) \\
&= e^{2\alpha x}\beta \\
&\neq 0 \quad \text{when} \quad \beta \neq 0,
\end{aligned}$$

and $\beta \neq 0$ is necessary for the root $\alpha \pm i\beta$ to be complex.

Example 4.11M In *Mathematica* this proof may be done this way.

Solution.

```
In[60]:= y = {E^(a x) Cos[b x],E^(a x) Sin[b x]}

Out[60]=   a x              a x
         {E    Cos[b x], E     Sin[b x]}
```

```
In[61]:= w = {y, D[y,x]}

Out[61]=    a x              a x
         {{E    Cos[b x], E    Sin[b x]},

            a x                a x
          {E    a Cos[b x] - E    b Sin[b x],

            a x                a x
           E    b Cos[b x] + E    a Sin[b x]}}

In[62]:= Simplify[Det[w], Trig->True]

Out[62]=  2 a x
         E      b
```

This is nonzero if $b \neq 0$. ◇

Our findings are worth recording as a theorem, after which we will give some examples.

Theorem 4.16 *Suppose that the second-order linear differential operator $L(y) = ay'' + by' + cy$ has real coefficients with $a \neq 0$. The character of the roots of the characteristic equation $p(r) = ar^2 + br + c = 0$ determines the character of pairs of linearly independent solutions of the homogeneous equation $L(y) = 0$ in this way:*

1. *If $p(r) = 0$ has two different real roots, r_1 and r_2, then two solutions of $L(y) = 0$ are $y_1(x) = e^{r_1 x}$ and $y_2(x) = e^{r_2 x}$.*
2. *If $p(r) = 0$ has a double (real) root $r = r_1$, then two solutions of $L(y) = 0$ are $y_1(x) = e^{r_1 x}$ and $y_2(x) = xe^{r_1 x}$.*
3. *If $p(r) = 0$ has two conjugate complex roots $r_1, r_2 = \alpha \pm \beta i$, then two real solutions of $L(y) = 0$ are $y_1(x) = e^{\alpha x} \cos \beta x$ and $y_2(x) = e^{\alpha x} \sin \beta x$.*

The first case was illustrated in Example 4.1. Here are examples of the remaining two cases.

Example 4.12 Find two linearly independent solutions to the differential equation $L_2(y) = y'' + 4y' + 4y = 0$.

Solution. Here when we substitute $y = e^{rx}$, we find that $L_2(e^{rx}) = (r^2 + 4r + 4)e^{rx} = p(r)e^{rx}$. In order for this to be 0, $p(r) = r^2 + 4r + 4 = (r+2)^2 = 0$. This means that there is a double root $r = -2$. From the theory, $y_1(x) = e^{-2x}$ and $y_2(x) = xe^{-2x}$ are two solutions. So a complete description of the kernel is $y = c_1 e^{-2x} + c_2 xe^{-2x}$. ◇

Example 4.12M Solve the differential equation of example 4.10 in *Mathematica* and demonstrate the linear independence of the two solutions.

Solution.

```
In[63]:= DSolve[y''[x]+4y'[x]+4y[x] == 0, y[x], x]

Out[63]=             C[1] + x C[2]
         {{y[x] -> ---------------}}
                       2 x
                      E

In[64]:= soln[x_] = Expand[y[x]/.%[[1]]]

Out[64]= C[1]     x C[2]
         ----  +  ------
          2 x      2 x
         E        E

In[65]:= y1[x_] = Coefficient[soln[x],C[1]]

Out[65]=  -2 x
         E

In[66]:= y2[x_] = Coefficient[soln[x],C[2]]

Out[66]=  x
         ----
          2 x
         E

In[67]:= basis = {y1[x],y2[x]}

Out[67]=   -2 x   x
         {E    , ----}
                  2 x
                 E

In[68]:= WronskianDet = Det[{basis,D[basis,x]}]

Out[68]=  -4 x
         E
```

◇

Example 4.13 Completely describe the kernel of the differential equation $L_2(y) = y'' - 6y' + 25y = 0$.

Solution. Substitute $y = e^{rx}$, to get $L_2(e^{rx}) = (r^2 - 6r + 25)e^{rx} = p(r)e^{rx}$. In order for this to be 0, $p(r) = r^2 - 6r + 25 = 0$. The quadratic equation reveals the roots to be $r_1, r_2 = 3 \pm 4i$, and the theory in the third case, complex roots, says that $y_1(x) = e^{3x}\cos 4x$ and $y_2(x) = e^{3x}\sin 4x$ are linearly independent solutions. The kernel is completely described by the set of functions $y(x) = c_1e^{3x}\cos 4x + c_2e^{3x}\sin 4x$. ◇

Example 4.13M Solve the differential equation of Example 4.13 in *Mathematica* and demonstrate the linear independence of the two solutions.

Solution.

```
In[69]:= DSolve[y''[x]-6y'[x]+25y[x] == 0,y[x],x]

Out[69]=            3 x
         {{y[x] -> E    (C[2] Cos[4 x] - C[1] Sin[4 x])}}
```

```
In[70]:= soln[x_] = Expand[y[x]/.%[[1]]]

Out[70]= 3 x                  3 x
        E    C[2] Cos[4 x] - E    C[1] Sin[4 x]

In[71]:= y1[x_] = Coefficient[soln[x],C[1]]

Out[71]=    3 x
        -(E    Sin[4 x])

In[72]:= y2[x_] = Coefficient[soln[x],C[2]]

Out[72]= 3 x
        E    Cos[4 x]

In[73]:= basis = {y1[x],y2[x]}

Out[73]=     3 x                 3 x
        {-(E    Sin[4 x]), E    Cos[4 x]}

In[74]:= WronskianDet = Simplify[Det[{basis,D[basis,x]}]]

Out[74]=    6 x
        4 E
```

◇

These examples are worked and checked in *Mathematica* in the notebook named *Second-Order Linear.* This notebook develops each detail of the theory; it does not use `DSolve`. The three cell groupings in the notebook can be changed to have the notebook solve other problems for you. Just decide the case into which a problem falls and modify the operator defined for that case. Then execute the cells as a group or one at a time until the solution has been obtained and checked. Once you understand what must be done to solve problems like these, you can make regular use of `DSolve`.

Initial Value Problems

Quite often one seeks a solution of a differential equation that has special properties. Theorem 4.2 states that a linear initial value problem has a unique solution. Here is a sample *Mathematica* session that illustrates how to use *Mathematica* to solve initial-value problems. It also illustrates the behavior of a family of solutions that pass through a fixed point with unrestricted slopes, and a family that has specified slope when $x = 0$, but no specific starting point is given. Observe that in each case the solution is expressed in terms of a single parameter since only one of two initial conditions has been given. The differential equation in each case is that of Example 4.13.

Example 4.14M Solve the differential equation of Example 4.13 given the condition $y(0) = 1$.

Solution.

```
In[75]:= DSolve[{y''[x]-6y'[x]+25y[x] == 0, y[0] == 1},y[x],x]

Out[75]=              3 x
         {{y[x] -> E    (Cos[4 x] - C[1] Sin[4 x])}}
```

The C[1] is changed to c1 because C[1] cannot be used as an iterator in the Table function.

```
In[76]:= soln[x_] = Expand[y[x]/.%[[1]]]/.C[1]->c1

Out[76]= 3 x                  3 x
         E    Cos[4 x] - c1 E    Sin[4 x]

In[77]:= curves = Table[soln[x], {c1,-2,2}]

Out[77]=  3 x                 3 x
         {E    Cos[4 x] + 2 E    Sin[4 x],

           3 x              3 x               3 x
          E    Cos[4 x] + E    Sin[4 x], E    Cos[4 x],

           3 x              3 x
          E    Cos[4 x] - E    Sin[4 x],

           3 x                3 x
          E    Cos[4 x] - 2 E    Sin[4 x]}

In[78]:= Plot[Evaluate[curves],{x,-1,1},PlotRange->{-15,15}];
```


◇

Example 4.15M Solve the differential equation of Example 4.13 given the condition $y'(0) = 2$. No initial point is specified.

Solution.

```
In[79]:= DSolve[{y''[x]-6y'[x]+25y[x] == 0,  y'[0] == 2}, y[x], x]

Out[79]=
                  3 x                     (2 - 3 C[2]) Sin[4 x]
         {{y[x] -> E    (C[2] Cos[4 x] + ---------------------)}}
                                                    4
```

The C[2] is changed to c2 for the same reason as above.

```
In[80]:= soln[x_] = Expand[y[x]/.%[[1]]]/.C[2]->c2

Out[80]=
                             3 x               3 x
             3 x            E    Sin[4 x]   3 c2 E    Sin[4 x]
         c2 E    Cos[4 x] + -------------- - -----------------
                                  2                  4
```

```
In[81]:= curves = Table[soln[x], {c2,-2,2}]

Out[81]=
              3 x                  3 x
         {-2 E    Cos[4 x] + 2 E     Sin[4 x],

                                   3 x              3 x
             3 x              5 E     Sin[4 x]  E     Sin[4 x]
          -(E    Cos[4 x]) + ----------------, ---------------,
                                     4                2

                          3 x
           3 x           E    Sin[4 x]
          E    Cos[4 x] - -------------,
                               4

             3 x            3 x
          2 E    Cos[4 x] - E    Sin[4 x]}

In[82]:= Plot[Evaluate[curves],{x,-1,1.1}, PlotRange->{-25,25}];
```

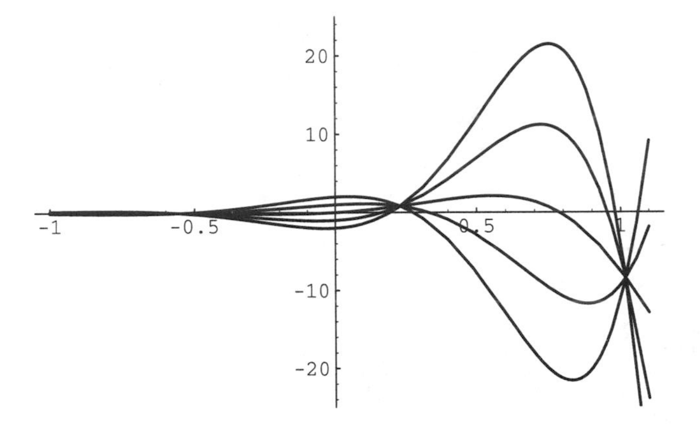

◇

In both of the plots it appears that all of the curves pass through a single point to the right of $x = 0$. In the first plot, all of the curves do pass through the point $(0, 1)$. Is it the case that there are other common points on the graphs? You can investigate these matters in the exercises.

EXERCISES 4.3

PART I. In each differential equation in problems 1–8, make the substitution $y = e^{rx}$ to find the characteristic polynomial. Solve the characteristic polynomial. Describe the kernel of the operator and write down a complete solution to the differential equation. Do this manually and by *Mathematica.*

1. $y'' - 4y' + 3y = 0$.
2. $y'' - 3y' = 0$.
3. $y'' + 4y' + 3y = 0$.
4. $y'' - 2y = 0$.
5. $y'' - 2y' + y = 0$.

6. $y'' + 2y' + y = 0.$

7. $y'' - 4y' + 13y = 0.$

8. $y'' - 2y' + 2y = 0.$

PART II. For each solution given in problems 1–8 above, evaluate the Wronskian determinant of the indicated basis for the kernel of the differential operator. Show that in each case, the Wronskian determinant satisfies the differential equation

$$W'(x) = -\left(\frac{\text{Coefficient}(y')}{\text{Coefficient}(y'')}\right) W(x),$$

where the coefficients come from the original differential equation. The problems in this part are problems 1b–8b.

PART III. For each of problems 1–8 above, find the solution for which $y(0) = 1, y'(0) = 0$. Name the solution to problem $i, y_i(x)$. Plot each of these solutions on a separate plot. The problems in this part are problems 1c–8c.

9. Plot all of the solutions $y_{1c}(x) - y_{8c}(x)$ on a common set of axes over the interval $-3 \leq x \leq 3$. In *Mathematica* this is done by the function call
```
Plot[Evaluate[y1c[x], y2c[x], y3c[x], y4c[x],
  y5c[x], y6c[x], y7c[x], y8c[x]], x,-3,3].
```
These functions have nothing in common but the initial condition.

PART IV. Theory.

10. Prove Theorem 4.14 by expanding $p(x)$ in a Taylor series about r and interpreting why some of the leading terms are missing. [The value $p(r)$, and the appropriate derivatives $p^{(k)}(r)$ are zero.] Observe that the remaining terms have a common factor.

11. In the two plots just before these exercises, it appears that all of the curves pass through a single point to the right of $x = 0$. Show that this is the case and find all of the points where the curves have a common crossing. The differential equation is

$$y'' - 6y' + 25y = 0.$$

 (a) In the first plot, all solutions pass through $y(0) = 1$.
 (b) In the second plot, all solution have slope 2 when they cross the y-axis: $y'(0) = 2$.

Each of these polynomials has a multiple root at the indicated point x_0. Find the order of multiplicity in each case. In *Mathematica* use the function `Factor`. Also in *Mathematica* expand the polynomial in a power series about x_0.

12. $-2 + 5x - 3x^2 - x^3 + x^4; \qquad x_0 = 1.$

13. $16 - 8x^2 + x^4; \qquad x_0 = 2.$

14. $81 + 81x - 18x^2 - 18x^3 + x^4 + x^5; \qquad x_0 = -3.$

15. $-81 - 27x + 54x^2 - 6x^3 - 5x^4 + x^5; \qquad x_0 = 3.$

4.4 Higher-Order Constant Coefficients (Homogeneous)

Recall that the Wronskian $W(y_1, y_2, \ldots, y_n)$ of a set of functions having $n-1$ continuous derivatives on some interval I is the determinant

$$W(y_1, y_2, \ldots, y_n) = \begin{vmatrix} y_1 & y_2 & \cdots & y_n \\ y_1' & y_2' & \cdots & y_n' \\ \vdots & \vdots & \ddots & \vdots \\ y_1^{(n-1)} & y_2^{(n-1)} & \cdots & y_n^{(n-1)} \end{vmatrix}.$$

Theorem 4.17 makes this very useful to us, both in theory and in practice.

Theorem 4.17 *Given a set $\{y_1, y_2, \ldots, y_n\}$ of functions having $n-1$ continuous derivatives on an interval I. If the functions are not linearly independent on I, then $W(y_1, y_2, \ldots, y_n)(x) = 0$ for every x in I. If the functions $\{y_1, y_2, \ldots, y_n\}$ are all solutions of the same linear homogeneous differential equation, then $W(y_1, y_2, \ldots, y_n) \neq 0$ on I if and only if the set $\{y_1, y_2, \ldots, y_n\}$ is linearly independent.*

Proof. Theorem 4.17 follows from Theorem 4.6. The proof requires several properties of determinants. It is not difficult to understand, but it is complicated to write out. It was outlined in exercises 4.1. □

Example 4.16 The functions $\cos x$ and $\sin x$, two solutions of the linear differential equation $y'' + y = 0$, have the Wronskian $W[\cos x, \sin x] = 1$, which means that they are linearly independent. On the other hand, $W[\cos(x - \pi/2), \sin x] = 0$, indicates that the functions $\cos(x - \pi/2)$ and $\sin x$, which are solutions of the same differential equation, $y'' + y = 0$, are linearly dependent. In fact, $\cos(x - \pi/2) = \sin x$. ◇

Theorem 4.18 gives the relationship between the order of the differential operator and the dimension of its kernel.

Theorem 4.18 *The dimension of the kernel of an nth-order linear differential operator is n. That is, suppose that I is an interval and the functions $a_n(x), a_{n-1}(x), \ldots, a_1(x), a_0(x)$ are continuous on I and $a_n(x)$ is nonzero on I, then a homogeneous linear differential equation of the form*

$$L_n(y) = a_n(x)\frac{d^n y}{dx^n} + a_{n-1}(x)\frac{d^{n-1} y}{dx^{n-1}} + \cdots + a_1(x)\frac{dy}{dx} + a_0(x)y = 0$$

has n linearly independent solutions that form a basis for the set of all solutions on I.

Here is an example of the use of the Wronskian to define a differential equation with known solutions. This is rarely done manually because the computations can be so involved, but we can easily use *Mathematica* to help us.

Example 4.17 Suppose that we want a linear differential equation whose solutions are the span of the two functions $y_1(x) = x$ and $y_2(x) = e^x$. Consider

$$\begin{aligned} W[x, e^x, y] &= \begin{vmatrix} x & e^x & y \\ 1 & e^x & y' \\ 0 & e^x & y'' \end{vmatrix} \\ &= e^x \begin{vmatrix} x & 1 & y \\ 1 & 1 & y' \\ 0 & 1 & y'' \end{vmatrix} \\ &= e^x[(x-1)y'' - xy' + y]. \end{aligned}$$

Then $W[x, e^x, y] = 0$ provided we choose as our differential equation $L(y) = (x-1)y'' - xy' + y = 0$.

You may wish to verify that both $y_1(x) = x$ and $y_2(x) = e^x$ are solutions. The kernel is then the set of their linear combinations. ◇

We often want to find a differential equation whose solutions are the span of the linearly independent set of functions $\{y_1(x), y_2(x), \ldots, y_n(x)\}$. This is an easy task, now that we have seen the last example. Up to a nonzero factor, an equation that has these properties is simply $L_n(y) = W(y_1, y_2, \ldots, y_n, y) = 0$. Why? Because W(collection of functions) $= 0$ is a condition for the linear dependence of the functions in the collection, and we want y to depend on $y_1, y_2, \ldots, y_n$. For certain choices of $y_1, y_2, \ldots, y_n$, the differential equation may even have constant coefficients. See the exercises.

Higher-Order Differential Equations with Constant Coefficients

We now state the theorems that enable us to completely describe the kernel of any higher order differential equation having constant coefficients. This discussion goes well in theory, but N.H. Abel showed that it is not possible to solve polynomial equations of degree five or greater in terms of radicals (like the quadratic formula, for instance). This means that we cannot expect *Mathematica* to be able to give us theoretically exact roots to every higher order polynomial equation that we try to solve. *Mathematica* can give us numerical answers, and we often will have to be content with these unless we or someone else has contrived the problem so that the roots can be found in terms of radicals. If you have ever seen the formal solution of the cubic or the quartic, you will know that sometimes a theoretically exact solution is so incredibly complicated as to be essentially useless. We will not hesitate to seek numerical roots of polynomials from *Mathematica* in either of these cases.

Distinct Real Roots

Theorem 4.19 *If the characteristic polynomial of the nth-order linear differential equation $L_n(y) = a_n y^{(n)} + a_{n-1} y^{(n-1)} + \cdots + a_1 y' + a_0 y = 0$ has n distinct real roots $r_1, r_2, \ldots, r_n$, then the n functions $y_1(x) = e^{r_1 x}, y_2(x) = e^{r_2 x}, \ldots, y_n(x) = e^{r_n x}$ form a basis for the kernel of the differential equation $L_n(y) = 0$. In this case, every solution of $L_n(y) = 0$ can be written as*

$$y(x) = c_1 e^{r_1 x} + c_2 e^{r_2 x} + \cdots + c_n e^{r_n x}.$$

Proof. This is proved the same way that Theorem 4.16 was proved for the case of two functions. □

Example 4.18 The third-order constant coefficients differential equation

$$L_3(y) = y''' - 6y'' + 11y' - 6y = 0$$

has characteristic polynomial $p(r) = r^3 - 6r^2 + 11r - 6 = (r-1)(r-2)(r-3)$, which has the distinct roots $r_1 = 1, r_2 = 2$ and $r_3 = 3$. The three linearly independent solutions are $y_1(x) = e^{r_1 x} = e^x, y_2(x) = e^{r_2 x} = e^{2x}$, and $y_3(x) = e^{r_3 x} = e^{3x}$, so every solution in the kernel has the form $y(x) = c_1 e^x + c_2 e^{2x} + c_3 e^{3x}$. ◇

Example 4.18M We can reproduce these steps in *Mathematica* this way:
Solution.

```
In[83]:= L[x_,y_] = y'''[x]-6y''[x]+11y'[x]-6y[x]

Out[83]=                                    (3)
         -6 y[x] + 11 y'[x] - 6 y''[x] + y    [x]

In[84]:= p[r_] = Coefficient[L[x, Function[t,Exp[r t]]],Exp[r x]]

Out[84]=              2    3
         -6 + 11 r - 6 r  + r

In[85]:= roots = r/.Solve[p[r] == 0,r]

Out[85]= {3, 2, 1}

In[86]:= solns = Map[Function[m,Exp[m x]],roots]

Out[86]=   3 x   2 x   x
         {E   , E   , E }

In[87]:= y[x_] = solns.{c3,c2,c1}

Out[87]=  x        2 x         3 x
         E  c1 + E     c2 + E     c3
```

These same results can be obtained directly using the built-in function `DSolve`.

```
In[88]:= Clear[x,y]
         L[x_,y_] = y'''[x]-6y''[x]+11y'[x]-6y[x]

Out[88]=                                    (3)
         -6 y[x] + 11 y'[x] - 6 y''[x] + y    [x]
```

```
In[89]:= DSolve[L[x,y] == 0,y[x],x]

Out[89]=
                x          x          2 x
         {{y[x] -> E  (C[1] + E  C[2] + E    C[3])}}

In[90]:= y[x_] = Expand[y[x]/.%[[1]]]

Out[90]= x            2 x            3 x
         E  C[1] + E     C[2] + E     C[3]
```

A basis for the kernel of L can be obtained simply this way:

```
In[91]:= Table[Coefficient[y[x],C[i]],{i,1,3}]

Out[91]=  x    2 x    3 x
         {E , E    , E   }
```

◇

Repeated Real Roots

Theorem 4.20 *If the characteristic polynomial $p(r)$ of the nth-order linear differential equation $L_n(y) = a_n y^{(n)} + a_{n-1}y^{(n-1)} + \cdots + a_1 y' + a_0 y = 0$ has a single root r repeated n times, then the n functions $y_1(x) = e^{rx}, y_2(x) = xe^{rx}, \ldots, y_n(x) = x^{n-1}e^{rx}$ form a basis for the kernel of the differential equation $L_n(y) = 0$.*

In this case, every solution of $L_n(y) = 0$ can be written as

$$y(x) = (c_1 + c_2 x + \cdots + c_n x^{n-1})e^{rx}.$$

If the characteristic polynomial has a single root r repeated m times and $n-m$ distinct roots $r_{m+1}, r_{m+2}, \ldots, r_n$, then the n functions $y_1(x) = e^{rx}, y_2(x) = xe^{rx}, \ldots, y_m(x) = x^{m-1}e^{rx}; y_{m+1}(x) = e^{r_{m+1}x}, \ldots, y_n(x) = e^{r_n x}$ form a basis for the kernel of the differential equation $L_n(y) = 0$, and every solution can be written in the form

$$y(x) = (c_1 + c_2 x + \cdots + c_m x^{n-1})e^{rx} + c_{m+1}e^{r_{m+1}x} + \cdots + c_n e^{r_n x}.$$

If the characteristic polynomial has a single root r repeated m times and other roots that are also repeated, then the other roots are handled just as the root r was, and the process is continued.

Proof. This is proved the same way that Theorem 4.16 was proved for the case of two functions. The essential observation is that derivatives with respect to r and with respect to x can be interchanged. This, combined with the representation of polynomials having repeated roots, does the job. □

Example 4.19 The fifth-order constant coefficients differential equation

$$L_5(y) = y^{(5)} - 10y^{(4)} + 39y''' - 74y'' + 68y' - 24y = 0$$

has characteristic polynomial $p(r) = r^5 - 10r^4 + 39r^3 - 74r^2 + 68r - 24 = (r-1)(r-3)(r-2)^3$, which has the distinct roots $r_1 = 1, r_2 = 3$ and the repeated root $r_3 = r_4 = r_5 = 2$, having multiplicity 3. The five linearly independent solutions are $y_1(x) = e^{r_1 x} = e^x, y_2(x) = e^{r_2 x} = e^{3x}, y_3(x) = e^{r_3 x} = e^{2x}, y_4(x) = xe^{r_3 x} = xe^{2x}$, and $y_5(x) = x^2 e^{r_3 x} = x^2 e^{2x}$, so every solution in the kernel has the form

$$y(x) = c_1 e^x + c_2 e^{3x} + c_3 e^{2x} + c_4 x e^{2x} + c_5 x^2 e^{2x}.$$

Use the Wronskian to check these five functions for linear independence. ◇

Example 4.19M Do Example 4.19 in *Mathematica*.

Solution.

```
In[92]:= Clear[x,y]
         L[x_,y_] = y'''''[x]-10y''''[x]+39y'''[x]-74y''[x]+
                        68y'[x]-24y[x]

Out[92]=                                              (3)
         -24 y[x] + 68 y'[x] - 74 y''[x] + 39 y    [x] -

               (4)        (5)
           10 y    [x] + y    [x]

In[93]:= DSolve[L[x,y] == 0,y[x],x]

DSolve::dsdeg:
   Warning: Differential equation of order higher than four
     encountered. DSolve may not be able to find the
     solution.

Out[93]=              x          x          x             x  2
         {{y[x] -> E  (C[1] + E  C[2] + E  x C[3] + E  x  C[4] +

                 2 x
                E    C[5])}}

In[94]:= y[x_] = Expand[y[x]/.%[[1]]]

Out[94]=  x          2 x         2 x            2 x  2
         E  C[1] + E    C[2] + E    x C[3] + E    x  C[4] +

            3 x
           E    C[5]

In[95]:= basis = Table[Coefficient[y[x],C[i]],{i,1,5}]

Out[95]=   x   2 x   2 x      2 x  2   3 x
         {E , E   , E    x, E    x , E   }

In[96]:= Factor[Coefficient[L[x,Function[t,Exp[r t]]],Exp[r x]]]

Out[96]=                     3
         (-3 + r) (-2 + r)  (-1 + r)
```

Note how this function calculates successive derivatives of a function.

```
In[97]:= W[x_] = NestList[Function[t,D[t,x]],z[x],4]

Out[97]=                           (3)       (4)
         {z[x], z'[x], z''[x], z    [x], z    [x]}
```

When applied to the list of functions, basis, the Wronskian matrix is produced:

```
In[98]:= W[x_] = NestList[Function[t,D[t,x]],basis,4]

Out[98]=    x   2 x   2 x      2 x  2   3 x
         {{E , E   , E    x, E    x , E   },

            x     2 x   2 x        2 x        2 x          2 x  2
          {E , 2 E   , E    + 2 E    x, 2 E    x + 2 E    x ,

             3 x     x     2 x     2 x      2 x
           3 E   }, {E , 4 E   , 4 E    + 4 E    x,

             2 x      2 x        2 x  2    3 x
           2 E    + 8 E    x + 4 E    x , 9 E   },

            x     2 x      2 x      2 x
          {E , 8 E   , 12 E    + 8 E    x,

              2 x       2 x        2 x  2     3 x
           12 E    + 24 E    x + 8 E    x , 27 E   },

            x      2 x      2 x       2 x
          {E , 16 E   , 32 E    + 16 E    x,

              2 x       2 x         2 x  2     3 x
           48 E    + 64 E    x + 16 E    x , 81 E   }}
```

The Wronskian determinant is the determinant of the Wronskian matrix.

```
In[99]:= Det[W[x]]

Out[99]=    10 x
         4 E
```

◇

Example 4.20 The fifth-order constant coefficients differential equation

$$L_5(y) = y^{(5)} - 8y^{(4)} + 25y''' - 38y'' + 28y' - 8y = 0$$

has characteristic polynomial

$$p(r) = r^5 - 8r^4 + 25r^3 - 38r^2 + 28r - 8 = (r-1)^2(r-2)^3,$$

which has the repeated roots $r_1 = r_2 = 1$ and $r_3 = r_4 = r_5 = 2$, having multiplicities 2 and 3, respectively. The five linearly independent solutions are $y_1(x) = e^{r_1 x} = e^x, y_2(x) = xe^{r_1 x} = xe^x, y_3(x) = e^{r_3 x} = e^{2x}, y_4(x) = xe^{r_3 x} = xe^{2x}$, and $y_5(x) = x^2 e^{r_3 x} = x^2 e^{2x}$, so every solution in the kernel has the form

$$y(x) = c_1 e^x + c_2 x e^x + c_3 e^{2x} + c_4 x e^{2x} + c_5 x^2 e^{2x}.$$

Use the Wronskian to check these five functions for linear independence. ◇

Example 4.20M Example 4.20 can be worked in *Mathematica* this way.

Solution.

```
In[100]:=Clear[x,y]
         L[x_,y_] = y'''''[x]-8y''''[x]+25y'''[x]-
                        38y''[x]+ 28y'[x]-8y[x]

Out[100]=
                                                       (3)           (4)
         -8 y[x] + 28 y'[x] - 38 y''[x] + 25 y     [x] - 8 y     [x] +

            (5)
          y    [x]

In[101]:=DSolve[L[x,y] == 0,y[x],x]

DSolve::dsdeg:
   Warning: Differential equation of order higher than four
     encountered. DSolve may not be able to find the
     solution.

Out[101]=
                          x                    x          x
         {{y[x] -> E  (C[1] + x C[2] + E  C[3] + E  x C[4] +

                  x  2
                 E  x  C[5])}}

In[102]:=y[x_] = Expand[y[x]/.%[[1]]]

In[103]:= x          x            2 x          2 x
         E  C[1] + E  x C[2] + E    C[3] + E    x C[4] +

           2 x  2
          E    x  C[5]

In[104]:=basis = Table[Coefficient[y[x],C[i]],{i,1,5}]

Out[104]= x    x       2 x    2 x      2 x  2
         {E , E  x, E    , E    x, E    x }

In[105]:=Factor[Coefficient[L[x,Function[t,Exp[r t]]], Exp[r x]]]

Out[105]=          3          2
         (-2 + r)  (-1 + r)

In[106]:=W[x_] = NestList[Function[t,D[t,x]],basis,4]

Out[106]=  x    x       2 x    2 x      2 x  2
         {{E , E  x, E    , E    x, E    x },

            x    x    x         2 x   2 x        2 x
          {E , E  + E  x, 2 E    , E    + 2 E    x,

             2 x          2 x  2
           2 E    x + 2 E    x },

            x      x    x         2 x      2 x        2 x
          {E , 2 E  + E  x, 4 E    , 4 E    + 4 E    x,

             2 x       2 x          2 x  2
           2 E    + 8 E    x + 4 E    x },

            x      x    x         2 x       2 x        2 x
          {E , 3 E  + E  x, 8 E    , 12 E    + 8 E    x,

              2 x         2 x          2 x  2
           12 E    + 24 E    x + 8 E    x },
```

```
  x     x     x          2 x        2 x          2 x
{E , 4 E  + E  x, 16 E    , 32 E    + 16 E    x,

     2 x          2 x            2 x  2
  48 E    + 64 E    x + 16 E    x }}
```

```
In[107]:=Det[W[x]]

Out[107]=   8 x
          2 E
```

◇

We are only beginning to take advantage of the considerable power built into *Mathematica* for finding these solutions.

Complex Roots

We use the theory from the section on second-order equations as it applies in this case. The proofs are essentially the same, but some of the ideas are combined. We still take a pair of complex-valued functions and break them into their real and imaginary parts, which we proceed to retain as the functions we seek. Repeated roots are handled just like they were in the preceding case. You might think of using Theorem 4.20 on a complex-valued exponential and then breaking the product into its real and imaginary parts which we want here.

Theorem 4.21 *If the characteristic polynomial $p(r)$ of the nth-order linear differential equation $L_n(y) = a_n y^{(n)} + a_{n-1}y^{(n-1)} + \cdots + a_1 y' + a_0 y = 0$ is real and has a pair of complex roots $r = a \pm bi$, then the functions $y_1(x) = e^{ax}\cos bx$ and $y_2(x) = e^{ax}\sin bx$ are linearly independent solutions of the differential equation $L_n(y) = 0$.*

If the characteristic polynomial has a pair of complex conjugate roots $r = a \pm bi$ repeated m times, then the $2m$ functions

$$y_1(x) = e^{ax}\cos bx, y_2(x) = xe^{ax}\cos bx, \ldots, y_m(x) = x^{m-1}c^{ax}\cos bx,$$

$$y_{m+1}(x) = e^{ax}\sin bx, y_{m+2}(x) = xe^{ax}\sin bx, \ldots, y_{2m}(x) = x^{m-1}e^{ax}\sin bx$$

are linearly independent solutions of the differential equation $L_n(y) = 0$, and every function of the form

$$y(x) = (c_1 + c_2 x + \cdots + c_m x^{m-1})e^{ax}\cos bx$$
$$+ (c_{m+1} + c_{m+2}x + \cdots + c_{2m}x^{m-1})e^{ax}\sin bx.$$

is a solution.

If the characteristic polynomial has a complex conjugate pair repeated m times and other roots that are either distinct or repeated, then the other roots are handled individually as the theory suggests, and the process is continued.

Example 4.21M Solve $y'''' - 8y''' + 74y'' - 232y' + 841y = 0$ completely and evaluate the Wronskian determinant.

Solution.

```
In[108]:=Clear[x,y]
         L[x_,y_] = y''''[x]-8y'''[x]+74y''[x]-232y'[x]+841y[x]

Out[108]=
                                                    (3)         (4)
         841 y[x] - 232 y'[x] + 74 y''[x] - 8 y    [x] + y    [x]

In[109]:=DSolve[L[x,y] == 0,y[x],x]

Out[109]=            2 x
         {{y[x] -> E     (C[2] Cos[5 x] + x C[4] Cos[5 x] -

                C[1] Sin[5 x] - x C[3] Sin[5 x])}}

In[110]:=y[x_] = Expand[y[x]/.%[[1]]]

Out[110]= 2 x                    2 x
         E    C[2] Cos[5 x] + E    x C[4] Cos[5 x] -

           2 x                2 x
          E    C[1] Sin[5 x] - E    x C[3] Sin[5 x]

In[111]:=basis = Table[Coefficient[y[x],C[i]],{i,1,4}]

Out[111]=    2 x                 2 x                 2 x
         {-(E    Sin[5 x]), E    Cos[5 x], -(E    x Sin[5 x]),

           2 x
          E    x Cos[5 x]}

In[112]:=Factor[Coefficient[L[x,Function[t,Exp[r t]]], Exp[r x]]]

Out[112]=                 2 2
         (29 - 4 r + r )

In[113]:=W[x_] = NestList[Function[t,D[t,x]],basis,3];
         (* Suppress output *)

In[114]:=Det[W[x]]

Out[114]=       8 x           4      8 x          2          2
         2500 (E    Cos[5 x]  + 2 E    Cos[5 x]  Sin[5 x]  +

             8 x          4
            E    Sin[5 x] )

In[115]:=Simplify[Det[W[x]]]

Out[115]=      8 x
         2500 E
```

◇

Solving higher-order constant coefficients linear differential equations is easy in theory. With the help of *Mathematica* even very large problems are not difficult, so long as they have been properly constructed. If the characteristic equation does not have exact solutions representable by radicals, then numerical solutions may have to suffice. However, the formal character of the solutions can change dramatically when solutions have to be obtained numerically, but formulas are still desired. If all that is needed is an accurate picture or a list of reliable points that lie on a solution, then `NDSolve` may give you what you need. On the other hand you can inadvertently cause messy roots by copying the operator incorrectly. It is wise to verify the operator.

Using DKernel.m

There is a package that you can read in that readily finds the kernel for essentially any linear differential equation with constant numeric coefficients. It works by following the theory that has been outlined. Example 4.22 shows how to use DKernel.m . You will notice that DKernel.m returns a basis for the kernel. If you want solutions, you can form linear combinations by Dotting with a vector of constants. The first cell loads the package DKernel.m .

Example 4.22M Load DKernel.m and find the kernel of the linear differential equation $L(y) = y^{(4)} - 8y^{(3)} + 74y^{(2)} - 232y' + 841y = 0$.

Solution.

```
In[116]:=<<DKernel.m (* the actual location of this package *)
                     (* depends on where your system supervisor *)
                     (* put it.)

In[117]:=Clear[y]
         L[x_,y_] = y''''[x]-8y'''[x]+74y''[x]-232y'[x]+841y[x]

Out[117]=
                                                      (3)       (4)
         841 y[x] - 232 y'[x] + 74 y''[x] - 8 y    [x] + y    [x]

In[118]:=DKernel[L[x,y] == 0,y[x],x]

Out[118]= 2x          2x           2x                2x
         {E   Cos[5x], E   Sin[5x], E   x Cos[5x], E   x Sin[5x]}
```

◇

This is simple, fast and direct. DKernel handles integer or rational coefficients, as well as reals, but not differential equations that contain a parameter in the coefficients. Both DSolve and DKernel may have trouble with differential equations of order five or greater. This is because of Abel's theorem which says that polynomial equations of degree five and above are not all solvable in radicals. Even in cases where an exact solution can be obtained, it may be too complicated for you to be able to grasp. This is another situation where NDSolve has great value.

The form and order of the parameters to DKernel is the same as for DSolve. DKernel.m does not accept initial conditions (they are not needed).

Exercises 4.4

Part I. Use the determinant method of Example 4.17 to find a differential equation of second-order whose kernel has as basis the two given functions. Do this by *Mathematica*. Simplify your differential equation. Note that sometimes the differential equation has constant coefficients and sometimes not. The determinant you want may be generated by the function call

```
Det[NestList[Function[t, D[t, x]], y1[x], y2[x], y[x], 2]].
```

(Remove the `Det[` and the last `]` to see what `NestList` does.)

1. $y_1(x) = x^3$; $\quad y_2(x) = x^2$.
2. $y_1(x) = e^{2x}$; $\quad y_2(x) = e^{-3x}$.
3. $y_1(x) = e^x \cos 4x$; $\quad y_2(x) = e^x \sin 4x$.
4. $y_1(x) = e^{2x}$; $\quad y_2(x) = xe^{3x}$.

You can use this method in Chapter 7 to generate problems with nonconstant coefficients that have recognizable solutions. Chapter 7 concentrates on differential equations with nonconstant coefficients.

PART II. Modify the function call of problems 1–4 to find a differential equation of third-order whose kernel has as basis the given functions. The determinant you want may be generated by the function call

```
Det[NestList[Function[t, D[t, x]], y1[x], y2[x], y3[x], y[x], 3]].
```

5. $y_1(x) = x^3$; $\quad y_2(x) = x^2$; $\quad y_3(x) = x$.
6. $y_1(x) = e^{2x}$; $\quad y_2(x) = e^{-3x}$; $\quad y_3(x) = e^{-2x}$.
7. $y_1(x) = e^x \cos 4x$; $\quad y_2(x) = e^x \sin 4x$; $\quad y_3(x) = x^2$.
8. $y_1(x) = e^{2x}$; $\quad y_2(x) = xe^{3x}$; $\quad y_3(x) = x$.

You may need to `Simplify` in order to make your operator more reasonable.

PART III. Applying initial conditions to problems 2–4 .

For each of problems 2–4 above, find a linear combination of the given functions such that $y(0) = 1, y'(0) = 0$. Name the solution to problem $i, z_i(x)$. Plot each of these functions on a separate plot. **These are problems 2b–4b.** (Is it possible to find a combination of the functions in problem 1 which satisfies these initial conditions?)

PART IV. Applying initial conditions to problems 6–8.

For each of problems 6–8 above, find a linear combination of the given functions such that $y(0) = 1, y'(0) = 0, y''(0) = 0$. Name the solution to problem $i, u_i(x)$. Plot each of these functions on a separate plot. **These are problems 6b–8b.** (Is it possible to find a combination of the functions in problem 5 which satisfies these initial conditions?)

PART V. Solving homogeneous differential equations which have constant coefficients.

For the following differential equations, find a basis for the kernel and a description of all solutions. You may find that the package `DKernel.m` gives you the kernel in the most straightforward manner. (It goes step-by-step through the theory and merely does what you would do by hand.) You may get partial help from *Mathematica* by using `Solve` to find the roots of a manually supplied characteristic polynomial (which you can write down by inspection).

9. $y''' + y'' - 2y = 0.$
10. $y''' - 6y'' - y' + 30y = 0.$
11. $2y''' - y'' - 200y' + 100y = 0.$
12. $y''' + y'' - 3y' - 3y = 0.$
13. $y''' - 18y'' + 65y = 0.$
14. $y^{(4)} - 19y''' + 81y'' - 29y' - 130y = 0.$
15. $2y^{(5)} - 33y^{(4)} + 147y''' - 158y'' - 60y' = 0.$
16. $y^{(6)} + 2y^{(5)} - 425y^{(4)} + 1450y''' + 22484y'' - 47432y' = 0.$
17. $y''' - 5y'' + 8y' - 4y = 0.$
18. $y''' + 15y'' + 75y' + 125y = 0.$
19. $y^{(4)} - 2y''' - 12y'' + 40y' - 32y = 0.$
20. $y^{(5)} - 15y^{(4)} + 75y''' - 125y'' = 0.$
21. $y^{(8)} + 7y^{(7)} + 6y^{(6)} - 50y^{(5)} - 95y^{(4)} + 63y''' + 216y'' + 108y' = 0.$
22. $y'' - 4y' + 13y = 0.$
23. $y^{(4)} - 8y''' + 56y'' - 192y' + 468y = 0.$
24. $y^{(4)} - 6y''' + 54y'' - 126y' + 493y = 0.$
25. $y^{(6)} - 20y^{(5)} + 224y^{(4)} - 1616y''' + 7668y'' - 23376y' + 35360y = 0.$
26. $y^{(4)} - 12y''' + 87y'' - 380y' + 600y = 0.$
27. $y^{(8)} - 12y^{(7)} + 202y^{(6)} - 636y^{(5)} + 5829y^{(4)} + 24{,}280y''' + 15{,}200y'' + 696{,}000y' + 1{,}800{,}000y = 0$

4.5 The Method of Undetermined Coefficients

Now that we are able to completely describe the kernel of a homogeneous linear differential equation with constant coefficients, our next task is to begin to solve nonhomogeneous problems. Theorem 4.3 told us that nonhomogeneous problems have solutions and it even told us the form those solutions would take. At this point all we are missing is some method of actually finding the particular solutions we know to exist.

There are two approaches we can take. One is appealing because it appears to be simple, but solves only certain special kinds of problems, and one is completely general, but appears to be complicated. The simple technique is known as the method of **undetermined coefficients**. It is the subject of this section. The other method is the method of **variation of parameters**. It will be covered in the next section.

An example will set the tone of the discussion. Here we will use the fact that if $L(y)$ is a linear differential operator with constant coefficients, then $L(\text{polynomial}) = \text{polynomial}$. In other words constant coefficients operators take polynomials into polynomials. This suggests that if the function on the right hand side of the equation is a polynomial, we might profitably guess that $y_p(x)$, the particular solution we want, is a polynomial.

Example 4.23 Find a particular solution to the nonhomogeneous differential equation $L(y) = y'' - 5y' + 6y = 12x^3 - 30x^2 - 18x + 73$.

Solution. We observe that $L(\text{cubic polynomial}) = \text{cubic polynomial}$. (Why?) It therefore seems reasonable to assume that

$$y_p(x) = ax^3 + bx^2 + cx + d,$$

the most general cubic. We want to find out what a, b, c, and d should be in order for

$$L(y_p)(x) = 12x^3 - 30x^2 - 18x + 73.$$

Substituting the cubic $y_p(x)$ into the operator yields the cubic

$$\begin{aligned}(6ax + 2b) &- 5(3ax^2 + 2bx + c) + 6(ax^3 + bx^2 + cx + d)\\ &= 6ax^3 + (-15a + 6b)x^2 + (6a - 10b + 6c)x + (2b - 5c + 6d).\end{aligned}$$

For this to agree with the right-hand side,

$$12x^3 - 30x^2 - 18x + 73$$

we must equate coefficients, which gives the equations

$$\begin{aligned}(x^3):&\quad 6a = 12\\ (x^2):&\quad -15a + 6b = -30\\ (x):&\quad 6a - 10b + 6c = -18\\ (1):&\quad 2b - 5c + 6d = 0\end{aligned}$$

These equations are easily solved from the top down, giving $a = 2, b = 0, c = -5$, and $d = 8$.

Thus the particular solution we have found is $y_p(x) = 2x^3 - 5x + 8$. You are encouraged to check it.

The kernel for $L(y) = y'' - 5y' + 6y$ was obtained in example 4.10, so here is a complete solution of the problem:

$$\begin{aligned} y(x) &= c_1 e^{2x} + c_2 e^{3x} + y_p(x) \\ &= c_1 e^{2x} + c_2 e^{3x} + 2x^3 - 5x + 8. \end{aligned}$$

This is a solution of the nonhomogeneous equation for every choice of c_1 and c_2. ◇

Example 4.23M We could have done Example 4.23 this way in *Mathematica*.

Solution.

```
In[119]:=Clear[L,x,y]
         L[x_,y_] = y''[x]-5y'[x]+6y[x];
         rhs = 12x^3-30x^2-18x+73;
```

Verify that this is correct.

```
In[120]:=L[x,y] == rhs

Out[120]=
                                               2        3
         6 y[x] - 5 y'[x] + y''[x] == 73 - 18 x - 30 x  + 12 x

In[121]:=assume[x_] = a x^3 + b x^2 + c x + d

Out[121]=
                      2      3
         d + c x + b x  + a x

In[122]:=result = Collect[L[x,assume],x]

Out[122]=
                                                              2
         2 b - 5 c + 6 d + (6 a - 10 b + 6 c) x + (-15 a + 6 b) x  +

              3
           6 a x

In[123]:=eqn = CoefficientList[result,x] == CoefficientList[rhs,x]

Out[123]={2 b - 5 c + 6 d, 6 a - 10 b + 6 c, -15 a + 6 b, 6 a} ==

           {73, -18, -30, 12}

In[124]:=yp[x_] = assume[x]/.Solve[eqn][[1]]

Out[124]=
                    3
         8 - 5 x + 2 x
```
◇

It is the possibility that one can guess the form of a solution and then actually find the solution by solving a system of linear algebraic equations (not a hard task) that makes this so appealing. The problem is that sometimes it is difficult to guess a proper form to assume for the proposed solution. We will see by example that this is so. Theorem 4.24 will say exactly what to guess and why. When working with the class of problems where this method is appropriate, the theory says that all you need to know is how to solve homogeneous linear differential equations with constant coefficients. It is reasonable to claim that you know how to do this since we just finished in the last section. We first need to develop some supplementary theory.

Composition of Linear Differential Operators

Definition 4.10 *The* **composition** L_2L_1 *of the linear differential operators* L_2 *and* L_1 *is defined to be* $(L_2L_1)(y) = L_2(L_1(y))$ *for any function* y *that meets the combined differentiability requirements of the composition.*

It is important to know the order of the composition of two linear differential operators.

Theorem 4.22 *If the linear differential operators* L_2 *and* L_1 *have orders* m *and* n, *respectively, then the order of either composition of* L_1 *and* L_2 *is* $m+n$.

If L_2 *and* L_1 *have constant coefficients, then* $L_1L_2 = L_2L_1$, *but this is not necessarily true if either operator has variable coefficients.*

The proof of the first part is an exercise in manipulating summations, as is the commutativity. An example suffices to show noncommutativity. (Find such an example. First-order will do.)

Here is an example of the commutativity and the sum of orders properties. It is repeated in *Mathematica.*

Example 4.24 Given the constant coefficient differential operators

$$L_3(y) = a_3y''' + a_2y'' + a_1y' + a_0y$$

and

$$L_2(y) = b_2y'' + b_1y' + b_0y$$

show that $L_3(L_2(y)) = L_2(L_3(y))$ for all functions y that have five continuous derivatives.

Solution.

$$\begin{aligned} L_3(L_2(y)) &= a_3(b_2y'' + b_1y + b_0y)''' + a_2(b_2y'' + b_1y' + b_0y)'' \\ &\quad + a_1(b_2y'' + b_1y' + b_0y)' + a_0(b_2y'' + b_1y' + b_0y) \\ &= a_0b_0y + (a_1b_0 + a_0b_1)y' + (a_2b_0 + a_1b_1 + a_0b_2)y'' \\ &\quad + (a_3b_0 + a_2b_1 + a_1b_2)y^{(3)} + (a_3b_0 + a_2b_1)y^{(4)} + a_3b_2y^{(5)}. \end{aligned}$$

Similarly,

$$\begin{aligned} L_2(L_3(y)) &= b_2(a_3y''' + a_2y'' + a_1y' + a_0y)'' \\ &\quad + b_1(a_3y''' + a_2y'' + a_1y' + a_0y)' + b_0(a_3y''' \\ &\quad + a_2y'' + a_1y' + a_0y) \\ &= a_0b_0y + (a_1b_0 + a_0b_1)y' + (a_2b_0 + a_1b_1 + a_0b_2)y'' \\ &\quad + (a_3b_0 + a_2b_1 + a_1b_2)y^{(3)} + (a_3b_0 + a_2b_1)y^{(4)} + a_3b_2y^{(5)}. \end{aligned}$$

These are equal. ◇

Example 4.24M Here is how to do example 4.24 in *Mathematica*. Note the use of D rather than primes (′) for differentiation. This is because composition works better with D.

Solution.

```
In[125]:=L3[x_,y_] := a[3]D[y[x],{x,3}]+a[2]D[y[x],{x,2}]+
                       a[1]D[y[x],{x,1}]+a[0]y[x]

In[126]:=L2[x_,y_] := b[2]D[y[x],{x,2}]+b[1]D[y[x],{x,1}]+b[0]y[x]
```

Compose L3 with L2.

```
In[127]:=L3L2Result = Collect[L3[x, Function[z,L2[z,y]]], y]

Out[127]=a[0] b[0] y[x] + (a[1] b[0] + a[0] b[1]) y'[x] +
         (a[2] b[0] + a[1] b[1] + a[0] b[2]) y''[x] +
                                               (3)
         (a[3] b[0] + a[2] b[1] + a[1] b[2]) y    [x] +
                                  (4)                    (5)
         (a[3] b[1] + a[2] b[2]) y    [x] + a[3] b[2] y    [x]
```

Compose L2 with L3.

```
In[128]:=L2L3Result = Collect[L2[x, Function[z,L3[z,y]]], y]

Out[128]=a[0] b[0] y[x] + (a[1] b[0] + a[0] b[1]) y'[x] +
         (a[2] b[0] + a[1] b[1] + a[0] b[2]) y''[x] +
                                               (3)
         (a[3] b[0] + a[2] b[1] + a[1] b[2]) y    [x] +
                                  (4)                    (5)
         (a[3] b[1] + a[2] b[2]) y    [x] + a[3] b[2] y    [x]
```

Are they equal?

```
In[129]:=Simplify[L2L3Result == L3L2Result, Trig->False]

Out[129]=True
```

◇

This example in *Mathematica* could be modified by making one coefficient or more in either operator contain x as a variable.

Annihilators

In order to be able to use the method of undetermined coefficients effectively, you need to be able to look at a function, f, *just like those we have been producing as the members of the kernel of a linear differential equation with constant coefficients*, and write down an operator A that maps f to zero: $A(f) = 0$.

Definition 4.11 *Given a function f, an* **annihilator** *of f is a linear operator A such that $A(f) = 0$.*

For our purposes, an annihilator A should have constant coefficients. The choice of A is not unique, but smaller order is preferable to larger order.

Example 4.25 Here are some annihilators (and their characteristic polynomials).

1. $A_1(y) = y' - 3y$ is an annihilator of e^{3x}.
2. $A_2(y) = y'' - 5y' + 6y$ is also an annihilator of e^{3x}. (But not obviously so.)
3. $A_3(y) = y'' - 6y' + 9y$ is an annihilator of xe^{3x}. It also annihilates e^{3x}.
4. $A_4(y) = y''' - 9y'' + 27y' - 27y$ is an annihilator of x^2e^{3x}. It also annihilates e^{3x} and xe^{3x}.
5. $A_5(y) = y'' + 9y$ is an annihilator of $\sin 3x$. It also annihilates $\cos 3x$.
6. $A_6(y) = y'' - 6y' + 13y$ is an annihilator of $e^{3x}\sin 2x$. It also annihilates $e^{3x}\cos 2x$.
7. $A_7(y) = y^{(6)} - 18y^{(5)} + 147y^{(4)} - 684y''' + 1911y'' - 3042y' + 2197y$ annihilates $x^2e^{3x}\cos 2x$. It also annihilates $xe^{3x}\cos 2x$, $e^{3x}\cos 2x$, $x^2e^{3x}\sin 2x$, $xe^{3x}\sin 2x$, and $e^{3x}\sin 2x$.

The secret here is to look at the characteristic polynomial of the annihilator. For the operators above they are

1. $(r - 3)$;
2. $r^2 - 5r + 6 = (r - 3)(r - 2)$;
3. $(r - 3)^2$;
4. $(r - 3)^3$;
5. $(r^2 + 9) = (r + 3i)(r - 3i)$;
6. $(r^2 - 6r + 13) = (r - 3 + 2i)(r - 3 - 2i)$; and
7. $(r^2 - 6r + 13)^3$. ◇

As Example 4.25 demonstrates, we need not construct an annihilator. The characteristic polynomial in factored form gives all of the information we need! Now consider Theorem 4.23 which makes the observation more powerful.

Theorem 4.23 *If L_1 is an annihilator of f_1 and L_2 is an annihilator of f_2, and L_1 and L_2 commute, then L_1L_2 annihilates $f_1 + f_2$. If $p_1(r)$ is the characteristic polynomial of L_1 and $p_2(r)$ is the characteristic polynomial of L_2, then $p_1(r)p_2(r)$ is the characteristic polynomial of L_1L_2.*

Proof.

$$\begin{aligned} L_1L_2(f_1 + f_2) &= L_1(L_2(f_1) + L_2(f_2)) \\ &= L_1(L_2(f_1) + 0) \\ &= L_1(L_2(f_1)) \\ &= L_2(L_1(f_1)) \end{aligned}$$

$$= L_2(0)$$
$$= 0$$

Also, $L_1L_2(e^{rx}) = L_1(p_2(r)e^{rx}) = p_1(r)p_2(r)e^{rx}$. □

Theorem 4.23 tells us how to construct an annihilator for a sum of functions: Obtain the characteristic polynomials (in factored form) of annihilators of the individual terms and multiply them. This is the characteristic polynomial of an annihilator of the sum. It is helpful, but not necessary, to remember that a given annihilator may annihilate more than one term of your sum. Don't construct unnecessarily high-order annihilators. They increase the size of the linear system that you have to solve to find your particular solution.

Here is how to construct an annihilator for a function having several terms. Suppose that our function is

$$f(x) = x^2e^{2x} + 5\sin(3x) + x + xe^{2x}\sin(3x).$$

In every case we concentrate on the factored form of the characteristic polynomial, rather than on the actual differential operator that is the annihilator. This table summarizes the considerations for each term and for the entire function:

Function	Characteristic polynomial	Basis of functions annihilated
e^{2x}	$(r-2)$	$\{e^{2x}\}$
x^2e^{2x}	$(r-2)^3$	$\{e^{2x}, xe^{2x}, x^2e^{2x}\}$
$\sin(3x)$	(r^2+3^2)	$\{\sin(3x), \cos(3x)\}$
x	r^2	$\{1, x\}$
$e^{2x}\sin(3x)$	$(r-(2+3i))(r-(2-3i))$ $= (r^2-4r+13)$	$\{e^{2x}\sin(3x), e^{2x}\cos(3x)\}$
$xe^{2x}\sin(3x)$	$(r^2-4r+13)^2$	$\{e^{2x}\sin(3x), e^{2x}\cos(3x),$ $xe^{2x}\sin(3x), xe^{2x}\cos(3x)\}$
$f(x)$	$(r-2)^3(r^2+3^2)$ $\times\, r^2(r^2-4r+13)^2$	$\{e^{2x}, xe^{2x},$ $x^2e^{2x}, \sin(3x), \cos(3x), 1, x,$ $e^{2x}\sin(3x), e^{2x}\cos(3x),$ $xe^{2x}\sin(3x), xe^{2x}\cos(3x)\}$

Theorem 4.24 is stated in the form of an algorithm to guide you as you use the method of undetermined coefficients.

Theorem 4.24 *To solve $L(y) = f(x)$ by the method of undetermined coefficients:*

1. *Factor the characteristic polynomial $p_L(r)$ of L into linear and irreducible quadratic factors, using powers to show repeated factors, and form B_L, a basis for the kernel of L.*

2. *Separate $f(x)$ into groups of terms that are annihilated by a single annihilator.*
3. *Form $p_A(r)$, the characteristic polynomial for an annihilator of $f(x)$, factored into linear and irreducible quadratic factors, using powers to show repeated factors.*
4. *Using $p_A(r)p_L(r)$, the characteristic polynomial of AL, write out a basis B_{AL} for the kernel of the operator AL.*
5. *Find B, the complement of B_L in B_{AL}. That is, B consists of those functions in B_{AL} that are not in B_L.*
6. *Form $y_p(x)$ as a linear combination of the functions in B.* **This is the form that the candidate particular solution takes**. *The coefficients are as yet undetermined.*
7. *Equate coefficients of $L(y_p(x)) = f(x)$. The functions that appear in $L(y_p(x))$ are all in the kernel of A. The function $f(x)$ is there, too. This is why we know there is a match for everything and that the coefficients are unique.*
8. *Solve the resulting system of equations to determine values for the coefficients.*
9. *Use these values to produce an actual particular solution $y_p(x)$.*
10. *Form $y(x) = y_c(x) + y_p(x)$.*

Notice that essentially everything is done through solving homogeneous equations by simply inspecting the characteristic polynomials! This is why it is so important for you to understand completely how to solve homogeneous equations.

Steps (1), (3), (7), and (8) can be assisted greatly by *Mathematica.* These steps are the ones that are simple in theory, but are difficult in practice. The notebook *Undetermined Coefficients* performs all of these steps (with your help). Here are three examples.

Example 4.26 Solve the nonhomogeneous differential equation

$$L(y) = y''' - 3y'' + 4y' - 12y = e^{2x} - 5\sin(3x) = f(x)$$

by the method of undetermined coefficients.

Solution. We follow the steps presented in Theorem 4.24.

1. The characteristic polynomial of L is $r^3 - 3r^2 + 4r - 12 = (r-3)(r^2+4)$, so a basis for the kernel of L is $B_L = \{e^{3x}, \sin(2x), \cos(2x)\}$.
2. e^{2x} is annihilated by an operator having characteristic polynomial $(r-2)$; $-5\sin(3x)$ is annihilated by an operator having characteristic polynomial $(r^2 + 3^2)$.
3. $(r-2)(r^2+3^2)$ is the characteristic polynomial for A, an annihilator of $f(x)$. A basis for the kernel of A is $B_A = \{e^{2x}, \sin(3x), \cos(3x)\}$.
4. $(r-3)(r^2+4)(r-2)(r^2+3^2)$ is the characteristic polynomial of A_L, so a basis for the kernel of A_L is $B_{AL} = \{\boxed{e^{3x}}, e^{2x}, \boxed{\sin(2x), \cos(2x)}, \sin(3x), \cos(3x)\}$. The boxed entries are the members of B_L.
5. The complement of B_L in B_{AL} is $B = \{e^{2x}, \sin(3x), \cos(3x)\} = B_A$.

6. Assume a particular solution of the form

$$y_p(x) = a_1 e^{2x} + a_2 \sin(3x) + a_3 \cos(3x).$$

7. Form $L(y_p)(x) = -8a_1e^{2x} + (15a_2 - 15a_3)\cos(3x) + (15a_2 + 15a_3)\sin(3x)$. Equate coefficients with $f(x) = e^{2x} - 5\sin(3x)$ to get the system of equations

$$\begin{aligned}(e^{2x}):&\quad -8a_1 = 1\\ (\cos(3x)):&\quad 15a_2 - 15a_3 = 0\\ (\sin(3x)):&\quad 15a_2 + 15a_3 = -5.\end{aligned}$$

8. Solve these equations to get $a_1 = -1/8, a_2 = -1/6$, and $a_3 = -1/6$.
9. Produce

$$y_p(x) = (-1/8)e^{2x} - (1/6)\sin(3x) - (1/6)\cos(3x).$$

The complete solution of $L(y) = f$ is then

$$\begin{aligned}y(x) &= c_1e^{3x} + c_2\sin(2x) + c_3\cos(2x)\\ &\quad - (1/8)e^{2x} - (1/6)\sin(3x) - (1/6)\cos(3x),\end{aligned}$$

where c_1, c_2, and c_3 are arbitrary constants. ◇

We now solve another example that looks superficially like Example 4.24, but turns out to be quite different because the right-hand side is in the kernel of L. In the last example B_L and B_A were disjoint. In this example they are not.

Example 4.27 Solve the nonhomogeneous differential equation

$$L(y) = y''' - 3y'' + 4y' - 12y = e^{3x} - 5\sin(2x) = f(x)$$

by the method of undetermined coefficients.

Solution. We again follow the steps presented in Theorem 4.24.

1. The characteristic polynomial of L is $r^3 - 3r^2 + 4r - 12 = (r-3)(r^2+4)$, so a basis for the kernel of L is $B_L = \{e^{3x}, \sin(2x), \cos(2x)\}$.
2. e^{3x} is annihilated by an operator having characteristic polynomial $(r-3)$; $-5\sin(2x)$ is annihilated by an operator having characteristic polynomial $(r^2 + 2^2)$.
3. $(r-3)(r^2+2^2)$ is the characteristic polynomial for A, an annihilator of $f(x)$. A basis for the kernel of A is $B_A = \{e^{3x}, \sin(2x), \cos(2x)\} = B_L$.
4. $(r-3)^2(r^2+4)^2$ is the characteristic polynomial of AL, so a basis for the kernel of AL is $B_{AL} = \{\boxed{e^{3x}}, xe^{3x}, \boxed{\sin(2x), \cos(2x)}, x\sin(2x), x\cos(2x)\}$. The boxed entries are the members of B_L.
5. The complement of B_L in B_{AL} is $B = \{xe^{3x}, x\sin(2x), x\cos(2x)\} \neq B_A$.
6. Assume a particular solution of the form

$$y_p(x) = a_1xe^{3x} + a_2x\sin(2x) + a_3x\cos(2x).$$

7. Form

$$L(y_p)(x) = 13a_1e^{3x} + (-8a_2 - 12a_3)\cos(2x) + (12a_2 - 8a_3)\sin(2x).$$

Equate coefficients with $f(x) = e^{3x} - 5\sin(2x)$ to get the system of equations

$$\begin{aligned}(e^{3x})&: \quad 13a_1 = 1\\ (\cos(2x))&: \quad -8a_2 - 12a_3 = 0\\ (\sin(2x))&: \quad 12a_2 - 8a_3 = -5.\end{aligned}$$

8. Solve these equations to get $a_1 = 1/13, a_2 = -15/52$, and $a_3 = 5/26$.
9. Produce

$$y_p(x) = (1/13)xe^{3x} - (15/52)x\sin(2x) + (5/26)x\cos(2x).$$

The complete solution of $L(y) = f$ is

$$\begin{aligned}y(x) &= c_1e^{3x} + c_2\sin(2x) + c_3\cos(2x)\\ &\quad + (1/13)xe^{3x} - (15/52)x\sin(2x) + (5/26)x\cos(2x),\end{aligned}$$

where c_1, c_2, and c_3 are arbitrary constants. ◇

Example 4.28 Consider the linear operator L whose characteristic polynomial is $p_L(r) = (r-4)(r+3)^3(r-2)$. (Write out $L(y)$. It is fifth-order.) We wish to solve the nonhomogeneous problem $L(y) = 5x^3e^{-3x} + 7xe^{-3x} + 36e^{2x}$.

Solution. The first two terms are annihilated by an operator whose characteristic polynomial is $(r+3)^4$ and the third by an operator whose characteristic polynomial is $(r-2)$. So the characteristic polynomial for A, an annihilator of the right-hand side is $p_A(r) = (r-2)(r+3)^4$. This means that the characteristic polynomial of AL is $p_A(r)p_L(r) = ((r+3)^4(r-2))((r-4)(r+3)^3(r-2)) = (r-4)(r+3)^7(r-2)^2$.

We see that the kernel of AL has as basis

$$\{\boxed{e^{4x}, e^{2x}}, xe^{2x}, \boxed{e^{-3x}, xe^{-3x}, x^2e^{-3x}}, x^3e^{-3x}, x^4e^{-3x}, x^5e^{-3x}, x^6e^{-3x}\},$$

and a basis for the kernel of L is contained in boxes. This means that the form of our particular solution is a linear combination of the remaining functions and is therefore

$$y_p(x) = a_1xe^{2x} + a_2x^3e^{-3x} + a_3x^4e^{-3x} + a_4x^5e^{-3x} + a_5x^6e^{-3x}.$$

When we have equated coefficients, we will have a set of five equations in five unknowns. We now need the definition for $L(y)$. It is

$$L(y) = y^{(5)} + 3y^{(4)} - 19y''' - 63y'' + 54y' + 216y,$$

and we equate coefficients to make

$$\begin{aligned}L(y_p(x)) &= -250a_1e^{2x} + (210a_2 - 288a_3 + 120a_4)e^{-3x}\\ &\quad + (840a_3 - 1440a_4 + 720a_5)xe^{-3x}\\ &\quad + (2100a_4 - 4320a_5)x^2e^{-3x} + 4200a_5x^3e^{-3x}\\ &= 5x^3e^{-3x} + 7xe^{-3x} + 36e^{2x}.\end{aligned}$$

The resulting system of equations is

$$\begin{aligned}
(x^3e^{-3x})&: \quad 4200a_5 = 5 \\
(x^2e^{-3x})&: \quad 2100a_4 - 4320a_5 = 0 \\
(xe^{-3x})&: \quad 840a_3 - 1440a_4 + 720a_5 = 7 \\
(e^{-3x})&: \quad 210a_2 - 288a_3 + 120a_4 = 0 \\
(e^{2x})&: \quad -250a_1 = 36
\end{aligned}$$

This system, with the help of *Mathematica*, has the solution

$$\begin{aligned}
&a_1 = -18/125, a_2 = 4318/300125, a_3 = 2369/205800, \\
&a_4 = 3/1225, a_5 = 1/840.
\end{aligned}$$

From these, we find a particular solution to be

$$\begin{aligned}
y_p(x) &= -(18/125)xe^{2x} + (4318/300125)x^3e^{-3x} \\
&\quad + (2369/205800)x^4e^{-3x} + (3/1225)x^5e^{-3x} + (1/840)x^6e^{-3x}. \qquad \diamond
\end{aligned}$$

You can modify the *Undetermined Coefficients* notebook to have it work this problem for you. It would be instructive to do so.

Exercises 4.5

Part I. Given an annihilator, what is annihilated? In each problem 1–10, you are given a constant coefficients linear operator. State a basis for the functions that are annihilated by the operator. Look at the factored form of the characteristic polynomial. The use of `Factor` in *Mathematica* may be helpful.

1. $A_1(u) = u''' - u'$.
2. $A_2(u) = u'' - 7u' + 10u$.
3. $A_3(u) = u'' - 4u' + 5u$.
4. $A_4(u) = u''' - 11u'' + 33u' - 35u$.
5. $A_5(u) = u''' + u'' - 16u' + 20u$.
6. $A_6(u) = u''' - 900u'$.
7. $A_7(u) = u^{(5)} - 5u^{(4)} - 5u''' + 45u'' - 108u$.
8. $A_8(u) = u^{(4)} - 20u''' + 167u'' - 770u' + 1666u$.
9. $A_9(u) = u^{(6)} - 4u^{(5)} + 7u^{(4)} - 6u''' + 2u''$.
10. $A_{10}(u) = u^{(8)} - 34u^{(7)} + 538u^{(6)} - 5272u^{(5)} + 335,188u^{(4)}$
    $- 161{,}992u''' + 491{,}816u'' - 881{,}504u' + 703{,}040u$

PART II. Construct the characteristic polynomial of an annihilator for each of the following functions. Remember that if $L_1(u_1) = 0$ and $L_2(u_2) = 0$ then $L_1L_2(u_1 + u_2) = 0$. The characteristic polynomial of L_1L_2 is the product of the two characteristic polynomials. Have *Mathematica* multiply out your product polynomial. You may use the function `MakeOperator` that is defined in the notebook *Undetermined Coefficients* to check your annihilator. It builds a differential operator from its characteristic polynomial. As in problems 1–10, list all of the functions that are annihilated by your operator.

11. e^x

12. e^{4x}

13. xe^{3x}

14. x^2e^{6x}

15. xe^{-2x}

16. $e^x + xe^{3x}$

17. $e^{4x}\cos 5x$

18. $e^{3x} + 3e^{4x}\cos 5x$

19. $e^{3x} + 3e^{4x}\cos 5x + x^3$

20. $e^{3x} + 3e^{4x}\cos 5x + x^3 + x^2e^{6x}$

PART III. Given these pairs of differential operators, what operator is the composition of the operators? Multiply the two characteristic polynomials and construct the composite operator by inspection.

21. $L_1(y) = y' - 4y$; $L_2(y) = y'' + y$

22. $L_1(y) = y'' + y' - 4y$; $L_2(y) = y'' + y'$

23. $L_1(y) = y'' - 5y'$; $L_2(y) = y'' + 2y$

24. $L_1(y) = y'' + y$; $L_2(y) = y'' + y$

PART IV. Nonhomogeneous differential equations with constant coefficients. For the following differential equations, factor the characteristic polynomials of the operator and of an annihilator of the right hand side. Multiply these two polynomials in factored form. By inspection write down a kernel of the composite operator that corresponds to the product polynomial and delete from this kernel the kernel of the operator of the differential equation. Finally, state the form to assume for a candidate particular solution by undetermined coefficients.

25. $y'' - 4y' - 12y = 15e^x - 4\cos x - 13\sin x$.

26. $y'' + 9y = 10e^x - 9x$.

27. $y''' - 4y'' + 14y' - 20y = -4 - 3e^x + 14x - 10x^2$.

28. $y''' - 3y'' + 4y' - 2y = 8 - 8x + 2x^2$.

29. $y''' - 8y'' - 10y' - 100y = -117e^x - 11\cos x - 92\sin x$.

30. $y''' - 3y'' + 3y' - y = e^{9x} - \sin x$.

PART V. Modify the notebook *Undetermined Coefficients* to do the manipulations required to solve each of problems 25–30. **These are problems 31–36**.

4.6 Variation of Parameters

We have seen that the method of undetermined coefficients, which works only in those special cases where the right-hand side has a special form, gets complicated to execute even though the steps are straightforward in theory. The method of variation of parameters is also straight-forward in theory, but its complexity is on two fronts: how to invert the Wronskian matrix, and how to integrate the functions that appear. When this method is being done manually, these are truly daunting problems. But our salvation is in having *Mathematica* to do these difficult steps for us. Even *Mathematica* can have difficulty with the integrals that it encounters, but except in rare circumstances you can be assured that if *Mathematica* has trouble with an integral, you are not likely to evaluate it easily yourself!

Given the general nonhomogeneous linear nth-order differential equation

$$L_n(y) = a_n(x)\frac{d^n y}{dx^n} + a_{n-1}(x)\frac{d^{n-1}y}{dx^{n-1}} + \cdots + a_1(x)\frac{dy}{dx} + a_0(x)y = f(x), \quad (4.6)$$

where f and all of the coefficient functions are continuous on some interval over which $a_n(x) \neq 0$. We want to solve the nonhomogeneous linear differential equation $L_n(y) = f(x)$, where all we know about $f(x)$ is that it is continuous.

The idea behind the method is that linear operators send linear combinations of functions to linear combinations of their images. We always find the kernel of L_n first. So we have a set of functions $\{y_1, y_2, \ldots, y_n\}$ that is a basis for the kernel of L_n. But L_n sends every linear combination of these functions to 0. How can we distort them in some way, so that $L_n(\text{distorted functions}) = f(x)$? The way to do the distorting is to think of

$$c_1y_1(x) + c_2y_2(x) + \cdots + c_ny_n(x)$$

as

$$v_1(x)y_1(x) + v_2(x)y_2(x) + \cdots + v_n(x)y_n(x),$$

where we have changed the parameters $\{c_1, c_2, \ldots, c_n\}$ into $\{v_1(x), v_2(x), \ldots, v_n(x)\}$. We have made the parameters vary by turning them into functions. Hence the name **variation of parameters**.

Since we have n coefficients to determine, we need to impose n conditions. To determine the first $n-1$ conditions, we will let $n-1$ functions be 0. The differential equation will then determine the last condition for us.

Recall from multidimensional calculus that if $f(x)$ and $g(x)$ are differentiable and their values are n-vectors, then their dot product $f(x) \cdot g(x)$ is differentiable and the derivative looks like the usual product rule:

$$(f(x) \cdot g(x))' = f'(x) \cdot g(x) + f(x) \cdot g'(x).$$

We will use this rule to reduce the notation in the derivation of the method of variation of parameters. To do this let $Y(x) = \{y_1(x), y_2(x), \ldots, y_n(x)\}$ and $V(x) = \{v_1(x), v_2(x), \ldots, v_n(x)\}$. Then the particular solution we seek can be written as $y_p(x) = Y(x) \cdot V(x)$, or $y_p = Y \cdot V$ for short. We begin our search for the n conditions that are to be imposed.

Calculate $y_p' = Y' \cdot V + Y \cdot V'$. We take as our first condition $Y \cdot V' = 0$. Then $y_p' = Y' \cdot V$.

Calculate $y_p'' = Y'' \cdot V + Y' \cdot V'$. We take as our second condition $Y' \cdot V' = 0$. Then $y_p'' = Y'' \cdot V$.

Using the same reasoning, for $1 \leq k \leq n-1, y_p^{(k)} = Y^{(k)} \cdot V$ and the kth condition is $Y^{(k-1)} \cdot V' = 0$. This means that $y_p^{(n)} = Y^{(n)} \cdot V + Y^{(n-1)} \cdot V'$, from which we need to obtain the nth condition. We do this by substituting all of these derivatives back into L_n and requiring that $L_n(y_p) = f$. This is what results:

$$\begin{aligned} L_n(y_p) &= L_n(Y) \cdot V + a_n(Y^{(n-1)} \cdot V') \\ &= 0_n \cdot V + a_n(Y^{(n-1)} \cdot V') \\ &= a_n(Y^{(n-1)} \cdot V') \\ &= f. \end{aligned}$$

This gives as the nth condition $Y^{(n-1)} \cdot V' = f/a_n$, where $a_n(x)$ is the coefficient of the highest-order derivative in the differential equation. The notation $L_n(Y)$ is short for the vector $\{L_n(y_1), L_n(y_2), \ldots, L_n(y_n)\} = \{0, 0, \ldots, 0\} = 0_n$.

In summary, the n conditions are:

$$\begin{cases} Y \cdot V' = 0 \\ Y' \cdot V' = 0 \\ \quad \vdots \\ Y^{(n-1)} \cdot V' = f/a_n \end{cases} \tag{4.7}$$

Equations 4.7 are just the row equations of this matrix equation:

$$W(x)V' = \begin{pmatrix} y_1 & y_2 & \cdots & y_n \\ y_1' & y_2' & \cdots & y_n' \\ \vdots & \vdots & \ddots & \vdots \\ y_1^{(n-1)} & y_2^{(n-1)} & \cdots & y_n^{(n-1)} \end{pmatrix} \begin{pmatrix} v_1' \\ v_2' \\ \vdots \\ v_n' \end{pmatrix} = \begin{pmatrix} 0 \\ 0 \\ \vdots \\ f/a_n \end{pmatrix} = F(x)$$

Observe that the coefficient matrix is just the Wronskian matrix (not determinant). Since $\{y_1, y_2, \ldots, y_n\}$ is a linearly independent set, the Wronskian determinant $W(y_1, y_2, \ldots, y_n) \neq 0$, and the matrix $W(x)$ has an inverse. This means that we can solve for V', getting

$$V'(x) = (W(x))^{-1}F(x).$$

If we know $V'(x)$, we can find $V(x)$ by integrating this vector equation. The integration is done component by component, giving

$$V(x) = \int (W(x))^{-1}F(x))\,dx + c,$$

where c is a vector of constants.

Once we have $V(x)$, we obtain $y_p(x)$ from either $Y(x) \cdot V(x)$ or $V(x) \cdot Y(x)$:

$$\begin{aligned} y_p(x) &= V(x) \cdot Y(x) \\ &= \left(\int (W(x))^{-1}F(x)\,dx + c \right) \cdot Y(x) \\ &= \left(\int (W(x))^{-1}F(x)\,dx \right) \cdot Y(x) + c \cdot Y(x) \\ &= \left(Y(x) \cdot \int (W(x))^{-1}F(x)\,dx \right) + c \cdot Y(x). \end{aligned}$$

Notice that our arbitrary constant from the integration has produced a full description of the kernel, $c \cdot Y(x)$, as well, so that what we have here is actually the complete solution of the nonhomogeneous equation $L_n(y) = f$.

Theorem 4.25 *To find a particular solution $y_p(x)$ of equation 4.6, find a basis for the kernel of L_n. Let such a basis be $\{y_1, y_2, \ldots, y_n\}$. We seek functions $\{v_1, v_2, \ldots, v_n\}$ such that $y_p(x) = v_1(x)y_1(x) + v_2(x)y_2(x) + \cdots + v_n(x)y_n(x)$. Form the system*

$$\begin{pmatrix} y_1 & y_2 & & y_n \\ y_1' & y_2' & \cdots & y_n' \\ \vdots & \vdots & \ddots & \vdots \\ y_1^{(n-1)} & y_2^{(n-1)} & \cdots & y_n^{(n-1)} \end{pmatrix} \begin{pmatrix} v_1' \\ v_2' \\ \vdots \\ v_n' \end{pmatrix} = \begin{pmatrix} 0 \\ 0 \\ \vdots \\ f/a_n \end{pmatrix}.$$

The coefficient matrix has an inverse, so integrate

$$\begin{pmatrix} v_1' \\ v_2' \\ \vdots \\ v_n' \end{pmatrix} = \begin{pmatrix} y_1 & y_2 & \cdots & y_n \\ y_1' & y_2' & \cdots & y_n' \\ \vdots & \vdots & \ddots & \vdots \\ y_1^{(n-1)} & y_2^{(n-1)} & \cdots & y_n^{(n-1)} \end{pmatrix}^{-1} \begin{pmatrix} 0 \\ 0 \\ \vdots \\ f/a_n \end{pmatrix}. \tag{4.8}$$

to find the vector of functions $\{v_1, v_2, \ldots, v_n\}$. Then form the solution

$$y_p(x) = v_1(x)y_1(x) + v_2(x)y_2(x) + \cdots + v_n(x)y_n(x).$$

This is a particular solution of equation 4.6. The complete solution of equation 4.6 is therefore

$$y(x) = c_1y_1(x) + c_2y_2(x) + \cdots + c_ny_n(x) + y_p(x).$$

Example 4.29 Find a particular solution of the differential equation

$$L_2(y)(x) = x^2y''(x) - 4xy'(x) + 6y(x) = (x-1)^3$$

by the method of variation of parameters, given that $y_1(x) = x^2$ and $y_2(x) = x^3$ are linearly independent functions in the kernel of L_2.

Solution. We seek functions $\{v_1(x), v_2(x)\}$ such that $y_p(x) = v_1(x)y_1(x) + v_2(x)y_2(x)$ is a particular solution. Form the system

$$\begin{pmatrix} x^2 & x^3 \\ 2x & 3x^2 \end{pmatrix}\begin{pmatrix} v_1' \\ v_2' \end{pmatrix} = \begin{pmatrix} 0 \\ (x-1)^3/x^2 \end{pmatrix}.$$

Invert the coefficient matrix to get

$$\begin{aligned}\begin{pmatrix} v_1' \\ v_2' \end{pmatrix}(x) &= \begin{pmatrix} x^2 & x^3 \\ 2x & 3x^2 \end{pmatrix}^{-1}\begin{pmatrix} 0 \\ (x-1)^3/x^2 \end{pmatrix} \\ &= \begin{pmatrix} 3x^{-2} & -x^{-1} \\ -2x^{-3} & x^{-2} \end{pmatrix}\begin{pmatrix} 0 \\ (x-1)^3/x^2 \end{pmatrix} \\ &= \begin{pmatrix} -(x-1)^3/x^3 \\ (x-1)^3/x^4 \end{pmatrix}.\end{aligned}$$

Integrate this to get

$$\begin{pmatrix} v_1 \\ v_2 \end{pmatrix}(x) = \begin{pmatrix} (-1/2x^2) + 3/x - x + 3\ln x \\ 1/3x^3 - 3/2x^2 + 3/x + \ln x \end{pmatrix}.$$

Then

$$\begin{aligned}y_p(x) &= x^2\left(\frac{-1}{2x^2} + \frac{3}{x} - x + 3\ln x\right) + x^3\left(\frac{1}{3x^3} - \frac{3}{2x^2} + \frac{3}{x} + \ln x\right) \\ &= -\frac{1}{6} + \frac{3x}{2} + 3x^2 - x^3 + 3x^2\ln x + x^3\ln x\end{aligned}$$

is the desired particular solution. ◇

Example 4.29M Solve the problem of Example 4.29 in *Mathematica*. Notice how these steps parallel the ones above.

Solution.

```
In[130]:=basis = {x^2,x^3}

Out[130]=  2   3
         {x , x }
```

```
In[131]:=W[x_] = Table[D[basis,{x,i}],{i,0,1}]

Out[131]=   2   3            2
         {{x , x }, {2 x, 3 x }}

In[132]:=F[x_] = {0,(x-1)^3/x^2}

Out[132]=                  3
                 (-1 + x)
         {0,  ────────────}
                    2
                   x

In[133]:=Inverse[W[x]].F[x]

Out[133]=                3                3
              (-1 + x)        (-1 + x)
         {-(────────────), ────────────}
                   3               4
                  x               x

In[134]:=v[x_] = Integrate[Inverse[W[x]].F[x],x]

Out[134]= -1      3                        1       3      3
         {──── + - - x + 3 Log[x], ──── - ──── + - + Log[x]}
             2   x                      3       2   x
          2 x                        3 x    2 x

In[135]:=yp[x_] = Simplify[basis.v[x]]

Out[135]= 1     3 x        2    3       2            3
         -(-) + ─── + 3 x  - x  + 3 x  Log[x] + x  Log[x]
           6     2

In[136]:=Simplify[L[x,yp] == (x-1)^3]

Out[136]=True
```

◇

The notebook *Variation of Parameters* contains the machinery necessary to solve problems such as these. The integrations can be complicated, but that is why we have *Mathematica*. The notebook *Complete Solution* uses `DKernel` to find a basis for the kernel of the operator part of a differential equation and variation of parameters to find a particular solution. It then returns the complete solution as an n-parameter family of functions.

You can then take this solution and begin to impose conditions on it in order to select the solution from the family that satisfies some identifying set of criteria. Since there are n parameters that can be determined, n conditions need to be imposed. When those n conditions are in the form of standard initial conditions, the unique solution is easy to obtain: just solve the n linear algebraic equations in n unknowns that the initial conditions produce. It is possible that the n conditions do not all occur at a single point in the interval over which you want a solution, but are given at several (even infinitely many) places on the interval. In this case you are said to have a **boundary value problem**. Boundary value problems are not different from initial value problems to *Mathematica*, because all we present to *Mathematica* is a system of linear algebraic equations. *Mathematica* does not know or care where those equations came from. As with any system of equations, the linear algebraic system we have may have a unique solution, or no solution, or an infinite number of solutions. These latter two eventualities are what makes most people think boundary value problems are a totally different type of problem from initial value problems. But they can be essentially the same, except that initial value problems in standard form always have unique solutions, whereas a perfectly reasonably stated boundary value problem may lose either existence or uniqueness. This loss is not a property of the differential equation itself, but of the form of the boundary conditions. We will see in Chapter 8, systems of differential equations, an existence and uniqueness theorem for boundary value problems.

Example 4.30M Solve the boundary value problem $y'' - y' - 12y = 6x$ along with the boundary conditions: $y(0) = 3$ and $\int_0^2 y(x)\,dx = 1$.

Solution. Note that there is a point condition at 0 and that every point in the interval $[0, 2]$ is involved in the integral. Therefore this problem has its boundary conditions at an infinite number of points!

```
In[137]:=Clear[x,y]
         L[x_,y_] = y''[x]-y'[x]-12y[x]

Out[137]=-12 y[x] - y'[x] + y''[x]

In[138]:=s[x_] = y[x]/.DSolve[L[x,y] == 6x,y[x],x][[1]]

Out[138]=1    x   C[1]    4 x
         -- - - + ---- + E    C[2]
         24   2   3 x
                  E

In[139]:=const = Solve[{s[0] == 3, Integrate[s[x],{x,0,2}] == 1},
                 {C[1],C[2]}]

Out[139]=
                         6                 6      14
                    -23 E           71 (-3 E  + 3 E  )
         {{C[1] -> --------------- + -------------------,
                        6      14            6      14
                   4 - 7 E  + 3 E    24 (4 - 7 E  + 3 E  )

                         6                   6
                     23 E             71 (4 - 4 E )
           C[2] -> --------------- + ---------------------}}
                        6      14            6      14
                   4 - 7 E  + 3 E    24 (4 - 7 E  + 3 E  )

In[140]:=z[x_] = Simplify[s[x]/.const[[1]]]

Out[140]=
                                                           6
         1    4 x           71                        67 E
         -- + E    (----------------------- + -----------------------) +
         24                 6      14                 6      14
                    6 (4 - 7 E  + 3 E  )      6 (4 - 7 E  + 3 E  )

                      6                          14
                -255 E                       71 E
           ---------------------- + ----------------------
                    6      14                6      14
           8 (4 - 7 E  + 3 E  )     8 (4 - 7 E  + 3 E  )      x
           ------------------------------------------------ - -
                                3 x                           2
                               E
```

This is so complicated that its form gets totally lost. Let's turn these complicated expressions into numbers and then look at what we have. Here is the numeric value of the constants:

```
In[141]:={d1,d2} = N[{C[1],C[2]}/.const[[1]]]

Out[141]={2.95708, 0.00125293}
```

Here is the solution with numeric coefficients.

```
In[142]:=zn[x_] = s[x]/.{C[1]->d1,C[2]->d2}

Out[142]=1      2.95708                     4 x   x
         -- + ------- + 0.00125293 E      - -
         24     3 x                           2
               E

In[143]:=Plot[zn[x],{x,0,2}];
```

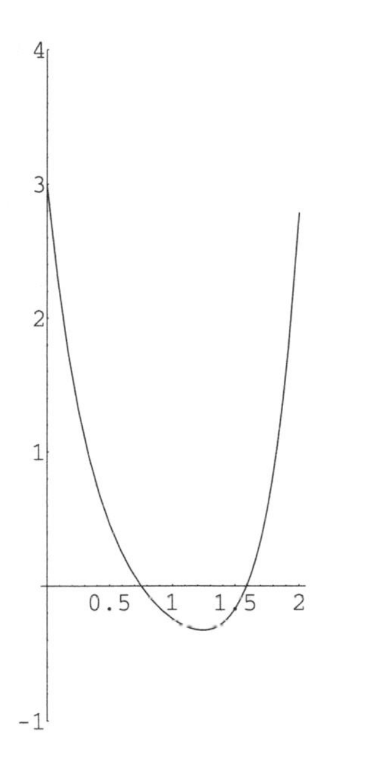


```
In[144]:=NIntegrate[zn[x],{x,0,2}]

Out[144]=1.
```

The form of zn[x] is simply that of s[x]: the sum of a polynomial and two exponentials. Now go back and look at z[x] again through enlightened eyes. Doesn't it have the same form? ◇

You will find boundary value problems such as these to be immensely powerful and exciting. When you have an existence and uniqueness theorem to work with, and a collection of acceptable forms for boundary conditions, those of you who have a primary interest in applications will have a truly powerful tool in your possession, because your intuition will suggest wonderful boundary conditions that have a real physical significance. We will look again at boundary value problems in Chapter 8 in the context of systems of differential equations.

Exercises 4.6

These exercises consist of 10 problems to be solved by variation of parameters. Parts I, II, and III are steps that guide you through the solution. **You may take any problem and perform the instructions given in the three parts to complete the solution.**

Part I. Wronskian. After each differential equation below is a basis for the kernel. Construct the Wronskian matrix $W(x)$ of the basis set. Calculate the determinant of $W(x)$ to verify linear independence.

1. $y''' - y'' + y' - y = 1;$ $\{e^x, \sin x, \cos x\}.$
2. $y'' - 5y' + 6y = x^2;$ $\{e^{3x}, e^{2x}\}.$
3. $y'' + 4y' - 5y = \sin x;$ $\{e^{-5x}, e^x\}.$
4. $y'' - 2y' + 2y = e^{3x};$ $\{e^x \sin x, e^x \cos x\}.$
5. $y'''' - 2y'' + 2y' = \cos x;$ $\{e^x \sin x, e^x \cos x, 1\}.$
6. $y'''' - 2y'' + 2y' = x^3 - 1;$ $\{e^x \sin x, e^x \cos x, 1\}.$
7. $y'''' - 2y'' + y' = e^{-x};$ $\{1, e^x, xe^x\}.$
8. $y''' - 5y'' + 9y' - 5y = 10e^{-x} \cos x;$ $\{e^{2x} \sin x, e^{2x} \cos x, e^x\}.$
9. $y''' - 6y'' + 12y' - 8y = 2e^x \sin x;$ $\{e^{2x}, xe^{2x}, x^2e^{2x}\}.$
10. $y''' - 3y' + 2y = \sin x;$ $\{e^x, xe^x, e^{-2x}\}.$

Part II. For each differential equation in problems 1–10, construct the linear equation 4.7, $W(x)V'(x) = Q(x)$, which the coefficient functions $V_i(x)$ must satisfy. **These problems are continuations of problems 1–10.**

Part III. With the help of *Mathematica* solve each of the linear equations for the vector $V'(x)$. If you need further help, use *Mathematica* to perform the integrations necessary to find the vector $V(x)$. Complete the solution of each of the original problems by variation of parameters by forming the dot product of the kernel with $V(x)$ as suggested by Theorem 4.25. Write out the complete solution of the original problem and check. **These problems are also continuations of problems 1–10.**

5 Applications of Second-Order Equations

5.0 Introduction

Once again, second-order differential equations with constant coefficients serve as a special topic for our study. This time we study them not just because it is easy to explain everything that is going on, but because of the special interests one can have in the physical situation that these differential equations are modeling. We study several cases of the motion of a single weight hanging from a spring:

- where the motion is free;
- where the motion is damped;
- when there is a forcing function present.

This second-order setting for differential equations also applies to series electronic circuits that contain only resistance, capacitance, and inductance. Such electronic circuits can also be forced by an oscillating voltage. The behavior of many of these electronic circuits is analogous to that of systems of springs.

Resonance phenomena are also investigated, especially in the exercises.

Several notebooks are provided that enable you to experiment with the behavior of a system without having to painstakingly repeat a manual sequence of calculations. For instance, if an idea you are working with suggests further investigations of theoretical or applied interest, you should not have to endure intrusive manual calculations. You should be free to think and experiment. Do so.

It is worth remembering that *Mathematica* can often do a symbolic computation as easily as it can do a numerical one.

5.1 Simple Harmonic Motion

To study simple harmonic motion, we need to review Hooke's[1] law. Hooke's law states that the restoring force of a spring is proportional to the displacement from the length of the spring when it is at rest. The spring produces a force $F = -kx$, when it is x units from its rest position. The number k is called the **spring constant**. If the spring is stretched, then x is considered to be positive. If the spring is compressed, then x is considered to be negative. When the spring is hanging this seems counterintuitive, but $dx/dt > 0$ means that the length of the spring is increasing, and this increasing length is downward.

Newton's second law of motion is $F = ma$, where m is the mass of the object in motion and a is the acceleration on the object. In our setting, we will use $a = d^2x/dt^2$, where x is displacement. When the spring is hanging vertically, there is an additional force on the mass, that due to gravity. In these problems it is important to choose an appropriate system of units [foot-pound-second (English), or gram-centimeter-second (cgs), for instance] and not mix units. It is also important to remember that weight is a unit of force. A calculation is needed to obtain an object's mass from its weight.

Suppose that the natural length of a massless spring is L and that when an object of mass m is suspended from the end of the spring, the spring extends d units and is in its equilibrium position. The spring is stretched further and released. What is the differential equation of the motion? When the mass is in equilibrium, as in the center of Figure 5.1, the upward restoring force of the spring and the downward force due to gravity are in balance: $kd = mg$. Thus, when the spring is extended x units further, for a total extension of $d + x$, the force on the system is the sum of the force produced by the spring and the force due to gravity.

$$m\frac{d^2x}{dt^2} = -k(d + x) + mg = -kx + (-kd + mg) = -kx,$$

since $mg = kd$. Thus

$$m\frac{d^2x}{dt^2} + kx = 0.$$

We simplify further to get

$$\frac{d^2x}{dt^2} + \frac{k}{m}x = \frac{d^2x}{dt^2} + \omega^2 x = 0, \tag{5.1}$$

where $\omega^2 = k/m$.

---

[1] Robert Hooke (1635–1703) was a contemporary of Isaac Newton. Newton had ideas, and Hooke did experiments for the Royal Society to verify them. Their association was long and rancorous. Hooke thought that Newton's best ideas were stolen from him. When Newton was thinking about universal gravitation, Hooke communicated an (incorrect) idea that caused Newton to redirect his thoughts toward a valid theory. Hooke thought that the whole theory was stolen from him.

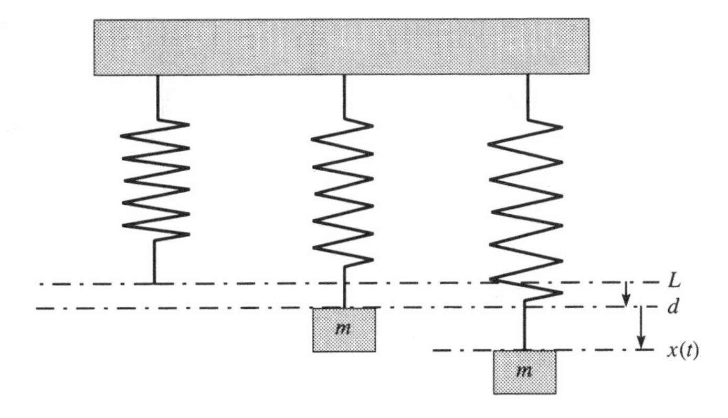

Figure 5.1: Positions of a simple spring system.

This is the differential equation of **free harmonic motion**. The parameter ω is called the **angular velocity**.

From the section on second-order differential equations with constant coefficients, you will recognize one form of the solution to be

$$x(t) = c_1 \cos \omega t + c_2 \sin \omega t.$$

This is often called the **equation of motion** of the free vibrations described by the differential equation. The **period** of the free vibrations so described is the number $T = 2\pi/\omega$, and the **frequency** of the free vibrations is $f = (1/T) = \omega/(2\pi)$. For instance, if the equation of motion is $x(t) = 4\cos 5t + 3\sin 5t$, then the period of the free vibrations is $T = (2\pi/5)$, and the frequency is $f = 5/(2\pi)$. This period means that the motion repeats every $5/(2\pi)$ units of time, and that during each 2π units of time the motion is repeated five times. During each unit of time, $5/(2\pi)$ of the motion is completed.

Summary of Consistent Basic Units

For your use in the exercises, here is a summary of three standard consistent sets of units that are in common use.

System	Distance	Time	Mass	Acceleration	Force
	$s, x, y, \ldots$	t	m	a	F
SI	meter (m)	second (sec)	kilogram (kg)	m/sec^2	Newton (Nt)
cgs	centimeter (cm)	second (sec)	gram (g)	cm/sec^2	dyne
English	foot (ft)	second (sec)	slug	ft/sec^2	pound (lb)

You can find an astonishingly complete list of conversion factors in the package `Miscellaneous`Units``. Once you have loaded this package you can invoke

`Convert[ old, newunits]`. There are also functions `SI[expr]`, `MKS[expr]`, `CGS[expr]`, which convert *expr* to the International System, or to the mks or cgs systems. In this package there are not only units of distance, time, mass, acceleration, and force, but also electrical, information, inverse length, volume, viscosity, luminosity, radiation, angles, power, area, amount, pressure, energy, frequency, and more.

Here is a short list of conversion factors.

Length	1 km = 1000 m;	1 mi = 5280 ft;
	1 m = 100 cm;	1 ft = 30.48 cm;
	1 m = 3.281 ft;	1 mi = 1.609 km;
	1 km = 0.6214 mi;	
Mass	1 kg = 1000 g = 0.06852 slug;	
	1 slug = 14.59 kg = 32 lb.;	
Force	1 kg wt = 2.205 lb = 9.807 kg;	
	1 Nt = 105 dynes = 0.2248 lb;	
	1 lb = 4.448 Nt.	

Example 5.1 Suppose that a long spring hanging vertically is stretched to an equilibrium position of 2 feet by a 10.24-pound object. If the spring is stretched 3 additional feet and released from rest, what is the equation of motion of the system?

Solution. First we need to calculate the mass of the object. Since the force of gravity in the foot-pound-second system is 32 feet per second per second, and the force is $mg = 32m = 10.24$, it follows that $m = (10.24/32) = 0.32$ slug. Next, we need to calculate the spring constant k. The force $F = -10.24$ pounds $= -k$ (2 feet), so $k = 5.12$ pounds per foot. Next, formulate the differential equation as

$$m\frac{d^2x}{dt^2} + kx = 0.32\frac{d^2x}{dt^2} + 5.12x = 0,$$

or

$$\frac{d^2x}{dt^2} + (5.12/0.32)x = \frac{d^2x}{dt^2} + 16x = 0,$$

with initial conditions $x(0) = 3$ and $x'(0) = 0$.

This initial value problem, which describes the motion, has as solutions

$$x(t) = c_1 \cos 4t + c_2 \sin 4t.$$

The initial conditions give the equations

$$x(0) = c_1 \cos 0 + c_2 \sin 0 = c_1 = 3$$

and

$$x'(0) = -4c_1 \sin 0 + 4c_2 \cos 0 = 4c_2 = 0.$$

From these we find that $c_1 = 3$ and $c_2 = 0$. So the equation of motion is

$$x(t) = 3 \cos 4t,$$

which is a pure cosine curve, with **amplitude** 3. This means that the mass rises 3 feet and falls 3 feet in its excursion about the rest position. This is a 6-foot total excursion, which is why it was stated that the spring was very long. At its minimum excursion the mass must not compress the spring fully. The frequency of this motion is $f = 4/(2\pi) = (2/\pi)$ and the period is $T = (\pi/2)$. ◇

Example 5.1M Solve the initial value problem of example 5.1 in *Mathematica* by using DSolve.

Solution.

```
In[1]:=  soln = DSolve[{x''[t]+16x[t] == 0, x[0] == 3,
                x'[0] == 0},x[t],t]

Out[1]=  {{x[t] -> 3 Cos[4 t]}}
```
◇

Example 5.2 Suppose that a spring hanging vertically is stretched to a rest position of 2 inches by a 10.24-pound weight. If the spring is stretched 3 additional inches and released from rest, what is the equation of motion of the system?

Solution. Notice that the only difference between this example and the last example is that the units have been changed from feet to inches. The mass is still 0.32 slug. Put the length units into feet, since the force was given in pounds. The rest position is 2 inches = (2 inches) (1/12) (foot/inch) =(1/6) feet. The release point is an additional

$$3 \text{ inches} = (3 \text{ inches})\ (1/12)\ (\text{foot/inch}) = (1/4)\text{feet}.$$

Calculate the spring constant. $F = -10.24$ pounds $= -k$ (1/6 feet), so $k = 61.44$ pounds per foot. Next, formulate the differential equation as

$$0.32\frac{d^2x}{dt^2} + 61.44x = 0,$$

or

$$\frac{d^2x}{dt^2} + (61.44/0.32)x = \frac{d^2x}{dt^2} + 192x = 0;$$

with initial conditions

$$x(0) = 1/4, \qquad x'(0) = 0.$$

This differential equation of motion has as solutions

$$x(t) = c_1 \cos 8\sqrt{3}t + c_2 \sin 8\sqrt{3}t.$$

The initial conditions give the equations

$$x(0) = c_1 \cos 0 + c_2 \sin 0 = c_1 = 1/4$$

and

$$x'(0) = -8\sqrt{3}c_1 \sin 0 + 8\sqrt{3}c_2 \cos 0 = 8\sqrt{3}c_2 = 0.$$

From these we find that $c_1 = 1/4$ and $c_2 = 0$. So the equation of motion is

$$x(t) = (1/4)\cos 8\sqrt{3}t,$$

which is a pure cosine curve with amplitude 1/4 . This means that the mass rises and falls with a total excursion of 1/2 foot. The spring need not be nearly as long as before. ◇

Example 5.2M Solve the differential equation of Example 5.2 in *Mathematica* by using `DSolve`.

Solution.

```
In[2]:=  soln = DSolve[{x''[t]+192x[t] == 0, x[0] == 1/4,
                  x'[0] == 0}, x[t], t]

Out[2]=
                    -8 I Sqrt[3] t        16 I Sqrt[3] t
                   E               (1 + E               )
         {{x[t] -> ------------------------------------------}}
                                  8
```

Our solution is real: convert to real form (the imaginary part is 0).

```
In[3]:=  x[t_] = Simplify[ComplexExpand[x[t]/.soln[[1]] ] ]

Out[3]=  Cos[8 Sqrt[3] t]
         ----------------
                4
```

This is the same result that we got in Example 5.2. ◇

Sometimes it is not nearly so easy to determine the amplitude of the motion that is obtained. This happens when the mass is not released from rest, but is given an upward (negative) or downward (positive) initial velocity. The motion is still a single sine (or cosine) wave, but there is an additive adjustment to the angle.

To clarify, suppose that we have the general equation of motion,

$$x(t) = c_1 \cos \omega t + c_2 \sin \omega t$$

from a differential equation $d^2x/dt^2 + \omega^2 x = 0$ that describes free harmonic motion. Recall from your trigonometry studies that $\sin(a + b) = \sin a \cos b + \sin b \cos a$. We wish to use this formula in the context

$$\sin(a + \omega t) = c_1 \cos \omega t + c_2 \sin \omega t = x(t).$$

We can take any point in the plane, different from the origin, and write it uniquely as

$$(c_2, c_1) = A(\cos\phi, \sin\phi), \quad \text{for} \quad 0 \le \phi < 2\pi \quad \text{and} \quad A > 0.$$

This is a polar representation of the point (c_2, c_1). Using the association, $c_1 = A\sin\phi, c2 = A\cos\phi$ we can write

$$
\begin{aligned}
x(t) &= c_1 \cos \omega t + c_2 \sin \omega t \\
&= A \sin \phi \cos \omega t + A \cos \phi \sin \omega t \\
&= A \sin(\omega t + \phi).
\end{aligned}
$$

This expresses the equation of free harmonic motion as a pure sine curve, from which the **amplitude** A can be immediately determined by inspection. But this does introduce a different parameter ϕ, the **phase angle**. In the first two examples, the phase angle was 0. Notice that we still have a two-parameter family of solutions of $x'' + \omega^2 x = 0$, but expressed in terms of two different parameters. The basis $\{\cos \omega t, \sin \omega t\}$ for the kernel is not nearly as clear in this representation, but we can see other things such as the amplitude of the motion much more clearly. Figure 5.2 is a picture of the amplitude and the phase angle.

Example 5.3 Suppose that a spring hanging vertically is stretched to an equilibrium position of 15.36 inches by a 1.6-pound weight. If the spring is stretched 4 additional inches and released with an upward velocity of 15 inches per second, what is the equation of motion of the system? What is the velocity of the mass the first time it passes downward through the rest position?

Solution. Again we have to find the mass $m = (1.6/32) = (1/20)$ slug, and put the length units into feet, since the force was given in pounds. The rest position is 15.36 inches = (15.36 inches) (1/12) (foot/inch) = 1.28 feet. The release point $x(0)$ is an additional 1/3 feet, and the initial velocity $x'(0) = -(15(\text{inch/second}))$ $(1/12)$ (foot/inch) = 5/3 (foot/second).

We calculate the spring constant. $F = -1.6$ pounds $= -k(1.28$ foot), so $k = (1.6/1.28) = 1.25 = 5/4$ pounds per foot. Next, formulate the differential equation as

$$
\left(\frac{1}{20}\right)\frac{d^2x}{dt^2} + \left(\frac{5}{4}\right)x = 0,
$$

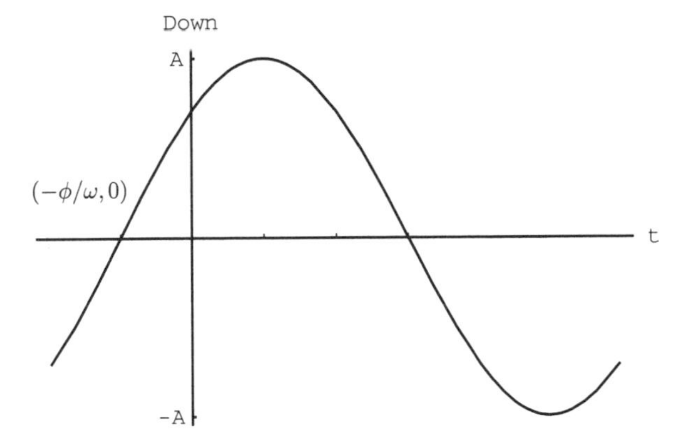

Figure 5.2: The amplitude and phase angle when the solution is expressed as a sine function.

or

$$\frac{d^2x}{dt^2} + \left(\frac{100}{4}\right)x = \frac{d^2x}{dt^2} + 25x = 0;$$

with initial conditions $x(0) = 1/3$ foot and $x'(0) = -5/4$ foot per second.

This differential equation has as solutions $x(t) = c_1 \cos 5t + c_2 \sin 5t$. The initial conditions give the equations

$$x(0) = c_1 \cos 0 + c_2 \sin 0 = c_1 = 1/3$$

and

$$x'(0) = -5c_1 \sin 0 + 5c_2 \cos 0 = 5c_2 = -5/4.$$

From these we find that $c_1 = 1/3$ and $c_2 = -1/4$. Now $\sqrt{c_1^2 + c_2^2} = 5/12$. So the equation of motion is

$$\begin{aligned}
x(t) &= \frac{1}{3}\cos 5t - \frac{1}{4}\sin 5t \\
&= \frac{5}{12}\left(\frac{\frac{1}{3}}{\frac{5}{12}}\cos 5t - \frac{\frac{1}{4}}{\frac{5}{12}}\sin 5t\right) \\
&= \frac{5}{12}\left(\frac{4}{5}\cos 5t - \frac{3}{5}\sin 5t\right) \\
&= \frac{5}{12}\left(\sin\left(\pi + \arctan\left(-\frac{4}{3}\right)\right)\cos 5t\right. \\
&\quad \left. + \left(\cos\left(\pi + \arctan\left(-\frac{4}{3}\right)\right)\sin 5t\right)\right) \\
&= \frac{5}{12}\sin\left(5t + \left[\pi + \arctan\left(-\frac{4}{3}\right)\right]\right)
\end{aligned}$$

which is a sine curve with period $2\pi/5$, amplitude $5/12$, and phase angle $\pi + \arctan(-4/3)$. This means that the mass rises and falls with a total excursion of $5/6$ foot.

The mass passes through the rest position when

$$x(t) = \left(\frac{5}{12}\right)\sin\left(5t + \pi + \arctan\left(-\frac{4}{3}\right)\right) = 0,$$

which is when

$$5t + \pi + \arctan\left(-\frac{4}{3}\right) = k\pi.$$

Since the mass started below the rest point, the first time through, when $k = 1$, the mass is rising. The next time through, when $k = 2$, is the first time through moving downward. This is when $5t + \pi + \arctan(-4/3) = 2\pi$, or when $t = (1/5)(\pi + \arctan(-4/3)) \approx 0.813778$. The velocity is therefore $x'(0.813778) \approx 2.08333$. Positive means downward. ◇

Example 5.3M Solve the preceding exercise in *Mathematica*.
Solution.

```
In[4]:=  soln = DSolve[{x''[t]+25x[t] == 0,
                        x[0] == 1/3, x'[0] == -5/4},
                        x[t], t]

Out[4]=            Cos[5 t]   Sin[5 t]
         {{x[t] -> -------- - --------}}
                      3          4

In[5]:=  s2[t_] = x[t]/.soln[[1]]

Out[5]=  Cos[5 t]   Sin[5 t]
         -------- - --------
            3          4
```

Look at a plot of the solution.

```
In[6]:=  Plot[s2[t],{t,0,2},AxesLabel->{"t","Down"}];
```

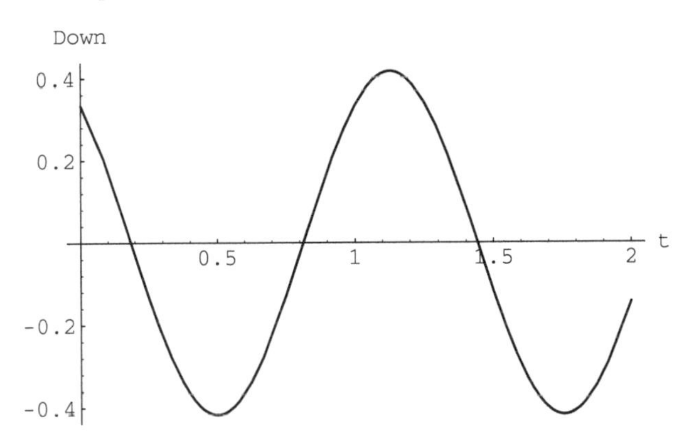

Calculate the time of the first downward excursion through equilibrium.

```
In[7]:=  N[(Pi-ArcTan[-4/3])/5]

Out[7]=  0.813778
```

The downward velocity at the equilibrium position is:

```
In[8]:=  s2'[%]

Out[8]=  2.08333
```

This agrees with the manual solution. ◇

Exercises 5.1

Part I. Convert the following functions having the form

$$c_1 \cos \omega t + c_2 \sin \omega t$$

to the forms $A \sin(\omega t + \phi)$ and $A \cos(\omega t + \phi)$. Do these both manually and by *Mathematica*. You may find these transformation rules to be helpful:

```
amplitude = y[t]/.c1_. Cos[a_] + c2_. Sin[a_]->Sqrt[c1^2 + c2^2]
SinePhaseAngle = y[t]/.c1_. Cos[a_] + c2_. Sin[a_]->ArcTan[c2,c1]
CosinePhaseAngle = y[t]/.c1_. Cos[a_] + c2_. Sin[a_]->ArcTan[c1,-c2]
```

1. $2 \sin 3t - \cos 3t$.

2. $\frac{1}{2} \sin \frac{3}{2}t + \frac{1}{4} \cos \frac{3}{2}t$.

3. $\frac{\sqrt{3}}{2} \cos 4t - \frac{1}{2} \sin 4t$.

4. $5 \sin 6t$.

5. $\frac{1}{2} \cos 4t + \frac{\sqrt{3}}{2} \sin 4t$.

6. $-3 \cos 5t$.

7. $-\frac{\sqrt{2}}{2} \cos \pi t - \frac{\sqrt{2}}{2} \sin \pi t$.

8. $2 \cos \frac{\pi t}{3} - \sin \frac{\pi t}{3}$.

Part II. These initial-value problems model simple harmonic motion. Solve each of the problems and convert each of the solutions into the single sine and single cosine forms of Part I.

9a. $\frac{1}{5} \frac{d^2x}{dt^2} + 10x = 0;$

$x(0) = 0,\ x(0) = 1.$

10a. $\frac{1}{5} \frac{d^2x}{dt^2} + 20x = 0;$

$x(0) = 0,\ x'(0) = -1.$

11a. $\frac{d^2x}{dt^2} + 100x = 0;$

$x(0) = 1/2,\ x'(0) = 0.$

12a. $\dfrac{d^2x}{dt^2} + 25x = 0;$

$x(0) = -1/2, \ x'(0) = 0.$

13a. $\dfrac{d^2x}{dt^2} + 40x = 0;$

$x(0) = 1/2, \ x'(0) = 1.$

14a. $\dfrac{d^2x}{dt^2} + 40x = 0;$

$x(0) = -1/2, \ x'(0) = 1.$

15a. $5\dfrac{d^2x}{dt^2} + 10x = 0;$

$x(0) = 1/2, \ x'(0) = -1.$

16a. $5\dfrac{d^2x}{dt^2} + 10x = 0;$

$x(0) = -1/2, \ x'(0) = -1.$

PART III. Each of the problems in Part II describes a spring that is undamped and is set in motion from some initial conditions. Describe in words the positioning of the spring at the moment of release. Use such terms as above or below equilibrium, and upward or downward velocity. Recall that downward is positive. **These are problems 9b–16b.**

PART IV. For each problem in Part II give the coordinates of the first maximum and the first minimum of the motion of the spring. **These are problems 9c–16c.**

PART V. Simple harmonic motion: problems.

17. Suppose that an object moving in a straight line has the equation of motion

$$mx'' + 6x' + 3x = 0.$$

For what values of m is the motion oscillatory?

18. A weight of 4 pounds stretches a spring 6 inches. A 5-pound weight is attached to the spring as the only weight, is pulled downward 8 inches below equilibrium, and released with an upward velocity of 3 inches per second. Find a formula for the position of the weight thereafter.

19. A mass of 30 grams stretches a spring 3 centimeters. An object of unknown mass is attached to the spring and set in motion. The period is observed to be 0.5 second. What is the mass of the object?

PART VI. Simple harmonic motion: theory.

20. Verify mathematically the physical observation that if the motion of the mass starts at rest from its rest position, then it remains at the rest position.
21. Show that if the mass is started moving from its rest position by being given a nonzero initial velocity, then the motion is a pure sine curve (the phase angle is zero).
22. Show that if the mass starts its motion away from its rest position, but with initial velocity 0, the motion is a pure cosine curve (the phase angle is zero).
23. Show that if the mass is not at rest when it starts x_0 units above or below its rest position, then the motion of the mass eventually takes it further than x_0 units away from the rest position. That is, the amplitude of the motion is greater than $|x_0|$.

5.2 Damped Harmonic Motion

Having a situation where simple harmonic motion is possible is essentially unknown in nature. If it were possible, then such a device would demonstrate perpetual motion, and this does not occur. What actually happens is that a spring system such as is described in the last section always experiences some form of damping of its motion. Usually such a system is operating in the air. Objects moving in air have their motion opposed by the air itself. For slowly moving systems such as spring systems, the force of resistance due to the air, or **air resistance**, is proportional to the speed of the object through the air. In other words, air resistance is $-\beta(dx/dt)$, for an object whose position is $x(t)$. The number β is called the **damping constant**. We will assume that any **damping** of motion that is present is of this form, though damping proportional to $(dx/dt)^2$ does occur in certain situations.

The differential equation of an object of mass m attached to a spring with spring constant k and damping proportional to velocity is therefore

$$m\frac{d^2x}{dt^2} = -kx - \beta\frac{dx}{dt},$$

or

$$m\frac{d^2x}{dt^2} + \beta\frac{dx}{dt} + kx = 0.$$

This can be simplified to yield the differential equation of **free damped motion**

$$\frac{d^2x}{dt^2} + 2\lambda\frac{dx}{dt} + \omega^2 x = 0, \tag{5.2}$$

where $2\lambda = \beta/m$ and $\omega^2 = k/m$. The "2" in the expression 2λ is present to simplify some algebra that will be encountered.

One can artificially control the amount of damping by attaching a paddle (of low mass) to the object at the end of the spring or by attaching a piston that rides in an enclosed tube and thus provides resistance to the motion. The attachment should be by a stiff rod in order for it to resist on the way down as well as on the way up. Some very complicated behavior is exhibited when the resistance is a "semi-resistor" and resists motion in only one direction [See Lazer & McKenna, 1990].

The differential equation of free damped motion can be solved by standard methods. The characteristic equation is $r^2 + 2\lambda r + \omega^2 = 0$. Solving for r by the quadratic equation yields $r_1, r_2 = -\lambda \pm \sqrt{\lambda^2 - \omega^2}$. The solutions are real and distinct if $\lambda^2 - \omega^2 > 0$; repeated if $\lambda^2 - \omega^2 = 0$, or $\lambda = \pm\omega$; and complex if $\lambda^2 - \omega^2 < 0$. These three situations are all important, and will be discussed separately. Note that these situations depend on the magnitude of λ, since ω^2 is determined by the spring itself and not by external conditions. Here are the formal solutions in the three cases.

CASE I. $\lambda^2 - \omega^2 > 0$. The motion is said to be **overdamped**. The two roots $r_1, r_2 = -\lambda \pm \sqrt{\lambda^2 - \omega^2}$ are real and both are negative since $\sqrt{\lambda^2 - \omega^2} < \sqrt{\lambda^2} = \lambda$. This means that the solution $x(t) = c_1 e^{r_1 t} + c_2 e^{r_2 t}$ will head quickly for 0 and the motion will die out rapidly. In this case, there will be at most one critical point on the curve. The motion is not oscillatory.

CASE II. $\lambda^2 - \omega^2 = 0$. The motion is said to be **critically damped**. The two roots $r_1, r_2 = -\lambda \pm \sqrt{\lambda^2 - \omega^2} = -\lambda \pm 0 = -\lambda, -\lambda$ are equal, so the solution is

$$x(t) = c_1 e^{-\lambda t} + c_2 t e^{-\lambda t} = e^{-\lambda t}(c_1 + c_2 t).$$

Here again, the motion is not oscillatory and there will be at most one critical point.

CASE III. $\lambda^2 - \omega^2 < 0$. The motion is said to be **underdamped**. The two roots $r_1, r_2 = -\lambda \pm \sqrt{\lambda^2 - \omega^2} = -\lambda \pm i\sqrt{\omega^2 - \lambda^2}$. These roots are complex and the solution is

$$\begin{aligned} x(t) &= e^{-\lambda t}(c_1 \cos \sqrt{\omega^2 - \lambda^2} t + c_2 \sin \sqrt{\omega^2 - \lambda^2} t) \\ &= A e^{-\lambda t} \sin(\sqrt{\omega^2 - \lambda^2} t + \phi). \end{aligned}$$

The motion is oscillatory because of the second factor, but the first factor $Ae^{-\lambda t}$ causes the motion to decay. This is referred to as **damped oscillation**. The quantity $Ae^{-\lambda t}$ is sometimes called the **damped amplitude**, and $2\pi/\sqrt{\omega^2 - \lambda^2}$ is called the **quasi period**. The quantity $\sqrt{\omega^2 - \lambda^2}/(2\pi)$ is called the **quasi frequency**. These latter quantities are not a true period or frequency because the motion is damped and not periodic.

Example 5.4 (Case I) The solution of the initial value problem

$$x'' + 26x' + 25x = 0,$$

$$x(0) = 4,$$

$$x'(0) = 20$$

is readily shown to be

$$x(t) = -e^{-25t} + 5e^{-t}.$$

Where does the system make its greatest excursion from equilibrium?

Solution. We need to solve $x'(t) = 25e^{-25t} - 5e^{-t} = 0$. This can be written $5 - e^{24t} = 0$, and this has as solution, $t = (\ln 5)/24 \approx 0.0670599$. ◇

Example 5.4M In *Mathematica* Example 5.4 can be done this way:

Solution.

```
In[9]:=  soln = DSolve[{x''[t]+26x'[t]+25x[t] == 0,
                  x[0] == 4,
                  x'[0] == 20},
                  x[t],t]

Out[9]=
                     -25 t    5
         {{x[t] -> -E       + --}}
                               t
                              E

In[10]:= s1[t_] = x[t]/.soln[[1]]

Out[10]=   -25 t    5
         -E       + --
                     t
                    E

In[11]:= Plot[s1[t],{t,0,1}];
```


When does the maximum excursion occur?

```
In[12]:= FindRoot[s1'[t] == 0,{t,0}]

Out[12]= {t -> 0.0670599}
```

What is the maximum excursion?

```
In[13]:= s1[t/.%]

Out[13]= 4.48867
```

Example 5.5 (Case II) The solution of

$$x'' + 6x' + 9x = 0,$$
$$x(0) = -2,$$
$$x'(0) = 15$$

for $t > 0$ is easily verified to be $x(t) = e^{-3t}(-2 + 9t)$. Find the location of the single extremum.

Solution. We find that $x'(t) = e^{-3t}(15 - 27t)$, which is 0 when $15 - 27t = 0$. This is when $t = 5/9$. The x-position is therefore $x(5/9) = 3e^{-5/3}$. This is the only extremum. Since $x > 0$, this is a maximum downward excursion. Had the zero occurred for $t < 0$, there would have been no extremum for t positive. ◇

Example 5.5M In *Mathematica* this looks like:

Solution.

```
In[14]:= DSolve[{x''[t]+6x'[t]+9x[t] == 0,
                                   x[0] == -2,
                                  x'[0] == 15},x[t],t]

Out[14]=
                 -2     9 t
         {{x[t] -> ---- + ----}}
                  3 t    3 t
                 E      E

Out[14]= s2[t_] = x[t]/.%[[1]]

Out[14]=  -2     9 t
         ---- + ----
          3 t    3 t
         E      E

In[15]:= Plot[s2[t],{t,0,3}, PlotRange->{-2,1}];
```

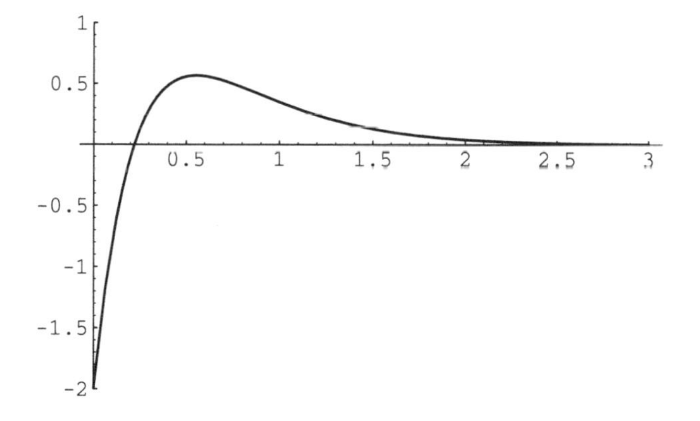

When does the extremum occur?

```
In[16]:= Solve[s2'[t] == 0,t]

Out[16]=          5
         {{t -> -}}
                  9
```

Where is the extremum?

```
In[17]:= s2[t/.%[[1]]]

Out[17]=  3
         ----
          5/3
         E

In[18]:= N[%]

Out[18]= 0.566627
```

◇

Example 5.6 (Case III) An 8-pound weight stretches a spring 9.6 inches. The damping on the system is half the velocity of the motion. The weight is set in motion from a point 6 inches above the equilibrium position with an initial velocity of 1 foot per second upward. What is the equation of motion?

Solution. The mass of the weight is $m = 8/32 = 1/4$ slug. The initial position is $x(0) = (-6 \text{ inch})(1/12)(\text{foot/inch}) = -1/2$ foot, and the spring constant is found from -8 pounds $= -k$ (9.6 inch)(1/12) (foot/inch) $= -4/5k$ foot.

Thus $k = (5/4)8 = 10$ pounds/foot. The damping constant was given as 1/2, so the differential equation of motion is

$$\frac{1}{4}x'' + \frac{1}{2}x' + 10x = 0,$$

or

$$x'' + 2x' + 40x = 0,$$

with initial conditions

$$x(0) = -1/2 \quad \text{and} \quad x'(0) = -1.$$

◇

Example 5.6M The solution of Example 5.6 by *Mathematica* is:

Solution.

```
In[19]:= soln = DSolve[{x''[t]+2x'[t]+40x[t] == 0,
             x[0] == -1/2,
            x'[0] == -1},
             x[t],t]

Out[19]=
                    I   -t - I Sqrt[39] t
         {{x[t] -> -- E
                   52

                                          2 I Sqrt[39] t
               (13 I - Sqrt[39] + 13 I E                +

                               2 I Sqrt[39] t
                 Sqrt[39] E                   )}}

In[20]:= s2[t_] = Simplify[ComplexExpand[x[t]/.soln[[1]]]]

Out[20]= -(13 Cos[Sqrt[39] t] + Sqrt[39] Sin[Sqrt[39] t])
         -------------------------------------------------
                                   t
                               26 E
```

```
In[21]:= Plot[s2[t],{t,0,3}, PlotRange->{-1,1}];
```

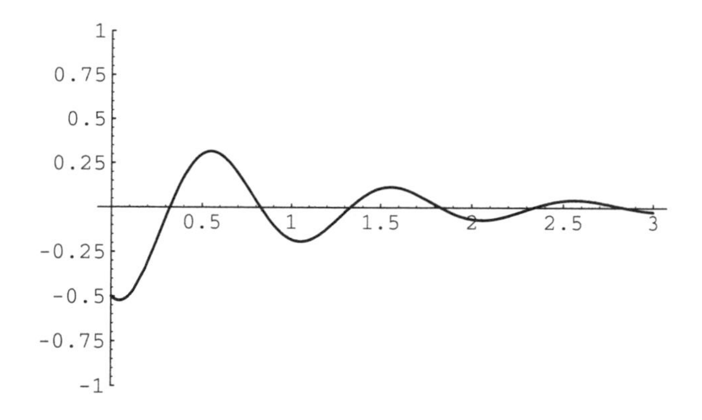

Use "pattern matching" to find the amplitude.

```
In[22]:= amplitude = s2[t]/.
              c3_ Exp[c4_](c1_ Cos[a_] + c2_ Sin[a_])->
                                      c3*Sqrt[c1^2+c2^2]

Out[22]=     -2
         ---------
         Sqrt[13]
```

The solution and the exponential envelopes plot as:

```
In[23]:= Plot[{amplitude*Exp[-t],-amplitude*Exp[-t], s2[t]},
                        {t,0,3}];
```

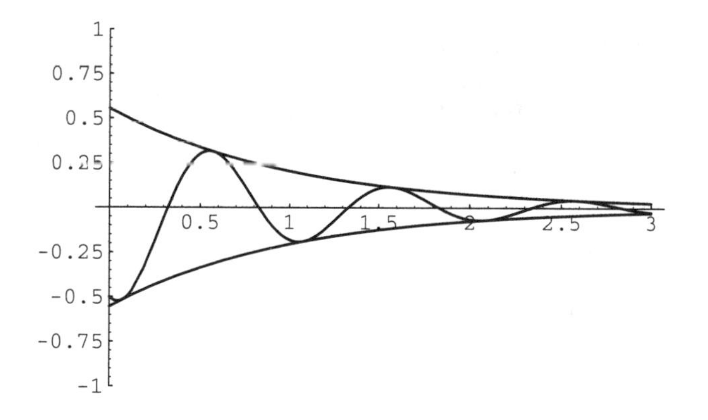

◇

Exercises 5.2

Part I. Convert the following functions having the form

$$c_1 e^{-\lambda t}\cos\omega t + c_2 e^{-\lambda t}\sin\omega t$$

to the form

$$e^{-\lambda t}\sin(\omega t+\phi)$$

or

$$Ae^{-\lambda t}\cos(\omega t+\theta).$$

Do these both manually and by *Mathematica*. State the upper and lower envelope curves in each case. Modify the transformation rules given in Exercises 5.1 to include the exponential factor.

1a. $2e^{-t}\sin 3t - e^{-t}\cos 3t.$

2a. $\frac{1}{2}e^{-3t}\sin\frac{3}{2}t + \frac{1}{4}e^{-3t}\cos\frac{3}{2}t.$

3a. $\frac{\sqrt{3}}{2}e^{-5t}\cos 4t - \frac{1}{2}e^{-5t}\sin 4t.$

4a. $5e^{-2t}\sin 6t.$

5a. $\frac{1}{2}e^{-2t}\cos 4t + \frac{\sqrt{3}}{2}e^{-2t}\sin 4t.$

6a. $-3e^{-4t}\cos 5t.$

7a. $-\frac{\sqrt{2}}{2}e^{-4t}\cos 8t - \frac{\sqrt{2}}{2}e^{-4t}\sin 8t.$

8a. $-\frac{\sqrt{2}}{2}e^{-4t}\cos 8t - \frac{\sqrt{2}}{2}e^{-4t}\sin 8t.$

PART II. For each function given in Part I, plot it and the two envelopes on a common set of axes. Make the orientation of the plots reflect the fact that downward is positive. **These are problems 1b–8b.**

PART III. For each problem in Part I give the coordinates of all relative maxima and all relative minima. **These are problems 1c–8c.**

PART IV. Solve these initial value problems. They model free (unforced) damped harmonic motion. Convert each of the solutions to the single sine and single cosine forms of Part I. Repeat Parts II and III for these solution functions. Describe in words each of these problems as a spring system.

9. $y'' + 10y' + 41y = 0;\quad y(0) = \frac{\sqrt{3}}{2}, y'(0) = -\frac{4+5\sqrt{3}}{2}.$

10. $y'' + 8y' + 41y = 0;\quad y(0) = -3, y'(0) = 12.$

11. $4y'' + 24y' + 45y = 0;\quad y(0) = \frac{1}{4}, y'(0) = 0.$

12. $y'' + 8y' + 80y = 0; \quad y(0) = -\frac{\sqrt{2}}{2}, y'(0) = -2\sqrt{2}.$

13. $9y'' + 54y' + 14y = 0; \quad y(0) = 2, y''(0) = -\frac{26}{3}.$

14. $y'' + 2y' + 10y = 0; \quad y(0) = -1, y'(0) = 7.$

15. $y'' + 4y' + 40y = 0; \quad y(0) = 0, y'(0) = 30.$

16. $y'' + 4y' + 20y = 0; \quad y(0) = \frac{1}{2}, y'(0) = -1 + 2\sqrt{3}.$

PART V. Problems about free damped harmonic motion.

17. A mass weighing 16 pounds stretches a spring 3 inches. The mass undergoes damping that is proportional to velocity. When the velocity is 1/2 foot per second, the damping is 1 pound. Determine the motion of the mass if the mass is given an upward velocity of 3 inches per second from the equilibrium position.

18. A closed sheet metal cube of side 3 feet is floating in still water of density 62.4 pounds per cubic foot. It is weighted so the top stays parallel to the water surface. It floats at equilibrium 2 feet being submerged. If the cube is pressed downward 6 inches and released from rest, it bobs up and down. Suppose the buoyancy is proportional to the net volume of water displaced as compared to equilibrium. See Figure 5.3.
    (a) What is the period of the motion assuming no damping?
    (b) What is the quasi period of the motion if the motion is damped with magnitude of damping equal to the velocity?

19. A 20-pound weight stretches a spring 4 feet. The damping on the spring system can be adjusted. Damping is numerically equal to b ($b > 0$) times the velocity. Determine b so that the motion is (a) underdamped, (b) critically damped, (c) overdamped.

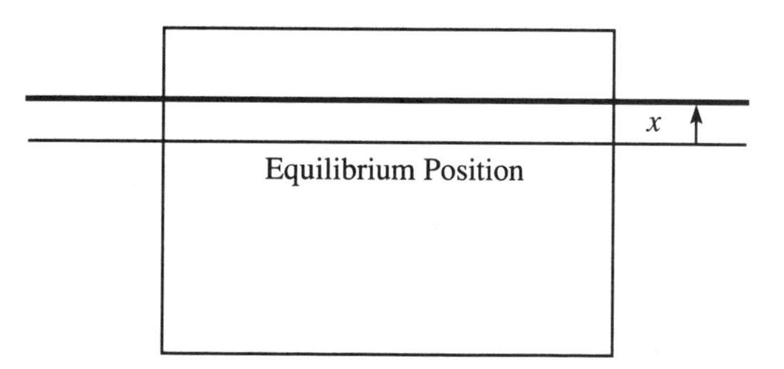

Figure 5.3: Floating cube.

PART VI. The theory of damped harmonic motion. The initial value problem for this part is $x'' + 2\lambda x' + \omega^2 x = 0, x(0) = \alpha, x'(0) = \beta$, with λ and ω chosen so that the roots are complex.

20. In the case of free, damped motion, find the location of each point where the solution curve is tangent to one of the envelopes.

21. In the case of free, damped motion, find the location of each point where the solution crosses the equilibrium position. Distinguish between up-going and down-going crossings.

22. Show that the time interval between two successive maxima (minima) is constant. This is a measurable quantity. Call it Δt.

23. Show that if $t_1 < t_2$ are the abscissas of successive maxima of $x(t)$, then $\Delta x = (x(t_1)/x(t_2))$ is constant. It is a measurable quantity. The number $\Delta = \ln(\Delta x)$ is called the **logarithmic decrement**.

24. Suppose that you have observed (and recorded) the behavior of a system undergoing damped harmonic motion, and have measured the time interval Δt between successive maxima and have calculated the logarithmic decrement Δ. Show how to determine λ and the angular velocity ω from Δt and Δ. If the mass m has also been measured, the damping factor $\beta = 2m\lambda$ can be calculated. This observation thus allows the straight-forward calculation of a quantity that is difficult to measure accurately.

5.3 Forced Oscillation

One often has an external force acting on a damped system. For instance the differential equation might be

$$m\frac{d^2x}{dt^2} = -kx - \beta\frac{dx}{dt} + f(t),$$

or

$$m\frac{d^2x}{dt^2} + \beta\frac{dx}{dt} + kx = f(t).$$

In the notation we used before this becomes

$$\frac{d^2x}{dt^2} + 2\lambda\frac{dx}{dt} + \omega^2 x = F(t), \tag{5.3}$$

where $F(t) = f(t)/m, 2\lambda = \beta/m$ and $\omega^2 = k/m$. The function $F(t)$ is called the **forcing function**. This is just a nonhomogeneous second-order differential equation with constant coefficients, and is easily solved both in theory and in practice when the function $F(t)$ is simple. Usually the problem is posed as an **initial value problem**, with the initial conditions $x(0) = x_0$ and $x'(0) = x_1$ included.

Example 5.7M Interpret and solve the initial-value problem

$$\frac{1}{10}\frac{d^2x}{dt^2} + 0.6\frac{dx}{dt} + x = \frac{1}{5}\cos 3t, \quad x(0) = \frac{1}{2}, x'(0) = 0.$$

Solution. Here we can say that we have a mass of 1/10 (unit of mass), a damping constant of 0.6, and a spring constant of 1. The initial position of the mass is 1/2 (length unit) below the equilibrium position, and initially the mass is at rest.

Multiply the differential equation by 10 to get

$$\frac{d^2x}{dt^2} + 6\frac{dx}{dt} + 10x = 2\cos 3t$$

as the differential equation to solve. We let *Mathematica* do the work. See the notebook *Forced Oscillations.*

```
In[24]:= Clear[L,t,x]
         L[t_,x_] = x''[t]+6x'[t]+10 x[t]

Out[24]= 10 x[t] + 6 x'[t] + x''[t]

In[25]:= solnRule = DSolve[{L[t,x] == 2 Cos[3t]}, x[t], t]

Out[25]=
                C[2] Cos[t]   2 Cos[3 t]   C[1] Sin[t]
        {{x[t] -> ----------- + ---------- - ----------- +
                    3 t          325          3 t
                   E                          E
            36 Sin[3 t]
            -----------}}
               325

In[26]:= Clear[soln]
         soln[t_] = x[t]/.solnRule[[1]]

Out[26]= C[2] Cos[t]   2 Cos[3 t]   C[1] Sin[t]   36 Sin[3 t]
         ----------- + ---------- - ----------- + -----------
            3 t          325           3 t           325
           E                          E

In[27]:= steadyState[t_] = soln[t]/.{C[1]->0, C[2]->0}

Out[27]= 2 Cos[3 t]   36 Sin[3 t]
         ---------- + -----------
            325          325

In[28]:= c1c2 = Solve[{soln[0] == 1/2, soln'[0]== 0}]

Out[28]=
                      747              321
         {{C[1] -> -(---), C[2] -> ---}}
                      650              650

In[29]:= x[t_] = soln[t]/.c1c2[[1]]

Out[29]= 321 Cos[t]   2 Cos[3 t]   747 Sin[t]   36 Sin[3 t]
         ---------- + ---------- + ---------- + -----------
             3 t         325           3 t          325
         650 E                     650 E
```

```
In[30]:= Simplify[L[t,x] == 2Cos[3t] && x[0] == 1/2 && x'[0] == 0]

Out[30]= True

In[31]:= Plot[x[t],{t,0,5}];
```

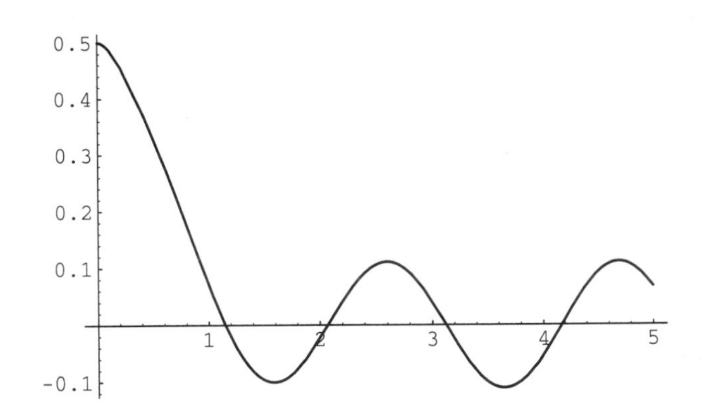

This is what the transient solution looks like:

```
In[32]:= Plot[x[t]-steadyState[t],{t,0,5},PlotRange->{-1/2,1/2}];
```

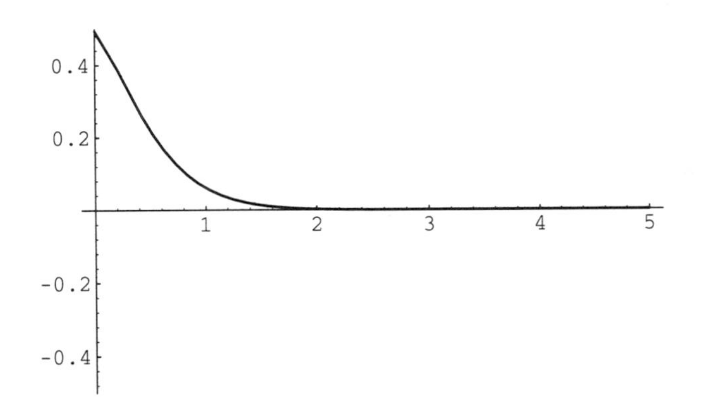

◇

Notice that the solution itself eventually looks almost exactly like simple harmonic motion. There is, however, a small **transient** from the strongly overdamped (homogeneous) system, whose plot is just above. This affects the solution noticeably only for two to three seconds. After that, the system is effectively in its **steady-state** condition. The transient solution is obtained from the homogeneous system and the steady-state solution is a particular solution of the nonhomogeneous system. Both are used when satisfying the initial conditions, but their effects on the solution differ.

Resonance Phenomena

The solution that satisfies the initial-value problem

$$\begin{cases} 5\dfrac{d^2x}{dt^2} + 2\lambda\dfrac{dx}{dt} + \omega^2 x = F_0 \cos\gamma t \\ x(0) = x_0 \\ x'(0) = x_1 \end{cases} \tag{5.4}$$

can be expressed as

$$x(t) = Ae^{-\lambda t}\sin(\sqrt{\omega^2 - \lambda^2}t + \phi) + B\cos\gamma t + C\sin\gamma t, \tag{5.5}$$

where the first term, $Ae^{-\lambda t}\sin(\sqrt{\omega^2 - \lambda^2}t + \phi)$, is a solution of the unforced system, and the two terms $B\cos\gamma t + C\sin\gamma t$ are due to the forcing function. The constants B and C, which can easily be found by undetermined coefficients, depend on the parameters λ and ω, and hence on the basic parameters m, β, and k, as well as on the parameter F_0. Using the same reasoning as before, we can write the sum of a sine and a cosine as

$$B\cos\gamma t + C\sin\gamma t = g(\gamma)\sin(\gamma t + \theta),$$

where the amplitude function $g(\gamma)$ depends not only on γ but on λ, ω, and F_0, and hence on the basic parameters m, β, and k. The angle θ is a new phase angle.

The solution is thus expressible as

$$x(t) = Ae^{-\lambda t}\sin(\sqrt{\omega^2 - \lambda^2}t + \phi) + g(\gamma)\sin(\gamma t + \theta),$$

and it is easily seen that in the presence of damping, that is, when $\lambda \neq 0$, the first term decays away and leaves the second term as the dominant term in the solution. For these reasons, the first term is called the transient solution and the second term is called the steady-state solution. We saw this situation in the example above.

Since $g(\gamma)$ is not constant, what does it look like, and how does it behave? We understand everything else about the solution. It is left as an exercise for you to show that

$$g(\gamma) = \frac{F_0}{\sqrt{(\omega^2 - \gamma^2)^2 + 4\lambda^2\gamma^2}}. \tag{5.6}$$

Several things are immediately clear. This expression does explicitly depend on the parameters we indicated it would. The expression inside the radical is always positive, so we have no need to worry about complex numbers. The expression is not constant as a function of γ. You have as an exercise to show that $g(\gamma)$ has a maximum at $\gamma_1 = \sqrt{\omega^2 - 2\lambda^2}$. The notebook *Resonance Phenomena* examines the maximum value and shows the effects of λ on the graph of $g(\gamma)$. As $\lambda \to 0$ the system begins to exhibit resonance, which means that the solution has an amplitude that is unbounded as $t \to \infty$.

Exercises 5.3

Part I. Convert the following functions having the form

$$c_1e^{-\lambda t}\cos\alpha t + c_2e^{-\lambda t}\sin\alpha t + d_1\cos\gamma t + d_2\sin\gamma t$$

to the form

$$Ae^{-\lambda t}\sin(\alpha t+\phi) + B\sin(\gamma t+\delta)$$

or

$$Ae^{-\lambda t}\cos(\alpha t+\theta) + B\cos(\gamma t+\tau).$$

Do these both manually and by *Mathematica*. You may wish to add new transformation rules that distinguish between those terms which have an exponential factor and those that do not. These problems are combinations of problems from Part I of Exercises 5.1 and 5.2. They do not necessarily correspond to forced spring systems of the form in the text.

1. $2e^{-t}\sin 3t - e^{-t}\cos 3t + \dfrac{\sqrt{3}}{2}\cos 4t - \dfrac{1}{2}\sin 4t.$

2. $\dfrac{1}{2}e^{-2t}\cos 4t + \dfrac{\sqrt{3}}{2}e^{-2t}\sin 4t + 5\sin 6t.$

3. $2e^{-3t}\cos\dfrac{\pi t}{3} - e^{-3t}\sin\dfrac{\pi t}{3} - 3\cos 5t.$

4. $5e^{-2t}\sin 6t + \dfrac{1}{2}\sin\dfrac{3}{2}t + \dfrac{1}{4}\cos\dfrac{3}{2}t.$

5. $-\dfrac{\sqrt{2}}{2}e^{-4t}\cos\pi t - \dfrac{\sqrt{2}}{2}e^{-4t}\sin\pi t + \dfrac{1}{2}\cos 4t + \dfrac{\sqrt{3}}{2}\sin 4t.$

6. $-3e^{-\pi t}\cos 5t + 2\cos\dfrac{\pi t}{3} - \sin\dfrac{\pi t}{3}.$

7. $2e^{-3t}\cos\dfrac{\pi t}{3} - e^{-3t}\sin\dfrac{\pi t}{3} - \dfrac{\sqrt{2}}{2}\cos\pi t - \dfrac{\sqrt{2}}{2}\sin\pi t.$

8. $5e^{-2t}\sin 6t + 2\sin 3t - \cos 3t.$

Part II. Solve these initial value problems that model forced harmonic motion. Convert each of the solutions to the single sine and single cosine forms of Part I. The operators are from the same numbered problems of Part IV of Exercises 5.2.

9. $y'' + 10y' + 41y = 3\cos 2t; \qquad y(0) = \dfrac{\sqrt{3}}{2}, y'(0) = -\dfrac{1}{2}.$

10. $y'' + 8y' + 41y = 5\cos\dfrac{t}{2}; \qquad y(0) = -3, y'(0) = 12.$

11. $4y'' + 24y' + 45y = 4\cos t;$ $\quad y(0) = \frac{1}{4}, y'(0) = 0.$

12. $y'' + 8y' + 80y = 5\cos 4t;$ $\quad y(0) = -\frac{\sqrt{2}}{2}, y'(0) = -2\sqrt{2}.$

13. $9y'' + 54y' + 14y = \cos\frac{3t}{2};$ $\quad y(0) = 2, y'(0) = -\frac{26}{3}.$

14. $y'' + 2y' + 10y = \cos 3t;$ $\quad y(0) = -1, y'(0) = 7.$

15. $y'' + 4y' + 40y = 4\cos 3t;$ $\quad y(0) = 0, y'(0) = 30.$

16. $y'' + 4y' + 20y = 3\cos 4t;$ $\quad y(0) = \frac{1}{2}, y'(0) = -1.$

PART III. Problems on forced oscillation.

17. In problem 18 of the exercises for Section 5.2, it was assumed that the water was still. This made the motion of the cube unforced. Suppose that there are waves that impart a force of $(3/2)\cos 4t$. What is a formula for the vertical displacement of the cube at any positive time t?
    (a) Assume no damping.
    (b) Assume damping with magnitude equal to that of the velocity.

18. A 16-pound weight stretches a spring 5/32 foot. The spring is initially compressed 1 foot and given a further upward velocity of 1 foot per second. The system undergoes damping numerically equal to twice the velocity.
    (a) Describe the motion if the system is driven by a force of $\cos 2t - 3\sin 2t$.
    (b) Identify the transient motion and the steady-state motion.
    (c) Plot the motion, the transient motion, and the steady-state motion.

19. Suppose that a spring is stretched 6 inches by a 12-pound weight. The weight is pulled downward 3 inches below equilibrium and released from rest. If no damping is present, but there is an impressed force of $9\sin 4t$ pounds, describe the motion.

PART IV. Theory of forced oscillation. Throughout this part of the exercises we are concerned with the nonhomogeneous differential equation

$$\frac{d^2x}{dt^2} + 2\lambda\frac{dx}{dt} + \omega^2 x = F_0\cos\gamma t. \tag{N}$$

20. Derive equation 5.6,

$$g(\gamma) = \frac{F_0}{\sqrt{(\omega^2 - \gamma^2)^2 + 4\lambda^2\gamma^2}}$$

from equation 5.5

$$x(t) = Ae^{-\lambda t}\sin(\sqrt{\omega^2 - \lambda^2}t + \phi) + B\cos\gamma t + C\sin\gamma t,$$

where the first term, $Ae^{-\lambda t}\sin(\sqrt{\omega^2 - \gamma^2}t + \phi)$ is a solution of the unforced system, and the two terms $B\cos\gamma t + C\sin\gamma t$ are a particular solution of the forced equation (N). You first need to determine the values of B and C.

21. Show that the maximum of the function $g(\gamma) = F_0/\sqrt{(\omega^2-\gamma^2)^2 + 4\lambda^2\gamma^2}$ occurs when $\gamma = \gamma_1 = \sqrt{\omega^2 - 2\lambda^2}$. There is also a critical point when $\gamma = 0$. To what does it correspond?

22. Find the maximum value $g(\gamma_1)$ attained by $g(\gamma)$. How does this maximum value behave as $\lambda \to 0$? How does γ_1 behave as $\lambda \to 0$? As $\gamma \to \omega$?

23. What is $\lim_{\gamma\to\infty} g(\gamma)$? Is there a physical interpretation for this?

24. Suppose that the spring system being modeled happens to be the suspension system of your car.

    (a) Explain what the speed with which you traverse a lengthy "washboard" defect in a roadway has to do with our investigations of $g(\gamma)$.
    (b) What is the relationship between speed and γ?
    (c) What differing effects do you as driver feel for slow speeds, intermediate speeds, and high speed?
    (d) Which speed strategy is best? Which is worst?
    (e) Explain how such a "washboard" defect might come into being. [Assume that the roadway can be made to change shape by applying vertical forces to it.]

5.4 Simple Electronic Circuits

A Serial Circuit

As we will see, the differential equations that model simple electronic circuits are directly analogous to those that model the oscillatory motion of springs. Here we will use **time** t (in seconds), **charge** q (in coulombs[2]), **voltage** v (in volts[3]), **current** I (in amperes[4]), **capacitance** C (in farads[5]), **resistance** R (in

---

[2]Charles-Augustin de Coulomb (1736–1806), French physicist. Studied forces, including forces produced by electrical charges (1777).

[3]Alessandro Giuseppe Volta (1745–1827), Italian physicist. Constructed the first batteries (known as Voltaic cells) that would produce electricity continuously (1800).

[4]André Marie Ampère (1775–1836), French physicist. Studied current and electromechanical effects in parallel (1820) and helically wound (1823) wires.

[5]Michael Faraday (1791–1867), English physicist. Discovered electromagnetic lines of force (1821); liquefied gases (1823); built transformers and discovered electromagnetic induction (1831); built an electric motor and generator (1831); stated laws of electronics (1832).

Table 5.1: Analogies between a mechanical spring system and a series electric circuit.

Mechanical Spring System $m\frac{d^2x}{dt^2}+b\frac{dx}{dt}+kx=F(t).$		Series Electric Circuit $L\frac{d^2q}{dt^2}+R\frac{dq}{dt}+\frac{1}{C}q=E(t).$	
Displacement	x	Charge	q
Velocity	(dx/dt)	Current	$i=(dq/dt)$
Mass	m	Inductance	L
Damping factor	b	Resistance	R
Spring constant	k	1/Capacitance	$(1/C)$
Forcing function	$F(t)$	Impressed voltage	$E(t)$

ohms[6]), and **inductance** L (in henrys[7]). See Table 5.1 for the correspondence between these quantities and the analogous quantities for springs. Here are the fundamental relationships among these quantities:

A charge q on a capacitor of capacitance C produces a **voltage**: $v = q/C$.

The time rate of change of charge is called **current**: $I = (dq/dt)$.

In a circuit, voltage drops across a component because of the action of the component on the charge present. Typical components are **resistors**, **capacitors**, and **inductors**. These produce resistance, capacitance, and inductance in the circuit, respectively. Resistors, capacitors, and inductors act *linearly*. As always, this is important to us.

Voltage drops are produced across our three types of components in these ways:

$$\left(\frac{q}{C}\right), \quad IR\left(=R\frac{dq}{dt}\right), \quad \text{and} \quad L\frac{dI}{dt}\left(=L\frac{d^2q}{dt^2}\right).$$

Furthermore, in the simple circuit of Figure 5.4, if we consider the applied voltage source $E(t)$ to be introducing voltage into the circuit, then R, L, and C, are "consuming" voltage. **Kirchhoff's voltage law**[8] says that the total voltage drop around a circuit is 0. For our circuit this gives

$$E(t)-\frac{q}{C}-R\frac{dq}{dt}-L\frac{d^2q}{dt^2}=0. \tag{5.7}$$

[6]Georg Simon Ohm (1789–1854), German physicist. Studied the relationships between current and resistance (1827) known as Ohm's law.

[7]Joseph Henry (1797–1878), American physicist. Studied and produced useful electromagnets (1823); studied inductive effects on electrical currents (1823).

[8]Gustav Robert Kirchhoff (1824–1887), German physicist. Studied spectroscopy (1859), blackbody radiation (1860), stated the fundamental laws of behavior of voltage and current known now as Kirchhoff's laws; discovered cesium (1860) and rubidium (1861) from their spectra.

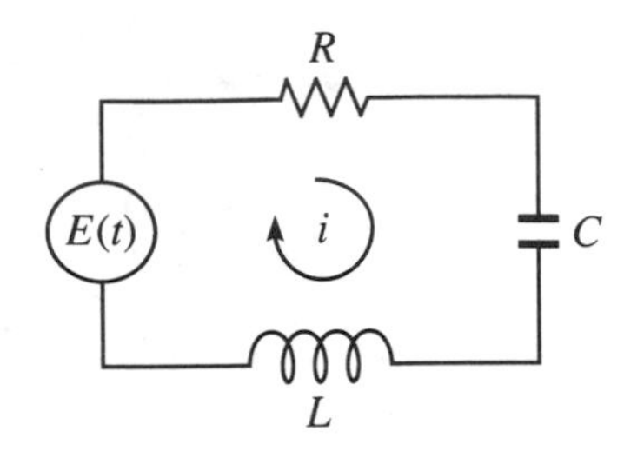

Figure 5.4: A simple passive serial circuit.

Recall that $I = (dq/dt)$, so that $(dI/dt) = (d^2q/dt^2)$. We can rewrite equation 5.7 as

$$L\frac{d^2q}{dt^2} + R\frac{dq}{dt} + \frac{1}{C}q = E(t). \tag{5.8}$$

This allows us to restate Kirchoff's voltage law as the sum of the voltage drops across the components of a circuit equals the applied voltage. If we differentiate equation 5.8 which expresses relationships based on the charge q across the capacitor, we obtain

$$L\frac{d^2I}{dt^2} + R\frac{dI}{dt} + \frac{1}{C}I = E'(t). \tag{5.9}$$

This equation has the same form as equation 5.8, but it is expressed in terms of the current in the circuit. From current we can get voltage. Often we are given that $E(t) = f_0 \cos(\gamma t)$. When this is the case, divide equation 5.8 through by L to get

$$\frac{d^2q}{dt^2} + \frac{R}{L}\frac{dq}{dt} + \frac{1}{LC}q = \frac{f_0}{L}\cos\gamma t,$$

or

$$\frac{d^2q}{dt^2} + 2\lambda\frac{dq}{dt} + \omega^2 q = F_0 \cos\gamma t, \tag{5.10}$$

where $2\lambda = R/L, \omega^2 = 1/(LC)$, and $F_0 = f_0/L$. If we include the initial conditions $q(0) = q_0$ and $q'(0) = i(0) = q_1$, we have reproduced equation 5.4 which models the motion of a spring system. This means that we already know how simple circuits such as that in Figure 5.4 behave, because we analyzed equation 5.4 completely, and the same analysis holds.

Resonance Phenomena

The solution that satisfies the initial-value problem

$$\begin{cases} 5\dfrac{d^2x}{dt^2} + 2\lambda\dfrac{dx}{dt} + \omega^2 x = F_0\cos\gamma t \\ x(0) = x_0 \\ x'(0) = x_1, \end{cases} \tag{5.11}$$

which was studied in the last section, can be expressed as

$$x(t) = Ae^{-\lambda t}\sin(\sqrt{\omega^2 - \lambda^2}t + \phi) + g(\gamma)\sin(\gamma t + \theta), \tag{5.12}$$

and in the presence of damping, that is, when $\lambda \neq 0$, the first term decays away and leaves the second term as the dominant term in the solution. As before, the first term is called the transient solution and the second term is called the steady-state solution. The transient solutions for voltage and current have the same form. The steady-state amplitude is

$$g(\gamma) = \frac{F_0}{\sqrt{(\omega^2 - \gamma^2)^2 + 4\lambda^2\gamma^2}},$$

and $g(\gamma)$ has a maximum at $\gamma_1 = \sqrt{\omega^2 - 2\lambda^2}$. As $\lambda \to 0$ and $\gamma \to \omega$ the system begins to exhibit **resonance**. This unbounded resonance effect occurs in an undamped system when the period of the forcing function approaches the period of the free undamped system. This phenomenon is examined in the exercises. We have all heard this effect when a public address system goes into oscillation. The oscillation starts small and then grows to deafening proportions. In extreme cases the public address system—either the amplifier or speakers, or both—can be damaged or destroyed from the extreme current to which the components are exposed during such an event.

Exercises 5.4

Throughout these exercises we are concerned with either the homogeneous differential equation

$$L\frac{d^2q}{dt^2} + R\frac{dq}{dt} + \frac{1}{C}q = 0, \tag{H}$$

or with the nonhomogeneous differential equation

$$L\frac{d^2q}{dt^2} + R\frac{dq}{dt} + \frac{1}{C}q = f_0\cos\gamma t, \tag{N}$$

where L is in henrys, R is in ohms, C is in farads (usually microfarads: μf), q is in coulombs, and f_0 is in volts.

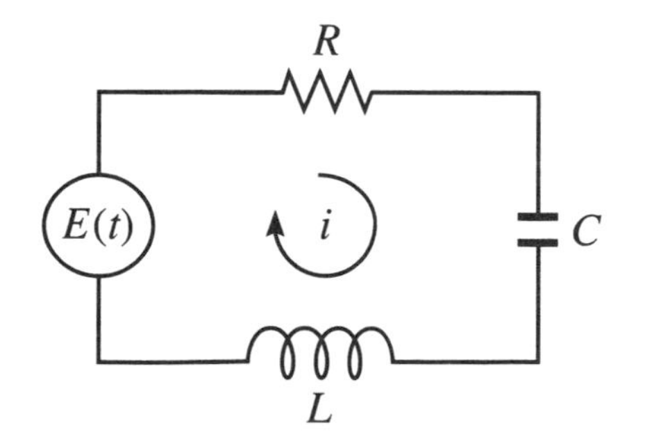

PART I. Solve these initial value problems that model the simple electronic circuit illustrated above. Convert each of the solutions to the forms of part I of exercises 5.3. Use $i = dq/dt$ to find the current.

1. $L = 0.05h$, $R = 20\Omega$, $C = 100\mu f$, $E(t) = 100V$; $q(0) = q'(0) = 0$.
2. $L = 0.05h$, $R = 20\Omega$, $C = 100\mu f$, $E(t) = (100\cos t)V$; $q(0) = q'(0) = 0$.
3. $L = 0.05h$, $R = 50\Omega$, $C = 0.0004\mu f$, $E(t) = 117V$; $q(0) = q'(0) = 0$.
4. $L = 0.05h$, $R = 50\Omega$, $C = 0.0004\mu f$, $E(t) = (120\cos 120\pi t)V$; $q(0) = q'(0) = 0$.
5. $L = 0.2h.$, $R = 300\Omega$, $C = 10\mu f$, $E(t) = 0$; $q(0) = 10 - 6$, $q'(0) = 0$.
6. $L = 0.2h.$, $R = 300\Omega$, $C = 10\mu f$, $E(t) = (120\cos 120\pi t)V$; $q(0) = 10^{-6}$, $q'(0) = 0$.
7. $L = 2h.$, $R = 400\Omega$, $C = 10\mu f$, $E(t) = (10\cos 120\pi t)V$; $q(0) = q'(0) = 0$.
8. $L = 2h.$, $R = 400\Omega$, $C = 10\mu f$, $E(t) = (10\cos 120\pi t)V$; $q(0) = 1$, $q'(0) = 0$.

PART II. Theory. Throughout this part of the exercises we are concerned with either the homogeneous differential equation (H) or with the nonhomogeneous differential equation (N), above, where L is in henrys, R is in ohms, C is in farads, q is in coulombs, and f_0 is in volts.

9. Express $g(\gamma)$, the amplitude of the steady-state term, in terms of L, R, and C.
10. Find where the maximim and minimum of the function $g(\gamma)$, occur as functions of L, R, and C.
11. In equation (N), found on page 217, take $R = 0$.
    (a) Show that the solution which also satisfies the initial conditions $q(0) = q'(0) = 0$ is $q(t) = (F_0/\omega^2 - \gamma^2)(\cos\gamma t - \cos\omega t)$, where $\omega = (1/\sqrt{LC})$ and $F_0 = f_0/L$.
    (b) Evaluate $\lim_{\gamma\to\omega} m(F_0/\omega^2 - \gamma^2)(\cos\gamma t - \cos\omega t)$.
    (c) Show that the limit is the solution of the initial-value problem

$$\frac{d^2q}{dt^2} + \omega^2 q = F_0\cos\omega t, \quad q(0) = q(0) = 0.$$

12. Rewrite the solution of the last problem as

$$\begin{aligned} q(t) &= \frac{F_0}{\omega^2 - \gamma^2}(\cos\gamma t - \cos\omega t) \\ &= \frac{-2F_0}{\omega^2 - \gamma^2}\sin\frac{1}{2}(\gamma - \omega)t\sin\frac{1}{2}(\gamma + \omega)t. \end{aligned}$$

When ω is positive, $\gamma - \omega < \gamma + \omega$. Define $\varepsilon = \frac{1}{2}(\gamma - \omega)$. When γ is near ω, then ε is small, and

$$q(t) \approx \frac{-F_0}{2\varepsilon\gamma} \sin \varepsilon t \sin \gamma t.$$

Calculate

$$\lim_{\varepsilon \to 0} \frac{-F_0}{2\varepsilon\gamma} \sin \varepsilon t \, \sin \gamma t.$$

13. Using *Mathematica*, plot the three functions

$$\frac{-F_0}{2\varepsilon\gamma}(\sin \varepsilon t)(\sin \gamma t)$$

and

$$\frac{\pm F_0}{2\varepsilon\gamma} \sin \varepsilon t$$

on the same axes for $\varepsilon = 1/2, \gamma = 5$, and $F_0 = 1$, using $0 \leq t \leq 30$. What you see illustrates the phenomenon known as **beats**. You plotted three curves on the same axes. Explain which is which. The effect you see also lies at the heart of AM (**amplitude modulated**) radio broadcasting, which includes short-wave, where a signal of the form $s(t) \sin \gamma t$ is broadcast. In this context, $s(t)$ is called the **signal** and $\sin \gamma t$ is called the **carrier**. The frequency γ is the frequency to which you tune your dial. (The actual situation is somewhat more complicated than this, but this is the basic idea.) [By contrast, **FM** signals are broadcast in the form $\sin(\gamma + s(t))t$, where $s(t)$ modulates the frequency, rather than the amplitude.]

5.5 Two Nonlinear Examples (Optional)

Even though this chapter and the previous one as well were devoted to linear differential equations, these two examples are of interest because almost everyone has heard of them. In each case we make some headway toward solving the equations using techniques that are new to us, and we obtain qualitative information from the differential equations themselves.

Escape Velocity

Newton's **law of gravitation** states that the mutual force of gravitational attraction acting between two bodies is proportional to the reciprocal of the square of the distance between them. Suppose that we are interested in the force acting on a small body of mass m_s due to the presence of a large body of mass m_l. If y is the distance separating the two centers with positive being in the direction of greater separation, then

$$m_s \frac{d^2y}{dt^2} = -\frac{Gm_s m_l}{y^2},$$

where G is Newton's **universal gravitation constant**. It is positive. The negative indicates that the force acts to oppose the separation of the two bodies. To solve this differential equation, divide through by m_s. Then let $v = dy/dt$ and multiply through by dy/dt to get

$$v\frac{dv}{dt} = -\frac{Gm_l}{y^2}\frac{dy}{dt}.$$

The variables are separated. Integrate both sides of the equation to get

$$\frac{1}{2}v^2 = \frac{Gm_l}{y} + c.$$

If the initial separation between the centers is $y_0 = R > 0$, and the initial velocity is v_0, then

$$c = \frac{1}{2}v_0^2 - \frac{Gm_l}{R}.$$

This means that

$$\frac{1}{2}v^2 = \frac{Gm_l}{y} + \frac{1}{2}v_0^2 - \frac{Gm_l}{R}.$$

We see from this that if the initial velocity v_0 is such that $\frac{1}{2}v_0^2 < Gm_l/R$, then $|v|$ decreases to 0 as y increases to

$$\frac{Gm_l}{(Gm_l/R) - (1/2){v_0}^2}.$$

After v reaches 0, then $|v|$ begins to increase again, and the smaller body picks up speed as it falls back toward the larger body. On the other hand, if $\frac{1}{2}v_0^2 \geq Gm_l/R$, then v can never become 0 since the right-hand side is always positive. The two bodies therefore continue to separate forever, though at a smaller and smaller rate. The smaller body is said to **escape** from the gravitational control of the larger. The initial velocity v_0 that permits this escape is

$$v_0 = \sqrt{\frac{2Gm_l}{R}}$$

and is called the **escape velocity**. Notice that the escape velocity depends on the mass m_l of the larger body, but not on the mass of the smaller one. It is often the case that R is the radius of the larger body, so that the larger body completely determines the escape velocity.

Planetary Orbits

In Chapter 8 we consider the equation(s) of motion of a small body about a larger body under the assumptions of Newtonian mechanics. We show that the motion lies in a plane and that if the larger body is located at the origin of a rectangular coordinate system then in polar coordinates (r, θ), the smaller body orbits according to the equation of motion:

$$\left(\frac{H^2}{r^2}\right)\frac{d^2r}{d\theta^2} - \left(\frac{2H^2}{r^3}\right)\left(\frac{dr}{d\theta}\right)^2 = \frac{H^2}{r} - G, \tag{5.13}$$

which is a second-order nonlinear differential equation for r in terms of θ. We take H and G to be positive constants. Their physical significance is explained in Chapter 8. One fascinating thing about this equation is that a simple transformation reduces it to a linear differential equation that we can solve by the methods of Chapter 4.

Let $r = 1/u$. Then

$$\frac{dr}{d\theta} = \frac{-1}{u^2}\frac{du}{d\theta}$$

and

$$\frac{d^2r}{d\theta^2} = \frac{2}{u^3}\left(\frac{du}{d\theta}\right)^2 - \frac{1}{u^2}\frac{d^2u}{d\theta^2}.$$

When these three expressions are substituted into equation 5.13 there results

$$u^2H^2\left(\frac{2}{u^3}\left(\frac{du}{d\theta}\right)^2 - \frac{1}{u^2}\frac{d^2u}{d\theta^2}\right) - 2H^2u^3\left(\frac{-1}{u^2}\frac{du}{d\theta}\right)^2 = H^2u - G,$$

or, after simplification,

$$\frac{d^2u}{d\theta^2} + u = \frac{G}{H^2}.$$

This is **Binet's equation**. It is a nonhomogeneous second-order differential equation with constant coefficients. Once u is known, we find $r(\theta) = 1/u(\theta)$. From our studies in Chapter 4 we know that

$$u(\theta) = \frac{G}{H^2} + c\cos(\theta + d).$$

Let $k = 1/c$, and $\varepsilon = cH^2/G$. Then

$$r(\theta) = \frac{1}{(G/H^2) + c\cos(\theta + d)} = \frac{k\varepsilon}{1 + \varepsilon\cos(\theta)} \tag{5.14}$$

if the initial conditions are chosen so that $d = 0$. This is a polar equation for a conic with one focus at the origin. See Figure 5.5. When the orbit is an ellipse the other focus is at $(\varepsilon a, 0)$, where $a = (k\varepsilon/(1 - \varepsilon^2))$ is the length of the semimajor axis. (See a standard calculus book such as Thomas/Finney.) The line $x = k$ is the **directrix** of the conic and the parameter $\varepsilon > 0$, called the **eccentricity** of the conic, determines the nature of the conic in accordance with this table:

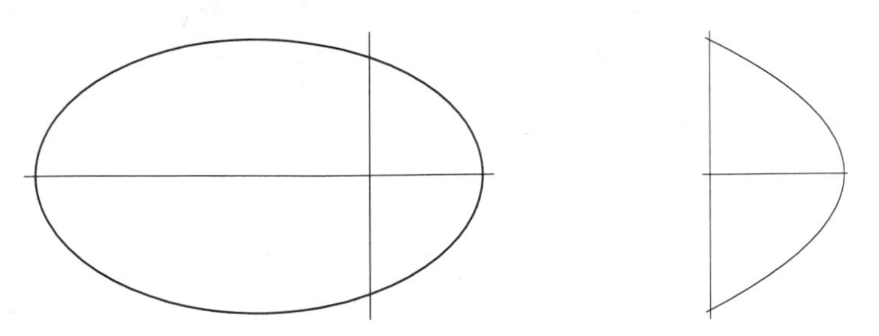

Figure 5.5: Elliptical and hyperbolic orbits.

$$\begin{aligned} \varepsilon > 1&: \quad \text{hyperbola} \\ \varepsilon = 1&: \quad \text{parabola} \\ 0 < \varepsilon < 1&: \quad \text{ellipse.} \end{aligned}$$

In your studies of physics you will learn that $\varepsilon > 1$ corresponds to a high-energy **hyperbolic orbit** that will pass by the larger body once with enough energy never to return, $\varepsilon < 1$ corresponds to a low-energy **elliptical orbit** such as planets and periodic comets (**Halley's Comet**, for instance) exhibit within our solar system. A **parabolic orbit**, $\varepsilon = 1$, is statistically so rare as to be considered impossible. At a minimum, measurement errors are likely to mis-classify it. The orbit would require so long to return, if it were truly elliptical, that it would be impossible to say that it had returned, and if it were properly hyperbolic we would not see it again, anyway, as would also be the case if it were parabolic.

6 The Laplace Transform

6.0 Introduction

In this chapter we introduce the Laplace[1] transform and show how to use it to solve differential equations of the kind discussed in Chapters 4 and 5—linear differential equations with constant coefficients. The Laplace transform really shines when a nonhomogeneous constant coefficients initial value problem must be solved. The Laplace transform package **LaplaceTransform.m** provided in *Mathematica* (versions 2.1 and later) is full-featured and easy to use. If you have a nonhomogeneous constant coefficients initial value problem to solve, the **LaplaceTransform.m** package should do it very effectively.

The Laplace transform is a linear transformation that, when applied to a linear initial value problem, converts the problem from a differential problem into an algebraic problem. The algebraic problem is easy to solve for the transform of the unknown function, and the solution to the differential problem can be obtained by reversing the process: the inverse Laplace transform. Traditionally, Laplace transforms have been applied by using a table of transforms to obtain the desired algebraic system, and then manipulating the algebraic system into a form where the terms of the solution can be identified in a table of inverse Laplace transforms. We will demonstrate this method, but will concentrate on letting *Mathematica* perform all of the manipulations for us.

The Laplace transform gives us a direct way to solve a constant coefficients initial value problem where the nonhomogeneous part is not continuous. The

[1]Pierre-Simon de Laplace (1749–1827), French man of science, was born of peasant parents. He sought out centers of power and found his in Paris, where he was a contemporary of d'Alembert and Lagrange. Best known by many for his work *Mèchanique Céleste* which exploited Newton's ideas to give a mathematical basis to the motions in the solar system, he is also recognized as the founder of the theory of probability. He accompanied Napoleon on many expeditions as principal scientific adviser.

methods of Chapter 4 also work, but they are tedious to apply because several different solutions on abutting subintervals have to be obtained and "glued" together to force continuity of the solution across points where the nonhomogeneous part is not continuous.

The Laplace transform takes problems such as this in easy stride. There are simple methods for expressing certain discontinuous functions in terms of step functions. The resulting functions have straightforward Laplace transforms and inverse Laplace transforms. For example, a stable spring system may be given a momentary shock that introduces motion. What is the nature of that motion? A switch might be thrown (open or closed) to stimulate an electronic circuit to change behavior. What is the nature of the behavior before and after the event?

Since the **LaplaceTransform.m** package provided by *Mathematica* is so general and powerful, its demands for computer resources and time are great. Thus, for our purposes in this chapter, a Laplace transform package, **LPT.m**, with fewer features, has been provided. It is *much* faster and requires fewer computer resources. **LPT.m** provides you with only enough functionality to do the problems in this chapter (and in Chapters 4, 5, and 8). It is, however, smaller and much faster than the general package. You should consider **LPT.m** to be a package for the beginner; the package **LaplaceTransform.m** is for professionals who need its power. You will not need its power unless you pursue problems of greater complexity than those in this text.

6.1 The Laplace Transform

The Laplace transform is an example of an integral transform. Here is its definition.

Definition 6.1 *Suppose that the function $y(t)$ is defined for $0 \le t < \infty$ and the improper integral $\int_0^\infty e^{-st} f(t)\,dt$ exists for $s > s_0$. Then the* **Laplace transform** *of $y(t)$ is said to exist for $s > s_0$ and we denote it by $Y(s) = \mathscr{L}(y(t)) = \int_0^\infty e^{-st} f(t)\,dt$.*

We will soon develop some theory to make our work easier, but for now we obtain the Laplace transform of some elementary functions. Before doing so, let us agree upon the notation

$$g(t)\,\Big|_a^\infty = \lim_{b\to\infty}\left(g(t)\,\Big|_a^b\right) = \left(\lim_{b\to\infty} g(b)\right) - g(a).$$

This will make things much simpler to write. Recall from your study of l'Hôpital's rule in calculus that $\lim_{t\to\infty}(t^m/e^{st}) = 0$ whenever $s > 0$. This is used several times in the derivations below.

$$\mathscr{L}(1) = \int_0^\infty e^{-st}\,1\,dt = \left.\frac{e^{-st}}{-s}\right|_0^\infty = \frac{1}{s} \text{ for } s > 0,$$

since $\lim_{b\to\infty} e^{-sb} = 0$ when $s > 0$.

$$\begin{aligned}
\mathcal{L}(t) &= \int_0^\infty e^{-st} t\,dt \\
&= \frac{te^{-st}}{-s}\bigg|_0^\infty + \frac{1}{s}\int_0^\infty e^{-st} 1\,dt \\
&= 0 + \frac{1}{s}L(1) \\
&= \frac{1}{s}\left(\frac{1}{s}\right) \\
&= \frac{1}{s^2} \text{ for } s > 0.
\end{aligned}$$

$$\begin{aligned}
\mathcal{L}(t^2) &= \int_0^\infty e^{-st} t^2\,dt \\
&= \frac{t^2 e^{-st}}{-s}\bigg|_0^\infty + \frac{2}{s}\int_0^\infty e^{st} t\,dt \\
&= 0 + \frac{2}{s}L(t) \\
&= \frac{2}{s^3} \quad \text{for} \quad s > 0.
\end{aligned}$$

It is left as an exercise in mathematical induction to show that

$$\mathcal{L}(t^n) = \frac{n!}{s^{n+1}}, \quad \text{for} \quad s > 0.$$

We also have

$$\begin{aligned}
\mathcal{L}(e^{kt}) &= \int_0^\infty e^{-st} e^{kt}\,dt \\
&= \int_0^\infty e^{(k-s)t}\,dt \\
&= \frac{e^{(k-s)t}}{k-s}\bigg|_0^\infty \\
&= \frac{1}{s-k}, \text{ for } s > k.
\end{aligned}$$

It is left as an exercise for you to find the Laplace transforms of $\sin t$ and $\cos t$ from their complex representations

$$\sin t = \frac{e^{it} - e^{-it}}{2i} \text{ and } \cos t = \frac{e^{it} + e^{-it}}{2}$$

by using the above transform (formally). We find each of these Laplace transforms directly using integration by parts twice.

$$\begin{aligned}\mathcal{L}(\sin t) &= \int_0^\infty e^{-st}\sin t\,dt\\ &= -\frac{e^{-st}\sin t}{s}\bigg|_0^\infty + \frac{1}{s}\int_0^\infty e^{-st}\cos t\,dt\\ &= 0 + \frac{1}{s}\mathcal{L}(\cos t).\end{aligned}$$

We can get $\mathcal{L}(\cos t)$ from this equation in a moment, but we continue with $\mathcal{L}(\sin t)$.

$$\begin{aligned}&= \frac{1}{s}\left[-\frac{e^{-st}\cos t}{s}\bigg|_0^\infty - \frac{1}{s}\int_0^\infty e^{-st}\sin t\,dt\right]\\ &= \frac{1}{s}\left[\frac{1}{s} - \frac{1}{s}\mathcal{L}(\sin t)\right]\\ &= \frac{1}{s^2} - \frac{1}{s^2}\mathcal{L}(\sin t), \quad \text{for} \quad s > 0.\end{aligned}$$

Thus, $s^2\mathcal{L}(\sin t) = 1 - \mathcal{L}(\sin t)$. So, $(s^2+1)\mathcal{L}(\sin t) = 1$, and

$$\mathcal{L}(\sin t) = \frac{1}{1+s^2}, \quad \text{for} \quad s > 0.$$

From $\mathcal{L}(\sin t) = \frac{1}{s}\mathcal{L}(\cos t)$ we see that

$$\mathcal{L}(\cos t) = s\,\mathcal{L}(\sin t) = \frac{s}{1+s^2}, \quad \text{for} \quad s > 0.$$

Once the package **LPT.m** is loaded, these transformations are very easy to obtain by *Mathematica* , using `Transform[y[t],t,s]` to transform `y[t]`. Here is how to load the package and obtain these results by *Mathematica.*

```
In[1]:=  <<LPT.m          (* Load LPT.m *)

In[2]:=  Transform[1,t,s]          (* transform from t to s *)

Out[2]=  1
         -
         s

In[3]:=  Transform[t,t,s]

Out[3]=   -2
         s

In[4]:=  Transform[t^2,t,s]

Out[4]=  2
         --
          3
         s
```

```
In[5]:=  Transform[Exp[k t],t,s]

Out[5]=    1
         ------
         -k + s

In[6]:=  Transform[Sin[t],t,s]

Out[6]=    1
         ------
              2
         1 + s

In[7]:=  Transform[Cos[t],t,s]

Out[7]=    s
         ------
              2
         1 + s
```

Being able to transform a given function may be an interesting exercise in calculus, but it is not as useful as we would like. What we need are some theorems that provide general rules about the behavior of Laplace transforms. These are provided in the next section.

Exercises 6.1

Load the package **LPT.m**. Use `Transform[ , t, s]` to reproduce these Laplace transforms that were found in the text.

1. 1.
2. t.
3. t^2.
4. e^{kt}.
5. $\sin t$.
6. $\cos t$.
7. Find $\mathcal{L}(t^n)$ for $t = 3, 4, 5$.
8. Using the definition, find $\mathcal{L}(\sinh t)$ and $\mathcal{L}(\cosh t)$.
9. Use mathematical induction to prove that for each positive integer n,

$$\mathcal{L}(t^n) = n!/s^{n+1}, \text{ for } s > 0.$$

10. Use the definition of the Laplace transform to formally find the Laplace transforms of $\sin t$ and $\cos t$ from their complex representations

$$\sin t = \frac{e^{it} - e^{-it}}{2i} \text{ and } \cos t = \frac{e^{it} + e^{-it}}{2}.$$

11. Use l'Hôpital's rule and mathematical induction to prove that for each positive integer m,

$$\lim_{t\to\infty} \frac{t^m}{e^{st}} = 0 \text{ whenever } s > 0.$$

6.2 Properties of the Laplace Transform

The Laplace transform can be applied to functions that are not continuous and to functions that are large for large values of their argument, but they cannot be too discontinuous or get too large too fast. These definitions clarify these thoughts.

Definition 6.2 *A function $f(t)$ is said to be* **piecewise continuous** *on the closed and bounded interval $[a, b]$, provided that either $f(t)$ is continuous, or there is an increasing sequence $a_0, a_1, \ldots, a_n$ such that on each open subinterval $a_{k-1} < t < a_k$, $f(t)$ is continuous and has a finite limit at each end of the subinterval. Such a function $f(t)$ is said to be piecewise continuous on $[0, \infty)$ if it is piecewise continuous on each closed and bounded subinterval of $[0, \infty)$.*

This explains "not too discontinuous." Here is how to insure that a function does not grow too fast.

Definition 6.3 *The function $f(t)$ defined for $0 \le t < \infty$ is said to be of* **exponential order** *provided that there is a number k and nonnegative numbers T and M such that for $t > T, |f(t)| \le Me^{kt}$.*

It is often convenient to verify the exponential order of a function $f(t)$ by calculating

$$\lim_{t\to\infty} \frac{|f(t)|}{e^{kt}} = L$$

for some suitably chosen number k. If L is finite, then M can be chosen to be any number larger than L (and this determines T). Otherwise, $f(t)$ is not of exponential order.

It is easy to verify that polynomials of any finite degree n, exponentials of the form e^{bt} with b constant, $\sin bt$, and $\cos bt$ are all of exponential order. So are finite sums and products of them. The set of functions just described is precisely the set of functions that can be obtained as solutions from the differential equations discussed in Chapters 4 and 5. It is for this reason that the Laplace transform is useful—it is applicable to this set of functions (and more).

Example 6.1 The polynomial t^3 is of exponential order.

Solution. To prove this, use l'Hôpital's rule three times:

$$\lim_{t\to\infty} \frac{t^3}{e^{kt}} = \lim_{t\to\infty} \frac{3t^2}{ke^{kt}} = \lim_{t\to\infty} \frac{6t}{k^2e^{kt}} = \lim_{t\to\infty} \frac{6}{k^3e^{kt}} = 0$$

for any positive number k. Therefore, if t is sufficiently large, $|t^3| < e^t$, and t^3 is of exponential order. ◇

It is clear that an extension of this argument will show that t^n is of exponential order for any positive integer n: just apply l'Hôpital's rule n times.

Example 6.2 Any exponential $f(t) = e^{bt}$ is of exponential order.

Solution. We have immediately that

$$\lim_{t\to\infty} \frac{e^{bt}}{e^{kt}} = \lim_{t\to\infty} e^{(b-k)t} = 0$$

for $k > b$. So $e^{bt} < e^{kt}$ for large t. ◇

The function $f(t) = e^{t^2}$ is not of exponential order. Note that

$$\lim_{t\to\infty} \frac{e^{t^2}}{e^{kt}} = \lim_{t\to\infty} e^{t^2-kt} = \lim_{t\to\infty} e^{t(t-k)} = \infty$$

since $t(t-k) > 0$ for large t. This means that there is no acceptable M.

Example 6.3 Any bounded function is of exponential order.

Solution. Call our bounded function $f(t)$ and assume that for all nonnegative $t, |f(t)| \le B$. Then

$$\lim_{t\to\infty} \frac{|f(t)|}{e^{kt}} \le \lim_{t\to\infty} \frac{B}{e^{kt}} = 0$$

for $k > 0$, which is what we need. Take $M = 1$, for instance. ◇

Since both $\sin t$ and $\cos t$ are bounded, they are of exponential order. It is left as an exercise for you to show that if $f(t)$ and $g(t)$ are both of exponential order, then so are $f(t) + g(t)$ and $f(t)g(t)$. This will verify that any function that is a solution to a differential equation from Chapters 4 or 5 is of exponential order.

The Fundamental Existence Theorem for Laplace Transforms

This theorem is the reason for our definitions of piecewise continuous and exponential order.

Theorem 6.1 *If the function $y(t)$ is piecewise continuous and of exponential order for $0 \le t < \infty$, then the Laplace transform of $y(t)$ exists. That is, there is a number s_0 such that $Y(s) = \mathscr{L}(y(t))$ exists for $s > s_0$.*

Proof. There are nonnegative numbers T, k, and M such that $|y(t)| \le Me^{kt}$ for $t > T$. Thus

$$\lim_{t\to\infty} \int_0^b e^{-st} y(t)\, dt = \int_0^T e^{-st} y(t)\, dt + \lim_{t\to\infty} \int_T^b e^{-st} y(t)\, dt.$$

The first integral exists. So does the second, from this calculation:

$$\begin{aligned}\left|\int_T^b e^{-st} y(t)\,dt\right| &\le \int_T^b e^{-st} |y(t)|\,dt \\ &\le \int_T^b e^{-st} M e^{kt}\,dt \\ &= \int_T^b M e^{(k-s)t}\,dt \\ &= M \frac{e^{(k-s)b} - e^{(k-s)T}}{k-s}.\end{aligned}$$

The limit of this latter function as $b \to \infty$ exists, provided that $s > k$. Therefore the Laplace transform of $y(t)$ exists. □

Linearity of Laplace Transforms

Theorem 6.2 $\mathscr{L}(f(t)+g(t)) = \mathscr{L}(f(t)) + \mathscr{L}(g(t))$ *and* $\mathscr{L}(cf(t)) = c\mathscr{L}(f(t))$ *when* c *is constant provided that the Laplace transform of each of* $f(t)$ *and* $g(t)$ *exists.*

Proof. When we apply to the improper integrals our convention about limits, we have

$$\begin{aligned}\mathscr{L}(f(t)+g(t)) &= \int_0^\infty e^{-st}(f(t)+g(t))\,dt \\ &= \int_0^\infty e^{-st} f(t)\,dt + \int_0^\infty e^{-st} g(t)\,dt \\ &= \mathscr{L}(f(t)) + \mathscr{L}(g(t)).\end{aligned}$$

Also,

$$\mathscr{L}(cf(t)) = \int_0^\infty e^{-st}(cf(t))\,dt = c\int_0^\infty e^{-st} f(t)\,dt = c\mathscr{L}(f(t)).$$ □

This example illustrates how to use the linearity of the Laplace transform.

Example 6.4 Determine the Laplace transform of $f(t) = 1 + 4t + 2e^{3t} - 5\sin t$.
Solution.

$$\begin{aligned}\mathscr{L}(1+4t+2e^{3t}-5\sin t) &= \mathscr{L}(1) + 4\mathscr{L}(t) + 2\mathscr{L}(e^{3t}) - 5\mathscr{L}(\sin t) \\ &= \frac{1}{s} + \frac{4}{s^2} + \frac{2}{s-3} - \frac{5}{1+s^2}.\end{aligned}$$ ◇

Laplace Transforms of Derivatives

In order to apply Laplace transforms to the solution of differential equations, we need to be able to find the Laplace transform of the derivative of a function. This theorem says what it is.

Theorem 6.3 *If $y'(t)$ has a Laplace transform and $\mathcal{L}(y(t)) = Y(s)$, then $\mathcal{L}(y'(t)) = sY(s) - y(0)$.*

Proof. Integrate by parts.

$$\begin{aligned}\mathcal{L}(y'(t)) &= \int_0^\infty e^{-st} y'(t)\, dt \\ &= e^{-st} y(t)\Big|_0^\infty + s\int_0^\infty e^{-st} y(t)\, dt \\ &= -y(0) + s\mathcal{L}(y(t)) \\ &= sY(s) - y(0).\end{aligned}$$

A similar argument shows that

$$\begin{aligned}\mathcal{L}(y''(t)) &= -y'(0) + s\mathcal{L}(y'(t)) \\ &= -y'(0) + s(-y(0) + s\mathcal{L}(y(t))) \\ &= -y'(0) - sy(0) + s^2\mathcal{L}(y(t)) \\ &= s^2Y(s) - y'(0) - sy(0).\end{aligned}$$ □

The following theorem is proved by mathematical induction.

Theorem 6.4 *For each positive integer n, if $y^{(n)}(t)$ exists and is piecewise continuous and of exponential order and $y^{(k)}(t)$ is continuous for $0 \le k \le n-1$, then*

$$\begin{aligned}\mathcal{L}(y^{(n)}(t)) - s^nY(s) - \sum_{j=0}^{n-1} s_j y^{(n-1-j)}(0)) \\ = s^nY(s) - y^{(n-1)}(0) - sy^{(n-2)}(0) - \cdots - s^{n-1}y(0).\end{aligned}$$ □

The Solution of Differential Equations

To illustrate the use of the linearity of Laplace transforms and of Theorem 6.4, consider this example, whose solution is $y(t) = e^{-3t}$.

Example 6.5 Use Laplace transforms to solve the first-order initial-value problem

$$y'(t) + 3y(t) = 0, \quad y(0) = 1.$$

Solution. We have

$$\begin{aligned} \mathscr{L}(y'(t) + 3y(t)) &= \mathscr{L}(0) = 0 \\ \mathscr{L}(y'(t)) + 3\mathscr{L}(y(t)) &= 0 \\ s\mathscr{L}(y(t)) - y(0) + 3\mathscr{L}(y(t)) &= 0. \end{aligned}$$

Use $y(0) = 1$ to get

$$(s+3)\mathscr{L}(y(t)) - 1 = 0.$$

Solve this for

$$\mathscr{L}(y(t)) = \frac{1}{s+3} = \frac{1}{s-(-3)} = \mathscr{L}(e^{-3t}).$$

Therefore $y(t) = e^{-3t}$. ◇

It is worth pointing out that if both $y(t)$ and $z(t)$ are continuous and $\mathscr{L}(y(t)) = \mathscr{L}(z(t))$, then $y(t) = z(t)$. We used this in the above example to get $y(t) = e^{-3t}$ from $\mathscr{L}(y(t)) = \mathscr{L}(e^{-3t})$.

Example 6.5M Use Laplace transforms from the package **LPT.m** to solve the first-order initial value problem

$$y'(t) + 3y(t) = 0, \quad y(0) = 1.$$

Solution.
Load the small Laplace transforms package **LPT.m**.

```
In[8]:= <<:RossDE:LPT.m
```

Transform the differential equation.

```
In[9]:= Transform[y'[t]+3y[t] == 0, t, s]

Out[9]= 3 LPT[y[t], t, s] + s LPT[y[t], t, s] - y[0] == 0
```

Substitute the initial condition into the previous result.

```
In[10]:= %/.y[0]->1

Out[10]= -1 + 3 LPT[y[t], t, s] + s LPT[y[t], t, s] == 0
```

Isolate LPT[y[t], t, s].

```
In[11]:= Solve[%,LPT[y[t], t, s]]

Out[11]=
                                 1
         {{LPT[y[t], t, s] -> -----}}
                               3 + s
```

The solution can now be identified.

Here is another familiar differential equation. It has solution $y(t) = \sin t$.

Example 6.6 Use Laplace transforms to solve the second-order initial value problem $y''(t) + y(t) = 0; \quad y(0) = 0, y'(0) = 1$.

Solution. Use linearity to get

$$\begin{aligned} \mathcal{L}(y''(t) + y(t)) &= \mathcal{L}(0) = 0 \\ \mathcal{L}(y''(t)) + \mathcal{L}(y(t)) &= 0 \\ s^2\mathcal{L}(y(t)) - y'(0) - sy(0) + \mathcal{L}(y(t)) &= 0. \end{aligned}$$

Substitute the initial values $y(0) = 0$ and $y'(0) = 1$ to get

$$(s^2 + 1)\mathcal{L}(y(t)) - 1 = 0.$$

Solve this for $\mathcal{L}(y(t)) = 1/(s^2 + 1) = \mathcal{L}(\sin t)$. Therefore $y(t) = \sin t$. ◇

Example 6.6M Use the function LPT in the package **LPT.m** to solve the second-order initial value problem $y''(t) + y(t) = 0; y(0) = 0, y'(0) = 1$.

*Solution.*Transform the differential equation. (We assume that **LPT.m** is still loaded.)

```
In[12]:= Transform[y''[t]+y[t] == 0, t, s]

Out[12]= LPT[y[t], t, s] + s (s LPT[y[t], t, s] - y[0]) - y'[0] == 0
```

Substitute the initial conditions.

```
In[13]:= %/.{y[0]->0,y'[0]->1}

Out[13]=                         2
         -1 + LPT[y[t], t, s] + s  LPT[y[t], t, s] == 0
```

Isolate LPT[y[t], t, s].

```
In[14]:= Solve[%,LPT[y[t], t, s]]

Out[14]=                            1
         {{LPT[y[t], t, s] -> ------}}
                                   2
                              1 + s
```

The solution can now be identified. ◇

Needless to say, not every problem will produce a function that you can immediately identify as the transform of a known function, but the idea is correct: transform an initial value problem (with variable t) and then manipulate the resulting algebraic function of s so that it is not (too) difficult to recover a function of t that has the same transform. This function of t is the desired solution.

The Scaling Property

We can often determine the Laplace transform of some function from the transform of a simpler function. Here is one way.

Theorem 6.5 *If $c \neq 0$ and $\mathscr{L}(f(t)) = F(s)$, then*

$$\mathscr{L}(f(ct)) = \frac{1}{c}F\left(\frac{s}{c}\right).$$

Proof. In $L(f(ct)) = \int_0^\infty e^{-st} f(ct)\,dt$ make the change of variables $\tau = ct$, to get

$$\begin{aligned}\mathscr{L}(f(ct)) &= \frac{1}{c}\int_0^\infty e^{-(s/c)\tau} f(\tau)\,d\tau \\ &= \frac{1}{c}F\left(\frac{s}{c}\right)\end{aligned}$$ □

Example 6.7 Apply Theorem 6.5 to obtain $\mathscr{L}(\sin bt)$ and $\mathscr{L}(\cos bt)$ from $\mathscr{L}(\sin t)$ and $\mathscr{L}(\cos t)$.

Solution. From the theorem,

$$\mathscr{L}(\sin bt) = \frac{1}{b}\left(\frac{1}{1+(s/b)^2}\right) = \frac{b}{b^2+s^2},$$

and

$$\mathscr{L}(\cos bt) = \frac{1}{b}\left(\frac{s/b}{1+(s/b^2)}\right) = \frac{s}{b^2+s^2},$$ ◇

Example 6.7M Obtain $\mathscr{L}(\sin bt)$ and $\mathscr{L}(\cos bt)$ using *Mathematica*.

Solution.

```
In[1]:= <<:RossDE:LPT.m

In[2]:= Transform[Sin[b t],t,s]

Out[2]=     b
         -------
          2    2
         b  + s

In[3]:= Transform[Cos[b t],t,s]

Out[3]=     s
         -------
          2    2
         b  + s
```
◇

Example 6.8 Solve the initial value problem

$$y''(t) + b^2 y(t) = 0;\ y(0) = 1,\ y'(0) = 0$$

using Laplace transforms.

Solution. Transform the problem:

$$\begin{aligned}\mathscr{L}(y''(t) + b^2 y(t)) &= 0\\ \mathscr{L}(y''(t)) + b^2 \mathscr{L}(y(t))) &= 0\\ s^2 \mathscr{L}(y(t)) - y'(0) - sy(0) + b^2 \mathscr{L}(y(t)) &= 0.\end{aligned}$$

Substitute the initial values $y(0) = 1$ and $y'(0) = 0$ to get

$$(s^2 + b^2)\mathscr{L}(y(t)) - s = 0.$$

Solve this for

$$\mathscr{L}(y(t)) = \frac{s}{s^2 + b^2} = \mathscr{L}(\cos bt).$$

Therefore $y(t) = \cos bt$. ◇

Further Transformation Rules

It is instructive to differentiate the Laplace transform partially with respect to s:

$$\begin{aligned}\frac{\partial}{\partial s}(\mathscr{L}(y(t))) &= \frac{\partial}{\partial s}\int_0^\infty e^{-st} y(t)\, dt\\ &= \int_0^\infty \frac{\partial}{\partial s} e^{-st} y(t))\, dt\\ &= -\int_0^\infty e^{-st} t y(t)\, dt\\ &= -\mathscr{L}(ty(t)).\end{aligned}$$

This gives us a new transformation rule.

Theorem 6.6

$$\mathscr{L}(ty(t)) = -\frac{\partial}{\partial s}(L(y(t))) = -Y'(s).$$

Example 6.9 Find $\mathscr{L}(t \sin bt)$, $\mathscr{L}(t \cos bt)$, and $\mathscr{L}(te^{bt})$.

Solution. According to Theorem 6.6,

$$\begin{aligned}\mathscr{L}(t \sin bt) &= -\frac{\partial}{\partial s}(\mathscr{L}(\sin bt))\\ &= -\frac{\partial}{\partial s}\left(\frac{b}{b^2 + s^2}\right)\end{aligned}$$

$$= \frac{2bs}{(b^2+s^2)^2}.$$

$$\mathcal{L}(t\cos bt) = -\frac{\partial}{\partial s}(\mathcal{L}(\cos bt))$$

$$= -\frac{\partial}{\partial s}\left(\frac{s}{b^2+s^2}\right)$$

$$= \frac{s^2-b^2}{(b^2+s^2)^2}.$$

$$\mathcal{L}(te^{bt}) = -\frac{\partial}{\partial s}(\mathcal{L}(e^{bt}))$$

$$= -\frac{\partial}{\partial s}\left(\frac{1}{s-b}\right)$$

$$= \frac{1}{(s-b)^2}.$$

◇

Example 6.9M Find $\mathcal{L}(t\sin bt)$, $\mathcal{L}(t\cos bt)$, and $\mathcal{L}(te^{bt})$ using *Mathematica*.
Solution. We continue with the **LPT.m** package loaded.

```
In[4]:=  Transform[t Sin[b t],t,s]

Out[4]=    2 b s
         ----------
           2    2 2
         (b  + s )

In[5]:=  Transform[t Cos[b t],t,s]

Out[5]=        2
            2 s          1
         ---------- - -------
           2    2 2    2    2
         (b  + s )    b  + s

In[6]:=  Transform[t Exp[b t],t,s]

Out[6]=             -2
         (-b + s)
```

◇

An immediate corollary of Theorem 6.6 suggests how to transform $t^n y(t)$.

Corollary 6.1

$$\mathcal{L}(t^n y(t)) = (-1)^n \frac{\partial^n}{\partial s^n}(\mathcal{L}(y(t))) = (-1)^n Y^{(n)}(s).$$

The proof is by mathematical induction, and is left as an exercise.

Here is another common situation: what is the Laplace transform of $e^{kt}y(t)$? The theorem is called the **translation theorem**.

Theorem 6.7 *If $y(t)$ has a Laplace transform $\mathcal{L}(y(t)) = Y(s)$ for $s > s_0$, then for $s > k + s_0$,*

$$\mathcal{L}(e^{kt}y(t)) = Y(s-k).$$

Proof.

$$\begin{aligned}\mathcal{L}(e^{kt}y(t)) &= \int_0^\infty e^{-st}e^{kt}y(t)\,dt \\ &= \int_0^\infty e^{-(s-k)t}y(t)\,dt \\ &= Y(s-k), \quad \text{for} \quad s > k + s_0.\end{aligned}$$

□

We can now find $\mathcal{L}(t^2e^{kt})$ in two different ways. Using $\mathcal{L}(t^2y(t))$ with $y(t) = e^{kt}$ gives

$$\mathcal{L}(t^2e^{kt}) = (-1)^2\frac{\partial^2}{\partial s^2}\left(\frac{1}{s-k}\right) = \frac{2}{(s-k)^3}.$$

Using $\mathcal{L}(e^{kt}y(t))$ with $y(t) = t^2$ gives

$$\mathcal{L}(t^2e^{kt}) = \left.\frac{2}{s^3}\right|_{s\to s-k} = \frac{2}{(s-k)^3}.$$

It does not matter which way we choose to transform, though the latter way was simpler in this case. You should try this Laplace transform by *Mathematica*. We close this section with a summary of our results, stated as a theorem. You will want to refer to this theorem when doing problems by hand. There are extensive tables such as this for a large class of functions. See, for instance, the readily available sources Abramowitz and Stegun, or Spiegel. Our choice is to let *Mathematica* do the work for us whenever possible.

When you are working with *Mathematica* you need not worry about recalling these results. They can be obtained on a moment's notice by issuing the proper invocation of the function **Transform**. When we get into the next section, you will begin to appreciate the work that *Mathematica* is doing for you when it calculates these transforms. This is especially true when we get to chapter 8 and apply Laplace transforms to systems of equations.

Theorem 6.8 (Table of Laplace Transforms) *Denote $\mathcal{L}(f(t)) = F(s)$, $\mathcal{L}(g(t)) = G(s)$, and $\mathcal{L}(y(t)) = Y(s)$. Then*

1. $\mathcal{L}(f(t) + g(t)) = \mathcal{L}(f(t)) + \mathcal{L}(g(t))$;
2. $\mathcal{L}(cf(t)) = c\mathcal{L}(f(t))$;

3. $\mathcal{L}(y^{(n)}(t)) = s^n Y(s) - \sum_{j=0}^{n-1} s^j y^{(n-1-j)}(0)$
$= s^n Y(s) - y^{(n-1)}(0) - sy^{(n-2)}(0) - \cdots - s^{n-1}y(0);$

4. $\mathcal{L}(t^n y(t)) = (-1)^n \frac{\partial^n}{\partial s^n}(\mathcal{L}(y(t))) = (-1)^n Y^{(n)}(s);$

5. $\mathcal{L}(f(ct)) = \frac{1}{c} F\left(\frac{s}{c}\right)$, c is constant;

6. $\mathcal{L}(e^{kt}y(t)) = Y(s-k);$

7. $\mathcal{L}(c) = \frac{c}{s}$, c is constant;

8. $\mathcal{L}(t^n) = \frac{n!}{s^{n+1}};$

9. $\mathcal{L}(e^{kt}) = \frac{1}{s-k};$

10. $\mathcal{L}(\sin bt) = \frac{b}{b^2+s^2};$

11. $\mathcal{L}(\cos bt) = \frac{s}{b^2+s^2};$

12. $\mathcal{L}(\sinh bt) = \frac{b}{s^2-b^2};$

13. $\mathcal{L}(\cosh bt) = \frac{s}{s^2-b^2}.$

The proofs of the latter two Laplace transforms are left as exercises 5 and 6 below.

Exercises 6.2

Load the package **LPT.m**. Use `Transform[ ,t,s]` to find these Laplace transforms.

1. $\mathcal{L}(1+3t)$
2. $\mathcal{L}(2-7t^3)$
3. $\mathcal{L}(\sin kt - k\cos kt)$
4. $\mathcal{L}(\cos kt + k\sin kt)$
5. $\mathcal{L}(\sinh bt)$
6. $\mathcal{L}(\cosh bt)$
7. $\mathcal{L}(e^{5t}+7)$
8. $\mathcal{L}(t+3e^{2t})$

9. $\mathscr{L}(1 + 3t^3 - 6e^{4t})$

10. $\mathscr{L}(\sinh 3t + \cosh 3t)$

These problems involve powers of t.

11. $\mathscr{L}(t \sin bt)$

12. $\mathscr{L}(t \cos bt)$

13. $\mathscr{L}(te^{bt})$

14. $\mathscr{L}(t(e^{5t} + 7))$

15. $\mathscr{L}(t^2 \sin bt)$

16. $\mathscr{L}(t^2 \cos bt)$

17. $\mathscr{L}(t^2 e^{bt})$

18. $\mathscr{L}(t^2(e^{5t} + 7))$

19. $\mathscr{L}(t^3 \sin bt)$

20. $\mathscr{L}(t^3 \cos bt)$

21. $\mathscr{L}(t^3 e^{bt})$

22. $\mathscr{L}(t^3(e^{5t} + 7))$

These problems involve e^{kt}.

23. $\mathscr{L}(e^{kt} \sin bt)$

24. $\mathscr{L}(e^{kt} \cos bt)$

25. $\mathscr{L}(e^{kt}(t \sin bt))$

26. $\mathscr{L}(e^{kt}(t \cos bt))$

These problems involve derivatives. Collect all terms involving $\mathscr{L}(y)$.

27. $\mathscr{L}(y'(t) - 3y(t))$

28. $\mathscr{L}(y''(t) + 4y'(t) - 10y(t))$

29. $\mathscr{L}(y^{(4)}(t) - 16y(t))$

30. $\mathscr{L}(y'''(t) + y''(t) - 5y'(t) + 36y(t))$

31. Show that a piecewise continuous function is bounded on each closed and bounded subinterval of its domain.

32. Show that if $f(t)$ and $g(t)$ are both of exponential order, then so are $f(t)+g(t)$ and $f(t)g(t)$.

6.3 The Inverse Laplace Transform

The inverse Laplace transform $\mathscr{L}^{-1}$ is defined by $\mathscr{L}^{-1}(\mathscr{L}(y(t))) = y(t)$. If $y(t)$ is continuous, then there is no difficulty, but if $z(t)$ differs from $y(t)$ at a finite number of points, then $\mathscr{L}(y(t)) = \mathscr{L}(z(t))$, so $\mathscr{L}^{-1}(Y(s)) = \mathscr{L}^{-1}(Z(s))$ and we have a potential problem. However, we seek functions that are solutions to differential equations and are therefore necessarily continuous. Thus, this potential problem is not a problem for us.

There is a general definition of the inverse Laplace transform that involves integrals, but we leave that to you to pursue if you choose. In those situations where we need them, the inverse Laplace transform is easy to determine.

These properties of the inverse Laplace transform follow from Theorem 6.8:

Theorem 6.9 (Table of Inverse Laplace Transforms) *Denote* $\mathscr{L}(f(t)) = F(s)$, $\mathscr{L}(g(t)) = G(s)$, *and* $\mathscr{L}(y(t)) = Y(s)$. *Then*

1. $\mathscr{L}^{-1}(F(s) + G(s)) = \mathscr{L}^{-1}(F(s)) + \mathscr{L}^{-1}(G(s))$;
2. $\mathscr{L}^{-1}(cF(s)) = c\mathscr{L}^{-1}(F(s))$;
3. $\mathscr{L}^{-1}(s^n Y(s) - \sum_{j=0}^{n-1} s_j y^{(n-1-j)}(0))) = y^{(n)}(t)$
4. $\mathscr{L}^{-1}((-1)^n Y^{(n)}(s)) = t^n y(t)$;
5. $\mathscr{L}^{-1}\left(F\left(\frac{s}{c}\right)\right) = cf(ct), c$ is constant;
6. $\mathscr{L}^{-1}(Y(s-k)) = e^{kt}y(t)$;
7. $\mathscr{L}^{-1}\left(\frac{c}{s}\right) = c, c$ is constant;
8. $\mathscr{L}^{-1}\left(\frac{n!}{s^{n+1}}\right) = t^n$;
9. $\mathscr{L}^{-1}\left(\frac{1}{s-k}\right) = e^{kt}$;
10. $\mathscr{L}^{-1}\left(\frac{b}{b^2+s^2}\right) = \sin bt$;
11. $\mathscr{L}^{-1}\left(\frac{s}{b^2+s^2}\right) = \cos bt$;
12. $\mathscr{L}^{-1}\left(\frac{b}{s^2-b^2}\right) = \sinh bt$;
13. $\mathscr{L}^{-1}\left(\frac{s}{s^2-b^2}\right) = \cosh bt$.

In general, if we know $\mathscr{L}(y(t)) = Y(s)$ and $\mathscr{L}(z(t)) = Z(s)$ then we can find $\mathscr{L}^{-1}(Y(s)Z(s))$. This derivation tells us how:

$$\begin{aligned}
Y(s)Z(s) &= \left(\int_0^\infty e^{-sv} y(v)\,dv\right)\left(\int_0^\infty e^{-su} y(u)\,du\right)\\
&= \int_0^\infty \int_0^\infty e^{-s(v+u)} y(v)z(u)\,dv\,du\\
&= \int_0^\infty \left[\int_0^\infty e^{-s(v+u)} y(v)\,dv\right] z(u)\,du\\
&= \int_0^\infty \left[\int_u^\infty e^{-st} y(t-u)\,dt\right] z(u)\,du, \quad [\text{Let } v = t-u.]\\
&= \int_0^\infty \left[e^{-st}\int_0^t y(t-u)z(u)\,du\right] dt, \quad [\text{order of integration}]\\
&= \mathscr{L}\left(\int_0^t y(t-u)z(u)\,du\right).
\end{aligned}$$

We therefore have this theorem.

Theorem 6.10 *If* $\mathscr{L}(y(t)) = Y(s)$ *and* $\mathscr{L}(z(t)) = Z(s)$, *then*

$$\mathscr{L}^{-1}(Y(s)Z(s)) = \int_0^t y(t-u)z(u)\,du.$$

The integral $(y * z)(t) = \int_0^t y(t-u)z(u)\,du$ is called the **convolution** of y and z. It is a kind of generalized product. You can show that it is associative and commutative, and that it distributes over addition.

Example 6.10 Find $\mathscr{L}^{-1}((1/s)(1/(1+s^2)))$.

Solution. From Theorem 6.9 we recognize $\mathscr{L}^{-1}(1/s) = 1$ and $\mathscr{L}^{-1}(1/(1+s^2)) = \sin t$, so that

$$\begin{aligned}
\mathscr{L}^{-1}((1/s)(1/(1+s^2))) &= \int_0^t 1\sin(t-u)\,du\\
&= \cos(t-u)\,\Big|_{u=0}^{u=t}\\
&= 1-\cos t.
\end{aligned}$$

◇

Normally, we will avoid the use of the integral by using partial fractions to write

$$\frac{1}{s}\left(\frac{1}{1+s^2}\right) = \frac{1}{s} - \frac{s}{1+s^2}$$

so that

$$\begin{aligned}
\mathscr{L}^{-1}\left(\frac{1}{s}\left(\frac{1}{1+s^2}\right)\right) &= \mathscr{L}^{-1}\left(\frac{1}{s}\right) - \mathscr{L}^{-1}\left(\frac{s}{1+s^2}\right)\\
&= 1-\cos t.
\end{aligned}$$

You may recall from Calculus that partial fractions, though straightforward, were algebraically messy. To avoid this messiness, we will use the built-in *Mathematica* function `Apart`. Here is an illustration:

```
In[7]:=  Apart[1/(s(1+s^2)),s]

Out[7]= 1      s
        - - ------
        s        2
            1 + s
```

Example 6.10M Find $\mathcal{L}^{-1}((1/s)(1/(1+s^2)))$ using *Mathematica.*

Solution. Assume that **LPT.m** is loaded.

```
In[8]:=  InverseLaplaceTransform[(1/s)*(1/(1+s^2)),s,t]

Out[8]=  1 - Cos[t]
```

◇

It is of particular importance to recognize when Part 6 of Theorem 6.9 should be invoked. For instance

$$\begin{aligned}\mathcal{L}^{-1}\left(\frac{1}{5-2s+s^2}\right) &= \mathcal{L}^{-1}\left(\frac{1}{5-2s+s^2}\right)\\ &= \frac{1}{2}\mathcal{L}^{-1}\left(\frac{2}{2^2+(s-1)^2}\right)\\ &= \frac{1}{2}e^t\sin 2t.\end{aligned}$$

We will rely on *Mathematica* to make these observations for us. The results obtained should always be checked for correctness.

In the smaller Laplace transform package **LPT.m** the transform itself is named `LPT` and the inverse transform is named `ILPT`. In the standard package **LaplaceTransform.m**, the respective names are `LaplaceTransform` and `InverseLaplaceTransform`. Here is a partial comparison of the two packages.

	LPT.m	LaplaceTransform.m
Load	`<<LPT.m`	`<<LaplaceTransform .m`
Transform	`Transform[y[t],t,s]`	`LaplaceTransform [y[t], t, s]`
Inverse Transform	`ILPT[Y[s],s,t]`	`InverseLaplaceTransform[Y[s],s,t]`
Solve d.e.	`LPTSolve`: Like `DSolve`	Manual steps required
Initial Conditions	Like `DSolve`	Manual substitution

The functions `Transform` and `LPTSolve` defined in the package will accept initial value problems in the form acceptable to `DSolve`. That is, the differential equation and all of its initial values can be included in a `List` as the first argument. For example,

```
Transform[{y''[t]+9y[t] == 0, y[0] == 1, y'[0] == 0}, t, s]
```

or

```
LPTSolve[{y''[t]+9y[t] == 0, y[0] == 1, y'[0] == 0}, y[t], t, s].
```

Note that the form of `LPTSolve` is identical to that of `DSolve` except for the addition of the last argument. This will also be the case in Chapter 8 when we use Laplace transforms to solve first-order linear systems of differential equations with constant coefficients. There, when there are two or more unknown functions, the second argument will be a list of these arguments.

Here is the solution of a differential equation using the package **LPT.m**. The problem is solved twice — once with the steps done manually, using `Transform`, `Isolate`, and `IPLT`, and once automatically using `LPTSolve`. After ILPT is invoked, the substitution `ILPT-> InverseLaplaceTransform` invokes the inverse Laplace transform function `InverseLaplaceTransform` which is defined in the package **LaplaceTransform.m**. The latter two steps have been combined in `InverseTransform[IsolatedVars, s, t]` which can be invoked immediately after isolating the transform.

Example 6.11M Use the package **LPT.m** to manually solve the initial value problem $y''(t) - 5y'(t) + 6y(t) = e^{2t}; y(0) = 1, y'(0) = -2$.

Solution. Load the **LPT.m** package.

```
In[9]:= <<:RossDE:LPT.m
```

Transform the initial value problem.

```
In[10]:= Transform[{y''[t]-5y'[t]+6y[t] == Exp[2t],
                                y[0] == 1,
                               y'[0] == -2},
                               t, s]

Out[10]= {6 LPT[y[t], t, s] - 5 (s LPT[y[t], t, s] - y[0]) +

                                                         1
          s (s LPT[y[t], t, s] - y[0]) - y'[0] == ------,
                                                      -2 + s

         y[0] == 1, y'[0] == -2}
```

Isolate the transform of the unknown function $y[t]$.

```
In[11]:= Isolate[%,y[t],t,s]

Out[11]=                      -3              -2     4
         {LPT[y[t], t, s] -> ------ - (-2 + s)   + ------}
                             -3 + s                -2 + s
```

Prepare to find the inverse transform.

```
In[12]:= ILPT[%,s,t]

Out[12]=
                       -3                               -2
          {y[t] -> ILPT[------, s, t] + ILPT[-(-2 + s)   , s, t] +
                       -3 + s

                   4
              ILPT[------, s, t]}
                   -2 + s
```

Actually find the inverse transform.

```
In[13]:= %/.ILPT->InverseLaplaceTransform

Out[13]=
                       2 t      3 t    2 t
          {y[t] -> 4 E    - 3 E    - E    t}
```

Capture the solution.

```
In[14]:= w[t_] = y[t]/.%

Out[14]=
            2 t      3 t    2 t
          4 E    - 3 E    - E    t
```

Restate and name the initial-value problem.

```
In[15]:= sys[t_,y_] = {y''[t]-5y'[t]+6y[t] == Exp[2t],
                                  y[0] == 1,
                                  y'[0] == -2}

Out[15]=
                                               2 t
          {6 y[t] - 5 y'[t] + y''[t] == E    , y[0] == 1, y'[0] == -2}
```

The equation and both initial conditions are correct.

```
In[16]:= Simplify[sys[t,w]]

Out[16]= {True, True, True}
```

The problem can be solved completely in one step this way:

```
In[17]:= LPTSolve[{y''[t]-5y'[t]+6y[t] == Exp[2t],
                                  y[0] == 1,
                                  y'[0] == -2},
                            y[t], t, s]
```

The transformed system:

```
      {6 LPT[y[t], t, s] - 5 (s LPT[y[t], t, s] - y[0]) +

                                                     1
   s (s LPT[y[t], t, s] - y[0]) - y'[0] == ------,
                                                  -2 + s

y[0] == 1, y'[0] == -2}
```

The unknown(s) isolated:

```
                         -3                -2     4
{LPT[y[t], t, s] -> ------- - (-2 + s)   + -------}
                    -3 + s                  -2 + s

Out[17]=
                    2 t      3 t    2 t
     {y[t] -> 4 E     - 3 E    - E    t}
```

The solution can be identified from this expression. One could capture the solution and check it as we have done previously. ◇

The Laplace transform can be used in this mechanical way to solve differential equations. In certain circumstances this is permissible, and sometimes even standard practice. However, one should always be alert for errors of one kind or another. Even *Mathematica* still has errors in it, though fewer and fewer remain with each new release. On the other hand a new release can introduce new errors that had not been there previously. The packages are not nearly as completely tested as the main kernel itself, so do not be surprised to find an occasional error. The thing to do is to isolate the error as carefully as possible and report it. Bugs can and do get fixed.

Exercises 6.3

Load the package **LPT.m**. Use

```
ILPT[ , s, t]/.ILPT->InverseLaplaceTransform
```

to find these inverse Laplace transforms.

1. $\mathcal{L}^{-1}\left(\dfrac{1}{s^2}\dfrac{1}{1+s^2}\right)$
2. $\mathcal{L}^{-1}\left(\dfrac{1}{s+2}\dfrac{1}{1+s^2}\right)$
3. $\mathcal{L}^{-1}\left(\dfrac{1}{s+2}\dfrac{s}{1+s^2}\right)$
4. $\mathcal{L}^{-1}\left(\dfrac{5}{16+s^2}\right)$
5. $\mathcal{L}^{-1}\left(10\dfrac{1}{s-3}\dfrac{1}{9+s^2}\right)$
6. $\mathcal{L}^{-1}(\dfrac{1}{1+(s+4)^2})$
7. $\mathcal{L}^{-1}\left(\dfrac{s+6}{1+(s+4)^2}\right)$

8. $\mathscr{L}^{-1}\left(\frac{s-3}{25+(s+4)^2}\right)$

9. $\mathscr{L}^{-1}\left(\frac{10}{s}\frac{1}{1+(s-2)^2}\right)$

10. $\mathscr{L}^{-1}\left(\frac{10}{s}\frac{s}{16+(s-2)^2}\right)$

Follow the steps suggested in Example 6.10 to solve these initial value problems.

11. $y'(t) + 3y(t) = \sin 2t, y(0) = 4.$
12. $y''(t) + 4y'(t) + 4y(t) = 0, y(0) = 1, y'(0) = -3.$
13. $y''(t) + 4y'(t) + 4y(t) = 1, y(0) = 1, y'(0) = -3.$
14. $y''(t) + 4y'(t) + 4y(t) = e^{-2t}, y(0) = 1, y'(0) = -3.$
15. $y''(t) + 4y'(t) + 4y(t) = 1 + 5e^{-2t}, y(0) = 1, y'(0) = -3.$
16. $y'''(t) - 9y'(t) = 0, y(0) = -2, y'(0) = 4, y''(0) = 0.$
17. $y'''(t) - 9y'(t) = 7, y(0) = -2, y'(0) = 4, y''(0) = 0.$
18. $y'''(t) - 9y'(t) = 4e^{3t}, y(0) = -2, y'(0) = 4, y''(0) = 0.$
19. $y'''(t) - 9y'(t) = 3e^{-3t}, y(0) = -2, y'(0) = 4, y''(0) = 0.$
20. $y'''(t) - 9y'(t) = 7 + 4e^{3t} + 3e^{-3t}, y(0) = -2, y'(0) = 4, y''(0) = 0.$
21. $y''(t) + 36y(t) = 0, y(0) = a, y'(0) = b.$
22. $y''(t) + 36y(t) = 3e^{-3t}, y(0) = a, y'(0) = b.$
23. $y''(t) + 36y(t) = te^t \sin 3t, y(0) = a, y'(0) = b.$
24. $y''(t) + 36y(t) = t^2e^t \sin 3t, y(0) = a, y'(0) = b.$

6.4 Discontinous Functions and Their Transforms

The methods of Chapter 4 were fully capable of finding the solution of nonhomogeneous linear initial value problems for differential equations with constant coefficients and continuous nonhomogeneous parts. Some processes, however, are not continuous. For instance, a mechanical light switch goes from off to on without passing through intermediate values.

In calculus, one was often asked to discuss piecewise continuous functions such as

$$f(t) = \begin{cases} t^2, & 0 \le t \le 1 \\ \frac{1}{2}\left(t - \frac{3}{2}\right)^2, & 1 < t \le 2 \\ 6 - t^2, & 2 < t \end{cases} \tag{6.1}$$

In *Mathematica* one can define this function as

```
f[t_] := If[TrueQ[0 <= t <= 1],
   (*then*)
   t^2,
   (*else*)
 If[TrueQ[1 < t <= 2],
   (*then*)
   1/2 (t-3/2)^2,
   (*else*)
   If[TrueQ[2 < t],(*then*) 6-t^2]
   ]
  ]
```

This definition is even better:

```
f[t_] := t^2              /;TrueQ[N[0<=t<=1]]
f[t_] := 1/2(t-3/2)^2     /;TrueQ[N[1<t<=2]]
f[t_] := 6-t^2            /;TrueQ[N[2<t]]
```

This may look like three separate definitions, but it emphasizes the fact that there are three parts to the definition of $f(t)$. The `Condition` function, denoted `/;` , controls which of the three parts of the definition holds when t is given a value. The function `N`, which evaluates its arguments numerically, prevents problems when some of the constants are symbolic. An example of a symbolic constant is $1+\sqrt{2}$, or π.

Either of these two definitions is awkward to use to calculate the Laplace transform of $f(t)$, even though the Laplace transform is not difficult to express:

$$\mathscr{L}(f(t)) = \int_0^1 e^{-st}t^2\,dt + \int_1^2 e^{-st}\frac{1}{2}(t-32)^2\,dt + \int_3^\infty e^{-st}(6-t^2)\,dt.$$

Each of these integrals can be evaluated without unreasonable difficulty.

The function $f(t)$ has this appearance:

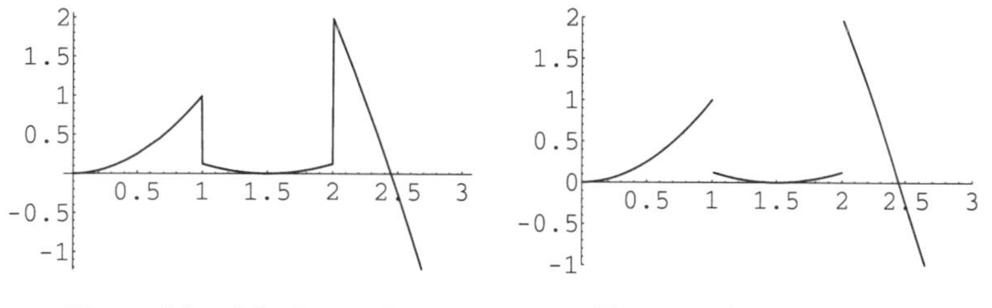

Plotted by *Mathematica*. The usual appearance.

You will notice that *Mathematica* connects the jumps in the definition of $f(t)$ with vertical line segments. One can force *Mathematica* to construct a graph as on the right which has the usual appearance, but the construction has to be done manually.

The evaluation of the Laplace transform of $f(t)$ made above is only an *ad hoc* procedure. It would be helpful to have a general procedure. The help we need is the Heaviside[2] **unit step function**, which has the definition

$$\mathcal{U}(t-c) = \begin{cases} 0, & t \leq c \\ 1, & c < t \end{cases},$$

and looks like this:

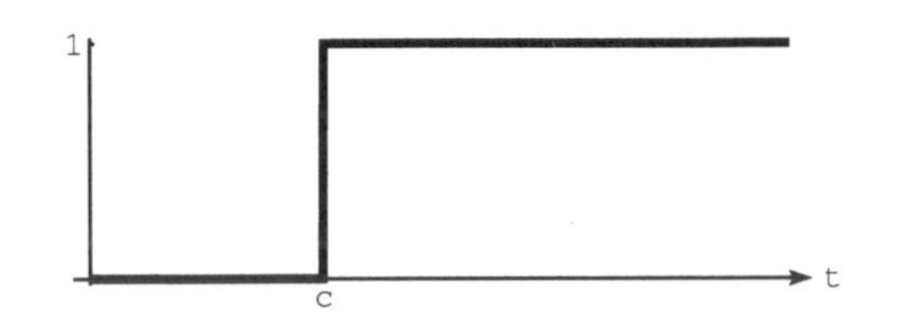

For purposes of the Laplace transform, if $c < 0$, $\mathcal{U}(t-c) = 1$ (since $t \geq 0$). The package **LPT.m** defines the unit step function `UnitStep` this way:

```
UnitStep[x_] := If[TrueQ[x > 0], 1, 0]/; NumberQ[N[x]]
```

As you can see, this is more general than the function $\mathcal{U}(t-c)$ defined above. This definition of the unit step function facilitates plotting, as in the case of this rectified half-wave:

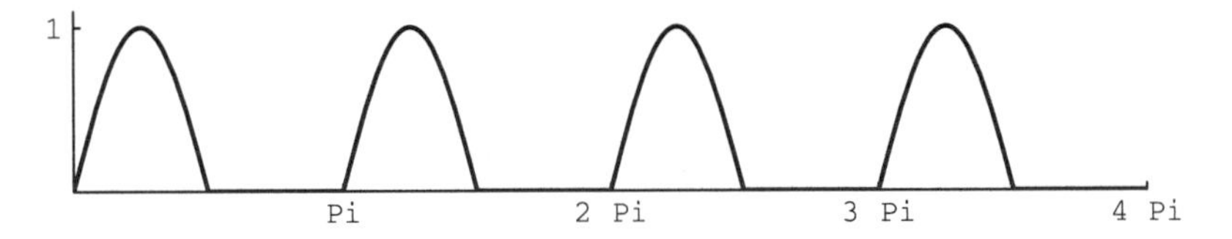

Figure 6.1: Rectified half-wave picture of `Sin[2t] UnitStep[Sin[2t]]`.

We will not use `UnitStep` in this way when using **LPT.m**, even though such use is permitted when using the general package **LaplaceTransform.m**.

The function $1 - \mathcal{U}(t-c)$ has a single down-going step.

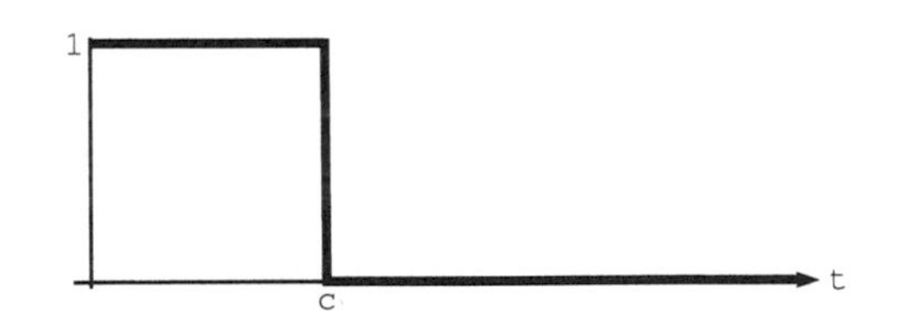

---

[2]Oliver Heaviside (1850–1925) was a British physicist. He became deaf at age 24. This forced him to quit his job with the Great Northern Telegraph company. Thereafter, he worked at his home in Devonshire on electromagnetic theory, where his work on the theory of the telephone made long-distance telephony possible. His nonstandard theoretical methods made it necessary for him to publish the results of his work privately.

More generally, we define the unit pulse function

$$\mathcal{P}(t,a,b) = \mathcal{U}(t-a) - \mathcal{U}(t-b), \text{for } a < b.$$

This function looks like:

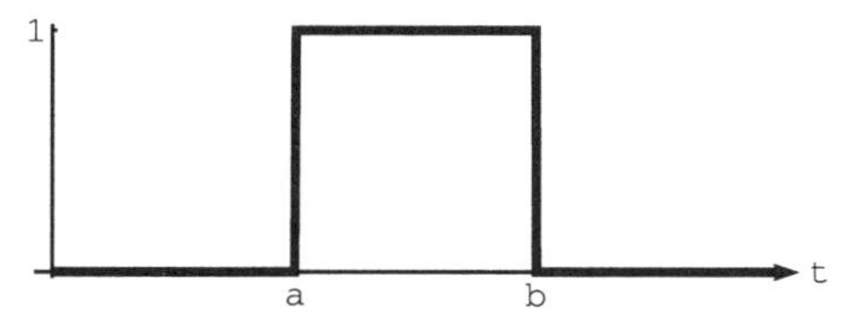

When $b < a$, as defined, this function is -1 between b and a. As a result, it is more useful to define

$$\mathcal{P}(t,a,b) = 1 + \mathcal{U}(t-a) - \mathcal{U}(t-b), \text{ for } b < a.$$

In both cases, the function rises at a and falls at b. When $b < a$ the function looks like:

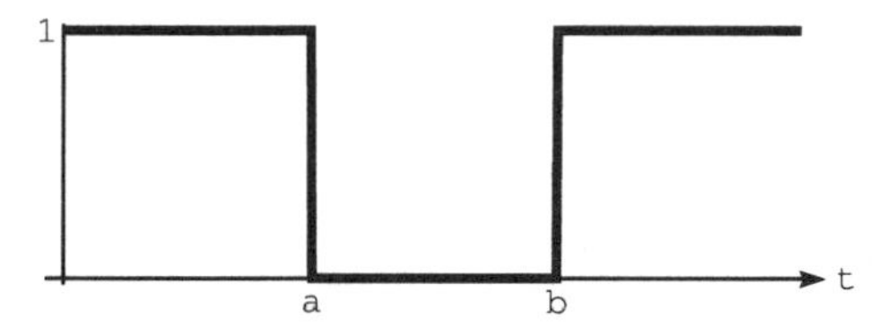

We can now describe the original function $f(t)$ in terms of the unit pulse function $\mathcal{P}$ and the unit step function $\mathcal{U}$ as

$$f(t) = t^2\mathcal{P}(t,0,1) + (1/2)(t-3/2)^2\mathcal{P}(t,1,2) + (6-t^2)\mathcal{U}(t-2).$$

This is progress, as we shall see in the paragraphs that follow, where the Laplace transform of $\mathcal{U}(t-c)$ is developed.

The Laplace Transform of $\mathcal{U}(t-c)$

For $c \geq 0$, we have

$$\begin{aligned}
\mathscr{L}(\mathcal{U}(t-c)) &= \int_0^c e^{-st}0\,dt + \int_c^\infty e^{-st}1\,dt \quad [\text{Let } \tau = t-c.] \\
&= 0 + \int_0^\infty e^{-s(\tau+c)}d\tau \quad [\text{Let } \tau = t.] \\
&= e^{-sc}\int_0^\infty e^{-st}1\,dt \\
&= e^{-sc}L(1) \\
&= \frac{e^{-sc}}{s}.
\end{aligned}$$

A similar calculation gives us this theorem and a corollary.

Theorem 6.11 *If $f(t)$ is defined for nonnegative t and has a Laplace transform, then for $c > 0$,*

$$\mathcal{L}(\mathcal{U}(t-c)f(t)) = e^{-sc}\mathcal{L}(f(t+c)).$$

Corollary 6.2 $\mathcal{L}(\mathcal{U}(t-c)f(t-c)) = e^{-sc}\mathcal{L}(f(t)) = e^{-sc}F(s)$.

Proof of the theorem: For $c \geq 0$, we have

$$\begin{aligned}
\mathcal{L}(\mathcal{U}(t-c)f(t)) &= \int_0^c e^{-st}0f(t)\,dt + \int_c^\infty e^{-st}f(t)\,dt \quad [\text{Let } \tau = t-c.] \\
&= 0 + \int_0^\infty e^{-s(\tau+c)}f(\tau+c)\,d\tau \quad [\text{Let } \tau = t.] \\
&= e^{-sc}\int_0^\infty e^{-st}f(t+c)\,dt \\
&= e^{-sc}\mathcal{L}(f(t+c)).
\end{aligned}$$

□

Aside from its utility in graphing, the simple effect of `UnitStep` when performing Laplace transforms is what makes it so useful to us. Though $e^{-sc}\mathcal{L}(f(t+c))$ looks superficially similar to $\mathcal{L}(e^{kt}f(t+c))$, it is significantly different, in that the factor e^{-sc} is the result of a transformation and the factor e^{kt} initiates a transformation.

We can now in effect add a new row to the bottom of Theorems 6.8 and 6.9.

Theorem 6.12 (6.8 and 6.9: last row) *If $f(t)$ has a Laplace transform then*

$$\mathcal{L}(\mathcal{U}(t-c)f(t)) = e^{-sc}\mathcal{L}(f(t+c))$$

and

$$\mathcal{L}^{-1}(e^{-sc}\mathcal{L}(f(t))) = \mathcal{U}(t-c)f(t-c).$$

Example 6.12M Find the Laplace transform of

$$\mathcal{U}(t-c)f(t)$$

and of

$$f(t) = t^2\mathcal{P}(t,0,1) + (1/2)(t-3/2)^2\mathcal{P}(t,1,2) + (6-t^2)\mathcal{U}(t-2)$$

by *Mathematica*. *Solution.*

```
In[18]:= <<:RossDE:LPT.m                    (* Load LPT.m *)

In[19]:= Transform[UnitStep[t-c]f1[t],t,s]

Out[19]= LPT[f1[c + t], t, s]
         --------------------
              c s
             E
```

```
In[20]:= Transform[t^2*UnitPulse[t,0,1]+
                   1/2 (t-3/2)^2*UnitPulse[t,1,2]+
                   (6-t^2)*UnitStep[t-2],
                          t,s]

Out[20]= 2       3        1         9          5          15
         -- - -------- - ----- - ---------- - -------- + -------- -
          3    2 s  3     s  3     2 s  2       s  2      2 s
         s    E    s     E  s   2 E    s    2 E   s    8 E    s

           7
         ------
            s
         8 E  s
```

Check this result.

```
In[21]:= InverseLaplaceTransform[%,s,t]

Out[21]=  2   15 UnitStep[-2 + t]   9 (-2 + t) UnitStep[-2 + t]
         t  + ------------------- - --------------------------- -
                       8                         2

                   2
          3 (-2 + t)  UnitStep[-2 + t]   7 UnitStep[-1 + t]
          ----------------------------- - ------------------ -
                       2                          8

                                                    2
          5 (-1 + t) UnitStep[-1 + t]   (-1 + t)  UnitStep[-1 + t]
          --------------------------- - ----------------------------
                       2                            2

In[22]:= Simplify[%-(t^2*UnitPulse[t,0,1]+
                   1/2(t-3/2)^2*UnitPulse[t,1,2]+
                   (6-t^2)*UnitStep[t-2])]

Out[22]=  2
         t  (1 - UnitStep[t])
```

This expression is 0 for $t >= 0$, so the transform is correct. ◇

The Dirac Delta

Before we proceed to solving differential equations with discontinuous nonhomogeneous parts, we can gain some more insight into $\mathcal{U}(t-c)$. In 1939 Dirac[3] proposed a "function" $\delta(t-c)$ that had these properties: (1) $\delta(t-c) = 0$ if $t \neq c$, and (2) $\int_{-\infty}^{\infty} \delta(t-c)\,dt = 1$.

[3]Paul A. M. Dirac (1902–1984), British physicist. He won a Nobel Prize at age 31 for his pioneering work in quantum mechanics. Though he recognized that his "delta-function" was not a function in the usual sense, his intuition was unerring, and his vision was vindicated when the theory of distributions or generalized functions was developed some years later. The Dirac δ was only one of the symbolic tools he created that enabled him to expand the theory of quantum mechanics so broadly.

It is clear that such an object is not a function in the usual sense. It does have its uses, however. One can find that $\int_{-\infty}^{\infty} \delta(t-c)\, f(t)\, dt = \int_{-\infty}^{\infty} \delta(t-c)\, f(c)\, dt = f(c), c \geq 0$. Also, $\mathscr{L}(\delta(t-c)) = e^{-sc}, c \geq 0$. What is most interesting to us is that $\int_{-\infty}^{t} \delta(\tau - c)\, d\tau = \mathcal{U}(t-c)$.

This tells us that, formally at least, $\mathcal{U}'(t-c) = \delta(t-c)$. This is worth pondering.

In *Mathematica* , $\delta(t-c)$ is typed as `DiracDelta[t-c]`, and is output as `DiracDelta[-c+t]`. The reason that the Dirac delta function $\delta(t-c)$ is mentioned is that if you solve a differential equation with nonhomogeneous part involving $\mathcal{U}(t-c)$ for some choice or choices of c, then the solution will be expressed in terms of $\mathcal{U}(t-c)$ for the same c. If you proceed to check your solution, you need to calculate derivatives of $\mathcal{U}(t-c)$. This means that $\delta(t-c)$ will appear. Higher-order derivatives of $\mathcal{U}(t-c)$ require higher derivatives of $\delta(t-c)$. This can be complicated. The most helpful observation you can make is that $\delta(t-c)$ and all derivatives of $\delta(t-c)$ (whatever they may mean) are zero if $t \neq c$.

If you have an expression, each term of which has a factor of $\delta(t-c)$ or any derivative of $\delta(t-c)$, then that expression is zero when $t \neq c$. The left- and right-hand limits of such a function are of interest, but we leave their investigation to a more advanced course.

The Solution of Differential Equations Having Discontinuous Nonhomogeneous Part

Since we can now easily represent discontinuous functions using $\mathcal{U}(t-c)$ so that the Laplace transforms are readily obtained, we can proceed to illustrate the process of solving differential equations that have a discontinuous nonhomogeneous part. Here is the first example.

Example 6.13M Using *Mathematica* solve the initial value problem

$$y''(t) + 25y(t) = \mathcal{U}(t-1),\ y(0) = 0,\ y'(0) = 0.$$

Observe that the solution is zero for $t < 1$ and only departs from zero for $t > 1$.

Solution. Here is how the solution process looks.

```
In[1]:=  <<:RossDE:LPT.m                    (* Load LPT.m *)
```

Perform the Laplace transform.

```
In[2]:=  Transform[{y''[t]+25y[t] == UnitStep[t-1],
                   y[0] == 0, y'[0] == 0},
                   t,s]

Out[2]=
                                    2
         {25 LPT[y[t], t, s] + s  LPT[y[t], t, s] - s y[0] -

                    1
           y'[0] == ----, y[0] == 0, y'[0] == 0}
                     s
                    E  s
```

Solve for the Laplace transform of y[t].

```
In[3]:=  Isolate[%,y[t],t,s]

Out[3]=
                                 1                s
         {LPT[y[t], t, s] -> -------- - ------------------}
                                s          s          2
                            25 E  s    25 E  (25 + s )
```

Perform the inverse Laplace transform.

```
In[4]:=  InverseTransform[%,s,t]

Out[4]=              UnitStep[-1 + t]
         {y[t] -> ---------------- -
                         25

            Cos[5 (-1 + t)] UnitStep[-1 + t]
            --------------------------------}
                           25
```

Capture the solution.

```
In[5]:=  w[t_] = y[t]/.%

Out[5]=  UnitStep[-1 + t]   Cos[5 (-1 + t)] UnitStep[-1 + t]
         ---------------- - --------------------------------
                25                         25
```

Plot the solution function.

```
In[6]:=  Plot[w[t],{t,0,3}];
```

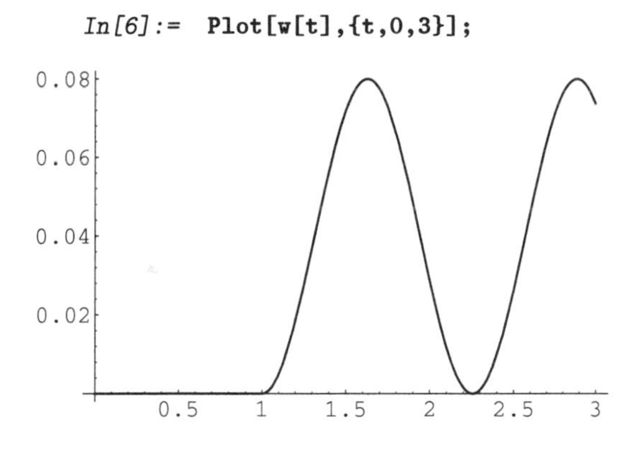

Restate the differential equation as an expression. This should evaluate to 0 when y->w. Note the use of D, rather than Derivative.

```
In[7]:=  L[t_,y_] := D[y[t],{t,2}]+25y[t]-UnitStep[t-1]
```

Check the solution.

```
In[8]:=  Simplify[L[t,w]]

Out[8]=
                   5 (1 - t)               5 (1 - t)
         (-2 Sin[---------] (10 Cos[---------] DiracDelta[-1 + t] -
                       2                       2

                   5 (1 - t)
              Sin[---------] DiracDelta'[-1 + t])) / 25
                       2
```

This expression is 0 when $t \neq 1$. The solution therefore checks. ◇

Here is a more substantial example using the function $f(t)$ defined in equation 6.1 by a three-part rule.

Example 6.14M Use *Mathematica* to solve the initial value problem

$$\begin{aligned} y''(t) + 5y'(t) + 6y(t) &= t^2\mathcal{P}(t,0,1) + (1/2)(t-3/2)^2\mathcal{P}(t,1,2) \\ &\quad + (6-t^2)\mathcal{U}(t-2), \\ y(0) &= 0, \\ y'(0) &= -2. \end{aligned}$$

Solution. Define the function f[t] having a three-part rule.

```
In[9]:=  f[t_] :=       t^2*UnitPulse[t,0,1]+
                        1/2 (t-3/2)^2*UnitPulse[t,1,2]+
                        (6-t^2)*UnitStep[t-2]
```

Perform the Laplace transform.

```
In[10]:= Transform[{y''[t]+5y'[t]+6y[t] == f[t],
                         y[0] == 0, y'[0] == -2},
                         t,s]

Out[10]=
                                   2
         {6 LPT[y[t], t, s] + s  LPT[y[t], t, s] +

            5 (s LPT[y[t], t, s] - y[0]) - s y[0] - y'[0] ==

            2      3         1          9           5          15
            -- - -------- - ------ - ---------- - --------- + -------- -
             3    2 s  3     s  3      2 s  2       s  2        2 s
            s    E    s     E  s    2 E    s     2 E  s     8 E    s

              7
            ------, y[0] == 0, y'[0] == -2}
              s
            8 E  s
```

Solve for the Laplace transform of y[t].

```
In[11]:= Isolate[%,y[t],t,s]

Out[11]= {LPT[y[t], t, s] ->

              s       2 s          s       2 s
        -3 - E  + 2 E         6 + 5 E  + 5 E
        ----------------- - ---------------- +
             2 s  3               2 s  2
          6 E    s            18 E    s

                s          2 s          s        2 s
         291 + 49 E  + 76 E        27 + E  + 36 E
         ---------------------- - ----------------- +
                  2 s                   2 s
             432 E    s            16 E    (2 + s)

                s          2 s
         219 - 11 E  + 448 E
         ----------------------}
                 2 s
            216 E    (3 + s)
```

Perform the inverse Laplace transform.

```
In[12]:= InverseTransform[%,s,t]

Out[12]=
                 76 + 291 UnitStep[-2 + t] + 49 UnitStep[-1 + t]
      {y[t] -> ------------------------------------------------ +
                                     432

         448    219 UnitStep[-2 + t]    11 UnitStep[-1 + t]
         --- + ---------------------- - --------------------
          3 t       3 (-2 + t)               3 (-1 + t)
         E         E                        E
         --------------------------------------------------- -
                                 216

          36    27 UnitStep[-2 + t]    UnitStep[-1 + t]
         --- + --------------------- + -----------------
          2 t       2 (-2 + t)           2 (-1 + t)
         E         E                    E
         ------------------------------------------------ -
                                16

         (5 t + 6 (-2 + t) UnitStep[-2 + t] +

            5 (-1 + t) UnitStep[-1 + t]) / 18 +

                        2
           2   3 (-2 + t)  UnitStep[-2 + t]
         (t  - ----------------------------- -
                            2

                    2
            (-1 + t)  UnitStep[-1 + t]
            ---------------------------) / 6}
                         2
```

Capture the solution. Suppress the output.

```
In[13]:= w[t_] = y[t]/.%;
```

Plot the solution function.

```
In[14]:= Plot[w[t],{t,0,3}];
```

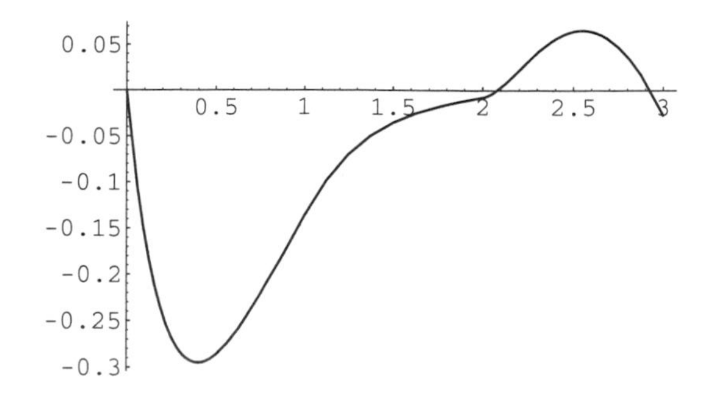

Restate the differential equation as an expression. This should evaluate to 0 when y->w. Note the use of D, rather than Derivative.

```
In[15]:= L[t_,y_] := D[y[t],{t,2}]+5D[y[t],t]+6y[t]-f[t]
```

Check the solution.

```
In[16]:= Simplify[L[t,w]]

Out[16]=  2   437 DiracDelta[-2 + t]
         t  + ---------------------- -
                       144

          6 - 3 t
       73 E        DiracDelta[-2 + t]
       ------------------------------ -
                     72

          4 - 2 t
       27 E        DiracDelta[-2 + t]
       ------------------------------ +
                     16

                                        2
       7 t DiracDelta[-2 + t]      5 t  DiracDelta[-2 + t]
       ---------------------- -    ----------------------- +
                 3                            4

       569 DiracDelta[-1 + t]
       ---------------------- +
                432

          3 - 3 t
       11 E        DiracDelta[-1 + t]
       ------------------------------ -
                    216

        2 - 2 t
       E        DiracDelta[-1 + t]   8 t DiracDelta[-1 + t]
       -------------------------- - ---------------------- -
                  16                           9
```

```
     2
5 t  DiracDelta[-1 + t]     2
------------------------ - t  UnitStep[t] +
          12

49 DiracDelta'[-2 + t]
---------------------- +
         144

    6 - 3 t
73 E         DiracDelta'[-2 + t]
-------------------------------- -
              72

    4 - 2 t
27 E         DiracDelta'[-2 + t]
-------------------------------- +
              16

                             2
2 t DiracDelta'[-2 + t]     t  DiracDelta'[-2 + t]
----------------------- - ----------------------- +
           3                         4

133 DiracDelta'[-1 + t]
----------------------- -
          432

    3 - 3 t
11 E         DiracDelta'[-1 + t]
-------------------------------- -
              216

 2 - 2 t
E         DiracDelta'[-1 + t]     t DiracDelta'[-1 + t]
----------------------------- - --------------------- -
             16                           9

 2
t  DiracDelta'[-1 + t]
----------------------
          12
```

These observations are in order: (1) there are two t^2 terms in the expression that combine into the single expression `t^2(1 - UnitStep[t])`, which is zero for $t \geq 0$; (2) every other term is zero except possibly at $t = 1$ or $t = 2$. The solution therefore satisfies the differential equation except possibly at $t = 1$, or 2. You can check that the initial values for the problem are satisfied. ◇

This example illustrates the power of the Laplace transform to solve complicated initial value problems. In so doing it also illustrates just how complex such solutions can be. Remember that if you need to solve a problem that involves more complicated situations than those illustrated, you should consider using the full implementation of Laplace transforms that is provided in the package **LaplaceTransform.m**.

Exercises 6.4

Load the package **LPT.m**. Use `Transform[ f[t],t,s]` to find these Laplace transforms. Express each function using `UnitStep` or `UnitPulse` individually or in combination before transforming.

1. $f_1(t) = \begin{cases} 1, & 0 \le t \le 2 \\ 0, & 2 < t \end{cases}$

2. $f_2(t) = \begin{cases} 0, & 0 \le t \le 1 \\ 1, & 1 < t \le 2 \\ 0, & 2 < t \end{cases}$

3. $f_3(t) = \begin{cases} 1, & 0 \le t \le 1 \\ 0, & 1 < t \le 2 \\ 1, & 2 < t \end{cases}$

4. $f_4(t) = \begin{cases} 2, & 0 \le t \le 1 \\ 1, & 1 < t \le 2 \\ 3, & 2 < t \end{cases}$

5. $f_5(t) = \begin{cases} t, & 0 \le t \le \pi \\ \sin t, & \pi < t \le 5 \\ e^{t-5}, & 5 < t \end{cases}$

6. $f_6(t) = \begin{cases} \sin t, & 0 \le t \le \pi/2 \\ 1, & \pi/2 < t \le 3\pi/2 \\ \cos t, & 3\pi/2 < t \end{cases}$

Use `Transform[ ,t s]` to find the Laplace transform of each system. Then use `Isolate[ , y[t],t,s]` to isolate the transform `LPT[ ,t,s]`. Finally, use `InverseTransform[ ,s,t]` to find the solution to each of these initial value problems. Capture each solution, plot it on an appropriate interval, and check it. Afterwards, solve each problem in a single step using `LPTSolve[ , y[t],t,s]`. The functions $f_i(t)$ are those defined in problems 1–6.

7. $y''(t) - 25y(t) = f_1(t), y(0) = 1, y'(0) = -1.$

8. $y''(t) - 25y(t) = f_2(t), y(0) = 1, y'(0) = -1.$

9. $y''(t) - 25y(t) = f_3(t), y(0) = 1, y'(0) = -1.$

10. $y''(t) + 7y'(t) + 12y(t) = f_4(t), y(0) = 0, y'(0) = 1.$

11. $y''(t) + 7y'(t) + 12y(t) = f_5(t), y(0) = 0, y'(0) = 1.$

12. $y''(t) + 7y'(t) + 12y(t) = f_6(t), y(0) = 0, y'(0) = 1.$

7 Higher-Order Differential Equations with Variable Coefficients

7.0 Introduction

In the previous two chapters we studied differential equations having constant coefficients. The theory was straightforward, and, with the help of *Mathematica*, the solutions were easy to obtain—even when the order of the differential equations was rather large. The persistent difficulty was, when we needed the roots of a polynomial of degree five or more, we could not be certain that they could be obtained exactly. This opened up the possibility of needing approximations to the roots, which opened up the need for knowing how accurately those approximations could be obtained, which led to numerical analysis, etc. The difficulties lay outside the field of differential equations, so we did not press for a resolution of them.

But when the differential equations no longer have constant coefficients, there begin to be real difficulties from the standpoint of differential equations. The earlier problems still remain, but new ones are added. Except in rare cases, the solutions we obtain will not be like they were before: combinations of polynomials, exponential functions, sines, and cosines. We still have all of these, but we add logarithms, and eventually will be forced to settle for solutions expressed as infinite series. Of course, polynomials, exponential functions, sines, cosines, and logarithms all have series expansions of one kind or another. We will obtain solution functions that are not readily identifiable as elementary functions, and indeed may not be elementary functions.

We start with simple problems and work our way up to more complicated problems. The simple problems in this chapter are of **Cauchy-Euler** type, where the method of solution is analogous to that of the constant coefficients problems we just studied. A simple change of variables can convert a problem of Cauchy-Euler type to one with constant coefficients, and vice versa.

Some problems with variable coefficients can be solved easily by series. These are the problems where solutions are centered on **ordinary points** of the equation. For some equations, certain points are not ordinary points and the simple methods do not apply. We therefore need to be able to determine when a point is ordinary and when it is not. Points that are not ordinary are called **singular**.

Some singular points can still be approached by straightforward methods. These are called **regular singular points**. Some cannot. These are called **irregular singular points**. We will concentrate our attention on ordinary points and regular singular points. The theorems will be stated for differential equations of order n, because using *Mathematica* it is essentially as easy to solve higher-order problems as it is to solve second-order problems. In Chapter 8 series solutions for vector and matrix differential equations will be presented, completing the argument that the general case can be approached (essentially) as easily as the special second-order case.

We will observe that the nature of solutions of a differential equation about a singular point can be inferred from the nature of solutions of Cauchy-Euler equations. Solutions of Cauchy-Euler equations are usually sought about a regular singular point. Since we will have a complete theory for Cauchy-Euler equations, we consider the general equation near a regular singular point to be merely a perturbation of a Cauchy-Euler equation. This enables us to make guesses about how to approach the general problem. The theorem, though complicated, vindicates this line of reasoning.

The investigation requires a review of **infinite series**. It demands skill in the technical manipulation of series. It introduces **product notation** as a companion to summation notation, and it introduces **difference equations** and their solutions. These difference equations are the **recursion relations** that say how any one coefficient in a series solution is related to one or more preceding coefficients.

Mathematica will do most of the manipulations for us. It will also check our answers for us. Without *Mathematica* the topics to be covered in this chapter demand both enormous amounts of thought and enormous amounts of manual manipulation. With *Mathematica*, the manipulative component, though not necessarily eliminated, is certainly reduced to manageable proportions. You are still encouraged to think.

The standard equations of mathematical physics such as the **Bessel** and **Legendre** equations, and equations of special functions in general, such as the **hypergeometric** equation, are covered as exercises.

7.1 Cauchy-Euler Differential Equations

The form of a **Cauchy[1]-Euler[2]** differential operator is

$$CE(y) = a_n x^n \frac{d^n y}{dx^n} + a_{n-1} x^{n-1} \frac{d^{n-1} y}{dx^{n-1}} + \cdots + a_1 x \frac{dy}{dx} + a_0 y, \tag{7.1}$$

[1]Augustin Louis Cauchy (1789–1875), French mathematician. Among his massive contributions to mathematics (some 800 books and articles) was the first truly acceptable theory of limits.

[2]Leonhard Euler (1707–1783), Swiss mathematician. Euler was the most prolific and versatile writer of mathematics to have ever lived. His influence can be seen throughout all of modern mathematics.

where $a_n, a_{n-1}, \ldots, a_1, a_0$ are numbers and $a_n \neq 0$. Examples of Cauchy-Euler differential equations are

$$3x^2 \frac{d^2y}{dx^2} - 7x\frac{dy}{dx} + 3y = 0$$

which is second-order homogeneous, and

$$x\frac{dy}{dx} + 4y = 5\cos(\ln x)$$

which is first-order nonhomogeneous. The key to identifying a Cauchy-Euler differential operator is to observe that each term has the form

$$a_k\, x^k \frac{d^k y}{dx^k},$$

where a_k is a number and the power of x and the order of the derivative agree. Missing terms are considered to have the correct form.

The central insight into solving Cauchy-Euler differential equations is that

$$CE(x^r) = p(r)\, x^r, \tag{7.2}$$

where $p(r)$ is a polynomial. This polynomial is the **characteristic polynomial** of the operator. A Cauchy-Euler operator maps a power of x into a polynomial times the same power of x. We will use this observation the same way we did the analogous observation in Chapter 4.

Example 7.1 Demonstrate equation 7.2 for the Cauchy-Euler differential equation

$$CE(y) = x^2y'' + 5xy' - 3y.$$

Solution. Consider

$$\begin{aligned} CE(x^r) &= x^2(r(r-1)x^{r-2}) + 5x(rx^{r-1}) - 3x^r \\ &= r(r-1)x^r + 5rx^r - 3x^r \\ &= [r(r-1) + 5r - 3]x^r \\ &= (r^2 + 4r - 3)x^r \\ &= p(r)x^r. \end{aligned}$$

◇

Observe that every term of CE takes x^r to a polynomial times x^r, because

$$\begin{aligned} a_k\, x^k \frac{d^k x^r}{dx^k} &= a_k x^k (r(r-1)(r-2)\cdots(r-k+1)x^{r-k}) \\ &= (a_k r(r-1)(r-2)\cdots(r-k+1))x^r \\ &= p_k(r)x^r. \end{aligned}$$

To get solutions to $CE(y) = 0$ we need only find the roots of the **characteristic equation** $p(r) = 0$. We have seen this before, and anticipate that we need to handle three cases.

Theorem 7.1 *The solutions of*

$$CE(y) = a_n x^n \frac{d^n y}{dx^n} + a_{n-1} x^{n-1} \frac{d^{n-1} y}{dx^{n-1}} + \cdots + a_1 x \frac{dy}{dx} + a_0 y,$$

are found this way. Substitute $y = x^r$ to get $CE(x^r) = p(r)x^r$. Then $CE(x^r) = 0$ when $p(r) = 0$.

1. *If $m \le n$ and $r_1, r_2, \ldots, r_m$ are distinct real characteristic roots, then x^{r_1}, $x^{r_2}, \ldots, x^{r_m}$ are linearly independent real solutions, and*

$$y(x) = c_1 x^{r_1} + c_2 x^{r_2} + \cdots + c_m x^{r_m}$$

   defines an m-dimensional subspace of the kernel of CE.
2. *If s is a characteristic root of multiplicity k, $1 < k \le m$, then $x^s, x^s \ln x, \ldots,$ $x^s (\ln x)^{k-1}$ are linearly independent solutions and*

$$y(x) = c_1 x^s + c_2 x^s \ln x + \cdots + c_k x^s (\ln x)^{k-1}$$

   defines a k-dimensional subspace of the kernel of CE.
3. *If $r = \alpha \pm \beta i$ is a complex conjugate pair of characteristic roots, then x^α $\cos(\beta \ln x)$ and $x^\alpha \sin(\beta \ln x)$ are real linearly independent solutions corresponding to the pair $\alpha \pm \beta i$ of complex roots. If $\alpha \pm \beta i$ is a pair of multiplicity k then*

$$x^\alpha \cos(\beta \ln x), x^\alpha \ln x \cos(\beta \ln x), \ldots, x^\alpha (\ln x)^{k-1} \cos(\beta \ln x)$$

   and

$$x^\alpha \sin(\beta \ln x), x^\alpha \ln x \sin(\beta \ln x), \ldots, x^\alpha (\ln x)^{k-1} \sin(\beta \ln x)$$

   are $2k$ real linearly independent solutions in the kernel of CE. Their linear combinations span a $(2k)$-dimensional subspace of the kernel of CE.
4. *A complete description of the kernel of CE is the set of all linear combinations of the functions found in the previous three cases.*

Proof (Indication).

1. The *Mathematica* notebook *Cauchy-Euler, Independence* shows that x^a, x^b, and x^c are linearly independent if a, b, and c are all different. A complete proof is analogous, but complicated.
2. Here we try an idea that worked before. If s is a double root of $p(r) = 0$, then $p(s) = p'(s) = 0$. Now $CE(x^s) = p(s)x^s = 0$. Taking r as a variable gives

$$\frac{\partial}{\partial r} CE(x^r) = CE\left(\frac{\partial}{\partial r} x^r\right) = CE(x^r \ln x)$$

$$= \frac{\partial}{\partial r}(p(r)x^r) = p'(r)x^r + p(r)x^r \ln x.$$

Thus, when $r = s$, $CE(x^s \ln x) = p'(s)x^s + p(s)x^s \ln x = 0x^s + 0x^s \ln x = 0$. So a second solution is $x^s \ln x$. That the solutions obtained from this pattern of reasoning are linearly independent is also illustrated in the notebook *Cauchy-Euler, Independence* in the case of multiplicity three.

3. Again we use an old technique. We have that

$$x^{\alpha+\beta i} = x^\alpha x^{\beta i} = x^\alpha e^{(\beta \ln x)i} = x^\alpha(\cos(\beta \ln x) + i \sin(\beta \ln x)).$$

Since CE is a real linear operator, it follows from Theorem 4.15 that

$$CE(x^\alpha \cos(\beta \ln x)) = 0$$

and

$$CE(x^\alpha \sin(\beta \ln x)) = 0. \qquad \square$$

The solutions $y_1(x) = x^\alpha \cos(\beta \ln x)$ and $y_2(x) = x^\alpha \cos(\beta \ln x)$ are shown to be linearly independent in the notebook *Cauchy-Euler, Independence.*

Example 7.2 Solve the homogeneous Cauchy-Euler differential equation

$$CE_3(y) = x^3 y''' - 2x^2 y'' - 2xy' + 8y = 0.$$

Solution. The substitution $y = x^r$ makes

$$\begin{aligned} CE_3(x^r) &= x^3 r(r-1)(r-2)x^{r-3} - 2x^2 r(r-1)x^{r-2} - 2xrx^{r-1} + 8x^r \\ &= r(r-1)(r-2)x^r - 2r(r-1)x^r - 2rx^r + 8x^r \\ &= [r(r-1)(r-2) - 2r(r-1) - 2r + 8]x^r \\ &= (r^3 - 5r^2 + 2r + 8)x^r \\ &= p(r)x^r \\ &= 0. \end{aligned}$$

This requires that $p(r) = r^3 - 5r^2 + 2r + 8 = (r+1)(r-4)(r-2) = 0$, which means that our characteristic roots are distinct: $r = -1, 4$, and 2. The three linearly independent solutions are $y_1 = x^{-1}, y_2 = x^4$, and $y_3 = x^2$. A complete description of the kernel is

$$y = c_1 x^{-1} + c_2 x^4 + c_3 x^2.$$

You are encouraged to check this. $\diamond$

Example 7.2M (From the *Mathematica* notebook *Solving Cauchy-Euler.*) Note how we follow the theory step-by-step.

Solution. Here is the operator.

```
In[1]:=  CEOp[x_,y_] = x^3*y'''[x]-2x^2y''[x]-2x y'[x]+8y[x]

Out[1]=                          2          3  (3)
          8 y[x] - 2 x y'[x] - 2 x  y''[x] + x  y   [x]
```

The function that we will substitute.

```
In[2]:=  ySub[x_] = x^r

Out[2]=   r
         x
```

The characteristic polynomial.

```
In[3]:=  poly[r_] = Coefficient[Expand[CEOp[x,ySub]],x^r]

Out[3]=                2    3
         8 + 2 r - 5 r  + r
```

Here are the distinct roots (in rule form).

```
In[4]:=  Solve[poly[r] == 0,r]

Out[4]=  {{r -> -1}, {r -> 4}, {r -> 2}}
```

Capture the roots.

```
In[5]:=  roots = r/.Solve[poly[r] == 0,r]

Out[5]=  {-1, 4, 2}
```

Produce the complete solution.

```
In[6]:=  s[x_] = (x^roots).{c1,c2,c3}

Out[6]=  c1         2        4
         -- + c3 x  + c2 x
         x
```

Check the solution.

```
In[7]:=  Simplify[CEOp[x,s] == 0]

Out[7]=  True
```

Of course, we can get the same result directly by using `DSolve`.

```
In[8]:=  DSolve[CEOp[x,y] == 0,y[x],x]

Out[8]=              C[1]     2          4
         {{y[x] -> ---- + x  C[2] + x  C[3]}}
                     x
```

This is the solution that we had before.

Example 7.3 Solve the differential equation

$$CE(y) = x^4 y^{(4)} + 9x^3 y''' + 10x^2 y'' - 30xy' - 24y = 0.$$

Solution. As before, substitute $y = x^r$ to get

$$\begin{aligned} CE(x^r) &= x^4 r(r-1)(r-2)(r-3)x^{r-4} + 9x^3 r(r-1)(r-2)x^{r-3} \\ &\quad + 10x^2 r(r-1)x^{r-2} - 30xrx^{r-1} - 24x^r \\ &= r(r-1)(r-2)(r-3)x^r + 9r(r-1)(r-2)x^r \\ &\quad + 10r(r-1)x^r - 30rx^r - 24x^r \\ &= [r(r-1)(r-2)(r-3) + 9r(r-1)(r-2) \\ &\quad + 10r(r-1) - 30r - 24]x^r \\ &= (r-3)(r+2)^3 x^r \\ &= 0 \end{aligned}$$

when $r = 3, -2, -2, -2$. Here we have an isolated root and a root of multiplicity three. The solution corresponding to the isolated root 3 is $y_1 = x^3$, and the solutions corresponding to the triple root -2 are $y_2 = x^{-2}, y_3 = x^{-2}\ln x$, and $y_4 = x^{-2}(\ln x)^2$. The complete description of the kernel is

$$y(x) = c_1 x^3 + c_2 x^{-2} + c_3 x^{-2}\ln x + c_4 x^{-2}(\ln x)^2. \qquad \diamond$$

Example 7.3M (From the notebook *Solving Cauchy-Euler.*) Solve the Cauchy-Euler differential equation

$$CE(y) = x^4 y^{(4)} + 9x^3 y''' + 10x^2 y'' - 30xy' - 24y = 0.$$

by *Mathematica.*

Solution. Here is the operator.

```
In[9]:=  CEOp[x_, y_] = x^4*y''''[x]+9x^3y'''[x]+10x^2y''[x]-
                                   30x y'[x]-24y[x]

Out[9]=
                                       2             3  (3)
         -24 y[x] - 30 x y'[x] + 10 x  y''[x] + 9 x  y    [x] +

            4  (4)
           x  y    [x]
```

The function that we will substitute.

```
In[10]:= ySub[x_] = x^r

Out[10]=  r
         x
```

The characteristic polynomial.

```
In[11]:= poly[r_] = Coefficient[Expand[CEOp[x,ySub]],x^r]

Out[11]=
                           2       3     4
         -24 - 28 r - 6 r  + 3 r  + r
```

The roots (in rule form).

```
In[12]:= Solve[poly[r] == 0,r]

Out[12]= {{r -> 3}, {r -> -2}, {r -> -2}, {r -> -2}}
```

Capture the roots.

```
In[13]:= roots = r/.Solve[poly[r] == 0,r]

Out[13]= {3, -2, -2, -2}
```

Produce the complete solution. Notice the way to account for the triple root.

```
In[14]:= s[x_] = (x^roots).{c1,c2,c3*Log[x],c4*Log[x]^2}

Out[14]=
                                              2
          c2         3    c3 Log[x]   c4 Log[x]
          -- + c1 x  +  --------- + ----------
           2               2             2
          x               x             x
```

Check the solution.

```
In[15]:= Simplify[CEOp[x,s] == 0]

Out[15]= True
```

Again, we can get the same result directly:

```
In[16]:= DSolve[CEOp[x,y] == 0,y[x],x]

Out[16]=
                          5                                      2
                  C[1] + x  C[4] + C[2] Log[x] + C[3] Log[x]
          {{y[x] -> -------------------------------------------- }}
                                     2
                                    x
```

◇

Example 7.4 Give a complete description of the kernel of the operator

$$CEOp(y) = x^2y'' + 5xy' + 29y.$$

Solution. Substitute $y = x^r$ to get

$$\begin{aligned} CEOp(x^r) &= x^2r(r-1)x^{r-2} + 5xrx^{r-1} + 29x^r \\ &= r(r-1)x^r + 5rx^r + 29x^r \\ &= [r(r-1) + 5r + 29]x^r \\ &= (r^2 + 4r + 29)x^r \\ &= 0 \end{aligned}$$

when $r^2 + 4r + 29 = 0$. This equation has the roots $-2 \pm 5i$, so the two real solutions are

$$y_1(x) = x^{-2} c_1 \cos(5 \ln x) \text{ and } y_2(x) = x^{-2} c_2 \sin(5 \ln x).$$

The complete description of the kernel can be written

$$y(x) = x^{-2}(c_1 \cos(5 \ln x) + c_2 \sin(5 \ln x)). \qquad \diamond$$

Example 7.4M (From the notebook *Solving Cauchy-Euler.*) Solve the Cauchy-Euler differential equation

$$CEOp(y) = x^2 y'' + 5xy' + 29y$$

by *Mathematica.*

Solution. Here is the operator.

```
In[17]:= CEOp[x_,y_] = x^2y''[x]+5x y'[x]+29y[x]

Out[17]=                    2
         29 y[x] + 5 x y'[x] + x  y''[x]
```

The function that we will substitute.

```
In[18]:= ySub[x_] = x^r

Out[18]= r
         x
```

The characteristic polynomial.

```
In[19]:= poly[r_] = Coefficient[Expand[CEOp[x,ySub]],x^r]

Out[19]=          2
         29 + 4 r + r
```

The distinct roots (in rule form).

```
In[20]:= Solve[poly[r] == 0,r]

Out[20]= {{r -> -2 + 5 I}, {r -> -2 - 5 I}}
```

Capture the first root.

```
In[21]:= CpxRoot = r/.Solve[poly[r] == 0,r][[1]]

Out[21]= -2 + 5 I

In[22]:= a = Re[CpxRoot]

Out[22]= -2
```

```
In[23]:= b = Im[CpxRoot]

Out[23]= 5
```

Produce the complete solution. This is a completely manual step.

```
In[24]:= s[x_] = x^a*{Cos[b Log[x]],Sin[b Log[x]]}.{c1,c2}

Out[24]= c1 Cos[5 Log[x]] + c2 Sin[5 Log[x]]
         -----------------------------------
                          2
                         x
```

Check the solution.

```
In[25]:= Simplify[CEOp[x,s] == 0]

Out[25]= True
```

Once again the use of DSolve is most direct (the constant names are different):

```
In[26]:= DSolve[CEOp[x,y] == 0,y[x],x]

Out[26]=
                 C[2] Cos[5 Log[x]] - C[1] Sin[5 Log[x]]
         {{y[x] -> ---------------------------------------}}
                                   2
                                  x
```

◇

Cauchy-Euler to Constant Coefficients

The observation that

$$\frac{d^k \ln x}{dx^k} = \frac{(-1)^{k+1}}{x^k}$$

suggests that it might be productive to attempt to change the variable in a Cauchy-Euler operator from x to t by using the substitution $t = \ln x$, or, equivalently, $x = e^t$. The reason is that these derivatives of $\ln x$ appear to cancel the powers of x in the coefficients. To make this change of variable we use the chain rule:

$$\frac{dy}{dx} = \frac{dy}{dt}\frac{dt}{dx} = \frac{dy}{dt}\frac{1}{x}.$$

This says that $x(dy/dx) = dy/dt$. This is promising: the coefficient is now constant!

Try the second derivative:

$$\begin{aligned}\frac{d^2y}{dx^2} &= \frac{d}{dx}\left(\frac{dy}{dx}\right)\\ &= \frac{d}{dx}\left(\frac{dy}{dt}\frac{1}{x}\right)\end{aligned}$$

$$
\begin{aligned}
&= \frac{d}{dx}\left(\frac{dy}{dt}\right)\frac{1}{x} + \frac{dy}{dt}\frac{d}{dx}\left(\frac{1}{x}\right) \\
&= \left(\frac{d^2y}{dt^2}\frac{1}{x}\right)\frac{1}{x} + \frac{dy}{dt}\left(\frac{-1}{x^2}\right) \\
&= \frac{1}{x^2}\left(\frac{d^2y}{dt^2} - \frac{dy}{dt}\right).
\end{aligned}
$$

This means that

$$x^2\frac{d^2y}{dx^2} = \frac{d^2y}{dt^2} - \frac{dy}{dt},$$

and again we have success: we get terms with constant coefficients. This is not an accident; the process continues through higher derivatives, as can be shown by mathematical induction.

Thus we can make the change of variable $t = \ln x$ in a Cauchy-Euler differential equation and arrive at a differential equation with constant coefficients. This is worth a theorem.

Theorem 7.2 *The change of variable $t = \ln x$ in a Cauchy-Euler differential equation $CE(y)(x) = f(x)$ converts it into a differential equation $L(y)(e^t) = g(t)$ with constant coefficients, where $g(t) = f(e^t)$, or equivalently, $f(x) = g(\ln x)$. These are equivalent:*

1. $y(x)$ is a solution of $CE(y)(x) = f(x)$.
2. $y(e^t)$ is a solution of $L(y)(e^t) = g(t)$. □

A comment is in order. We understand the sorts of functions that $g(t)$ can be in order for there to be easy solutions by undetermined coefficients. These are polynomials in t, exponentials linear in t, sines and cosines linear in t, and products of these. This says that the kinds of functions that $f(x)$ can be in order for there to be solutions that are obtainable (in x) by undetermined coefficients are polynomials in $\ln x$, exponentials linear in $(\ln x)$, sines and cosines linear in $(\ln x)$, and products of these. For example, these functions appearing on the right-hand side of a Cauchy-Euler differential equation would admit solutions (in x) by undetermined coefficients: $(\ln x)^3 - 4(\ln x)$, $7e^{5(\ln x)}$ $[= 7x^5]$, and $5\cos(2\ln x)$.

Example 7.5 Make the substitution $t = \ln x$ to change the Cauchy-Euler differential equation $x^2y'' + 5xy' + 4y = 0$ to a differential equation with constant coefficients. Solve the constant coefficients equation completely and recover the complete solution of the Cauchy-Euler equation.

Solution. From the work above, we know that

$$x\frac{dy}{dx} = \frac{dy}{dt} \text{ and } x^2\frac{d^2y}{dx^2} = \frac{d^2y}{dt^2} - \frac{dy}{dt},$$

so the differential equation becomes

$$x^2y'' + 5xy' + 4y = \left(\frac{d^2y}{dt^2} - \frac{dy}{dt}\right) + 5\left(\frac{dy}{dt}\right) + 4y$$
$$= \frac{d^2y}{dt^2} + 4\frac{dy}{dt} + 4y = 0.$$

This has characteristic polynomial $r^2+4r+4 = (r+2)^2 = 0$, and hence has a double root $r = -2$. The complete description of the kernel is $y(t) = e^{-2t}(c_1 + c_2t)$. This means that the solution as a function of x is $y = e^{-2\ln x}(c_1 + c_2 \ln x) = x^{-2}(c_1 + c_2 \ln x)$. You should verify this two ways: solve the Cauchy-Euler equation by those special techniques that apply to Cauchy-Euler equations, and by direct substitution and direct verification of linear independence. You may explore these ideas further in the notebook *Cauchy-Euler->Const Coeff.* ◇

Exercises 7.1

These problems should be done manually and by *Mathematica*. Feel free to use *Mathematica* to assist your manual solutions by finding the roots of high-order polynomials. Continue with the solution manually.

Part I. Solve these Cauchy-Euler differential equations. Their characteristic polynomials have distinct roots.

1. $x\dfrac{dy}{dx} + 2y = 0.$

2. $x^2\dfrac{d^2y}{dx^2} - 6y = 0.$

3. $x^3\dfrac{d^3y}{dx^3} + 2x^2\dfrac{d^2y}{dx^2} - 6x\dfrac{dy}{dx} = 0.$

4. $2x^3\dfrac{d^3y}{dx^3} - 5x^2\dfrac{d^2y}{dx^2} + 8x\dfrac{dy}{dx} - 6y = 0.$

5. $6x^3\dfrac{d^3y}{dx^3} - 13x^2\dfrac{d^2y}{dx^2} + 5x\dfrac{dy}{dx} - 8y = 0.$

6. $4x^3\dfrac{d^3y}{dx^3} - 12x^2\dfrac{d^2y}{dx^2} + 15x\dfrac{dy}{dx} - 12y = 0.$

7. $15x^3\dfrac{d^3y}{dx^3} - 2x^2\dfrac{d^2y}{dx^2} + 6x\dfrac{dy}{dx} - 8y = 0.$

8. $9x^3\dfrac{d^3y}{dx^3} + 36x^2\dfrac{d^2y}{dx^2} - 4x\dfrac{dy}{dx} - 8y = 0.$

PART II. Solve these Cauchy-Euler differential equations. Their characteristic polynomials have some or all roots repeated.

9. $x^2\dfrac{d^2y}{dx^2} + 5x\dfrac{dy}{dx} + 4y = 0.$

10. $x^3\dfrac{d^3y}{dx^3} + 4x^2\dfrac{d^2y}{dx^2} - 6x\dfrac{dy}{dx} - 12y = 0.$

11. $x^3\dfrac{d^3y}{dx^3} - x^2\dfrac{d^2y}{dx^2} - 6x\dfrac{dy}{dx} + 18y = 0.$

12. $x^4\dfrac{d^4y}{dx^4} + 4x^3\dfrac{d^3y}{dx^3} - 10x^2\dfrac{d^2y}{dx^2} - 6x\dfrac{dy}{dx} + y = 0.$

13. $27x^4\dfrac{d^4y}{dx^4} + 90x^3\dfrac{d^3y}{dx^3} - 12x^2\dfrac{d^2y}{dx^2} + 8x\dfrac{dy}{dx} + 8y = 0.$

14. $27x^4\dfrac{d^4y}{dx^4} + 45x^3\dfrac{d^3y}{dx^3} - 12x^2\dfrac{d^2y}{dx^2} + 28x\dfrac{dy}{dx} - 32y = 0.$

15. $x^4\dfrac{d^4y}{dx^4} + 6x^3\dfrac{d^3y}{dx^3} + 5x^2\dfrac{d^2y}{dx^2} - x\dfrac{dy}{dx} + y = 0.$

16. $20x^5\dfrac{d^5y}{dx^5} + 244x^4\dfrac{d^4y}{dx^4} + 761x^3\dfrac{d^3y}{dx^3} + 563x^2\dfrac{d^2y}{dx^2} + 25x\dfrac{dy}{dx} = 0.$

PART III. Solve these Cauchy-Euler differential equations. The characteristic polynomials have some complex roots. Convert the complex solutions to their equivalent real form.

17. $x^2\dfrac{d^2y}{dx^2} + 5x\dfrac{dy}{dx} + 13y = 0.$

18. $x^2\dfrac{d^2y}{dx^2} - 3x\dfrac{dy}{dx} + 20y = 0.$

19. $x^3\dfrac{d^3y}{dx^3} + 2x^2\dfrac{d^2y}{dx^2} - 15x\dfrac{dy}{dx} - 25y = 0.$

20. $x^4\dfrac{d^4y}{dx^4} - 10x^3\dfrac{d^3y}{dx^3} + 57x^2\dfrac{d^2y}{dx^2} - 189x\dfrac{dy}{dx} + 289y = 0.$

PART IV. Transformations to constant coefficients. We found that if $x = e^t$ then $x(dy/dx) = dy/dt$ and $x^2(d^2y/dx^2) = d^2y/dt^2 - dy/dt$.

21. Find an expression for $x^3(d^3y/dx^3)$ in terms of the derivatives of y with respect to t.

22. Find an expression for

$$x^4 \frac{d^4y}{dx^4}$$

in terms of the derivatives of y with respect to t.

23. Transform the Cauchy-Euler equations in Parts II and III into constant coefficients equations in t by the change of variables $x = e^t$.

7.2 Obtaining a Second Solution

We need to be able to find a second solution to a second-order linear differential equation given that one solution is known. Here is how to do this in the second-order case.

Suppose that $y_1(x)$ is a solution of $L_2(y) = a_2(x)y'' + a_1(x)y' + a_0(x)y = 0$.

How is a second solution to be found? Let $y = vy_1$. That is $y(x) = v(x)y_1(x)$. Then $y' = v'y_1 + vy_1'$ and $y'' = v''y_1 + 2v'y_1' + vy_1''$. Substitute into L_2:

$$\begin{aligned} L_2(y) &= L_2(v\, y_1) \\ &= a_2(x)(v''y_1 + 2v'y_1' + vy_1'') + a_1(x)(v'y_1 + vy_1') + a_0(x)vy_1 \\ &= (a_2(x)y_1)v'' + (2a_2(x)y_1' + a_1(x)y_1)v' \\ &\quad + (a_2(x)y_1'' + a_1(x)y_1' + a_0(x)y_1)v \\ &= (a_2(x)y_1)v'' + (2a_2(x)y_1' + a_1(x)y_1)v' + L_2(y_1)v \\ &= (a_2(x)y_1)v'' + (2a_2(x)y_1' + a_1(x)y_1)v', \end{aligned}$$

since $L_2(y_1) = 0$.

Let $w = v'$. Then $w' = v''$, and

$$\begin{aligned} a_2(x)y_1v'' + (2a_2(x)y_1' + a_1(x)y_1)v' &= a_2(x)y_1w' + (2a_2(x)y_1' + a_1(x)y_1)w \\ &= 0 \end{aligned}$$

is what we must solve. (If we know w, then we know v, and we can then find y.) The equation for w is merely a first-order linear differential equation having solution

$$\begin{aligned} w(x) &= \exp\left[-\int \frac{2a_2(x)y_1' + a_1(x)y_1}{a_2(x)y_1}\, dx\right] \\ &= \exp\left[-\int \left(2\frac{y_1'}{y_1} + \frac{a_1(x)}{a_2(x)}\right) dx\right] \\ &= \exp\left[-(\ln y_1^2) - \int \frac{a_1(x)}{a_2(x)}\, dx\right] \\ &= \frac{\exp\left[-\int a_1(x)/a_2(x)\, dx\right]}{y_1^2} \end{aligned}$$

This means that

$$v(x) = \int \frac{\exp\left[-\int (a_1(x)/a_2(x))dx\right]}{y_1(x)^2}\,dx$$

so we find the second solution in terms of y_1 to be

$$y_2(x) = v(x)y_1(x)$$

$$v(x) = y_1(x)\int \frac{\exp\left[-\int (a_1(x)/a_2(x))dx\right]}{y_1(x)^2}dx.$$

You should verify that $y_2(x)$ does satisfy $L_2(y) = a_2(x)y'' + a_1(x)y' + a_0(x)y = 0$ and that $W[y_1(x), y_2(x)] \neq 0$. See Example 7.14 for an application of this technique (to series).

More Thoughts on Substitutions

It is of interest to note here that the substitution of $y = vy_1$ into $L_2(y)$ giving $L_2(vy_1) = (a_2(x)y_1)v''+(2a_2(x)y_1{}'+a_1(x)y_1)v'+(a_2(x)y_1{}''+a_1(x)y_1{}'+a_0(x)y_1)v$ can be done without knowing in advance what role $y_1(x)$ is to play. One role is the one just discussed, where $L_2(y_1) = 0$. In this case, the order of the resulting equation can be decreased, as we saw. On the other hand, we can choose $y_1(x)$ to make the coefficient of $v'(x) = 0$. This means take $2a_2(x)y_1{}' + a_1(x)y_1 = 0$, so that

$$y_1(x) = e^{-\int (a_1(x)/2a_2(x))dx}.$$

Since $y_1 \neq 0$, the resulting differential equation has the form

$$L(v) = v'' + p(x)v = 0, \text{ where } p(x) = -\frac{L_2(y_1)(x)}{a_2(x)y_1(x)}.$$

This is one of the canonical forms for second-order differential equations.

Exercises 7.2

1. Write a few lines of code in *Mathematica* that finds this second solution, checks it, and verifies the linear independence. Then test your code on some second-order problems you make up using the Wronskian method of generating differential equations that was introduced in Example 4.17.

2. Project. Can one find a third solution of a third-order linear differential equation given two linearly independent solutions? With the help of *Mathematica* it can be done. The steps are interesting and require some careful thought. [*Hint:* Use your solutions one at a time to reduce the order from 3

to 2 to 1. Then, having solved the resulting first-order equation, perform the necessary two integrations to recover the third solution. Be careful to integrate the correct functions!]
Here is a sample problem. The differential equation

$$\begin{aligned} L_O(y) &= (-x^3 + 3x^2 - 3x)y''' + (x^3 - 3x + 3)y'' \\ &\quad + (-3x^2 + 3x)y' + (3x - 3)y = 0 \end{aligned}$$

has as two solutions $y_1(x) = x$ and $y_2(x) = x^3$. Find a third linearly independent solution.

7.3 Sums, Products and Recursion Relations

Manipulating Sums and Products

You are familiar with the summation symbol

$$\sum_{k=n_0}^{n_1} c_k = \begin{cases} c_{n_0} + c_{n_0+1} + \cdots + c_{n_1}, & n_1 \geq n_0 \\ 0, & n_1 = n_0 - 1 \end{cases}.$$

The condition on the second line is useful to reduce the necessity for specifying extra conditions on the indices and ranges of summation in complicated summation expressions. For instance, the sum

$$\sum_{\substack{k=n_0 \\ k \neq j}}^{n_1} c_k = \sum_{k=n_0}^{j-1} c_k + \sum_{k=j+1}^{n_1} c_k, \quad \text{for} \quad n_0 \leq j \leq n_1$$

makes sense without the need to state special cases when $j = n_0$ or $j = n_1$.
You may also be familiar with the product symbol

$$\prod_{k=n_0}^{n_1} c_k = \begin{cases} (c_{n_0})(c_{n_0+1}) \cdots (c_{n_1}) & n_1 \geq n_0 \\ 1 & n_1 = n_0 - 1 \end{cases}.$$

Once again the second line is needed for a product such as

$$\prod_{\substack{k=n_0 \\ k \neq j}}^{n_1} c_k = \left(\prod_{k=n_0}^{j-1} c_k \right) \left(\prod_{k=j+1}^{n_1} c_k \right), \quad \text{for} \quad n_0 \leq j \leq n_1$$

to make sense without the need to state special cases when $j = n_0$ or $j = n_1$.

Definition 7.1 *A* **difference equation** *is an equation of the form*

$$f(k, x_{k+m}, x_{k+m+1}, \ldots, x_{k+n}) = 0, \; m < n, \; k_0 \leq k \leq k_1. \qquad (7.3)$$

The upper limit k_1 may be infinite. The **order** *of the difference equation is $n - m$ $(= (k+n) - (k+m))$, the difference between the largest and smallest subscripts, which should both be nontrivially present in equation 7.3.*

A **solution** *of equation 7.3 is a sequence* $\{\phi_k\}_{k=k_0+m}^{k=k_1+n}$ *such that if* $k_0 \leq k \leq k_1$*, then*

$$f(k, \psi_{k+m}, \psi_{k+m+1}, \ldots, \psi_{k+n}) = 0.$$

A difference equation is sometimes called a **recursion relation** *or a* **recurrence relation**.

For example, from

$$\sum_{k=0}^{n+1} c_k = \sum_{k=0}^{n} c_k + c_{n+1}, \quad n \geq 0,$$

we see that if

$$s_n = \sum_{k=0}^{n} c_k \quad \text{for} \quad n \geq 0,$$

then

$$s_{n+1} = s_n + c_{n+1} \quad \text{for} \quad n \geq 0, \quad \text{and} \quad s_0 = c_0.$$

This is a **first-order difference equation initial-value problem**, and we know its unique solution, $s_n = \sum_{k=0}^{n} c_k$. It is first-order because the largest and smallest subscripts are one unit apart ($n+1-n = 1$). This solution is a sequence, each term of which is a sum.

Similarly, from

$$\prod_{k=0}^{n+1} c_k = (c_{n+1}) \prod_{k=0}^{n} c_k, \quad n \geq 0,$$

we see that if

$$p_n = \prod_{k=0}^{n} c_k \quad \text{for} \quad n \geq 0,$$

then

$$p_{n+1} = (p_n)(c_{n+1}) \quad \text{for} \quad n \geq 0. \quad \text{Furthermore,} \quad p_0 = c_0.$$

This again is a **first-order difference equation initial-value problem**, and we know its unique solution, $p_n = \prod_{k=0}^{n} c_k$. The equation is again first-order because the largest and smallest subscripts are one unit apart ($n + 1 - n = 1$). The solution is again a sequence, but this time each term of the sequence is a product.

The equation $p_{n+1} = (p_n)(c_{n+1})$ is homogeneous; $s_{n+1} = s_n + c_{n+1}$ is nonhomogeneous. In each case the operator is linear:

$$L_1(p) = p_{n+1} - c_{n+1}p_n$$

and

$$L_2(s) = s_{n+1} - s_n.$$

You should check both of these for linearity.

We first show how to do common manipulations with summations and products and then show how to solve some other more complicated difference equations. In other contexts, we call certain difference equations recursion relations because of the way they express how the terms of a sequence are related (recur). You should know the change of variables formula from calculus

$$\int_a^b f(g(x))g'(x)\,dx = \int_a^b f(y)\frac{dy}{dx}\,dx = \int_{g(a)}^{g(b)} f(y)\,dy.$$

Here we say that we have changed the variable from x to $y = g(x)$. This takes an especially simple form in the case where $y = g(x) = x - t$, a simple translation by an amount t. Here

$$\int_a^b f(x-t)\,dx = \int_{a-t}^{b-t} f(y)\,dy.$$

We illustrate this last transformation for sum and product

$$\sum_{k=m}^{n} c_{k-d} = \sum_{j=m-d}^{n-d} c_j,$$

and

$$\prod_{k=m}^{n} c_{k-d} = \prod_{j=m-d}^{n-d} c_j.$$

Here we let $j = k + d$. To verify these quickly, note that the summands (or multiplicands) are the same at the lower limit, they are the same at the upper limit, and the indices step by the same amount ($= 1$). This means

$$c_{k-d}\big|_{k=m} = c_{m-d} = c_j\big|_{j=m-d},$$

and

$$c_{k-d}\big|_{k=n} = c_{n-d} = c_j\big|_{j=n-d}.$$

This explanation works for both sums and products. In the latter sum and product (whose index is j) we can change the index back to k without changing the result, so we can say

$$\sum_{k=m}^{n} c_{k-d} = \sum_{k=m-d}^{n-d} c_k,$$

and

$$\prod_{k=m}^{n} c_{k-d} = \prod_{k=m-d}^{n-d} c_k,$$

and verify the result the same way. The process just illustrated is called **slipping the index**. We slipped the index up by d in each case.

Example 7.6M (a) Show how to use *Mathematica* to do the process illustrated in the last paragraph. Here the amount to be slipped is the fixed amount d. Notice that the substitution rule accommodates both sums and products, depending on which operator matches Op.

Solution.

```
In[27]:= Sum[c[k-d],{k,m,n}]/.
                Op_[p_,{k_,a_,b_}]:>
                Op[(p/.k->k+d),{k,a+d,b+d}]

Out[27]= Sum[c[k], {k, d + m, d + n}]

In[28]:= Product[c[k-d],{k,m,n}]/.
                Op_[p_,{k_,a_,b_}]:>
                Op[(p/.k->k+d),{k,a+d,b+d}]

Out[28]= Product[c[k], {k, d + m, d + n}]
```

◇

In case the upper limit is infinite it is still infinite after the index change, so an infinite limit does not appear to change when the index is slipped.

Our primary application of change of index manipulations is in situations such as

$$\sum_{k=2}^{\infty} k(k-1)c_k x^{k-2} = \sum_{k=0}^{\infty} (k+2)(k+1)c_{k+2}x^k.$$

Check to be sure that you understand completely how the transformation was made. We wanted to slip the index k up 2 units so that the exponent would be k rather than $k-2$.

Here is another example:

$$\sum_{k=1}^{\infty} kc_k x^{k+1} = \sum_{k=2}^{\infty} (k-1)c_{k-1}x^k.$$

Here we slipped the index down 1 for the same reason. Note that when $k-1=1$, $k=2$, so the summation starts at 2.

Example 7.6M (b) Illustrate on the last two summations how *Mathematica* can be made to automatically determine the amount by which the index should be slipped. The criterion chosen is to have the power of x be k.

Solution. First define a substitution rule that is to be applied to summations whose indices are to be slipped.

```
In[29]:= slipIndices = Op_[p_ x^(k_+e_.),{k_,a_,b_}]:>
                Op[(p/.k->k-e),{k,a+e,b+e}];

In[30]:= Sum[k(k-1)c[k]x^(k-2),{k,2,Infinity}]/.slipIndices

Out[30]= Sum[(1 + k) (2 + k) c[2 + k], {k, 0, Infinity}]

In[31]:= Sum[k c[k]x^(k+1),{k,1,Infinity}]/.slipIndices

Out[31]= Sum[(-1 + k) c[-1 + k], {k, 2, Infinity}]
```

◇

Example 7.7 Add the two sums $\sum_{k=2}^{\infty} k(k-1)c_k x^{k-2}$ and $\sum_{k=1}^{\infty} k c_k x^{k+1}$.

Solution. These two sums could not readily be added as they are, but can be added easily, once indices are slipped and the lower indices made equal by breaking the first two terms out of the first sum. This is how to add the two sums:

$$\sum_{k=2}^{\infty} k(k-1)c_k x^{k-2} + \sum_{k=1}^{\infty} k c_k x^{k+1}$$

$$= \sum_{k=0}^{\infty} (k+2)(k+1)c_{k+2}x^k + \sum_{k=2}^{\infty} (k-1)c_{k-1}x^k$$

$$= (2)(1)c_2 + (3)(2)c_3 x + \sum_{k=2}^{\infty} (k+2)(k+1)c_{k+2}x^k + \sum_{k=2}^{\infty} (k-1)c_{k-1}x^k$$

$$= (2)(1)c_2 + (3)(2)c_3 x + \sum_{k=2}^{\infty} [(k+2)(k+1)c_{k+2} + (k-1)c_{k-1}]x^k. \qquad \diamond$$

This sort of manipulation will occur often as we continue our studies in this chapter. To pursue this for a moment more, if we wanted the above sum to represent the zero function, then each coefficient of x to a power must be 0. So $c_2 = c_3 = 0$, and for $k \geq 2$,

$$(k+2)(k+1)c_{k+2} + (k-1)c_{k-1} = 0.$$

This is a third-order $(k + 2 - (k - 1) = 3)$ homogeneous linear difference equation. It is very similar to those we were looking at a short time ago.

Example 7.7M Do Example 7.7 in *Mathematica*.

Solution. The rule `slipIndices` was defined in Example 7.6.

```
In[32]:= Sum[k(k-1)c[k]x^(k-2),{k,2,Infinity}]+
         Sum[k c[k]x^(k+1),{k,1,Infinity}]/.slipIndices

Out[32]= Sum[(-1 + k) c[-1 + k], {k, 2, Infinity}] +

           Sum[(1 + k) (2 + k) c[2 + k], {k, 0, Infinity}]
```

Break out the first two terms.

```
In[33]:= %/.Sum[d_,{k,0,Infinity}]:>Sum[d,{k,0,1}]+
         Sum[d,{k,2,Infinity}]

Out[33]=                                  k
         2 c[2] + 6 x c[3] + Sum[(-1 + k) x  c[-1 + k],

                                                        k
            {k, 2, Infinity}] + Sum[(1 + k) (2 + k) x  c[2 + k],

            {k, 2, Infinity}]
```

Combine the two summations.

```
In[34]:= %/.Sum[c1_,{k,a_,b_}]+Sum[c2_,{k,a_,b_}]:>
                        Sum[c1+c2,{k,a,b}]

Out[34]=                                   k
         2 c[2] + 6 x c[3] + Sum[(-1 + k) x  c[-1 + k] +

                          k
           (1 + k) (2 + k) x  c[2 + k], {k, 2, Infinity}]
```

It is clear that the latter expression is the same as that obtained in Example 7.7. The terms inside the summation can be combined and simplified. This is done in the notebook *Formal Series, Ordinary Point.* ◇

Solving Recursion Relations

It is of some interest to solve the general first-order linear difference equation (recursion relation). The problem is to solve

$$y_{n+1} = c_n y_n + d_n, \quad n \geq 0, \quad y_0 = \alpha.$$

Notice that this combines the two first-order equations we saw before, but that the statement is slightly different: we have c_n and d_n, instead of c_{n+1} and d_{n+1}. This was done because it is often easier to see what a function of n is than to see what a function of $n+1$ is. There is no change in our methods: We evaluate the first several y_n until we see a pattern, guess the general form, and then prove that our guess is correct. The fundamental existence and uniqueness theorem, Theorem 7.3, says that for each initial point $y_0 = \alpha$, there is a unique solution. We proceed to find it.

$$\begin{aligned}
y_1 &= c_0 y_0 + d_0 \\
&= c_0 \alpha + d_0 \\
y_2 &= c_1 y_1 + d_1 \\
&= c_1(c_0\alpha + d_0) + d_1 \\
&= c_1 c_0 \alpha + (c_1 d_0 + d_1) \\
y_3 &= c_2 y_2 + d_2 \\
&= c_2(c_1 c_0 \alpha + (c_1 d_0 + d_1)) + d_2 \\
&= c_2 c_1 c_0 \alpha + c_2(c_1 d_0 + d_1) + d_2 \\
&= c_2 c_1 c_0 \alpha + (c_2 c_1 d_0 + c_2 d_1 + d_2) \\
y_4 &= c_3 y_3 + d_3 \\
&= c_3(c_2 c_1 c_0 \alpha + (c_2 c_1 d_0 + c_2 d_1 + d_2)) + d_3 \\
&= c_3 c_2 c_1 c_0 \alpha + c_3(c_2 c_1 d_0 + c_2 d_1 + d_2)) + d_3 \\
&= c_3 c_2 c_1 c_0 \alpha + (c_3 c_2 c_1 d_0 + c_3 c_2 d_1 + c_3 d_2 + d_3).
\end{aligned}$$

The coefficient of α is a product and the other term is a sum of products. From the patterns presented, we make the guess:

$$y_n = \alpha \prod_{i=0}^{n-1} c_i + \sum_{i=0}^{n-1} \left(\prod_{j=i+1}^{n-1} c_j \right) d_i. \tag{7.4}$$

If you do not understand what the expression in equation 7.4 says, write out what it means for $n = 0, 1$, and 4. This should help. You should check that when $n = 0, 1, 2, 3$, and 4 the expression agrees with the value of the corresponding y_n determined just above.

We now need to show that this expression satisfies $y_{n+1} = c_n y_n + d_n$ for every nonnegative integer n. You just showed it to be correct for $n = 0$. We now show that if it is correct for $n = m$, then it is correct for $n = m + 1$. Thus, by the principle of mathematical induction, it is true for every nonnegative integer n. Assume that

$$y_m = \alpha \prod_{i=0}^{m-1} c_i + \sum_{i=0}^{n-1} \left(\prod_{j=i+1}^{m-1} c_j \right) d_i.$$

is correct. Then

$$\begin{aligned}
c_m y_m + d_m &= c_m \left(\alpha \prod_{i=0}^{m-1} c_i + \sum_{i=0}^{m-1} \left(\prod_{j=i+1}^{m-1} c_j \right) d_i \right) + d_m \\
&= c_m \alpha \prod_{i=0}^{m-1} c_i + c_m \sum_{i=0}^{m-1} \left(\prod_{j=i+1}^{m-1} c_j \right) d_i + d_m \\
&= \alpha \prod_{i=0}^{m} c_i + \sum_{i=0}^{m-1} \left(c_m \prod_{j=i+1}^{m-1} c_j \right) d_i + d_m \\
&= \alpha \prod_{i=0}^{m} c_i + \sum_{i=0}^{m} \left(\prod_{j=i+1}^{m} c_j \right) d_i \\
&= y_{m+1}.
\end{aligned}$$

This completes the proof (by induction) that the stated expression for y_n is a solution of the difference equation for very nonnegative integer n.

In the coming sections we will have to solve difference equations of the form

$$f(n,k) y_{k+n} = \sum_{j=0}^{n-1} g_j(k) y_{k+j}, \quad k \geq k_0.$$

This is an nth-order difference equation so long as $f(n,k)$ and $g_0(k)$ are nonzero, because then the largest and smallest subscripts are $k + n$ and k, and their difference is n. In this case there is a solution function $F(k)$ such that for $k \geq k_0$,

$$f(n,k)F(k+n) = \sum_{j=0}^{n-1} g_j(k)F(k+j).$$

In general it is difficult to find a formula for such a solution function, but here is an existence and uniqueness theorem.

Theorem 7.3 (Existence and Uniqueness: Difference Equations) *Suppose that the function ϕ is a real-valued function defined throughout $\boldsymbol{R}^{n+1}$, and that $\{\alpha_i\}_{i=0}^{n-1}$ are numbers. If $f(n,k) \neq 0$ for $k \geq k_0$, then there is a unique solution of*

$$f(n,k)y_{k+n} = \phi(k, y_k, y_{k+1}, \ldots, y_{k+n-1}), \quad k \geq k_0, \tag{7.5}$$

having the property that

$$y_{k_0+j} = a_j \quad \textit{for} \quad 0 \leq j \leq n-1.$$

Equation 7.5 is a (possibly nonlinear) difference equation. If $\partial\phi(x,y,\ldots)/\partial y \neq 0$ then the order of the difference equation is n.

Proof. The proof hinges on the fact that if n consecutive values of y are known, then the $(n+1)$st is uniquely determined by the equation. Calculate

$$\begin{aligned} y_{k_0+n} &= \frac{1}{f(n,k_0)}\phi(k_0, y_{k_0}, y_{k_0+1}, \ldots, y_{k_0+n-1}) \\ &= \frac{1}{f(n,k_0)}\phi(k_0, \alpha_0, \alpha_1, \ldots, \alpha_{n-1}), \end{aligned}$$

so y_{k_0+n} is known uniquely. If $y_m, y_{m+1}, \ldots, y_{m+n-1}$ are known, then

$$y_{m+n} = \frac{1}{f(n,m)}\phi(m, y_m y_{m+1}, \ldots, y_{m+n-1})$$

is uniquely determined since $f(n,m) \neq 0$, and the proof is complete. □

Having the existence and uniqueness theorem puts us in a position analogous to where we were at the beginning of the course where we had an existence and uniqueness theorem for differential equations, but we did not know how to solve any of the equations whose solutions we could prove existed and were unique. In case $f(n,m) = 0$ for some $m \geq m_0$, we cannot determine y_{m+n} from

$$y_{m+n} = \frac{1}{f(n,m)}\phi(m, y_m y_{m+1}, \ldots, y_{m+n-1}.)$$

Two situations arise.

1. If $\phi(m, y_m, y_{m+1}, \ldots, y_{m+n-1}) = 0$, then y_{m+n} is arbitrary, and there may be a solution beyond m.

2. If $\phi(m, y_m, y_{m+1}, \ldots, y_{m+n-1}) \neq 0$ then we have

$$0 = f(n, m) y_{m+n} = \phi(m, y_m, y_{m+1}, \ldots, y_{m+n-1}) \neq 0.$$

Since this cannot happen, we have no solution that defines y_{k+n} for $k \geq m$.

Here is another example of a typical solution process. Solve the difference equation

$$\begin{aligned} a_{k+2} &= f(k) a_k, \quad k \geq 0; \\ a_0 &= \alpha, \\ a_1 &= \beta. \end{aligned}$$

Find out how the solution sequence starts:

$$\begin{aligned} a_2 &= f(0) a_0 = f(0)\alpha \\ a_3 &= f(1) a_1 = f(1)\beta \\ a_4 &= f(2) a_2 = f(2) f(0)\alpha \\ a_5 &= f(3) a_3 = f(3) f(1)\beta. \end{aligned}$$

This is enough to guess that the solution is

$$\left. \begin{aligned} a_{2m} &= \left(\prod_{j=0}^{m-1} f(2j) \right) \alpha \\ a_{2m+1} &= \left(\prod_{j=0}^{m-1} f(2j+1) \right) \beta \end{aligned} \right\} \quad \text{for} \quad m \geq 0, \tag{7.6}$$

and prove it by mathematical induction. (This is an exercise.)

Two-term variable coefficient difference equations such as this are easy to solve. Three-term difference equations can be very difficult, and more than three terms with variable coefficients are rarely solved—in the sense that a formula for the solution is given.

Exercises 7.3

Part I. Verify the equality of these sums or products:

1. $\displaystyle\sum_{i=0}^{5} 3^i = \sum_{i=4}^{9} 3^{i-4}.$

2. $\displaystyle(a + x) \sum_{k=0}^{n} \frac{n!}{k!(n-k)!} a^{n-k} x^k = \sum_{k=0}^{n+1} \frac{(n+1)!}{k!\,(n+1-k)!} a^{n+1-k} x^k.$

3. $\displaystyle\prod_{k=1}^{n} \frac{k+1}{k} = \prod_{k=0}^{n-1} \frac{k+2}{k+1} = n + 1.$

PART II. Verify that the given functions are solutions of the difference equation or difference initial value problem. Constants are represented by c, c_1, etc.

4. $a_{k+2} - 5a_{k+1} + 6a_k = 0; \quad a_k = c_1 2^k + c_2 3^k.$

5. $f_{k+1} = (k+1)f_k; \quad f_k = c\,k!\,.$

6. $r_{k+1} = \dfrac{1}{k+1} r_k; \quad r_k = \dfrac{c}{k!}\,.$

7. $b_{k+2} = \dfrac{b_k}{(k+2)(k+1)}; \quad b_k = \dfrac{c}{k!}\,.$

PART III. Theory.

8. Show that the difference equation $a_{k+2} = -a_k/((k+2)^2), \quad k \geq 0$ is second-order and has as solutions the two sequences, one even and one odd:

$$\left.\begin{aligned} a_{2m} &= \frac{(-1)^m a_0}{2^{2m}(m!)^2} \\ a_{2m+1} &= \frac{(-1)^m a_1}{\left(\prod_{j=0}^{m}(2j+1)\right)^2} \end{aligned}\right\} \quad \text{for} \quad m \geq 0,$$

   where a_0 and a_1 are arbitrary. Any linear combination of these is also a solution.

9. Prove that the two functions given in equation 7.6 define the only solution of the difference initial-value problem

$$\begin{aligned} a_{k+2} &= f(k)a_k, \quad k \geq 0; \\ a_0 &= \alpha, \\ a_1 &= \beta. \end{aligned}$$

10. The sequence a_n is defined by

$$\begin{aligned} a_n &= a_{n-1} - (n-1)a_{n-2}, \quad n \geq 2, \\ a_0 &= a_1 = 1. \end{aligned}$$

   Show that the solution sequence satisfies the equation

$$\sum_{n=0}^{\infty} a_n \frac{x^n}{n!} = \exp\left(x - \frac{x^2}{2}\right)$$

   by differentiating both sides of this equation and manipulating the result as a series expression. (You may need to slip the indices in the recursion relation, depending on just how you manipulate the series.) [From the solution of problem 91-9, *SIAM Review*, 34(2) pp. 315–317.]

7.4 Series Solutions of Differential Equations

You can easily verify that $y(x) = ce^{x^2}$ is a solution of $y' - 2xy = 0$ for each choice of the number c. From your study of Taylor[3] series you know that $e^t = \sum_{k=0}^{\infty} t^k/k!$, and this sum converges for every real (or complex) value of t. So you know that $ce^{x^2} = c\sum_{k=0}^{\infty}(x^2)^k/k! = c\sum_{k=0}^{\infty} x^{2k}/k!$ for every real x. This infinite series is a solution of $y' - 2xy = 0$. It is worth your while to verify this claim. The manipulations that you encounter are typical of those we will see throughout this study of series solutions of differential equations with variable coefficients.

How could we have found an infinite series solution directly from the differential equation without having to know the Taylor series for some function? This is the topic of this section. We will consider power series solutions of first-order equations now and defer second-order equations until later.

Review Power Series

A **power series** is an infinite series of the form $\sum_{k=0}^{\infty} a_k(x-c)^k$. The number c is called the **center**, and the numbers $\{a_k\}_{k=0}^{\infty}$ are the **coefficients**. Here are some of the properties of power series that we will use.

- The power series $\sum_{k=0}^{\infty} a_k(x-c)^k$ is **convergent** at z provided that $\lim_{n\to\infty}\sum_{k=0}^{n} a_k(z-c)^k$ exists.
- The power series $\sum_{k=0}^{\infty} a_k(x-c)^k$ is **absolutely convergent** at z if the power series $\sum_{k=0}^{\infty} |a_k||z-c|^k$ is convergent.
- If the power series $\sum_{k=0}^{\infty} a_k(x-c)^k$ is absolutely convergent, then it is convergent.
- A power series that is convergent, but not absolutely convergent, is called **conditionally convergent**.
- Either there is a number R such that if $|z-c| < R$, then $\sum_{k=0}^{\infty} a_k(z-c)^k$ is absolutely convergent or the series converges only for $z = c$, in which case we say $R = 0$. The largest such R is called the **radius of convergence** of the power series. The radius of convergence may be infinite, in which case the series is called **entire**.
- If R is finite, the convergence of $\sum_{k=0}^{\infty} a_k(x-c)^k$ at the two points where $|x-c| = R$ has to be examined specially.
- If the power series for $f(x) = \sum_{k=0}^{\infty} a_k(x-c)^k$ converges absolutely on an interval then the series can be integrated or differentiated term by term with the result that

$$f'(x) = \sum_{k=1}^{\infty} ka_k(x-c)^{k-1}$$

[3]Brooke Taylor (1685–1731), British mathematician. As secretary of the Royal Society, he supported Newton in the acrimonious battles with Leibniz over who invented calculus. Taylor series were invented by John Bernoulli, and Taylor knew it when he published his formula.

and

$$\int f(x)\,dx = \sum_{k=0}^{\infty} \frac{a_k(x-c)^{k+1}}{k+1}.$$

- The **ratio test** is of especial importance in the study of power series. It states that if $\lim_{k\to\infty} |(a_{k+1}/a_k)| = r$ and $r < 1$, then the series $\sum_{k=0}^{\infty} a_k x^k$ converges absolutely.

An Example of Solution by Series

Example 7.8 Solve $y' - 2xy = 0$ by assuming a solution of the form

$$y = \sum_{k=0}^{\infty} a_k x^k.$$

Solution. We have our solution as soon as we determine each of the coefficients $\{a_k\}_{k=0}^{\infty}$. This means that we have to determine an infinite collection of numbers.

We do this by attempting to find a formula for the coefficients in the form $a_k = f(k)$. It is not always possible to write down such a formula. But you can always establish a relationship between the coefficients that permits the calculation of any desired coefficient. This is essentially as good as a formula.

Assume that the series is absolutely convergent on a interval centered on 0 so that we can differentiate term by term. Thus, given

$$y = \sum_{k=0}^{\infty} a_k x^k,$$

we have

$$y' = \sum_{k=0}^{\infty} k a_k x^{k-1} = \sum_{k=1}^{\infty} k a_k x^{k-1}.$$

(The index of summation can start at 1 since the term corresponding to $k = 0$ is 0.)

Therefore

$$\begin{aligned} y' - 2xy &= \sum_{k=1}^{\infty} k a_k x^{k-1} - 2x \sum_{k=0}^{\infty} a_k x^k \\ &= \sum_{k=1}^{\infty} k a_k x^{k-1} - \sum_{k=0}^{\infty} 2 a_k x^{k+1} \\ &= \sum_{k=0}^{\infty} (k+1) a_{k+1} x^k - \sum_{k=1}^{\infty} 2 a_{k-1} x^k \end{aligned}$$

(slip $k \longrightarrow k+1$ in first sum)

$$= (0+1)a_{0+1}x^0 + \sum_{k=1}^{\infty}(k+1)a_{k+1}x^k - \sum_{k=1}^{\infty}2a_{k-1}x^k$$

$$= a_1 + \sum_{k=1}^{\infty}[(k+1)a_{k+1} - 2a_{k-1}]x^k$$

We want this sum to be 0 for all x in our interval centered on 0. For this to happen, two things must occur:

1. $a_1 = 0$, and
2. $(k+1)a_{k+1} - 2a_{k-1} = 0, \quad \text{for} \quad k \geq 1.$

We now have what is called a recursion relation (or recurrence relation). It shows the way coefficients in the assumed series are related for the series to be a solution.

Collections of coefficients have to cooperate in order for a power series to be a solution. Equation (2) expresses the form of that cooperation, namely $(k+1)a_{k+1} = 2a_{k-1}$ and equation (1) says that a_1 has no choice: it must be 0.

In order to get a function that says what each coefficient is, we need to "solve" the recursion relation as we did in the last section. The technique involves three steps:

1. look at some sample terms for early indices;
2. guess a formula that describes these terms;
3. use mathematical induction and prove that the guess is correct.

The recursion relation we have is $a_{k+1} = 2/(k+1)a_{k-1}$ for $k \geq 1$. By substituting values for k we can see that

$(k=1)$ $a_2 = 2/(1+1)a_0 = a_0$.
$(k=2)$ $a_3 = 2/(2+1)a_1 = (2/3)a_1 = (2/3)0 = 0$.
$(k=3)$ $a_4 = 2/(3+1)a_2 = (2/4)a_0 = (1/2)a_0$.

It is still too early to see a clear pattern, so we proceed.

$(k=4)$ $a_5 = 2/(4+1)a_3 = (2/5)a_3 = 0$.
$(k=5)$ $a_6 = 2/(5+1)a_4 = (1/3)a_4 = 1/((3)(2)(1))a_0 = 1/(3!)a_0$.

It is possible to make a reasonable guess at this point, especially given the way the denominator of a_6 has been expressed. If you are not ready to guess, try some more terms.

Here is my guess:

$$a_{2m+1} = 0, \quad m \geq 0. \quad (a_{\text{odd}} = 0)$$

and

$$a_{2m} = (1/m!)a_0, \quad m \geq 0.$$

This guess defines all of the coefficients. The question now is: is the guess correct?

Observe that the guess works for all of the coefficients that we had calculated:

$$a_1 = a_3 = a_5 = 0$$

and

$$a_0 = (1/0!)a_0, a_2 = (1/1!)a_0, a_4 = (1/2!)a_0,$$

and

$$a_6 = (1/3!)a_0.$$

This is not a proof, but if it had not worked we would have had proof that the guess was wrong.

Here is proof that the guess is correct. We use the recursion relation

$$a_{k+1} = \frac{2}{k+1}a_{k-1} \quad \text{for} \quad k \geq 1.$$

Start with the assumptions that a_0 is arbitrary, $a_1 = 0$, and that for some $m \geq 1$, we have the two adjacent coefficients $a_{2m+1} = 0$, and $a_{2m} = (1/m!)a_0$. We show that the next two coefficients (using $m + 1$) have the correct form.

$$a_{2(m+1)+1} = a_{2m+3} = \frac{2}{2(m+1)+1}a_{2(m+1)-1} = \frac{2}{2(m+1)+1}a_{2m+1} = 0,$$

so the next odd-numbered term has the correct form. For the next even-numbered coefficient,

$$a_{2(m+1)} = \frac{2}{2(m+1)}a_{2m} = \frac{1}{m+1}\frac{1}{m!}a_0 = \frac{1}{(m+1)!}a_0,$$

which also has the correct form. Since the correct form for the coefficients at $m + 1$ follows from the form for the coefficients at m, we have established the correctness of our formulas for all nonnegative integers k. (Note carefully where the recursion relation was used and where the assumptions were used. Both should be used in your proof.)

It is of interest to express our solution as a sum:

$$\begin{aligned} y &= \sum_{k=0}^{\infty} a_k x^k \\ &= \sum_{m=0}^{\infty} a_{2m}x^{2m} + \sum_{m=0}^{\infty} a_{2m+1}x^{2m+1}, \\ &\quad \text{(even terms, then odd terms)} \\ &= \sum_{m=0}^{\infty} \frac{1}{m!}a_0 x^{2m} + \sum_{m=0}^{\infty} 0x^{2m+1} \end{aligned}$$

$$= a_0 \sum_{m=0}^{\infty} \frac{1}{m!} x^{2m} + 0$$

$$= a_0 \sum_{m=0}^{\infty} \frac{1}{m!} x^{2m}.$$

Observe that, except for the names of the index and the arbitrary constant, this is the series that we obtained at the start of this investigation. We can change the name of the constant a_0 to c and that of the index from m to k without affecting the meaning of the sum. The terms of a series can be rearranged if the series converges absolutely. That is the case here since all of the terms of the series are nonnegative. ◇

Example 7.8M Work the previous example by *Mathematica*.

Solution. (From the notebook *Quick Series Solution.*) Define the operator

```
In[35]:= L[x_,y_] = y'[x]-2x y[x]

Out[35]= -2 x y[x] + y'[x]
```

Set n (the number of terms desired) to 10

```
In[36]:= n = 10

Out[36]= 10
```

The next cell says to create a sum of n terms and effectively turn it into a series by saying that the terms beyond n are indefinite. `O[x]^(n+1)` indicates we know nothing about terms of order `n+1` and beyond. Here is the series to be assumed.

```
In[37]:= y[x_] = Sum[a[i]x^i,{i,0,n}]+O[x]^(n+1)

Out[37]=
                                2          3          4          5
         a[0] + a[1] x + a[2] x  + a[3] x  + a[4] x  + a[5] x  +

                  6          7          8          9           10
           a[6] x  + a[7] x  + a[8] x  + a[9] x  + a[10] x   +

                11
           O[x]
```

Substitute into the differential operator.

```
In[38]:= LHS = Collect[L[x,y],x]

Out[38]=
                                                             2
         a[1] + (-2 a[0] + 2 a[2]) x + (-2 a[1] + 3 a[3]) x  +

                                   3                        4
           (-2 a[2] + 4 a[4]) x  + (-2 a[3] + 5 a[5]) x  +
```

```
                       5                          6
(-2 a[4] + 6 a[6]) x  + (-2 a[5] + 7 a[7]) x  +

                       7                          8
(-2 a[6] + 8 a[8]) x  + (-2 a[7] + 9 a[9]) x  +

                         9      10
(-2 a[8] + 10 a[10]) x  + O[x]
```

Equate the coefficients to 0. Notice that the recursion relation has appeared in the latter coefficients. (Its form remains to be guessed.) The operator `&&` means `And`.

```
In[39]:= sys = LogicalExpand[LHS == 0]

Out[39]= a[1] == 0 && -2 a[0] + 2 a[2] == 0 && -2 a[1] + 3 a[3] == 0 &&
           -2 a[2] + 4 a[4] == 0 && -2 a[3] + 5 a[5] == 0 &&
           -2 a[4] + 6 a[6] == 0 && -2 a[5] + 7 a[7] == 0 &&
           -2 a[6] + 8 a[8] == 0 && -2 a[7] + 9 a[9] == 0 &&
           -2 a[8] + 10 a[10] == 0
```

Find the coefficients. Using `Reverse[Table[a[i],i,0,n]]` makes the form of the coefficients agree with what you would find by hand: the latter terms are found in terms of the former.

```
In[40]:= coeff = Solve[sys,Reverse[Table[a[i],{i,0,n}]]]

Out[40]=
                   a[0]                         a[0]
         {{a[10] -> ----, a[9] -> 0, a[8] -> ----, a[7] -> 0,
                   120                          24

                   a[0]                         a[0]
           a[6] -> ----, a[5] -> 0, a[4] -> ----, a[3] -> 0,
                    6                            2

           a[2] -> a[0], a[1] -> 0}}
```

Get the solution series; take `a[0]==1`.

```
In[41]:= s[x_] = y[x]/.coeff[[1]]/.a[0]->1

Out[41]=
                   4    6    8    10
              2   x    x    x    x          11
         1 + x  + -- + -- + -- + --- + O[x]
                  2    6    24   120
```

The known solution has the same initial terms.

```
In[42]:= soln[x_] = Series[Exp[x^2],{x,0,n}]

Out[42]=
                   4    6    8    10
              2   x    x    x    x          11
         1 + x  + -- + -- + -- + --- + O[x]
                  2    6    24   120
```

Check the series that we have found.

```
In[43]:= Simplify[L[x,s]]

Out[43]=      10
         O[x]
```

What this says is that `L[x,s] == 0` through terms of order 9 in x. `O[x]^10` says the remainder starts at the x^{10} term. ◇

There are functions of great value in the sciences, among them the Bessel[4] functions, that can only be defined in terms of power series. One can argue that the familiar sine, cosine, exponential, and other functions with which we are familiar can also only be defined by power series. In fact when they need to be evaluated (in your calculator, for instance), this is done by calculating the value from a power series or from some other function that was obtained from a power series.

Exercises 7.4

Part I. Verify that the given series are solutions of the differential equation or initial value problem.

1. $\dfrac{dy}{dx} - 2xy(x) = 0; \qquad y(x) = \displaystyle\sum_{k=0}^{\infty} \frac{x^{2k}}{k!}.$

2. $\dfrac{d^2y}{dx^2} + y(x) = 0, \quad y(0) = 1, \quad y'(0) = 0; \quad y(x) = \displaystyle\sum_{k=0}^{\infty} \frac{(-1)^k x^{2k}}{(2k)!}.$

3. $\dfrac{d^2y}{dx^2} + y(x) = 0, \quad y(0) = 0, \quad y'(0) = 1; \quad y(x) = \displaystyle\sum_{k=0}^{\infty} \frac{(-1)^k x^{2k+1}}{(2k+1)!}.$

4. $\dfrac{d^3y}{dx^3} - \dfrac{d^2y}{dx^2} + \dfrac{dy}{dx} - y(x) = 0; \qquad y(x) = \displaystyle\sum_{k=0}^{\infty} \frac{x^k}{k!}.$

Part II. Solve the following differential equations in powers of x by the methods of this section. You may wish to modify the notebook *Quick Series Solution* to assist you by checking your calculations. Find the complete solution by the methods of Chapter 2 and compare the power series for the complete solution with the result you found. Use the built-in *Mathematica* function `Series` to compute the series for the complete solution.

5. $\dfrac{dy}{dx} - 3y = 0.$

---

[4]Friedrich Wilhelm Bessel (1784–1846), German astronomer. Used parallax measurement to determine the distance to a fixed star. He was a close personal friend of Gauss. His Bessel functions have many important uses in physics.

6. $\dfrac{dy}{dx} + (x+1)y = 0.$

7. $\dfrac{dy}{dx} + (x^2+1)y = 0.$

7.5 Series Solutions About Ordinary Points

An important application of formal series methods to the solution of linear ordinary differential equations is the case of second-order differential equations with variable coefficients. When second-order differential equations are solved by series methods, essentially every special case behavior is encountered or at least alluded to. There are enough opportunities for study in the second-order case for individuals to actually specialize in studying them. The fundamental concepts are applicable to higher-order equations as well, so we propose to introduce the theory in the general case.

Definition 7.2 *The function f is* **analytic** *at x_0 means that $f(x)$ is represented by its power series near x_0. That is,*

$$f(x) = \sum_{k=0}^{\infty} c_k (x - x_0)^k \quad \text{for} \quad |x - x_0| < R.$$

The differential equation we consider is

$$a_n(x)y^{(n)}(x) + a_{n-1}(x)y^{(n-1)}(x) + \cdots + a_0(x)y(x) = q(x). \tag{7.7}$$

We seek a solution near x_0 where $a_n(x), a_{n-1}(x), \ldots, a_0(x)$, and $q(x)$ are analytic at x_0. We will concentrate on the solution of the homogeneous equation, because a particular solution of the nonhomogeneous equation can often be obtained by techniques learned previously. Series methods can be used to find a particular solution by equating coefficients on the left-hand side with the corresponding coefficients of the series expansion about x_0 of the function $q(x)$.

Definition 7.3 *The point x_0 near which we seek a solution is called the* **center** *of the expansion. Such points are either ordinary or singular.*

The point x_0 is an **ordinary point** *of equation 7.7 provided that $a_n(x_0) \neq 0$.*

The point x_0 is a **regular singular point** *of equation 7.7 provided that equation 7.7 is expressible as*

$$\begin{aligned} &(x - x_0)^n p_n(x) y^{(n)}(x) + (x - x_0)^{n-1} p_{n-1}(x) y^{(n-1)}(x) \\ &\quad + \cdots + (x - x_0) p_1(x) y'(x) + p(x) y(x) = q(x), \end{aligned} \tag{7.7 RSP}$$

where the functions $p_n(x), p^{n-1}(x), \ldots, p_0(x)$ are analytic at x_0.

Otherwise the point x_0 is an **irregular singular point**.

The fundamental theorem for series expansions about ordinary points was proved in 1866 by L. Fuchs.[5] It was he who introduced the term **fundamental solution**.

Theorem 7.4 (Ordinary Points) *If the coefficients and right-hand side of the equation*

$$y^{(n)}(x) + a_{n-1}(x)y^{(n-1)}(x) + \cdots + a_0(x)y(x) = q(x) \tag{7.7 OP}$$

are all analytic at the point x_0, and if R is the smallest of the radii of convergence of the coefficient functions of the equation, then there are n linearly independent solutions of the homogeneous equation

$$y^{(n)}(x) + a_{n-1}(x)y^{(n-1)}(x) + \cdots + a_0(x)y(x) = 0$$

of the form $y(x) = \sum_{k=0}^{\infty} c_k(x-x_0)^k$. These are all (absolutely) convergent with radii of convergence at least R. There is also a particular solution of the same form having the same convergence properties with radius of convergence at least the smaller of R and the radius of convergence of the series that represents $q(x)$.

The theorem concerning regular singular points, due to Frobenius[6] is more complicated and its results are less definitive. There are powerful applications, especially in the second-order case.

Theorem 7.5 (Regular Singular Points) *Suppose that equation 7.7 can be written in the form (7.7 RSP) with all of the functions $p_n(x), p_{n-1}(x), \ldots, p_0(x)$ and $q(x)$ analytic at x_0. Suppose that R is the smallest of the radii of convergence of the coefficients of the equation, then there are at most n linearly independent solutions having the form $y(x) = \sum_{k=0}^{\infty} c_k(x-x_0)^{k+r}$ that are solutions of the homogeneous equation. The number r is called an* **index** *of the equation. There are at most n values that r can take; to each distinct value of r there results a solution, but not all are necessarily linearly independent.*

There is a particular solution of the same form having the same convergence properties with radius of convergence the smaller of R and the radius of convergence of the series that represents $q(x)$.

The indices are roots of the **indicial equation** which we will see in due time.

When there is a root r of the indicial equation of multiplicity m greater than 1, then, in addition to a solution of the homogeneous equation of the form $y_1(x) = \sum_{k=0}^{\infty} c_k(x-x_0)^{k+r}$, there are often solutions of the form $y_2(x) = y_1(x)\sum_{j=1}^{m-1} d_j(\ln(x-x_0))^j + \sum_{k=0}^{\infty} c_k(x-x_0)^{k+r}$. These are clearly not strictly power series and they are defined only for $x > x_0$. Determining when such

---

[5] Immanual Lazarus Fuchs (1833–1902), German mathematician. He made many advances in the theory of differential equations, and was instrumental in the creation of the modern theory of differential equations.

[6] Ferdinand Georg Frobenius (1849–1917), German mathematician.

solutions exist is a delicate task which we leave to those who are interested in the question. The Bessel equation will provide us with an example where the solutions corresponding to the two roots are not linearly independent when the indices are integers. When the indices of the Bessel equation are not integers, however, the solutions corresponding to the two indices are linearly independent.

Classifying the Point x_0

Given that x_0 is not an ordinary point of equation 7.7 how does one convert equation 7.7 into equation (7.7 RSP)? How can you tell that the form (7.7 RSP) is not possible? Here is the calculation. It provides criteria that insure x_0 is a regular singular point.

Divide the homogeneous equation 7.7 by $a_n(x)$ to get

$$y^{(n)}(x) + \left(\frac{a_{n-1}(x)}{a_n(x)}\right) y^{(n-1)}(x) + \cdots + \left(\frac{a_0(x)}{a_n(x)}\right) y(x) = 0. \tag{7.8}$$

Then the requirement is that each of the n functions

$$\begin{aligned}
(x-x_0)\left(\frac{a_{n-1}(x)}{a_n(x)}\right) &= p_{n-1}(x),\\
(x-x_0)^2\left(\frac{a_{n-2}(x)}{a_n(x)}\right) &= p_{n-2}(x),\\
&\vdots\\
(x-x_0)^{n-1}\left(\frac{a_1(x)}{a_n(x)}\right) &= p_1(x),\\
(x-x_0)^{n}\left(\frac{a_0(x)}{a_n(x)}\right) &= p_0(x)
\end{aligned} \tag{7.9}$$

be analytic at x_0. When the functions defined in equations 7.9 are each analytic at x_0, multiply equation 7.8 through by $(x-x_0)^n$. Then

$$\begin{aligned}
&(x-x_0)^n y^{(n)}(x) + (x-x_0)^n\left(\frac{a_{n-1}(x)}{a_n(x)}\right) y^{(n-1)}(x)\\
&\qquad + \cdots + (x-x_0)^n\left(\frac{a_0(x)}{a_n(x)}\right) y(x)\\
&\quad = (x-x_0)^n y^{(n)}(x)\\
&\qquad + (x-x_0)^{n-1}\left((x-x_0)\frac{a_{n-1}(x)}{a_n(x)}\right) y^{(n-1)}(x)\\
&\qquad + \cdots + (x-x_0)\left((x-x_0)^{n-1}\frac{a_1(x)}{a_n(x)}\right) y'(x)\\
&\qquad + \left((x-x_0)^n\frac{a_0(x)}{a_n(x)}\right) y(x)
\end{aligned}$$

$$= (x - x_0)^n p_n(x) y^{(n)}(x) + (x - x_0)^{n-1} p_{n-1}(x) y^{(n-1)}(x)$$

$$+ \cdots + (x - x_0) p_1(x) y'(x) + p_0(x) y(x)$$

$$= 0,$$

which is equation (7.7 RSP). The equations 7.9 constitute the test for whether or not the singular point x_0 is regular. If each function in 7.9 is analytic at x_0, then x_0 is a regular singular point; otherwise x_0 is an irregular singular point. Here are some examples of differential equations to see how to classify points as ordinary, regular singular, or irregular singular. If all of the coefficients are analytic at x_0, then any point x_0 where $a_n(x_0) \neq 0$ is an ordinary point. If x_0 is a point where $a_n(x_0) = 0$, then expand every coefficient in a power series about x_0. Check for the factorization indicated in equation (7.7 RSP). For example:

Example 7.9 Given the differential equation

$$(x^2 - 1)^2 y'' + (5x - 5) y' + (x^4 - 1) y = 0.$$

Classify all points as being either ordinary, regular singular, or irregular singular.

Solution. The only singular points are 1 and -1 since that is where $x^2 - 1 = 0$. Every other point is an ordinary point. Consider the singular point $x_0 = 1$. After dividing through by $(x^2 - 1)$ we have

$$y'' + \left(\frac{5x - 5}{(x^2 - 1)^2} \right) y' + \left(\frac{x^4 - 1}{(x^2 - 1)^2} \right) y = 0. \tag{7.10}$$

From here on we let *Mathematica* do the work. We have

```
In[44]:= x0 = 1

Out[44]= 1

In[45]:= Series[(x-x0)((5x-5)/((x^2-1)^2)),{x,x0,3}]

Out[45]=
                                    2              3
         5   5 (-1 + x)   15 (-1 + x)    5 (-1 + x)                4
         - - ---------- + ------------ - ----------- + O[-1 + x]
         4       4             16             8
```

Call this analytic function $p_1(x)$. Now consider

```
In[46]:= Series[(x-x0)^2((x^4-1)/((x^2-1)^2)),{x,x0,3}]

Out[46]=
                               2            3
                     (-1 + x)     (-1 + x)              4
         (-1 + x) + ---------- + ---------- + O[-1 + x]
                         2            4
```

Call this analytic function $p_0(x)$.

Therefore when multiplied through by $(x-1)^2$, the coefficients of equation 7.10 factor into

$$(x-1)^2y'' + (x-1)p_1(x)y' + p_0(x)y = 0,$$

so $x_0 = 1$ is a regular singular point.

Now consider the singular point $x_0 = -1$. Again let *Mathematica* do the work. From equation 7.10 we see that we must consider

```
In[47]:= x0 = -1

Out[47]= -1

In[48]:= Series[(x-x0)(5x-5)/((x^2-1)^2),{x,x0,3}]

Out[48]=
                                           2             3
           -5          5    5 (1 + x)   5 (1 + x)    5 (1 + x)
        ---------- + -(-) - --------- - ---------- - ---------- +
        2 (1 + x)      4        8           16           32

                  4
          O[1 + x]
```

and

```
In[49]:= Series[(x-x0)^2((x^4-1)/((x^2-1)^2)),{x,x0,3}]

Out[49]=
                              2          3
                      (1 + x)    (1 + x)            4
        -(1 + x) + ---------- - ---------- + O[1 + x]
                         2          4
```

The second of these is analytic at $x_0 = -1$, but the first is not. (It is undefined for $x = -1$.)

The conclusion is that 1 is a regular singular point, -1 is an irregular singular point, and all other points are ordinary points of the equation. ◇

The notebook *Classify Points* does these calculations for us. The main function, `Classify[x, x0, y, operator]`, will say what the status of the point `x0` is in the differential operator `operator`. The parameters `x` and `y` are symbols, `x0` is an expression, and `operator` is the name of the differential operator.

Example 7.9M Two function calls suffice to make the tests we just did.

Solution. Assume `Classify` to be loaded. Define the operator.

```
In[50]:= L[x_,y_] = (x^2-1)^2 y''[x] + 5(x+1) y'[x] + (x^2+1) y[x]

Out[50]=       2                                 2
        (1 + x ) y[x] + 5 (1 + x) y'[x] + (-1 + x ) y''[x]

In[51]:= Classify[x,1,y,L]

The point x0 = 1 is a regular singular point.
```

```
In[52]:= Classify[x,-1,y,L]

The point x0 = -1 is an irregular singular point.
```

◇

Predicting Properties of the Recursion Relation

When finding a power series solution for a first-order equation, we can expect a recursion relation to appear that says how all of the coefficients after a certain index are related. It is possible to determine many of the attributes of that recursion relation by inspection of the differential operator itself. First, express the coefficients of the differential equation in powers of $(x - x_0)$. The essential observation to make is that a term such as $(x-x_0)^m(d^k y/dx^k)$ changes the term $(x-x_0)^j$ of a series expansion into $(x-x_0)^{m+j-k}$. In other words, the exponent changes by the amount $m - k$. The set of all changes that can occur determine the terms c_{k+s} that occur in the recursion relation, the order of the recursion relation, and the index at which the relation takes effect. Here is an example: Consider the differential equation with solution to be expanded about $x_0 = 0$:

$$
\begin{array}{lccccccc}
 & L(y)(x) = (1)y'' & + (2x+ & x^2)y' & + (4+ & x)y = 0 \\
 & \Downarrow & \Downarrow & \Downarrow & \Downarrow & \Downarrow \\
\text{Changes} \rightarrow & -2 & 0 & 1 & 0 & 1
\end{array}
$$

Turn these changes into a set by eliminating duplicates: $DS(L) = \{-2, 0, 1\}$. This set is called the **determining set** of the operator, because it determines so much of the behavior of the recursion relation. From this one learns to see that the terms involved in the recursion relation are $c_{k-(-2)}, c_{k-0}$, and c_{k-1}, or c_{k+2}, c_k, and c_{k-1}. The recursion relation is a three-term recursion relation (involving the three terms shown). It is third- $(\max DS(L) - \min DS(L) = 1 - (-2) = 3)$ order, and is applicable for $k \geq 1 = \max DS(L)$. These conclusions are based on the requirement that all summation indices should be adjusted so that for ordinary series the common power of $(x - x_0)$ is $(x - x_0)^k$ and for Frobenius's method the common power of $(x - x_0)$ is $(x - x_0)^{k+r}$. There will always be as many initial equations as the starting index of the general summation that defines the recursion relation. The starting index of the general summation should cause c_0 to be involved, but no prior c_j. An mth-order (linear) recursion relation has a solution involving m arbitrary constants, but if the order of the differential equation is $n < m$, the desired solution will involve (at most) n arbitrary constants. Note that the determining set is useful when the coefficients of the differential equation are polynomials, but not when they are expressible as infinite series, because then the determining set is not bounded, and has infinitely many members.

Example 7.10 Solve $L(y)(x) = y'' + (2x + x^2)y' + (4 + x)y = 0$ by a power series with center at $x_0 = 0$ to illustrate the properties of the determining set.

Solution. We showed above that the determining set $DS(L) = \{-2, 0, 1\}$. Watch how these numbers play a central role in our calculations. Since $x_0 = 0$ is an ordinary point, assume

$$y(x) = \sum_{k=0}^{\infty} c_k x^k.$$

Then

$$y'(x) = \sum_{k=1}^{\infty} k c_k x^{k-1}$$

and

$$y''(x) = \sum_{k=2}^{\infty} k(k-1) c_k x^{k-2}.$$

Substitution gives

$$\begin{aligned}
y'' &+ (2x + x^2)y' + (4 + x)y \\
&= \sum_{k=2}^{\infty} k(k-1)c_k x^{k-2} + (2x + x^2)\sum_{k=1}^{\infty} k c_k x^{k-1} + (4 + x)\sum_{k=0}^{\infty} c_k x^k \\
&= \sum_{k=2}^{\infty} k(k-1)c_k x^{k-2} + 2x\sum_{k=1}^{\infty} k c_k x^{k-1} + x^2\sum_{k=1}^{\infty} k c_k x^{k-1} \\
&\quad + 4\sum_{k=0}^{\infty} c_k x^k + x\sum_{k=0}^{\infty} c_k x^k \\
&= \sum_{k=2}^{\infty} k(k-1)c_k x^{k-2} + 2\sum_{k=1}^{\infty} k c_k x^k + \sum_{k=1}^{\infty} k c_k x^{k+1} \\
&\quad + 4\sum_{k=0}^{\infty} c_k x^k + \sum_{k=0}^{\infty} c_k x^{k+1}.
\end{aligned}$$

Notice the changes in the exponents: $-2, 0, 1, 0, 1$; these form the determining set $\{-2, 0, 1\}$. Adjust the indices so that all sums involve x^k :

$$\begin{aligned}
&= \sum_{k=0}^{\infty} (k+2)(k+1)c_{k+2} x^k + 2\sum_{k=1}^{\infty} k c_k x^k + \sum_{k=2}^{\infty} (k-1)c_{k-1} x^k \\
&\quad + 4\sum_{k=0}^{\infty} c_k x^k + \sum_{k=1}^{\infty} c_{k-1} x^k
\end{aligned}$$

(The choices for the coefficients are now the ones expected: c_{k+2}, c_k, and c_{k-1}.)

$$= (2c_2 + 4c_0) + (c_0 + 6c_1 + 6c_3)x$$

$$+ \sum_{k=2}^{\infty} [(k+2)(k+1)c_{k+2} + (4+2k)c_k + kc_{k-1}]x^k \tag{7.11}$$

$$= (2c_2 + 4c_0) + \sum_{k=1}^{\infty} [(k+2)(k+1)c_{k+2} + (4+2k)c_k + kc_{k-1}]x^k$$

$$= 0.$$

What appears to be an initial term involving x^1 in equation 7.11 is really a general term when $k = 1$:

$$(1+2)(1+1)c_{1+2} + (4 + 2 \cdot 1)c_1 + 1c_{1-1} = c_0 + 6c_1 + 6c_3.$$

The recursion relation is therefore

$$(k+2)(k+1)c_{k+2} + (4+2k)c_k + kc_{k-1} = 0, \quad k \geq 1,$$

which has the form that was predicted. There are three terms, the order is $(k+2) - (k-1) = 3$, and the recursion relation is in effect for $k \geq 1$. Thus we knew a lot about what to expect before starting this rather complicated procedure. If we take c_0 and c_1 as arbitrary, all of the subsequent coefficients are defined in terms of these two, and we get the two linearly independent series, one the coefficient of c_0 and the other the coefficient of c_1.

Here is how to finish the problem: c_0 is arbitrary, as is c_1 (since no condition defines either c_0 or c_1); $2c_2 + 4c_0 = 0$ implies $c_2 = -2c_0$.

From

$$c_{k+2} = \frac{-2(2+k)c_k - kc_{k-1}}{(k+2)(k+1)}$$

for $k \geq 1$, we get

$$(k=1) \quad c_3 = \frac{-(2)(3)c_1 - c_0}{(3)(2)}$$

$$= -\frac{1}{6}c_0 - c_1,$$

$$(k=2) \quad c_4 = \frac{-2(4)c_2 - 2c_1}{(4)(3)}$$

$$= \frac{-2(4)(-2c_0) - 2c_1}{(4)(3)}$$

$$= \frac{16c_0 - 2c_1}{(4)(3)}$$

$$= \frac{4}{3}c_0 - \frac{1}{6}c_1,$$

and so on.

Since three consecutive terms c_0, c_1, and c_2 have been expressed in terms c. c_0 and c_1, all of the rest will also be expressed in terms of c_0 and c_1. (Why?) It is often quite difficult to guess the solutions to a three-term recursion relation. Sometimes you can see patterns by solving twice: once with $c_0 = 1$ and $c_1 = 0$, and a second time with $c_0 = 0$ and $c_1 = 1$. These can give manageable series. They will be the two series that we seek. Any linear combination of them is a solution, and they are linearly independent. (Why?) The *Complete Series, Ordinary* notebook gives the first several initial terms of the series to be

```
                            2    -a[0]          3
a[0] + a[1] x - 2 a[0] x  + (----- - a[1]) x  +
                                6

 4 a[0]   a[1]   4    23 a[0]   a[1]   5
(------ - ----) x  + (------- + ----) x  +
   3       6            60       2

 -23 a[0]   a[1]   6    -361 a[0]   37 a[1]   7
(-------- + ----) x  + (--------- - -------) x  +
    45       5            1260        252

 529 a[0]   31 a[1]   8    5503 a[0]   29 a[1]   9        10
(-------- - -------) x  + (--------- + -------) x  + O[x]
   5040       280           45360       1680
```

Since the coefficients in the differential equation are entire, these series are entire as well. (Can you find a general expression for the coefficients of these series?) The notation `O[x]^n` signifies the beginning of the remainder of the series. In this case the remainder begins with x^{10}. ◇

Exercises 7.5

PART I. For each of the following differential equations, classify every real number as being either an ordinary point, a regular singular point, or an irregular singular point.

1. $x^3y'' - 5x^2y' + 4y = 0.$
2. $(x^2 - 4)^2y'' + (x - 2)y' + 5y = 0.$
3. $x^2(x - 6)^2y'' + 7xy' - (x^2 - 36)y = 0.$
4. $(x^3 - 5x^2 + 6x)^2y'' + 3x(x - 3)^3y' + (x + 4)y = 0.$

PART II. By the methods of this section, solve the following differential equations in a power series centered on the ordinary point $x_0 = 0$. You may wish to modify the notebook *Quick Series Solution* to assist you by checking your calculations. The notebook *Formal Series, Ordinary* will actually reproduce each step that you make throughout the solution process.

5. $y'' + xy = 0.$ [This is the Airy[7] differential equation.]

6. $y'' - 2xy + 8y = 0.$

7. $y'' - (x+1)y' - y = 0.$

8. $\left(1 - \frac{1}{2}x^2\right) y'' + xy' - y = 0.$

9. $y'' + 5x^3 y = 0.$

10. $y'' - xy' + x^2 y = 0.$

7.6 Series Solution About Regular Singular Points

The Method of Frobenius

As we saw in the general introduction to solution in series, Georg Frobenius gave us the technique for solving in series about the regular singular point $x = x_0$. The differential equation

$$\begin{aligned} L_n(y)(x) &= (x - x_0)^n p_n(x) y^{(n)}(x) \\ &\quad + (x - x_0)^{n-1} p_{n-1}(x) y^{(n-1)}(x) \\ &\quad + \cdots + (x - x_0) p_1(x) y'(x) + p_0(x) y(x) = q(x), \end{aligned} \tag{7.12}$$

has x_0 as a regular singular point when the functions $p_n(x), p^{n-1}(x), \ldots, p_0(x)$ are analytic at x_0. Theorem 7.5 stated that there is at least one solution having the form

$$y(x) = \sum_{k=0}^{\infty} c_k (x - x_0)^{k+r}.$$

It is worth noting that near x_0 each of the (analytic) functions $p_n(x), p_{n-1}(x), \ldots, p_0(x)$ is nearly constant, so that for x near x_0, equation 7.12 is very much like

$$\begin{aligned} L_n(y)(x) &= (x - x_0)^n p_n(x_0) y^{(n)}(x) \\ &\quad + (x - x_0)^{n-1} p_{n-1}(x_0) y^{(n-1)}(x) \\ &\quad + \cdots + (x - x_0) p_1(x_0) y'(x) + p_0(x_0) y(x) = q(x), \end{aligned} \tag{7.12 Approx.}$$

where $p_n(x_0), p_{n-1}(x_0), \ldots, p_0(x_0)$ are constants and $p_n(x_0) \neq 0$. Look carefully at the form of this differential equation. It is a differential equation of the **Cauchy-Euler** type, suggesting that for x near x_0 equation (7.12 Approx.) should have a solution that looks very much like $(x - x_0)^r$. If this is so, then the possible choices for r can be found among the roots of a polynomial equation. We can indeed find the choices for r from among the roots of a polynomial equation,

[7] George Biddle Airy (1801–1892), British astronomer. He suffered from astigmatism, and was the the first, in 1825, to design eyeglasses that could correct for astigmatism.

and that equation is

$$\begin{aligned}
&[r(r-1)\cdots(r-n+1)]p_n(x_0) \\
&\quad + [r(r-1)\cdots(r-n+2)]p_{n-1}(x_0) \\
&\quad + \cdots + rp_1(x_0) + p_0(x_0) \\
&= \sum_{i=0}^{n} \left[\prod_{j=0}^{i-1} (r-j) \right] p_i(x_0) \\
&= 0. \qquad (7.13)
\end{aligned}$$

This is called the **indicial equation** for equation 7.12. It is the coefficient of $(x-x_0)^r$ in equation (7.12 Approx.) when $y(x) = \sum_{k=0}^{\infty} c_k (x-x_0)^{k+r}$.

Example 7.11 Let $x_0 = 0$. Find the indicial equation and the indices of the differential equation

$$6x^3 y^{(3)} - x^2(1+3x)y'' - 11x(1+x)y' + (3+2x)y = 0.$$

Solution. Here $p_3(x) = 6, p_2(x) = -(1+3x), p_1(x) = -11(1+x)$, and $p_0(x) = 3+2x$, so that $p_3(0) = 6, p_2(0) = -1, p_1(0) = -11$, and $p_0(0) = 3$. The indicial equation is

$$6r(r-1)(r-2) - r(r-1) - 11r + 3 = (r-3)(2r-1)(3r+1) = 0,$$

which has as roots $r_1 = 3, r_2 = 1/2$, and $r_3 = -1/3$. These are the indices.

Our statement was that the solution should look "very much like $(x-x_0)^r$." What does "very much like" mean here? It means that the solutions have the form

$$y(x) = (x-x_0)^r v(x),$$

where

$$v(x) = \sum_{k=0}^{\infty} c_k (x-x_0)^k,$$

and we find $v(x)$ by series methods. This makes the solutions ultimately have the form

$$y(x) = (x-x_0)^r \sum_{k=0}^{\infty} c_k (x-x_0)^k = \sum_{k=0}^{\infty} c_k (x-x_0)^{k+r},$$

the form that Frobenius suggested. ◇

Since the indices r are found to be the solutions of a polynomial equation, we expect the same three cases we saw before: distinct roots, repeated roots, and conjugate complex roots. The number of linearly independent solution we can get by Frobenius's method does depend on these three cases. However, the case of complex roots does not occur in practice (at least in the important second-order applications of the theory), so we will not elaborate on it. You therefore can expect to only find problems that have been so contrived that the indicial equations have no complex roots. We do get three cases, however.

The three cases of Frobenius's method are:

CASE I. Distinct real indices (roots), no two of which differ by a positive integer.
CASE II. Repeated real indices.
CASE III. Real indices that differ by a positive integer.

The notebook *FormalSeries, RSP* can be used to find n linearly independent solutions in case I. It will find linearly independent solutions corresponding to the indices that are distinct and do not differ by a positive integer from some other index. In cases II and III, you can look for solutions having one of the "$\ln x$" forms by assuming a solution of the proper form.

The theorem that will guide your search for solutions follows. A special version for second-order equations is provided. Both theorems assume the indicial equation has only real roots.

Theorem 7.6 *The differential equation*

$$\begin{aligned} L_n(y)(x) &= (x-x_0)^n p_n(x) y^{(n)}(x) \\ &\quad + (x-x_0)^{n-1} p_{n-1}(x) y^{(n-1)}(x) \\ &\quad + \cdots + (x-x_0) p_1(x) y'(x) + p(x) y(x) = 0, \end{aligned} \tag{7.14}$$

has x_0 as a regular singular point when the functions $p_n(x), p_{n-1}(x), \ldots, p_0(x)$ are analytic at x_0. The form of the solution is $y(x) = \sum_{k=0}^{\infty} c_k (x-x_0)^{k+r}$, where r is a (real) root of the **indicial equation**

$$\sum_{i=0}^{n} \left[\prod_{j=0}^{i-1} (r-j) \right] p_i(x_0) = 0.$$

Each such a root r is an index.

The distribution of the indices determines three cases:

CASE I. *The roots $r_1, r_2, \ldots, r_n$ of the indicial equation are distinct and no two differ by an integer. In this case there are n linearly independent solutions having the form*

$$y_i(x) = (x-x_0)^{r_i} \sum_{k=0}^{\infty} c_{i,k} (x-x_0)^k, \quad 1 \le i \le n.$$

CASE II. *There is an index r_i of multiplicity $k_i > 1$ which differs from no other index by an integer. In this case there is one solution of the form*

$$y_i(x) = (x-x_0)^{r_i} \sum_{k=0}^{\infty} c_{i,k} (x-x_0)^k,$$

and $k_i - 1$ *further linearly independent solutions of the form*

$$y_{i,m}(x) = y_i(x)\,(\ln(x-x_0))^m$$
$$+(x-x_0)^{r_i}\sum_{j=1}^{m-1}(\ln(x-x_0))^j\sum_{k=0}^{\infty}d_{m,j,k}(x-x_0)^k,$$
$$\text{for} \quad 2 \le m < k_i.$$

CASE III. *There is at least one collection* $C_q : r_1 > r_2 > \cdots > r_q$ *of indices that differ from each other by integer amounts and no other index differs from any of these by an integer amount. Suppose that the corresponding multiplicities are* $m_1, m_2, \ldots, m_q$. *Let* $p = \sum m_i$. *Then, corresponding to the largest index* r_1 *in* C_q, *there is one solution having the form*

$$y_1(x) = (x-x_0)^{r_1}\sum_{k=0}^{\infty}c_{1,k}(x-x_0)^k.$$

There may be other solutions having this form corresponding to other smaller indices in C_q *or there may not. To any index* r_s *in* C_q *of multiplicity greater than 1, there correspond* m_s *solutions having the form*

$$y(x) = (x-x_0)^{r_s}\sum_{j=0}^{i}(\ln(x-x_0))^j\sum_{k=0}^{\infty}d_{i,j,k}(x-x_0)^k$$

for some $i, 0 \le i \le p$.

For simplicity when you need to apply Theorem 7.6 to the second-order differential equations of mathematical physics, here is a statement of the theorem that applies specially to that case.

Theorem 7.7 *The second-order differential equation*

$$(x-x_0)^2p_2(x)y''(x) + (x-x_0)p_1(x)y'(x) + p_0(x)y(x) = 0, \tag{7.15}$$

has x_0 *as a regular singular point when the functions* $p_2(x), p_1(x)$, *and* $p_0(x)$ *are analytic at* x_0. *The form of the solution is* $y(x) = \sum_{k=0}^{\infty} c_k(x-x_0)^{k+r}$, *where* r *is a (real) root of the indicial equation* $p_2(x_0)r(r-1) + p_1(x_0)r + p_0(x_0) = 0$.

There are three cases:

CASE I. *The roots* r_1, r_2 *of the indicial equation are distinct and do not differ by an integer. In this case there are two linearly independent solutions having the form*

$$y_i(x) = (x-x_0)^{r_i}\sum_{k=0}^{\infty}c_{i,k}(x-x_0)^k, \quad i = 1, 2.$$

CASE II. *There is an index* r_1 *of multiplicity 2. In this case there is one solution of the form*

$$y_1(x) = (x - x_0)^{r_1} \sum_{k=0}^{\infty} c_{1,k}(x - x_0)^k,$$

and one further linearly independent solution of the form

$$y_2(x) = y_1(x) \ln(x - x_0) + (x - x_0)^{r_1} \sum_{k=0}^{\infty} c_{2,k}(x - x_0)^k.$$

CASE III. *The two indices $r_1 > r_2$ differ from each other by a positive integer. Then, corresponding to the larger index r_1 there is one solution having the form*

$$y_1(x) = (x - x_0)^{r_1} \sum_{k=0}^{\infty} c_{1,k}(x - x_0)^k.$$

There may be a second solution having this form corresponding to the smaller index or there may not. In any event, there is a second solution having the form

$$y_2(x) = C y_1(x) \ln(x - x_0) + (x - x_0)^{r_2} \sum_{k=0}^{\infty} c_{2,k}(x - x_0)^k$$

where $C = 0$ if there is a second series solution.

Examples of each of the three cases follow. Most are second-order, but case III is illustrated for a fifth-order problem as well.

Frobenius, Case I

The first case to consider when solving in series about a regular singular point is when the indices do not differ by a positive integer. In this case there are two linearly independent solutions, one for each index. See the notebook §*7.6 RSP Case I.*

Example 7.12 Consider the differential equation

$$3x^2y'' + (x^2 - x)y' + (1 + 2x)y = 0.$$

Find a pair of series solutions expanded about $x_0 = 0$.

Solution. The determining set is $\{0, 1\}$, so we expect a two term recursion relation involving c_k and c_{k-1} that is in effect for $k \geq 1$. Assume a series solution of the form $y(x) = \sum_{k=0}^{\infty} c_k x^{k+r}$. Then

$$y'(x) = \sum_{k=0}^{\infty} (k + r)c_k x^{k+r-1},$$

$$y''(x) = \sum_{k=0}^{\infty} (k + r)(k + r - 1)c_k x^{k+r-2},$$

and when these substitutions are made,

$$
\begin{aligned}
&3x^2y'' + (x^2 - x)y' + (1+2x)y \\
&\quad = 3x^2 \sum_{k=0}^{\infty} (k+r)(k+r-1)c_k x^{k+r-2} + (x^2 - x) \sum_{k=0}^{\infty} (k+r)c_k x^{k+r-1} \\
&\qquad + (1+2x) \sum_{k=0}^{\infty} c_k x^{k+r} \\
&\quad = 3 \sum_{k=0}^{\infty} (k+r)(k+r-1)c_k x^{k+r} + \sum_{k=0}^{\infty} (k+r)c_k x^{k+r+1} \\
&\qquad - \sum_{k=0}^{\infty} (k+r)c_k x^{k+r} + \sum_{k=0}^{\infty} c_k x^{k+r} + 2\sum_{k=0}^{\infty} c_k x^{k+r+1} \\
&\quad = \sum_{k=0}^{\infty} 3(k+r)(k+r-1)c_k x^{k+r} + \sum_{k=1}^{\infty} (k+r-1)c_{k-1} x^{k+r} \\
&\qquad - \sum_{k=0}^{\infty} (k+r)c_k x^{k+r} + \sum_{k=0}^{\infty} c_k x^{k+r} + \sum_{k=1}^{\infty} 2c_{k-1} x^{k+r} \\
&\quad = [3r(r-1) - r + 1]c_0 x^r \\
&\qquad + \sum_{k=1}^{\infty} [3(k+r)(k+r-1)c_k + (k+r-1)c_{k-1} \\
&\qquad - (k+r)c_k + c_k + 2c_{k-1}]x^{k+r} \\
&\quad = [3r^2 - 4r + 1]c_0 x^r \\
&\qquad + \sum_{k=1}^{\infty} [(k+r-1)(3(k+r)-1)c_k + (k+r+1)c_{k-1}]x^{k+r} \\
&\quad = 0.
\end{aligned}
$$

From the indicial equation $(3r^2 - 4r + 1)c_0 = (r-1)(3r-1)c_0 = 0$, we have the indices $r_1 = 1/3$ and $r_2 = 1$, and c_0 arbitrary. The recursion relation is

$$(k+r-1)(3(k+r)-1)c_k + (k+r+1)c_{k-1} = 0, \quad k \geq 1.$$

When $r = 1/3$, the recursion relation becomes

$$(3k-2)(3k)c_k + (3k+4)c_{k-1} = 0, \quad k \geq 1,$$

or, solved for c_k since the leading coefficient is never 0,

$$c_k = -\frac{3k+4}{(3k-2)(3k)}c_{k-1}, \quad k \geq 1.$$

A formula for the general coefficient is

$$c_k = (-1)^k \prod_{j=1}^{k} \frac{3j+4}{(3j-2)(3j)} c_0, \quad k \geq 1.$$

The first several values produced are

$$c_1 = -\frac{7}{3} c_0$$

$$c_2 = \frac{35}{36} c_0$$

$$c_3 = -\frac{65}{324} c_0$$

$$c_4 = \frac{13}{486} c_0,$$

so the series corresponding to $r_1 = 1/3$ is

$$y_1(x) = c_0 x^{1/3} \left(1 - \frac{7}{3}x + \frac{35}{36}x^2 - \frac{65}{324}x^3 + \frac{13}{486}x^4 - \cdots\right).$$

When $r = 1$ the recursion relation becomes

$$(k)(3k+2)c_k + (k+2)c_{k-1} = 0, \quad k \geq 1,$$

which can be written

$$c_k = -\frac{k+2}{(k)(3k+2)} c_{k-1}, \quad k \geq 1,$$

since the leading coefficient is never 0. A formula for the general coefficient is

$$c_k = (-1)^k \prod_{j=1}^{k} \frac{j+2}{(j)(3j+2)} c_0, \quad k \geq 1.$$

The first several values produced are

$$c_1 = -\frac{3}{5} c_0$$

$$c_2 = \frac{3}{20} c_0$$

$$c_3 = -\frac{1}{44} c_0$$

$$c_4 = \frac{3}{1232} c_0.$$

so the series corresponding to $r_2 = 1$ is

$$y_2(x) = c_0 x \left(1 - \frac{3}{5}x + \frac{3}{20}x^2 - \frac{1}{44}x^3 + \frac{3}{1232}x^4 - \cdots\right).$$

Since c_0 is intended to be arbitrary here and not related to the c_0 of the previous solution, it would be better to change its name (to c_1). Then the solution to our differential equation is

$$y(x) = c_0 x^{1/3}\left(1 - \frac{7}{3}x + \frac{35}{36}x^2 - \frac{65}{324}x^3 + \frac{13}{486}x^4 - \cdots\right)$$
$$+c_1 x\left(1 - \frac{3}{5}x + \frac{3}{20}x^2 - \frac{1}{44}x^3 + \frac{3}{1232}x^4 - \cdots\right),$$

where c_0 and c_1 are arbitrary. These solutions are valid for $0 < x < \infty$. ◇

Frobenius, Case II

Example 7.13 Find a series solution of this differential equation about the regular singular point $x_0 = 0$ where the two indices are equal:

$$x^2 y'' + x(2x - 5)y' + 9y = 0.$$

Solution. We actually seek two series centered on 0. As we will see, the indices are 3 and 3. This is the case of a double root of the indicial equation. We will find the only solution that Frobenius's method produces and then, in Example 7.14, find a second linearly independent solution by using the formula (see Section 7.2) for a second solution in series.

We substitute $y(x) = \sum_{k=0}^{\infty} c_k x^{k+r}$. The determining set for the equation is $\{0,1\}$ which says that the recursion relation will involve the two terms c_k and c_{k-1} and is in effect for $k \geq 1$. The indicial equation turns out to be

$$(r^2 - 6r + 9)c_0 = (r - 3)^2 c_0 = 0,$$

which says that $r_1 = r_2 = 3$ and c_0 is arbitrary. So we can find only one series solution by Frobenius's method. The recursion relation is

$$2(-1 + k + r)c_{k-1} + (-3 + k + r)^2 c_k = 2(k + 2)c_{k-1} + k^2 c_k = 0, \quad k \geq 1,$$

which can also be written

$$c_k = -\frac{2(k + 2)c_{k-1}}{k^2}, \quad k \geq 1.$$

This produces the coefficients of the solution

$$y_1(x) = c_0 x^3\left(1 - 6x + 12x^2 - \frac{40}{3}x^3 + 10x^4 - \frac{28}{5}x^5 + \frac{112}{45}x^6 - \frac{32}{35}x^7 + \cdots\right)$$

Note that the recursion relation has the solution

$$c_k = \frac{(-1)^k 2^{k-1}(k + 1)(k + 2)}{k!} c_0, \quad k \geq 1.$$

You may wish to verify each of these assertions. We still need to find a second linearly independent solution. ◇

Example 7.14 Find a second series solution using the formula for a second solution:

$$y_2(x) = y_1(x) \int \frac{e^{-\int (a_1(x)/a_2(x))\,dx}}{y_1(x)^2}\,dx.$$

Solution. Here $a_1(x) = x(2x-5)$ and $a_2(x) = x^2$, so that

$$g(x) = e^{-\int (a_1(x)/a_2(x))\,dx} = e^{-\int (x(2x-5)/x^2)\,dx} = e^{-2x+5\ln x} = x^5 e^{-2x}.$$

We find that according to *Mathematica* the series for $(1/(y_1(x)^2))$ is

```
In[53]:= 1/(y1[x]^2)

Out[53]= -6    12   84   1376   2188   48176   1814848
         x   + -- + -- + ---- + ---- + ----- + ------- + O[x]
                5    4      3     2     5 x      45
               x    x    3 x     x
```

It is worth noting that the coefficients seem to be getting larger. When the coefficients are approaching 0 we get an entire function from our series; when the coefficients are bounded, the series has a finite positive radius of convergence. But when the coefficients are unbounded the series may converge only at a single point. We watch to see what happens.

The factor

$$\int \frac{e^{-\int (a_1(x)/a_2(x))\,dx}}{y_1(x)^2} dx$$

is, according to *Mathematica*,

```
In[54]:= Normal[Integrate[g[x]/(y1[x]^2),x]]

Out[54]=
                            3          4           5            6
                   2   940 x     2135 x     91084 x     111946 x
        10 x + 31 x  + ------ + ------- + -------- + ---------
                         9          6          75          27

          + Log[x]
```

This means that the second solution is

```
y2(x)  = y1(x) ln x +
                              3           4            5             6
                     2   940 x      2135 x      91084 x      111946 x
y1(x)(10 x + 31 x  + ------ + ------- + -------- + ---------
                           9           6           75            27

 +...).
```

Note the size of the coefficients. The $y_1(x)\ln x$ term is convergent for $x > 0$, but the term on the second line looks like it may converge only for x small.

When substituted into the differential equation we would like for

$$x^2y'' + x(2x-5)y' + 9y = 0 = \text{small}(x).$$

This means that we should satisfy the differential equation except for terms that are small because the series that is left over should be an approximation for the correct right-hand side, 0. These terms left over constitute the **residual error** in our solution. When the substitution of $y_2(x)$ is made, *Mathematica* says that the residual error is

```
        10                                             2
(-8 x     (245804643 - 729842778 x + 1003339260 x   +

               3         3
     381837505 x  + 660 x  Log[x])) / 2835.
```

◇

The common factor of x^{10} says that we formally satisfied the differential equation through terms of order 9. But the huge coefficients say that this term (and hence the residual error) is not small for x large. In fact, on the interval $[0, 0.1]$, the error is at most 0.00005. We include 0 since $\lim_{x\to\infty} x^n \ln x = 0 = 0$ for $n \geq 1$, which makes the residual error continuous at 0. When plotted, the residual error is very slow to leave 0, but when it does, it rises rapidly. Look at the notebook *Second Series Solution* for a derivation of the second solution and a plot of the residual error. A residual error is plotted in Example 7.16.

Note that $y_2(x)$ as provided to you is not the solution. It is merely the initial terms of an infinite series. The difference between it and the actual solution is the **actual error in the solution**. This is not the same as the residual error. There is a relationship between the two errors, however. Since the differential operator is linear, the relationship is easy to show. Write $y_2(x) = \hat{y}_2(x) + E_2(x)$, where this means that the exact solution $y_2(x)$ is the sum of $\hat{y}_2(x)$, the approximate solution we have found, and the actual error $E_2(x) = y_2(x) - \hat{y}_2(x)$, the difference between the exact and the approximate solutions. We have

$$\begin{aligned}
0 &= L(y_2(x)) \\
&= L(\hat{y}_2(x) + E_2(x)) \\
&= L(\hat{y}_2(x)) + L(E_2(x)) \\
&= \text{Residual Error} + L(E_2(x)),
\end{aligned}$$

so

$$\text{Residual Error} = -L(E_2(x)).$$

This implies that the actual error $E_2(x)$ is a particular solution of the differential equation

$$L(y) = -\text{Residual Error},$$

which suggests that the actual error may be hard to find, since we are having trouble solving the homogeneous form of just such an equation. But this does define the relationship between the actual error and the residual error. In numerical analysis, such a relationship between the actual error and the residual error for linear algebraic systems is exploited to improve the approximate solution by using iterative techniques.

Frobenius, Case III

Example 7.15 Solve the differential equation

$$xy'' + (x-8)y' - 5y = 0$$

about the regular singular point $x_0 = 0$ where the two indices differ by a positive integer.

Solution. We seek series centered on 0. As we will see, the indices are 0 and 9. The role of the -5 is also of interest: there is a polynomial solution of order 5 (corresponding to the index 0). A second linearly independent solution (corresponding to the index 9) is an infinite series whose leading coefficient is x^9. The problem is a special case of the problem

$$xy'' + (x-n_1)y' - n_0 y = 0, \quad 0 < n_0 < n_1,$$

which has a polynomial solution of order n_0 and a series solution whose leading term is x^{n_1+1}.

The determining set is $\{-1, 0\}$, so we expect a two-term recursion relation that takes effect for $k \geq 0$. Assume a series solution of the form

$$y(x) = \sum_{k=0}^{\infty} c_k x^{k+r}.$$

Then

$$y'(x) = \sum_{k=0}^{\infty} (k+r)c_k x^{k+r-1},$$

$$y''(x) = \sum_{k=0}^{\infty} (k+r)(k+r-1)c_k x^{k+r-2},$$

and

$$\begin{aligned}
xy'' &+ (x-8)y' - 5y \\
&= x\sum_{k=0}^{\infty} (k+r)(k+r-1)c_k x^{k+r-2} \\
&\quad + (x-8)\sum_{k=0}^{\infty} (k+r)c_k x^{k+r-1} - 5\sum_{k=0}^{\infty} c_k x^{k+r} \\
&= \sum_{k=0}^{\infty} (k+r)(k+r-1)c_k x^{k+r-1} \\
&\quad + \sum_{k=0}^{\infty} (k+r)c_k x^{k+r} - 8\sum_{k=0}^{\infty} (k+r)c_k x^{k+r-1} - 5\sum_{k=0}^{\infty} c_k x^{k+r}
\end{aligned}$$

$$
\begin{aligned}
&= \sum_{k=-1}^{\infty} (k+1+r)(k+r)c_{k+1}x^{k+r} \\
&\quad + \sum_{k=0}^{\infty}(k+r)c_k x^{k+r} - 8\sum_{k=-1}^{\infty}(k+1+r)c_{k+1}x^{k+r} - 5\sum_{k=0}^{\infty} c_k x^{k+r} \\
&= [r(r-1) - 8r]c_0 x^{r-1} \\
&\quad + \sum_{k=0}^{\infty}[(k+1+r)(k+r)c_{k+1} + (k+r)c_k - 8(k+1+r)c_{k+1} - 5c_k]x^{k+r} \\
&= [r^2 - 9r]c_0 x^{r-1} + \sum_{k=0}^{\infty}[(k+1+r)(k+r-8)c_{k+1} + (k+r-5)c_k]x^{k+r} \\
&= 0.
\end{aligned}
$$

From $r^2 - 9r = 0$, we have the indices $r = 0$ and $r = 9$. The recursion relation is

$$(k+1+r)(k+r-8)c_{k+1} + (k+r-5)c_k = 0, \qquad k \geq 0.$$

When $r = 0$, the recursion relation becomes

$$(k+1)(k-8)c_{k+1} + (k-5)c_k = 0, \qquad k \geq 0.$$

Note that $k - 5 = 0$ when $k = 5$ and $k - 8 = 0$ when $k = 8$. These are important events.

The first several equations produced from the recursion relation are (according to *Mathematica*):

```
-5 c[0] - 8 c[1] == 0
-4 c[1] - 14 c[2] == 0
-3 c[2] - 10 c[3] == 0
-2 c[3] - 20 c[4] == 0
-c[4] - 20 c[5] == 0
0 c[5] - 18 c[6] == 0
1 c[6] - 14 c[7] == 0
2 c[7] - 8 c[8] == 0
3 c[8] - 0 c[9] == 0
4 c[9] + 10 c[10] == 0
5 c[10] + 22 c[11] == 0
6 c[11] + 36 c[12] == 0
```

The first five equations determine the polynomial solution. The next four **bold** equations say that $c_6 = c_7 = c_8 = 0$, and c_9 is arbitrary. The recursion relation defines all of the subsequent coefficients in terms of c_9, because after $k = 9$ we have passed the point where any factor in the recursion relation is zero. For $k \geq 9$ the recursion relation says

$$c_{k+1} = -\frac{k-5}{(k+1)(k-8)} c_k,$$

and one can write down the equation

$$\begin{aligned} c_{9+m} &= (-1)^m \frac{9!}{3!} \frac{c_9}{(m+9)(m+8)(m+7)(m+6)(m+5)(m+4)m!} \\ &= (-1)^m \frac{9!}{3!} \frac{(m+3)(m+2)(m+1)c_9}{(m+9)!}, \, m \geq 0. \end{aligned}$$

The case $r = 9$ reproduces these coefficients (with c_0 replacing c_9, and c_m replacing c_{9+m}). You should verify this statement.

The solution to the given differential equation is

$$\begin{aligned} y(x) = c_0 &\left(1 - \frac{5}{8}x + \frac{5}{28}x^2 - \frac{5}{168}x^3 + \frac{1}{336}x^4 - \frac{1}{6720}x^5\right) \\ &+ c_9 x^9 \left(1 - \frac{2}{5}x + \frac{1}{11}x^2 - \frac{1}{66}x^3 + \frac{7}{3432}x^4 - \frac{1}{4290}x^5 + \cdots\right). \end{aligned}$$

These solutions are valid for $0 < x < \infty$. ◇

Case III for Higher-Order Equations

In case III of Theorem 7.6, $i = 0$ corresponds to having a solution of Frobenius type corresponding to a smaller index. Also in case III, it is possible for i to be at least as large as m_s. This case is illustrated in the notebook *Example 7.16 M* where Frobenius solutions are found, and solutions having logarithms are found. In this notebook, the differential equation is the fifth-order equation of the next example.

Example 7.16M Find series solutions of the differential equation

$$x^5 y^{(5)} + 4x^4 y^{(4)} - 26x^3 y^{(3)} + 44x^2 y'' + x(144 + 2x)y' - 576y = 0$$

about the regular singular point $x_0 = 0$.

Solution. The differential equation has as indicial equation $(r+3)^2(r-4)^3 = 0$. Thus, $r_1 = -3$ has multiplicity 2 and $r_2 = 4$ has multiplicity 3. Furthermore, $r_2 - r_1 = 7$, an integer. So we have two indices, each of which is repeated, which differ by an integer. This is the third case, the statement of which appears to be so complicated. Since the order of the differential equation is five, we need to find five linearly independent solutions.

The following solutions were obtained from the notebook *Example 7.16 M* in three steps. We want to find two linearly independent solutions corresponding to the double root $r_1 = -3$ and three linearly independent solutions corresponding to the triple root $r_2 = 4$.

Step I. Find power series solutions by specifying { a } where you are to supply a list of coefficient names. After the next several cells are executed, find that corresponding to $r_1 = -3$ there is the power series solution

$$y_1(x) = \left(\frac{-1}{1296000} + \frac{1}{x^3} - \frac{1}{36\,x^2} + \frac{1}{4500\,x}\right)$$
$$= x^{-3}\left(1 - \frac{x}{36} + \frac{x^2}{4500} - \frac{x^3}{1296000}\right).$$

This solution is exact.

Corresponding to $r_2 = 4$ there is the power series solution

$$y_3(x) = x^4\left[1 - \frac{x}{8} + \frac{5\,x^2}{2592} - \frac{x^3}{116640} + \frac{7\,x^4}{451630080} - \frac{7\,x^5}{508083840000} + \cdots\right].$$

We now know that there are only two Frobenius solutions. The remainder of the solutions will involve $\ln x$ or powers of $\ln x$.

We need one more solution corresponding to $r_1 = -3$ and two more corresponding to $r_2 = 4$. We now look for solutions that involve $\ln x$ but no higher powers of $\ln x$.

Step II. Use the notebook *Example 7.16 M* to find simple "log series". These have the form

$$\text{(Frobenius series)} + (\ln x)\text{(Frobenius series)}.$$

To find these, specify $\{a, b\}$ in the list of coefficient names. This should enable us to reproduce $y_1(x)$ and $y_3(x)$ and find at least one additional log series corresponding to each index. From $r_1 = -3$, there results a log series having two arbitrary constants.

$$\begin{aligned} y_{1,2}(x) = {} & \frac{109\,a_0}{142560000} + \frac{a_0}{x^3} - \frac{3\,a_0}{27500\,x} - \frac{x\,a_0}{513216000} - \frac{x^2\,a_0}{51321600000} \\ & + \frac{73\,a_1}{1320000} + \frac{a_1}{x^2} - \frac{82\,a_1}{6875\,x} - \frac{x\,a_1}{14256000} - \frac{x^2\,a_1}{1425600000} + \cdots \\ & + \ln(x)\left(\frac{-a_0}{2376000} + \frac{6\,a_0}{11\,x^3} - \frac{a_0}{66\,x^2} + \frac{a_0}{8250\,x} - \frac{a_1}{66000}\right. \\ & \left. + \frac{216\,a_1}{11\,x^3} - \frac{6\,a_1}{11\,x^2} + \frac{6\,a_1}{1375\,x} + \cdots\right) \end{aligned}$$

The constants can be chosen (see exercises 12 and 13) to reproduce $y_1(x)$ or to create the new solution

$$\begin{aligned} y_2(x) = {} & \left(\frac{11\,x}{216} - \frac{41\,x^2}{67500} + \frac{73\,x^3}{25920000} - \frac{x^4}{279936000} - \frac{x^5}{27993600000} + \cdots\right) x^{-3} \\ & + \ln(x)(y_1(x)). \end{aligned}$$

Likewise, from the index $r_2 = 4$, there results a similar log series having two arbitrary constants which can be chosen (see exercises 14 and 15) either to reproduce $y_3(x)$ or to generate the new solution

$$y_4(x) = x^4 \left[\frac{3\,x}{8} - \frac{407\,x^2}{46656} + \frac{5\,x^3}{104976} - \frac{17587\,x^4}{178845511680} + \frac{193657\,x^5}{2012012006400000} + \cdots \right] + \log(x)\,[y_3(x)]\,.$$

This gives us four of the required five solutions. There remains one more solution corresponding to $r_2 = 4$ to find. This solution will contain $(\ln x)^2$.

Step III. Use the notebook *Example 7.16 M* to find a solution of the form

$$\text{(Frobenius series)} + (\ln x)\text{(Frobenius series)} + (\ln x)^2\text{(Frobenius series)}.$$

To find this solution, specify $\{a, b, c\}$ in the list of coefficient names. This should enable us to reproduce $y_1(x)$, $y_2(x)$, $y_3(x)$ and $y_4(x)$ and find at least one additional log series corresponding to $r_2 = 4$. From $r_2 = 4$, there results a $\log^2$ series having three arbitrary constants:

$$\begin{aligned}
&y_{3,4,5}(x) \\
&= x^4 \left[a_0 + x\,a_1 + x^2\,a_2 + x^3 \left(\frac{-228188\,a_0}{129929943375} - \frac{45890551\,a_1}{519719773500} - \frac{549886\,a_2}{59410125} \right) \right. \\
&\quad + x^4 \left(\frac{1793469419\,a_0}{243494950522032000} + \frac{14321509579\,a_1}{44271809185824000} + \frac{1401670709\,a_2}{55668712968000} \right) \\
&\quad + x^5 \left(\frac{-3868286359\,a_0}{365242425783048000000} - \frac{14281510597\,a_1}{33203856889368000000} \right. \\
&\quad \left. \left. - \frac{410880559\,a_2}{13917178242000000} \right) + \cdots \right] \\
&\quad + x^4 \ln(x) \left[\left(\frac{105280\,a_0}{158427} + \frac{1669520\,a_1}{158427} \frac{5956416\,a_2}{17603} \right) + \right. \\
&\quad + x \left(\frac{-3296\,a_0}{158427} + \frac{25742\,a_1}{158427} + \frac{375192\,a_2}{17603} \right) \\
&\quad + x^2 \left(\frac{-221\,a_0}{1327509} - \frac{37421\,a_1}{2655018} - \frac{502\,a_2}{607} \right) \\
&\quad + x^3 \left(\frac{766\,a_0}{346479849} + \frac{67633\,a_1}{692959698} + \frac{4102\,a_2}{792135} \right) \\
&\quad + x^4 \left(\frac{-889439\,a_0}{147572697286080} - \frac{3014507\,a_1}{13415699753280} - \frac{192919\,a_2}{16869306960} \right)
\end{aligned}$$

$$+ x^5 \left(\frac{11331209\, a_0}{1660192844468400000} + \frac{17705011\, a_1}{75463311112200000} \right.$$

$$\left. + \frac{1106891\, a_2}{94889851650000} \right) + \cdots \Bigg]$$

$$+ x^4 \ln(x)^2 \left[\left(\frac{4384\, a_0}{52809} + \frac{104192\, a_1}{52809} + \frac{1492992\, a_2}{17603} \right) \right.$$

$$+ x \left(\frac{-548\, a_0}{52809} - \frac{13024\, a_1}{52809} - \frac{186624\, a_2}{17603} \right)$$

$$+ x^2 \left(\frac{685\, a_0}{4277529} + \frac{16280\, a_1}{4277529} + \frac{2880\, a_2}{17603} \right)$$

$$+ x^3 \left(\frac{-137\, a_0}{192488805} - \frac{3256\, a_1}{192488805} - \frac{64\, a_2}{88015} \right)$$

$$+ x^4 \left(\frac{959\, a_0}{745316652960} + \frac{259\, a_1}{8469507420} + \frac{14\, a_2}{10649815} \right)$$

$$\left. + x^5 \left(\frac{-959\, a_0}{838481234580000} - \frac{259\, a_1}{9528195847500} - \frac{14\, a_2}{11981041875} \right) + \cdots \right]$$

The three constants can be chosen to reproduce $y_3(x)$ and $y_4(x)$, or to generate the last remaining solution (see also exercise 16):

$$y_5(x) = x^4 \left[\frac{52809\, x}{104192} - \frac{45890551\, x^3}{1025405568000} + \frac{14321509579\, x^4}{87348147904512000} \right.$$

$$\left. - \frac{14281510597\, x^5}{6551111092838400000} + \cdots \right]$$

$$+ x^4 \log(x) \left[\frac{104345}{19536} + \frac{12871\, x}{156288} - \frac{1085209\, x^2}{151911936} \right.$$

$$\left. + \frac{67633\, x^3}{1367207424} - \frac{3014507\, x^4}{26469135728640} + \frac{17705011\, x^5}{148888888473600000} + \cdots \right]$$

$$+ \log(x)^2 \left[y_3(x) \right].$$

We now have five linearly independent solutions. Any linear combination of these is also a solution.

The notebook *Example 7.16M* can be used to find these solutions through as many terms as you have patience to wait for (or have memory available). The process is not fast, but since a reassuring check of the solution is made, having confidence in the results is likely worth the wait. You will observe that when you request more than just a few terms, *Mathematica* spreads the various terms of the results quite widely apart. You may have to look carefully across several pages of output to pick out the terms you want.

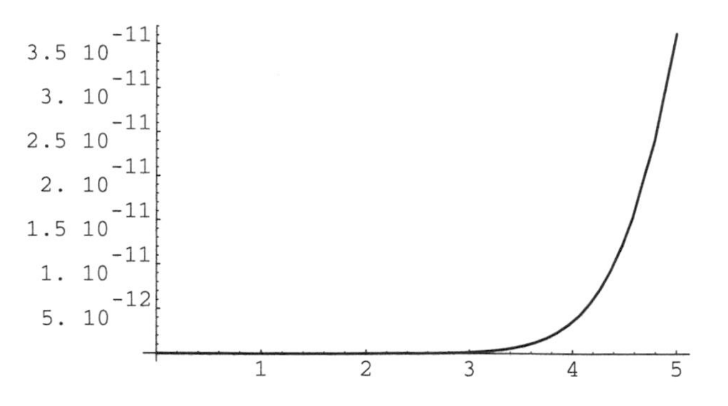

Figure 7.1: A plot of the residual error of the solution.

If you use the notebook *Example 7.16 M*, a check of the solution claims that when 15 terms are retained, then

$$L(x, y(x)) = \frac{x^{13} g(x, \ln x)}{\text{Large constant}} = \frac{x^{12}(x g(x, \ln x))}{\text{Large constant}}.$$

This is a **residual**. It measures how closely $L(x, y(x))$ approximates 0. (We wanted to solve a homogeneous equation, so we wanted the answer to be exactly 0. Instead, it is merely near 0.) Since

$$\lim_{x \to 0} x\, g(x, \ln x) = 0,$$

the factor x^{12} guarantees that our solution is a very good approximation to 0 near $x = 0$, because for x near 0, x^{13} is very small. In fact, a plot indicates that the residual is smaller than 4.0×10^{-11} for $0 < x \leq 5$, as Figure 7.1 shows. ◇

EXERCISES 7.6

Solve the following differential equations in power series centered on the regular singular point $x_0 = 0$ by the methods of this section. You may wish to modify the notebook *FormalSeries, RSP* to assist you by checking your calculations. The problems have been partitioned into those whose indices fall into case I or into case II or into case III of Theorem 7.6. *FormalSeries, RSP* permits you to assume series having several log parts.

PART I. These problems are in case I: the indices do not differ by an integer. The method of Frobenius gives you a linearly independent set of solutions.

1. $2x\dfrac{d^2y}{dx^2} - \dfrac{dy}{dx} + 3y = 0.$

2. $5x\dfrac{d^2y}{dx^2} + (3 + x)\dfrac{dy}{dx} + y = 0.$

3. $4x^2\dfrac{d^2y}{dx^2} + 9x\dfrac{dy}{dx} - (x - 3)y = 0.$

PART II. These problems are in case II: an index has multiplicity greater than 1; no unequal indices differ by an integer. The method of Frobenius may or may not give you a linearly independent set of solutions. If it does not, assume a solution with a log part. Add more and more log series until you have a full basis of solutions. Quite often the smallest index gives you the most linearly independent solutions, so you may want to try it first.

4. $x^2\dfrac{d^2y}{dx^2} + (5+2x)x\dfrac{dy}{dx} + 4y = 0.$

5. $2x^3\dfrac{d^3y}{dx^3} - 5x^2\dfrac{d^2y}{dx^2} + 3x(2x+1)\dfrac{dy}{dx} + 9y = 0.$

6. $9x^4\dfrac{d^4y}{dx^4} + 84x^3\dfrac{d^3y}{dx^3} + 166x^2\dfrac{d^2y}{dx^2} + 32x(2x+1)\dfrac{dy}{dx} + (4+x)y = 0.$

7. $135x^4\dfrac{d^4y}{dx^4} + 666x^3\dfrac{d^3y}{dx^3} + 444x^2\dfrac{d^2y}{dx^2} + 56x(x-1)\dfrac{dy}{dx} + 8y = 0.$

PART III. These problems are in case III: a pair of indices differ by an integer. The method of Frobenius may or may not give you a linearly independent set of solutions. If it does not, assume a solution with a log part. Add more and more log series until you have a full basis of solutions.

8. $4x^3\dfrac{d^3y}{dx^3} - 12x^2\dfrac{d^2y}{dx^2} + 15x\dfrac{dy}{dx} - (12+x)y = 0.$

9. $x^2\dfrac{d^2y}{dx^2} + x(2x+1)\dfrac{dy}{dx} - y = 0.$

10. $27x^4\dfrac{d^4y}{dx^4} + 45x^3\dfrac{d^3y}{dx^3} - 12x^2\dfrac{d^2y}{dx^2} + 14x(x+2)\dfrac{dy}{dx} - 32y = 0.$

11. $2x^4\dfrac{d^4y}{dx^4} + 7x^3\dfrac{d^3y}{dx^3} - 13x^2\dfrac{d^2y}{dx^2} - 5x(2x+3)\dfrac{dy}{dx} = 0.$

PART IV. Further examination of Example 7.16.

In the notebook *Example 7.16M* make certain that the operator defined is the operator of Example 7.16M. Execute each cell until you reach the boxed cell that reads:

```
y[x_]=MakeAssumedFunction[x,r,{a}]
(* {a}:     Power series *)
(* {a,b}:   Power series + Log series *)
(* {a,b,c}:Power series + Log series + Log^2 Series *)
```

Be sure that the function call is `MakeAssumedFunction[x,r,{a,b,c}]`, so that a $\log^2$ series will be produced when you execute the cell. Continue executing, using $n = 5$, until you have obtained the $\log^2$ series corresponding to r_1 and r_2.

12. In the series for `y12[x]` choose `a[0]` and `a[1]` to match the corresponding terms of the series for $y_1(x)$ in the solution of Example 7.16. The result you get should be equivalent to $y_1(x)$.

13. In the series for `y12[x]` make the three leading terms of the coefficient of `Log[x]` agree with the three leading terms of $y_1(x)$ that you just obtained. The result should agree with $y_2(x)$.

14. In the series for `y345[x]` create a system of equations by setting the coefficients in the lowest-order power series equal to the corresponding terms in $y_3(x)$. This should reproduce $y_3(x)$. (Alternatively, make the first three coefficients of `Log[x]` to be 0.)

15. Obtain $y_4(x)$ by making the low-order three coefficients of `Log[x]` agree with the corresponding coefficients of $y_3(x)$.

16. Obtain $y_5(x)$ by making the low-order three coefficients of `(Log[x])`^{2} agree with the corresponding coefficients of $y_3(x)$.

7.7 Important Classical Differential Equations and Functions

The definitions and problems that follow are but a quick glimpse at some very important material. You may find it productive to have *Mathematica* do some of the investigations. You may find the proofs to be difficult. For each of the functions defined, there is a special function built into *Mathematica*. You should investigate the properties of the functions *Mathematica* provides.

The Gamma Function

The **Gamma function** $\Gamma(p)$ is defined for positive p by

$$\Gamma(p) = \int_0^\infty t^{p-1} e^{-t} dt.$$

Mathematica calls $\Gamma(p)$ `Gamma[p]`.

G.1 The improper integral that defines $\Gamma(p)$ is convergent for $p > 0$.

G.2 $\Gamma(p+1) = p\Gamma(p)$. (Integrate by parts.) This is a recursive definition for the Gamma function.

G.3 $\Gamma(1) = 1; \Gamma(n+1) = n!$ for each positive integer n.

G.4 If p is not 0 or a negative integer, then $\Gamma(p) = \Gamma(p+1)/p$ can be used to extend the definition of $\Gamma(p)$ to negative p's.

G.5 Use *Mathematica* to plot $\Gamma(p)$ for $-3 \le p \le 4$.

G.6 Use *Mathematica* or otherwise show that $\Gamma(1/2) = \sqrt{\pi}$.

We define $p! = \Gamma(p+1)$ for all p except p a negative integer. This usage will be seen later in these problems.

The Bessel Function

Bessel's equation is

$$x^2\frac{d^2y}{dx^2} + x\frac{dy}{dx} + (x^2 - p^2)y = 0,$$

where p is a nonnegative constant. The **Bessel functions** of the first kind are traditionally denoted $J_p(x)$; those of the second kind, $Y_p(x)$. In *Mathematica* these are called `BesselJ[p,x]` and `BesselY[p,x]`, respectively.

J.1 Zero is a regular singular point of Bessel's equation.

J.2 The numbers p and $-p$ are the indices of Bessel's equation.

J.3 If a solution of the form $y(x) = \sum_0^\infty a_n x^{n+p}$ is assumed, the recursion relation is

$$a_n = -\frac{a_{n-2}}{n(2p+n)}, \quad \text{for} \quad n \geq 2.$$

J.4 One solution is

$$y(x) = a_0 x^p \sum_{n=0}^{\infty} (-1)^n \frac{x^{2n}}{2^{2n} n!(p+1)\cdots(p+n)}.$$

It is traditional to let $a_0 = 1/(2^p\Gamma(p+1))$ to get the **Bessel functions of the first kind of order p**,

$$J_p(x) = \sum_{n=0}^{\infty} (-1)^n \frac{(x/2)^{2n+p}}{n!\,\Gamma(p+n+1)}.$$

The use of Γ occurs because p need not be an integer. Using $p! = \Gamma(p+1)$, we have the alternate definition

$$J_p(x) = \sum_{n=0}^{\infty} (-1)^n \frac{(x/2)^{2n+p}}{n!\,(p+n)!}.$$

J.5 Write out several terms of $J_0(x)$ and of $J_1(x)$. Note that $J_0'(x) = -J_1(x)$.

J.6 If p is not an integer, then $J_{-p}(x)$ is linearly independent of $J_p(x)$ and is a second solution of Bessel's equation. Any solution of the form $y = c_1 J_p(x) + c_2 J_{-p}(x)$ with $c_2 \neq 0$ is called a **Bessel function of the second kind of order p**.

J.7 If p is an integer, then $J_{-p}(x) = (-1)^p J_p(x)$. In this case it is traditional to define

$$Y_p(x) = \lim_{q \to p} \frac{J_q(x)\cos q\pi - J_{-q}(x)}{\sin q\pi}.$$

$Y_p(x)$ is a **Bessel function of the second kind of order** p, and is linearly independent of $J_p(x)$.

J.8 Plot `BesselJ[0,x]`, `BesselJ[1,x]`, and `BesselY[0,x]` for $0 \le x \le 9$.

The Legendre Polynomials

The **Legendre equation** is

$$(1-x^2)\frac{d^2y}{dx^2} - 2x\frac{dy}{dx} + n(n+1)y = 0,$$

where n is constant.

P.1 Zero is a regular point of the Legendre equation.

P.2 If n is a nonnegative integer, then there is a polynomial solution of Legendre's equation

$$P_n(x) = \sum_{k=0}^{[n/2]} (-1)^n \frac{(2n-2k)!}{2^n k!\,(n-k)!\,(n-2k)!} x^{n-2k},$$

where $[n/2]$ denotes the greatest integer in $n/2$. These polynomials are the **Legendre polynomials**.

P.3 A second linearly independent solution is

$$Q_n(x) = \frac{1}{2} P_n(x) \ln \frac{1+x}{1-x} + \text{power series in } x.$$

Find the power series part of the solution. This solution is unbounded on $(-1, 1)$.

In *Mathematica*, $P_n(x)$ is called `LegendreP[n,x]` and $Q_n(x)$ is called `LegendreQ[n,x]`.

The Hypergeometric Equation

The equation

$$x(1-x)\frac{d^2y}{dx^2} + [c - (a+b+1)x]\frac{dy}{dx} - aby = 0$$

is called the **hypergeometric equation**. The numbers a, b, and c are constants. The numbers $x = 0$ and $x = 1$ are singular.

F.1 Zero is a regular singular point. The indices at $x = 0$ are 0 and 1. We concentrate on the solution about $x = 0$ produced by the index 0, and suppose that c is not 0 or a negative integer.

F.2 Substitute $y(x) = \sum a_n x^n$ to find that the recursion relation is

$$a_n = \frac{(a+n-1)(b+n-1)}{n(c+n-1)} a_{n-1} \quad \text{for} \quad n \ge 1; \; a_0 \text{ is arbitrary}.$$

F.3 Use the Gamma function to express

$$a_n = \frac{\Gamma(a+n)\Gamma(b+n)\Gamma(c)}{n!\,\Gamma(a)\Gamma(b)\Gamma(c+n)} a_0 \quad \text{for} \quad n \geq 0.$$

Let $a_0 = 1$, and define the solution function to be

$$F(a,b,c,x) = 1 + \sum_{n=1}^{\infty} \frac{\Gamma(a+n)\Gamma(b+n)\Gamma(c)}{n!\,\Gamma(a)\Gamma(b)\Gamma(c+n)} x^n.$$

This is called the **hypergeometric series** or the **hypergeometric function**. In *Mathematica*, $F(a,b,c,x)$ is called

Hypergeometric2F1[a, b, c, x].

By choosing the parameters a, b, and c properly, $F(a,b,c,x)$ can represent many functions. Here are some familiar ones.

F.4 $\dfrac{1}{(1-x)^a} = F(a,b,b,x)$.

F.5 $P_k(x) = F(-k, k+1, 1, (1-x, 2))$. (The Legendre polynomials.)

F.6 $\ln(1+x) = xF(1,1,2,-x)$.

F.7 $\arcsin x = xF\left(\frac{1}{2}, 1, \frac{3}{2}, x^2\right)$.

F.8 $\arctan x = xF\left(\frac{1}{2}, 1, \frac{3}{2}, -x^2\right)$.

F.9 $\cos x = \lim_{a\to\infty} F\left(a, a, \frac{1}{2}, \frac{x}{a}\right)$.

F.10 $\sin x = \lim_{a\to\infty} F\left(a, a, \frac{3}{2}, -\frac{x^2}{4a^2}\right)$.

F.11 $e^x = \lim_{a\to\infty} F\left(a, b, b, \frac{x}{a}\right)$.

F.12 Show that

$$\frac{d}{dx}F(a,b,c,x) = \frac{ab}{c}F(a+1, b+1, c+1, x).$$

Mathematica can give you an idea of what these statements are saying. To look at the first several terms of $F(a,b,c,x)$, define a shortened approximation like:

```
Fs[a_,b_,c_,x_] :=
          1+Sum[Gamma[a+1]Gamma[b+1]Gamma[c]/
                    (n! Gamma[a]Gamma[b]Gamma[c+n])*x^n, {n, 8}]
```

To use this to preview problem F.7, for instance, compute `x*Fs[1, 1, 2, -x]` and compare the result with `Series[Log[1+x],x,0,8]`. In problems F.9–F.11, you will need to use `ReplaceRepeated` like this:

```
Fs[...]//.Gamma[a+p_]:>(a+p-1)Gamma[a+p-1]
```

before taking the limits.

8 Differential Systems with Constant Coefficients

8.0 Introduction

This section serves to introduce both Chapter 8 and Chapter 9. In these two chapters, we study systems of ordinary differential equations. Our study will focus on first-order systems. These have the form

$$\begin{cases} y_1' = f_1(x, y_1, y_2, \ldots, y_n) \\ y_2' = f_2(x, y_1, y_2, \ldots, y_n) \\ \quad \vdots \\ y_n' = f_n(x, y_1, y_2, \ldots, y_n). \end{cases} \tag{8.1}$$

This is a system of n simultaneous first-order differential equations. We seek a solution on some interval $a \leq x \leq b$. We can make the notation more compact and suggestive without any loss of ability to express concepts clearly if we let $\boldsymbol{y} = (y_1, y_2, \ldots, y_n)$ be a vector whose components are the solution functions (dependent variables) that we seek. Then system 8.1 can be written as

$$\boldsymbol{y}' = \boldsymbol{f}(x, \boldsymbol{y}), \tag{8.2}$$

where the **bold type** denotes a vector object. The notation $\boldsymbol{f}(x, \boldsymbol{y})$ represents the entire n right-hand sides of system 8.1. The symbol $\boldsymbol{f}$ denotes $(f_1, f_2, \ldots, f_n)$, each component of which is a function from $\boldsymbol{R}^{n+1}$ to $\boldsymbol{R}$, so $\boldsymbol{f}$ is a function from $\boldsymbol{R}^{n+1}$ to $\boldsymbol{R}$. The arguments of each component of $\boldsymbol{f}$ consist of one dependent variable (x) and n dependent variables $\boldsymbol{y} = (y_1, y_2, \ldots, y_n)$. The notation of equation 8.2 is strongly suggestive of the notation $y' = f(x, y)$ of Section 1.1, where Theorem 1.1 gave us information about the existence and uniqueness of solutions to initial value problems. The existence and uniqueness theorem for system 8.2 will look very similar to Theorem 1.1.

In Section 8.1 we will discover that any nth-order differential equation

$$y^{(n)} = g(x, y, y', y'', \ldots, y^{(n-1)})$$

can be written in the form 8.1 as

$$\begin{cases} y_1' = y_2 \\ y_2' = y_3 \\ \quad \vdots \\ y_{n-1}' = y_n \\ y_n' = g(x, y_1, y_2, \ldots, y_n) \end{cases} \tag{8.3}$$

by writing

$$y_i = y^{(i-1)}, 1 \leq i \leq n.$$

This observation has very important and useful theoretical interest to us. It also tells us how to obtain accurate values for the solution and its derivatives when we want to use `NDSolve` to solve a higher-order differential equation numerically. If `NDSolve` is used for the calculation, the solution $y(x)$ and each of the derivatives $y'(x), y''(x), \ldots, y^{(n-1)}(x)$ which appear in equations 8.3 is obtained accurately. The differential equation itself provides the value for $y^{(n)}$ with the same accuracy. (How could one get values for the $(n+1)$st-derivative accurately if g is nice?)

Topics for Chapter 8

In Chapter 8 we will study systems of linear differential equations having constant coefficients. Such a system can be put in the vector form

$$\boldsymbol{y}'(x) - \boldsymbol{A}\boldsymbol{y}(x) = \boldsymbol{q}(x), \tag{8.1 v}$$

where $\boldsymbol{A}$ is a constant $n \times n$ matrix, $\boldsymbol{y}(x)$ is an n-vector, and $\boldsymbol{q}(x)$ is an n-vector of continuous functions. The system is called **homogeneous** if $\boldsymbol{q}(x) = 0$ and **nonhomogeneous** otherwise. Differential systems of this form arise naturally in physical situations. They can also be obtained from a single higher-order differential equation or from a system of higher-order differential equations. The differential operator in 8.1 v is the vector linear operator $\boldsymbol{L}(\boldsymbol{y}) = \boldsymbol{y}' - \boldsymbol{A}\boldsymbol{y}$.

Later in Chapter 8 we will consider the matrix system

$$\boldsymbol{Y}'(x) - \boldsymbol{A}\boldsymbol{Y}(x) = \boldsymbol{Q}(x), \tag{8.1 m}$$

where $\boldsymbol{A}$ is a constant $n\times n$ matrix, $\boldsymbol{Y}(x)$ is an $n\times n$ matrix, and $\boldsymbol{Q}(x)$ is an $n\times n$ matrix of continuous functions. The system is called **homogeneous** if $\boldsymbol{Q}(x) = 0$ and **nonhomogeneous** otherwise. It turns out that this is a natural way to pose such problems because of the favorable properties that matrices give us over vectors, the principal one being the ability to calculate inverses. The ability to calculate an inverse for a matrix has always been a powerful theoretical tool, but *Mathematica* now makes this a practical tool as well. In (8.1 m) we have the matrix linear differential operator $\boldsymbol{L}(\boldsymbol{Y}) = \boldsymbol{Y}' - \boldsymbol{A}\boldsymbol{Y}$. In this setting we even get a special kind of linearity where $\boldsymbol{L}(\boldsymbol{Y}\boldsymbol{K})(x) = \boldsymbol{L}(\boldsymbol{Y})(x)\boldsymbol{K}$ whenever K is a constant vector or matrix. This is a type of homogeneity or uniformity that is possessed by the operator. We still retain ordinary linearity.

The process for solving (8.1 v) relies heavily on algebraic techniques as it did in Chapter 4, but the ideas are somewhat more advanced. These algebraic ideas were reviewed in Section 4.2. Over the years, students have tended to get lost in the manipulations that are necessary to solve problems such as these, so an understanding of the method and the goals has been difficult to achieve. However, *Mathematica* will do these long and complicated manipulations for us. This enables us to concentrate on the process and gives us a real hope of understanding what is going on and why it is being done. Depending on the problem we are solving, these manipulations can be enormous, so it is very important to know what has to be done, how much of it has been completed, how much remains to be done, and how to check for correctness. All of this will be done within *Mathematica*, which makes it possible for you to keep long procedures in mind.

The key algebraic technique is the solution of **characteristic value problems**. These problems are computationally intensive, but *Mathematica* will be used to do the work.

The method of **variation of parameters** was introduced in Chapter 4. We are now prepared to explain the method. In Chapter 4 we used *Mathematica* to calculate the matrix inverses and the integrals involved, and we will do so here. But in the present chapter it will become clear what the process really involves, so you can go back to Chapter 4 and understand what was happening there.

The central goal is that of finding a **fundamental solution** (matrix) for a homogeneous system of equations with constant coefficients. This fundamental solution permits us to express every solution of both the homogeneous problem and any nonhomogeneous problem that has the same (vector or matrix) differential operator. Once a way has been found to obtain a fundamental solution, we can begin the study of initial value problems and quite general boundary value problems. *Mathematica* will carry the heavy computational burden for us.

Nonlinear systems such as system 8.1 are exceedingly difficult to "solve" in the sense of finding a formula for the solution. There exists a theory of such systems that attempts (with some real success) to explain the local and sometimes global behavior of the solutions without actually having access to the solutions themselves. We will begin to appreciate the value of this approach after finding that the solutions to simple linear problems are often so complicated as to defy understanding. We will discuss some of these ideas when we briefly consider the geometric theory of differential equations in Section 8.7. Perhaps a description of generic behavior can contribute to our understanding when actual solutions become too complicated. As the number of equations increases, even the clearest exposition of behavior also increases in complexity.

Sources of Differential Systems

One does not have to contrive systems such as system 8.1. They arise naturally from elementary applications as the setting of the problem is made more complicated. For instance, in the case of chemical mixing, if the number of containers increases from one to two or more cascaded containers where the outflow of one container is allowed to become part of the inflow of the next container, a system is required to track the amount of the solute in the various containers. (This type of system can also describe the dissemination of a pollutant throughout a chain of lakes.) If, in addition, there is a return of the solution from some of the later containers to some of the earlier containers, the equations become more complicated, but we will be able to actually solve them and give formulas for the amount of solute in each of the containers.

Other examples involve weights attached to spring systems where from the weight at the end of each spring, another spring and weight is attached. Such compound spring systems require a system of first-order linear differential equations to describe the motion. Also, it seems clear that the growth of rabbits that we studied in Chapter 3 is affected if wolves or other predators are introduced into the population. The growth of the rabbit population must slow down as predation takes place. When there are adequate prey, the predator population will increase, but what happens as the prey population begins to diminish? The rate of growth of the predator population is affected, and we have a system of differential equations, one describing the rate of growth of the prey population and one describing the rate of growth of the predator population, where each rate involves the population of both prey and predators. Such systems are typically nonlinear.

Newton's law of gravitation says that each two objects in the universe attract each other. This attraction is an acceleration: a force directed along the straight line joining the two bodies. A system of celestial objects such as our own solar system defines a set of differential equations that "determines" the motion of each object within the system. The determination is incomplete because the equations are nonlinear, their solutions may behave in wildly different ways when conditions differ slightly, and we are fundamentally unable to describe the system fully. Nonlinear systems are central to the study of the phenomenon of chaos: systems whose behavior is so unstable that it is exceedingly difficult to describe the behavior of solutions. Often these systems involving initial value problems fail to have unique solutions.

Electronic circuits give a rich source of examples of linear (and nonlinear) differential systems. Kirchoff's laws permit the various current paths in a complex electronic system to be separately described by a differential equation. Within any such device, the behavior of the resulting system of differential equations describes the currents within the various circuits comprising the device. A final linear application is radioactive decay series.

Section 8.8, which is optional, covers two nonlinear applications: planetary motion as a consequence of Newton's second law and Volterra-Lotka predator-prey equations. Section 8.9 looks briefly at defective systems of the form $\boldsymbol{A}\boldsymbol{y}' = \boldsymbol{P}\boldsymbol{y}$, where $\boldsymbol{A}$ is an $n \times n$ constant matrix that has no inverse. These systems are shown to be equivalent to differential systems with algebraic side conditions.

Topics from Chapter 9

The special case of system 8.1 on which we concentrate in Chapter 8 is the case where the system is linear. In this case $\boldsymbol{f}(x, \boldsymbol{y}) = \boldsymbol{P}(x)\boldsymbol{y} + \boldsymbol{q}(x)$, where $\boldsymbol{P}(x)$ is an $n \times n$ matrix of continuous functions and $\boldsymbol{q}(x)$ is an n-vector of continuous functions. Such a system looks like

$$\boldsymbol{y}' = \boldsymbol{P}(x)\boldsymbol{y} + \boldsymbol{q}(x).$$

We will solve this system when $\boldsymbol{P}(x)$ is analytic in a neighborhood of the **ordinary point** x_0. The series techniques to be used are directly analogous to the method we used previously in Chapter 6. We will even solve the system

$$(x - x_0)\boldsymbol{y}' = \boldsymbol{P}(x)\boldsymbol{y},$$

which is a regular singular point problem if $\boldsymbol{P}(x)$ is analytic in a neighborhood of the point x_0. The **method of Frobenius** for the solution of such equations by series techniques extends to this situation. We will see in Chapter 8 that a system of the form

$$(x - x_0)\boldsymbol{y}' = \boldsymbol{A}\boldsymbol{y},$$

where $\boldsymbol{A}$ is an $n \times n$ constant matrix is a generalization of differential equations of **Cauchy-Euler** type and the solution methods are analogous.

Existence and Uniqueness Theorem

Here is the fundamental existence and uniqueness theorem that guarantees that (most of) what we are attempting in Chapters 8 and 9 will work.

Theorem 8.1 (Existence and Uniqueness) *Let the function $\boldsymbol{f} : \boldsymbol{R}^{n+1}$ to $\boldsymbol{R}^n$ be continuous in the "box" $B = [a, b] \times [a_1, b_1] \times \cdots \times [a_n, b_n]$ in $\boldsymbol{R}^{n+1}$. Suppose that there is a number M such that each component function f_i of $\boldsymbol{f}$ satisfies*

$$|f_i(x, u_1, u_2, \ldots, u_n) - f_i(x, v_1, v_2, \ldots, v_n)| \leq M \sum_{j=1}^{n} |u_j - v_j|, \quad 1 \leq i \leq n,$$

whenever $(x, \boldsymbol{u})$ and $(x, \boldsymbol{v})$ are in B. If $(x_0, \alpha_1, \alpha_2, \ldots \alpha_n)$ is in the interior of B, then there is an open interval I containing x_0, and a unique vector function $\boldsymbol{y}(x) = (y_1(x), y_2(x), \ldots, y_n(x))$ such that $\boldsymbol{y}(x_0) = (\alpha_1, \alpha_2, \ldots, \alpha_n) = \boldsymbol{\alpha}$, and for each x in I, $(x, \boldsymbol{y}(x)) \in B$, and

$$\boldsymbol{y}'(x) = \boldsymbol{f}(x, \boldsymbol{y}(x)).$$

The proof is not difficult, but it involves concepts that are usually introduced in later courses, so we omit it. Two corollaries will be of value to us.

Corollary 8.1 (Linear Systems) *In the linear case M can be taken to be the maximum of the absolute values of the n^2 component functions in $\boldsymbol{P}(x)$.*

Therefore, every initial-value problem for the linear system

$$\boldsymbol{y}' = \boldsymbol{P}(x)\boldsymbol{y} + \boldsymbol{q}(x)$$
$$\boldsymbol{y}(x_0) = \boldsymbol{\alpha}$$

has a unique solution that passes through point $(x_0, \boldsymbol{\alpha})$.

The proof of Corollary 8.1 is left as an exercise.

Corollary 8.2 *In the case that* $\partial f_i/\partial y_j$ *exists, is bounded on* B*, and is continuous, for* $1 \leq i, j \leq n$*, then* M *can be taken to be the maximum absolute value of these* n^2 *functions over* B*. Therefore, if* $\boldsymbol{f}$ *is continuously differentiable on* B *in its last* n *places, then each of the initial value problems involving system 8.1 has a unique solution.*

The proof of Corollary 8.2 is also left as an exercise.

EXERCISES 8.0

1. Prove that Corollary 8.1 follows from Theorem 8.1.
2. Prove that Corollary 8.2 follows from Theorem 8.1.
3. Show that the system of equations

$$\begin{cases} \dfrac{dx}{dt} = x + \sin y \\ \dfrac{dy}{dt} = \cos x - y \\ x(0) = y(0) = 0 \end{cases}$$

   satisfies the conditions of Theorem 8.1.

4. Obtain a numerical approximation to the solution of the last problem using

```
NDSolve[{x'[t] == x[t] + Sin[y[t]],
         y'[t] == Cos[x[t]] - y[t],
          x[0] == 0,
          y[0] == 0},
          {x[t], y[t]}, {t, -Pi, Pi}].
```

   Capture the solution functions and plot them individually and parametrically as a pair `{x[t], y[t]}` in the plane. We have taken t to be in the interval $[-\pi, \pi]$. Notice that the parameters for using `NDSolve` to solve this initial-value problem are the same as those for `DSolve` except that instead of merely having the independent variable t in the last position, a range for t must be specified.

8.1 Reduction to First-Order Systems

In the introduction we indicated that a single nth-order differential equation can be reduced to a first-order system of differential equations. Here is an example of the technique. The equation in the example does not have constant coefficients, to illustrate that the technique is directly applicable to the problems of Chapter 9 as well as this chapter.

Example 8.1 Transform the third-order differential equation

$$5y''' - 3y'' + 7xy' - (\cos x)y = 4\sin 3x$$

with initial conditions $y(0) = 1, y'(0) = -1, y''(0) = 3$ into a first-order system of three equations.

Solution. To do this, let $y_1 = y, y_2 = y'$, and $y_3 = y''$. Then

$$y_1' = y' = y_2$$
$$y_2' = y'' = y_3$$

and

$$\begin{aligned} y_3' &= y''' \\ &= \frac{4\sin 3x + 3y'' - 7xy' + (\cos x)y}{5} \\ &= \frac{4}{5}\sin 3x + \frac{3}{5}y_3 - \frac{7}{5}xy_2 + \frac{\cos x}{5}y_1. \end{aligned}$$

The equivalent system we seek is

$$\begin{cases} y_1' = y_2 \\ y_2' = y_3 \\ y_3' = \dfrac{\cos x}{5}y_1 - \dfrac{7}{5}xy_2 + \dfrac{3}{5}y_3 + \dfrac{4}{5}\sin 3x. \end{cases}$$

This is a linear system with

$$\begin{pmatrix} y_1 \\ y_2 \\ y_3 \end{pmatrix}' = \begin{pmatrix} 0 & 1 & 0 \\ 0 & 0 & 1 \\ (\cos x)/5 & -(7/5)x & 3/5 \end{pmatrix} \begin{pmatrix} y_1 \\ y_2 \\ y_3 \end{pmatrix} + \begin{pmatrix} 0 \\ 0 \\ (4/5)\sin 3x \end{pmatrix}.$$

The initial conditions are

$$\begin{pmatrix} y_1 \\ y_2 \\ y_3 \end{pmatrix}(0) = \begin{pmatrix} y(0) \\ y'(0) \\ y''(0) \end{pmatrix} = \begin{pmatrix} 1 \\ -1 \\ 3 \end{pmatrix}.$$

We can express this as

$$\boldsymbol{y}'(x) = \boldsymbol{P}(x)\boldsymbol{y}(x) + \boldsymbol{q}(x)$$
$$\boldsymbol{y}(0) = \mathbf{c},$$

where

$$\boldsymbol{y}(x) = \begin{pmatrix} y_1(x) \\ y_2(x) \\ y_3(x) \end{pmatrix}, \boldsymbol{P}(x) = \begin{pmatrix} 0 & 1 & 0 \\ 0 & 0 & 1 \\ (\cos x)/5 & -(7/5)x & 3/5 \end{pmatrix},$$

$$\boldsymbol{q}(x) = \begin{pmatrix} 0 \\ 0 \\ (4/5)\sin 3x \end{pmatrix},$$

and

$$\boldsymbol{c} = \begin{pmatrix} 1 \\ -1 \\ 3 \end{pmatrix}.$$

It is comforting that a single *linear* differential equation transforms into a *linear* differential system. ◇

More generally, we can transform a system of higher-order equations into a single first-order system. An example is sufficient to enable you to make the transformation in these cases. Once again the example is more general than we require, but the point is still clear.

Example 8.2 Transform the system

$$\begin{cases} y''' = f(x, y, y', y'', z, z') \\ z'' = g(x, y, y', y'', z, z') \end{cases}$$

into a first-order system.

Solution. This is a nonlinear system of two equations in the two unknown functions y and z. It is third-order in y and second-order in z. We therefore need three dependent variables v_1, v_2, v_3 to transform y and two, v_4 and v_5, to transform z. We expect to have a system of five first-order differential equations. Let

$$\begin{cases} v_1 = y \\ v_2 = y' \\ v_3 = y'' \\ v_4 = z \\ v_5 = z'. \end{cases}$$

Then

$$\begin{cases} v_1' = y' = v_2 \\ v_2' = y'' = v_3 \\ v_3' = y''' = f(x, y, y, y'', z, z') = f(x, v_1, v_2, v_3, v_4, v_5) \\ v_4' = z' = v_5 \\ v_5' = z'' = g(x, y, y, y'', z, z') = g(x, v_1, v_2, v_3, v_4, v_5). \end{cases}$$

The first-order system that results is

$$\begin{cases} v_1' = v_2 \\ v_2' = v_3 \\ v_3' = f(x, v_1, v_2, v_3, v_4, v_5) \\ v_4' = v_5 \\ v_5' = g(x, v_1, v_2, v_3, v_4, v_5), \end{cases}$$

which is indeed a system of five first-order differential equations. Each of the two original equations has the highest-order derivative of one variable expressed explicitly in terms of lower-order derivatives of all of the variables. This made the transformation simple. ◇

Here is a more interesting example where the high-order derivatives are not explicitly expressed. This example does have constant coefficients, however.

Example 8.3 Transform the system

$$\begin{cases} y'' + 3z'' + 3w' + 3y' + 2y - z' + 2z + w = 0 \\ y'' + 4z'' + 3w' - y' - 5y + 2z' + z - w = 0 \\ y'' + 3z'' + 4w' + y' - 2z' + w = 0 \end{cases}$$

into a first-order system.

Solution. We concentrate on getting three differential equations in the three unknown functions y, z, and w. The highest-order derivatives are y'', z'', and w'. In order to solve for these derivatives explicitly, we express the system using matrices in such a way as to emphasize these highest-order derivatives:

$$\begin{pmatrix} 1 & 3 & 3 \\ 1 & 4 & 3 \\ 1 & 3 & 4 \end{pmatrix} \begin{pmatrix} y'' \\ z'' \\ w' \end{pmatrix} = - \begin{pmatrix} 2 & 3 & 2 & -1 & 1 \\ -5 & -1 & 1 & 2 & -1 \\ 0 & 1 & 0 & -2 & 1 \end{pmatrix} \begin{pmatrix} y \\ y' \\ z \\ z' \\ w \end{pmatrix}.$$

Now observe that

$$\begin{pmatrix} 1 & 3 & 3 \\ 1 & 4 & 3 \\ 1 & 3 & 4 \end{pmatrix}^{-1} = \begin{pmatrix} 7 & -3 & -3 \\ -1 & 1 & 0 \\ -1 & 0 & 1 \end{pmatrix},$$

so that multiplying through on the left by this matrix gives us the system

$$\begin{pmatrix} y'' \\ z'' \\ w' \end{pmatrix} = - \begin{pmatrix} 7 & -3 & -3 \\ -1 & 1 & 0 \\ -1 & 0 & 1 \end{pmatrix} \begin{pmatrix} 2 & 3 & 2 & -1 & 1 \\ -5 & -1 & 1 & 2 & -1 \\ 0 & 1 & 0 & -2 & 1 \end{pmatrix} \begin{pmatrix} y \\ y' \\ z \\ z' \\ w \end{pmatrix}$$

$$= - \begin{pmatrix} 29 & 21 & 11 & -7 & 7 \\ -7 & -4 & -1 & 3 & -2 \\ -2 & -2 & -2 & -1 & 0 \end{pmatrix} \begin{pmatrix} y \\ y' \\ z \\ z' \\ w \end{pmatrix}$$

which is expressed with matrix coefficients. It turns out to be convenient to retain this representation because when one lets $v_1 = y, v_2 = y', v_3 = z, v_4 = z'$, and $v_5 = w$, the coefficients for $v_2' = y'', v_4' = z''$, and $v_5' = w'$ can be obtained from:

$$\begin{pmatrix} v_2' \\ v_4' \\ v_5' \end{pmatrix} = -\begin{pmatrix} 29 & 21 & 11 & -7 & 7 \\ -7 & -4 & -1 & 3 & -2 \\ -2 & -2 & -2 & -1 & 0 \end{pmatrix} \begin{pmatrix} v_1 \\ v_2 \\ v_3 \\ v_4 \\ v_5 \end{pmatrix}.$$

The example is now easy to complete since we have explicitly solved for y'', z'', and w' as the derivatives v_2', v_4', and v_5'. The final form of the system is

$$\begin{pmatrix} v_1' \\ v_2' \\ v_3' \\ v_4' \\ v_5' \end{pmatrix} = -\begin{pmatrix} 0 & -1 & 0 & 0 & 0 \\ 29 & 21 & 11 & -7 & 7 \\ 0 & 0 & 0 & -1 & 0 \\ -7 & -4 & -1 & 3 & -2 \\ -2 & -2 & -2 & -1 & 0 \end{pmatrix} \begin{pmatrix} v_1 \\ v_2 \\ v_3 \\ v_4 \\ v_5 \end{pmatrix}.$$

◇

When the matrix of coefficients of the highest-order derivatives is singular (has no inverse), the system is somehow **defective**. The techniques for solving such defective systems are not as well defined as for normal systems, so they are rarely discussed. Defective systems are discussed briefly in Section 8.9 to demonstrate that they can be solved. Such systems are very interesting, and some work has been done on describing their solutions. They have applications in control theory and in reactor physics, among other places.

The following important example will provide us with several illustrations later.

Example 8.4 Transform the differential equation $y'' + \omega^2 y = 0$ into a first-order system. Find one solution of the system so that $y(0) = 1$, $y'(0) = 0$, and another so that $y(0) = 0$, $y'(0) = \omega$. Take $w \neq 0$.

Solution. To illustrate that reductions do not always have to be done the same way, let $y_1 = \omega y$ and $y_2 = y'$. Notice the additional ω. Then

$$y_1' = \omega y' = \omega y_2$$

and

$$y_2' = y'' = -\omega^2 y = -\omega y_1.$$

Written as a vector system this is

$$\begin{pmatrix} y_1 \\ y_2 \end{pmatrix}' = \begin{pmatrix} 0 & \omega \\ -\omega & 0 \end{pmatrix} \begin{pmatrix} y_1 \\ y_2 \end{pmatrix}.$$

The substitution we made has put ω in two places, rather than putting ω^2 in one place the way the standard substitution would have done. (Make the standard substitution to see that this is so.)

From Chapter 4 we know that the solution that satisfies $y(0) = 1, y'(0) = 0$ is $u(x) = \cos \omega x$. The solution to the vector system corresponding to the substitution $y_1 = \omega u, y_2 = u'$ is

$$\begin{pmatrix} y_1(x) \\ y_2(x) \end{pmatrix} = \begin{pmatrix} \omega u(x) \\ u'(x) \end{pmatrix} = \begin{pmatrix} \omega \cos \omega x \\ -\omega \sin \omega x \end{pmatrix} = \omega \begin{pmatrix} \cos \omega x \\ -\sin \omega x \end{pmatrix}.$$

Also from Chapter 4, the solution that satisfies $y(0) = 0, y'(0) = \omega$ is $v(x) = \sin \omega x$. The solution to the vector system corresponding to the substitution $y_1 = \omega v, y_2 = v'$ is

$$\begin{pmatrix} y_1(x) \\ y_2(x) \end{pmatrix} = \begin{pmatrix} \omega v(x) \\ v'(x) \end{pmatrix} = \begin{pmatrix} \omega \sin \omega x \\ \omega \cos \omega x \end{pmatrix} = \omega \begin{pmatrix} \sin \omega x \\ \cos \omega x \end{pmatrix}. \qquad \diamond$$

We can now see how to obtain a solution to the vector system from a solution to the second-order differential equation.

Form of the Solutions

Now that we can transform a single higher-order differential equation into a first-order system, how do we solve such systems? Our primary subject for this chapter is systems with constant coefficients. We can consider an example from Chapter 4 to motivate the technique for solving first-order systems with constant coefficients.

Consider the third-order differential equation

$$L(y) = 2y''' - 3y'' - 11y' + 6y = 0. \tag{8.4}$$

In Chapter 4 we learned that the substitution $y = e^{rx}$ led to the result $L(e^{rx}) = p(r)e^{rx} = 0$, so we want to have

$$p(r) = 2r^3 - 3r^2 - 11r + 6 = (2r - 1)(r + 2)(r - 3) = 0.$$

From this we obtain the three roots $r_1 = 1/2$, $r_2 = -2$, $r_3 = 3$ These produce the three linearly independent solutions

$$y_1(x) = e^{x/2}, y_2(x) = e^{-2x}, \text{ and } y_3(x) = e^{3x}.$$

Example 8.5 Transform equation 8.4 into a first-order system by the substitution $u_1 = y, u_2 = y', u_3 = y''$, and relate this system to the solution of equation 8.4 obtained above.

Solution. The substitution produces the system

$$\begin{pmatrix} u_1 \\ u_2 \\ u_3 \end{pmatrix}' = \begin{pmatrix} 0 & 1 & 0 \\ 0 & 0 & 1 \\ -3 & 11/2 & 3/2 \end{pmatrix} \begin{pmatrix} u_1 \\ u_2 \\ u_3 \end{pmatrix} = \boldsymbol{A} \begin{pmatrix} u_1 \\ u_2 \\ u_3 \end{pmatrix}, \tag{8.5}$$

where

$$\boldsymbol{A} = \begin{pmatrix} 0 & 1 & 0 \\ 0 & 0 & 1 \\ -3 & 11/2 & 3/2 \end{pmatrix}.$$

The function $y_1(x) = e^{x/2}$ found above should provide us with a solution in this vector environment if we use the definitions of u_1, u_2, and u_3 properly. Their definitions, respectively, were

$$u_1(x) = y_1(x) = e^{x/2},$$
$$u_2(x) = y_1'(x) = e^{x/2}/2,$$
$$u_3(x) = y_1''(x) = e^{x/2}/4.$$

Now we need to check whether or not the vector

$$\boldsymbol{U}_1(x) = \begin{pmatrix} u_1 \\ u_2 \\ u_3 \end{pmatrix}(x) = \begin{pmatrix} e^{x/2} \\ e^{x/2}/2 \\ e^{x/2}/4 \end{pmatrix} = \begin{pmatrix} 1 \\ 1/2 \\ 1/4 \end{pmatrix} e^{x/2}$$

is a solution of system 8.5. It certainly is, as the calculation

$$\begin{pmatrix} u_1 \\ u_2 \\ u_3 \end{pmatrix}'(x) = \begin{pmatrix} 1/2 \\ 1/4 \\ 1/8 \end{pmatrix} e^{x/2} = \begin{pmatrix} 0 & 1 & 0 \\ 0 & 0 & 1 \\ -3 & 11/2 & 3/2 \end{pmatrix} \begin{pmatrix} 1 \\ 1/2 \\ 1/4 \end{pmatrix} e^{x/2}$$

demonstrates.

Observe that the form of this vector solution is $\boldsymbol{U}_1(x) = \boldsymbol{k}_1 e^{rx}$, where $\boldsymbol{k}_1$ is a nonzero constant vector, and r is a number. This is also the case with the other two vector solutions that we obtain from the solutions of equation 8.5 this way:

$$\boldsymbol{U}_2(x) = \begin{pmatrix} y_2(x) \\ y_2'(x) \\ y_2''(x) \end{pmatrix} \begin{pmatrix} e^{-2x} \\ -2e^{-2x} \\ 4e^{-2x} \end{pmatrix} = \begin{pmatrix} 1 \\ -2 \\ 4 \end{pmatrix} e^{-2x},$$

and

$$\boldsymbol{U}_3(x) = \begin{pmatrix} y_3(x) \\ y_3'(x) \\ y_3''(x) \end{pmatrix} = \begin{pmatrix} e^{3x} \\ 3e^{3x} \\ 9e^{3x} \end{pmatrix} = \begin{pmatrix} 1 \\ 3 \\ 9 \end{pmatrix} e^{3x}.$$

You should verify as was done above that these two vector functions are indeed solutions of equation 8.5. We will soon have a way to easily test that these three solutions are linearly independent and hence that they form a basis for the kernel of the vector linear differential equation 8.5. We can form the Wronskian matrix of these solutions. It is

$$\boldsymbol{W}(x) = \{\boldsymbol{U}_1(x) \mid \boldsymbol{U}_2(x) \mid \boldsymbol{U}_3(x)\} = \begin{pmatrix} e^{x/2} & e^{-2x} & e^{3x} \\ e^{x/2}/2 & -2e^{-2x} & 3e^{3x} \\ e^{x/2}/4 & 4e^{-2x} & 9e^{3x} \end{pmatrix}.$$

This matrix will be used again later. It is worth noting that

$$\boldsymbol{W}'(x) = \boldsymbol{A}\boldsymbol{W}(x),$$

where the matrix $\boldsymbol{A}$ is given in system 8.5. ◇

From this example we have found the form of a vector solution to a homogeneous constant coefficients linear differential system: solutions should have the form $\boldsymbol{k}e^{rx}$, where $\boldsymbol{k}$ is a nonzero constant vector, r is a (possibly complex) number, and both are to be determined. It is this observation that leads us to consider the solution of vector characteristic value problems. We will do this later in the chapter.

EXERCISES 8.1

PART I. Transform these differential equations into first-order systems. If an initial-value problem is given, transform the initial conditions as well.

1. $y''' - y'' + y' - y = 1; \quad y(0) = -1,\ y'(0) = 0,\ y''(0) = 2.$

2. $\begin{cases} y'' - 5z + 6y = x^2 \\ z' + 4y = 1 - x \end{cases}.$

3. $y'' + 4y' - 5y = \sin x; \quad y(3) = 0,\ y'(3) = 1.$

4. $\begin{cases} y'' - 2y' + 2z = e^{3x} \\ z'' + y' - 2y + z = 0 \end{cases}.$

5. $y''' - 2y'' + 2y' = \cos x; \quad y(0) = 0,\ y'(0) = 0,\ y''(0) = -2.$

6. $\begin{cases} y''' - 2z'' + 2y' + z = x^3 - 1 \\ z'' + y' - 2y + z = 0 \end{cases}.$

7. $y''' - 2y'' + y' = e^{-x}; \quad y(0) = 1,\ y'(0) = -1,\ y''(0) = 0$

8. $\begin{cases} y''' - 5y'' + 9z' - 5y = 10e^{-x}\cos x \\ z' + 4y = 1 - x \end{cases}.$

9. $y''' - 6y'' + 12y' - 8y = 2e^{x}\sin x; \quad y(0) = 0,\ y'(0) = 1,\ y''(0) = 0.$

10. $\begin{cases} y''' - 3y' + 2z = \sin x \\ z' + 4y + 3z = 1 - x \\ y(0) = y'(0) = y''(0) = 0 \\ z(0) = 1. \end{cases}$

PART II. In each of these problems you have a differential equation and a basis for the kernel. Transform these differential equations into first-order systems by the transformation $y_i = y^{(i-1)}$. Show that each function in the given basis for the kernel transforms into a vector solution of the first-order system.

11a. $y''' - y'' + y' - y = 0; \qquad \{e^x, \sin x, \cos x\}.$

12a. $y'' - 5y' + 6y = 0; \qquad \{e^{3x}, e^{2x}\}.$

13a. $y'' + 4y' - 5y = 0; \qquad \{e^{-5x}, e^x\}.$

14a. $y'' - 2y' + 2y = 0; \qquad \{e^x \sin x, e^x \cos x\}.$

15a. $y''' - 2y'' + 2y' = 0; \qquad \{e^x \sin x, e^x \cos x, 1\}.$

16a. $y''' - 2y'' + 2y' = 0; \qquad \{e^x \sin x, e^x \cos x, 1\}.$

17a. $y''' - 2y'' + y' = 0;$ $\{1, e^x, xe^x\}.$

18a. $y''' - 5y'' + 9y' - 5y = 0;$ $\{e^{2x}\sin x, e^{2x}\cos x, e^x\}.$

19a. $y''' - 6y'' + 12y' - 8y = 0;$ $\{e^{2x}, xe^{2x}, x^2e^{2x}\}.$

20a. $y''' - 3y' + 2y = 0;$ $\{e^x, xe^x, e^{-2x}\}.$

PART III. For each problem in Part II, show that the set of vectors you constructed is a (Wronskian) matrix whose determinant is nonzero. If the problem was transformed into the vector system $\boldsymbol{y}' = \boldsymbol{A}\boldsymbol{y}$, show that $\boldsymbol{W}$ satisfies the matrix system $\boldsymbol{W}' = \boldsymbol{A}\boldsymbol{W}$. These are problems 11b–20b.

8.2 Theory of First-Order Systems

Recall the theory of solving linear equations and systems. To solve $\boldsymbol{L}(\boldsymbol{y}) = \boldsymbol{q}$ describe the kernel completely: call a typical member $\boldsymbol{u}$. Then find some particular solution $\boldsymbol{y}_0$ such that $\boldsymbol{L}(\boldsymbol{y}_0) = \boldsymbol{q}$. The complete solution of $\boldsymbol{L}(\boldsymbol{y}) = \boldsymbol{q}$ is then $\boldsymbol{y} = \boldsymbol{u} + \boldsymbol{y}_0$. This is how to solve linear equations. From here on the theory of differential equations provides techniques that enable us to find $\boldsymbol{u}$ and $\boldsymbol{y}_0$ when the operator $\boldsymbol{L}$ is a **first-order linear vector differential operator**. This is an operator of the form

$$\boldsymbol{L}(\boldsymbol{y}) = \boldsymbol{y}' - \boldsymbol{P}(x)\boldsymbol{y}, \tag{8.6}$$

where $\boldsymbol{L}$ is an n-vector of differentiable functions and $\boldsymbol{P}(x)$ is an $n \times n$ matrix of continuous functions. The function $\boldsymbol{q}$ is an n-vector of continuous functions. We will initially be interested in the homogeneous problem

$$\boldsymbol{L}(\boldsymbol{y}) = \boldsymbol{y}' - \boldsymbol{P}(x)\boldsymbol{y} = \boldsymbol{0} \quad \text{or} \quad \boldsymbol{y}' = \boldsymbol{P}(x)\boldsymbol{y}. \tag{8.7}$$

Given a vector $\boldsymbol{c}$, Corollary 8.1 says that system 8.7 has a unique solution $\boldsymbol{y}$ such that $\boldsymbol{y}(x_0) = \boldsymbol{c}$. We therefore can choose an independent set $\{\boldsymbol{c}_1, \boldsymbol{c}_2, \ldots, \boldsymbol{c}_n\}$ of vectors from $\boldsymbol{R}^n$, and on an interval I containing x_0, find unique solutions $\{\boldsymbol{y}_1, \boldsymbol{y}_2, \ldots, \boldsymbol{y}_n\}$ to the n homogeneous initial-value problems

$$\begin{cases} \boldsymbol{L}(\boldsymbol{y}_i) = \boldsymbol{y}_i' - \boldsymbol{P}(x)\boldsymbol{y}_i = \boldsymbol{0} \\ \boldsymbol{y}_i(x_0) = \boldsymbol{c}_i, \quad 1 \le i \le n. \end{cases} \tag{8.8}$$

Since $\boldsymbol{L}$ is linear, any linear combination of the $\boldsymbol{y}_i$ is also a solution of the homogeneous differential equation $\boldsymbol{L}(\boldsymbol{y}) = \boldsymbol{y}' - \boldsymbol{P}(x)\boldsymbol{y} = \boldsymbol{0}$ on I.

Theorem 8.2 (Linear Independence) *If the vectors $\{\boldsymbol{c}_1, \boldsymbol{c}_2, \ldots, \boldsymbol{c}_n\}$ are linearly independent, then the corresponding solutions $\{\boldsymbol{y}_1, \boldsymbol{y}_2, \ldots, \boldsymbol{y}_n\}$ to the homogeneous initial value problems*

$$\begin{cases} \boldsymbol{L}(\boldsymbol{y}_i) = \boldsymbol{y}_i' - \boldsymbol{P}(x)\boldsymbol{y}_i = \boldsymbol{0}, & x \in I, \\ \boldsymbol{y}_i(x_0) = \boldsymbol{c}_i, & 1 \le i \le n, \end{cases}$$

are linearly independent in the sense that $\{\boldsymbol{y}_1(x), \boldsymbol{y}_2(x), \ldots, \boldsymbol{y}_n(x)\}$ is a linearly independent set for each x in I.

Proof. We indicate a proof of Theorem 8.2 in the case of two equations. Compare this proof with that of Theorem 4.6. The proof follows from the fact that if $\boldsymbol{W}(x) = (\boldsymbol{y}(x)|\boldsymbol{z}(x))$ is the matrix whose columns are the vector solutions $\boldsymbol{y}$ and $\boldsymbol{z}$, then the determinant of $\boldsymbol{W}(x)$ is either never 0 or identically 0. This is a demonstration of this fact: Write

$$\boldsymbol{W}(x) = \begin{pmatrix} y_1(x) & z_1(x) \\ y_2(x) & z_2(x) \end{pmatrix}$$

where

$$\boldsymbol{y}(x) = \begin{pmatrix} y_1 \\ y_2 \end{pmatrix} \quad \text{and} \quad \boldsymbol{z}(x) = \begin{pmatrix} z_1 \\ z_2 \end{pmatrix}$$

are vector solutions of the homogeneous linear system

$$\begin{aligned} \boldsymbol{L}(\boldsymbol{y})(x) &= \boldsymbol{y}'(x) - \boldsymbol{P}(x)\boldsymbol{y}(x) \\ &= \boldsymbol{y}'(x) - \begin{pmatrix} p_{11}(x) & p_{12}(x) \\ p_{21}(x) & p_{22}(x) \end{pmatrix} \boldsymbol{y}(x) \\ &= \boldsymbol{0}. \end{aligned}$$

You can easily verify the identity

$$\left(\det \begin{pmatrix} y_1(x) & z_1(x) \\ y_2(x) & z_2(x) \end{pmatrix}\right)' = \det \begin{pmatrix} y_1'(x) & z_1'(x) \\ y_2(x) & z_2(x) \end{pmatrix} + \det \begin{pmatrix} y_1(x) & z_1(x) \\ y_2'(x) & z_2'(x) \end{pmatrix}.$$

We use this identity to show that

$$\begin{aligned} &(\det \boldsymbol{W}(x))' \\ &= \left(\det \begin{pmatrix} y_1(x) & z_1(x) \\ y_2(x) & z_2(x) \end{pmatrix}\right)' \\ &= \det \begin{pmatrix} y_1'(x) & z_1'(x) \\ y_2(x) & z_2(x) \end{pmatrix} + \det \begin{pmatrix} y_1(x) & z_1(x) \\ y_2'(x) & z_2'(x) \end{pmatrix} \\ &= \det \begin{pmatrix} p_{11}(x)y_1(x) + p_{12}(x)y_2(x) & p_{11}(x)z_1(x) + p_{12}(x)z_2(x) \\ y_2(x) & z_2(x) \end{pmatrix} \\ &\quad + \det \begin{pmatrix} y_1(x) & z_1(x) \\ p_{21}(x)y_1(x) + p_{22}(x)y_2(x) & p_{21}(x)z_1(x) + p_{22}(x)z_2(x) \end{pmatrix} \\ &= \det \begin{pmatrix} p_{11}(x)y_1(x) & p_{11}(x)z_1(x) \\ y_2(x) & z_2(x) \end{pmatrix} \\ &\quad + \det \begin{pmatrix} y_1(x) & z_1(x) \\ p_{22}(x)y_2(x) & p_{22}(x)z_2(x) \end{pmatrix} \\ &= p_{11}(x) \det \begin{pmatrix} y_1(x) & z_1(x) \\ y_2(x) & z_2(x) \end{pmatrix} + p_{22}(x) \det \begin{pmatrix} y_1(x) & z_1(x) \\ y_2(x) & z_2(x) \end{pmatrix} \\ &= (p_{11}(x) + p_{22}(x)) \det \boldsymbol{W}(x). \end{aligned}$$

This is a single homogeneous first-order linear differential equation that has the solution

$$\det \boldsymbol{W}(x) = \det \boldsymbol{W}(x_0)\, e^{\int (p_{11}(x)+p_{22}(x))dx}. \tag{8.9}$$

Therefore $\det \boldsymbol{W}(x)$ is never 0 if $\det \boldsymbol{W}(x_0) \neq 0$ and is identically 0 if $\det \boldsymbol{W}(x_0) = 0$. Equation 8.9 is known as **Abel's identity**. It is a more general form of the identity than that found during the proof of Theorem 4.6. The quantity $p_{11}(x) + p_{22}(x)$, the sum of the diagonal elements of the matrix, is known as the **trace** of the coefficient matrix. Abel's identity holds in higher dimensions as well, with the integrand being the trace of the coefficient matrix. In the $n \times n$ case, $\det \boldsymbol{W}(x) = \det \boldsymbol{W}(x_0) \exp(\int \operatorname{tr} \boldsymbol{P}(x)dx)$. □

Definition 8.1 *A linearly independent set of n vector-valued solutions to system 8.7 is called a* **fundamental set of solutions**.

The powerful property possessed by fundamental sets of solutions is that they form a basis for the kernel of the operator $\boldsymbol{L}$. Any function in the kernel of $\boldsymbol{L}$ is a solution of system 8.7 and is expressible as a linear combination of the members of the fundamental set, and the linear combination is unique. This means that if we have a fundamental set of solutions of system 8.7 we can uniquely describe any solution of system 8.7 as a linear combination of the members of that fundamental set. We state this as a theorem.

Theorem 8.3 (Fundamental Sets of Solutions) *If $\{\boldsymbol{y}_1, \boldsymbol{y}_2, \ldots, \boldsymbol{y}_n\}$ is a fundamental set of solutions of system 8.7 and $\boldsymbol{z}$ is a solution of system 8.7, then there is a unique linear combination of $\{\boldsymbol{y}_1, \boldsymbol{y}_2, \ldots, \boldsymbol{y}_n\}$ such that*

$$\boldsymbol{z}(x) = \sum_{j=1}^{n} a_i \boldsymbol{y}_i(x)$$

for each x in I.

Proof. Let $\boldsymbol{z}(x)$ be a vector solution. Consider the vector $\boldsymbol{z}(x_0)$. The constants $\{a_i\}_{i=1}^n$ required are the solutions of the linear algebraic system

$$\boldsymbol{z}(x_0) = \sum_{j=1}^{n} a_i \boldsymbol{y}_i(x_0).$$

This system has a unique solution $\{a_1, a_2, \ldots, a_n\}$ because the constant vectors $\{\boldsymbol{y}_1(x_0), \boldsymbol{y}_2(x_0), \ldots, \boldsymbol{y}_n(x_0)\}$ are linearly independent. The function $\boldsymbol{w}(x) = \sum_{j=1}^n a_i \boldsymbol{y}_i(x)$ is a solution of system 8.7 as we showed above, and the $\{a_1, a_2, \ldots, a_n\}$ were chosen so that $\boldsymbol{w}(x_0) = \boldsymbol{z}(x_0)$. The uniqueness property says that $\boldsymbol{w}(x) = \boldsymbol{z}(x)$ for each x in I. Therefore, $\boldsymbol{z}(x)$ is expressed uniquely as a linear combination of $\{\boldsymbol{y}_1(x), \boldsymbol{y}_2(x), \ldots, \boldsymbol{y}_n(x)\}$, as required. □

Here is a powerful observation about the linearly independent set $\{\boldsymbol{y}_1, \boldsymbol{y}_2, \ldots, \boldsymbol{y}_n\}$ of solutions of the vector system 8.7: Form the matrix $\boldsymbol{W}(x) =$

$(\boldsymbol{y}_1|\boldsymbol{y}_2|\ldots|\boldsymbol{y}_n)$ whose columns are the vectors $\{\boldsymbol{y}_1, \boldsymbol{y}_2, \ldots, \boldsymbol{y}_n\}$. Then

$$\begin{aligned}
\boldsymbol{W}'(x) &= (\boldsymbol{y}_1' \mid \boldsymbol{y}_2' \mid \cdots \mid \boldsymbol{y}_n') \\
&= (\boldsymbol{P}(x)\boldsymbol{y}_1 \mid \boldsymbol{P}(x)\boldsymbol{y}_2 \mid \cdots \mid \boldsymbol{P}(x)\boldsymbol{y}_n) \\
&= \boldsymbol{P}(x)(\boldsymbol{y}_1 \mid \boldsymbol{y}_2 \mid \cdots \mid \boldsymbol{y}_n) \\
&= \boldsymbol{P}(x)\boldsymbol{W}(x).
\end{aligned}$$

This means that $\boldsymbol{W}'(x) - \boldsymbol{P}(x)\boldsymbol{W}(x) = \boldsymbol{0}$. This is a differential equation with both coefficients and solution being $n \times n$ matrices. This is called a **matrix differential equation**. We first saw such a result in Example 8.5.

Properties of Matrices and of Matrix Differential Equations

Here are some properties of matrix differential equations and of matrices as they relate to these differential equations. Consider the $n \times n$ matrix differential operator $\boldsymbol{L}$ defined for each x in an interval I by

$$\boldsymbol{L}(\boldsymbol{Y})(x) = \boldsymbol{Y}'(x) - \boldsymbol{P}(x)\boldsymbol{Y}(x).$$

- $\boldsymbol{L}(\boldsymbol{Y})$ is defined for each $n \times n$ matrix $\boldsymbol{Y}$ of continuously differentiable functions defined on I. We say that $\boldsymbol{Y}$ is a **continuously differentiable matrix-valued function**. This means that each value of the function $\boldsymbol{Y}$ is a matrix, rather than that $\boldsymbol{Y}$ is a matrix, each entry of which is a function.
- The derivative of $\boldsymbol{Y}$ is the matrix of derivatives.
- The integral of $\boldsymbol{Y}$ is the matrix of integrals.
- If $\boldsymbol{Y}$ is a solution of $\boldsymbol{L}(\boldsymbol{Y}) = \boldsymbol{0}$, then $\det(\boldsymbol{Y})$ is either never 0 or identically 0.
- If $\det(\boldsymbol{Y}) \neq 0$, then $\boldsymbol{Y}$ is called a **fundamental matrix solution** or simply a **fundamental solution** of $\boldsymbol{L}(\boldsymbol{Y}) = \boldsymbol{0}$.
- If $\boldsymbol{Y}$ and $\boldsymbol{Z}$ are differentiable $n \times n$ matrix-valued functions, then the derivative of their product satisfies this rule: $(\boldsymbol{Y}\boldsymbol{Z})' = \boldsymbol{Y}'\boldsymbol{Z} + \boldsymbol{Y}\boldsymbol{Z}'$. Note that order of multiplication has been preserved: in each term, $\boldsymbol{Y}$ is the left factor and $\boldsymbol{Z}$ the right.
- $(\boldsymbol{Y} + \boldsymbol{Z})' = \boldsymbol{Y}' + \boldsymbol{Z}'$.
- If $\boldsymbol{K}$ is a constant matrix $(\boldsymbol{Y}\boldsymbol{K})' = \boldsymbol{Y}'\boldsymbol{K}$.
- This latter property guarantees the **homogeneity property**

$$\begin{aligned}
\boldsymbol{L}(\boldsymbol{Y}\boldsymbol{K}) &= (\boldsymbol{Y}\boldsymbol{K})' - \boldsymbol{P}(x)(\boldsymbol{Y}\boldsymbol{K}) \\
&= \boldsymbol{Y}'\boldsymbol{K} - \boldsymbol{P}(x)\boldsymbol{Y}\boldsymbol{K} \\
&= (\boldsymbol{Y}' - \boldsymbol{P}(x)\boldsymbol{Y})\boldsymbol{K} \\
&= \boldsymbol{L}(\boldsymbol{Y})\boldsymbol{K}
\end{aligned}$$

when $\boldsymbol{K}$ is a constant matrix. The property $\boldsymbol{L}(\boldsymbol{Y}\boldsymbol{K}) = \boldsymbol{L}(\boldsymbol{Y})\boldsymbol{K}$ is just another higher level way to pass constants out of a linear operator. Note that the matrix $\boldsymbol{K}$ does not come out on the left because $\boldsymbol{P}(x)$ is in the way: $\boldsymbol{L}(\boldsymbol{KY}) = \boldsymbol{KY}' + \boldsymbol{P}(x)\boldsymbol{KY} \neq \boldsymbol{KL}(\boldsymbol{Y})$, unless $\boldsymbol{P}(x)\boldsymbol{K} = \boldsymbol{KP}(x)$. So $\boldsymbol{L}$ has a one-sided homogeneity property toward constant matrices. This is still a very useful property.

- If $\boldsymbol{K}$ is constant and invertible and $\boldsymbol{Y}$ is a fundamental solution matrix, then $\boldsymbol{Z} = \boldsymbol{YK}$ is also a fundamental solution matrix because

$$\det(\boldsymbol{YK}) = \det(\boldsymbol{Y})\det(\boldsymbol{K}) \neq 0.$$

- If $\boldsymbol{A}$ is an $n \times n$ matrix, then there is only one solution of $\boldsymbol{L}(\boldsymbol{Y}) = \boldsymbol{0}$ such that $\boldsymbol{Y}(x_0) = \boldsymbol{A}$.
- If $\boldsymbol{Z}$ is a solution of $\boldsymbol{L}(\boldsymbol{Y}) = \boldsymbol{0}$ and $\boldsymbol{Y}$ is a fundamental solution, then $\boldsymbol{Z} = \boldsymbol{Y}(\boldsymbol{Y}(x_0))^{-1}\boldsymbol{Z}(x_0)$. The matrix function $\boldsymbol{U} = \boldsymbol{Y}(\boldsymbol{Y}(x_0))^{-1}$ is a solution of $\boldsymbol{L}(\boldsymbol{U}) = \boldsymbol{0}$ with $\boldsymbol{U}(x_0) = \boldsymbol{I}_n$, the $n \times n$ identity matrix. The proofs of these assertions are left as exercises.
- If $\boldsymbol{U}$ is a differentiable $n \times n$ matrix that is invertible on an interval $[a, b]$, then $\boldsymbol{U}$ is a fundamental solution of the matrix system $\boldsymbol{Y}' = \boldsymbol{PY}$ where $\boldsymbol{P}(x) = \boldsymbol{U}'(x)(\boldsymbol{U}(x))^{-1}$. This follows from the calculation

$$\boldsymbol{P}(x)\boldsymbol{U}(x) = \boldsymbol{U}'(x)(\boldsymbol{U}(x))^{-1}\boldsymbol{U}(x) = \boldsymbol{U}'(x).$$

Any other matrix $\boldsymbol{Z} = \boldsymbol{UK}$ that is obtained from $\boldsymbol{U}$ by right-multiplication by a constant $n \times n$ matrix $\boldsymbol{K}$ is also a solution of $\boldsymbol{Y}' = \boldsymbol{PY}$.

Variation of Parameters

Suppose that $\boldsymbol{Z}$ is a fundamental solution of $\boldsymbol{L}(\boldsymbol{Y}) = \boldsymbol{0}$. We can construct a particular solution of the nonhomogeneous equation $\boldsymbol{L}(\boldsymbol{Y}) = \boldsymbol{Q}$ in this manner: consider $\boldsymbol{Z} = \boldsymbol{YV}$, where the matrix function $\boldsymbol{V}$ is to be determined so that $\boldsymbol{Z}$ is a solution of $\boldsymbol{L}(\boldsymbol{Z}) = \boldsymbol{Q}$. Make the calculation

$$\begin{aligned}
\boldsymbol{L}(\boldsymbol{Z}) &= \boldsymbol{L}(\boldsymbol{YV}) \\
&= (\boldsymbol{YV})' - \boldsymbol{P}(x)(\boldsymbol{YV}) \\
&= \boldsymbol{Y}'\boldsymbol{V} + \boldsymbol{YV}' - \boldsymbol{P}(x)\boldsymbol{YV} \\
&= (\boldsymbol{Y}' - \boldsymbol{P}(x)\boldsymbol{Y})\boldsymbol{V} + \boldsymbol{YV}' \\
&= \boldsymbol{L}(\boldsymbol{Y})\boldsymbol{V} + \boldsymbol{YV}' \\
&= \boldsymbol{0V} + \boldsymbol{YV}' \\
&= \boldsymbol{YV}' \\
&= \boldsymbol{Q}.
\end{aligned}$$

From $\boldsymbol{YV}' = \boldsymbol{Q}$, invert $\boldsymbol{Y}$ to obtain $\boldsymbol{V}' = \boldsymbol{Y}^{-1}\boldsymbol{Q}$. Integrate this to get

$$\boldsymbol{V}(x) = \int_{x_0}^{x} \boldsymbol{Y}^{-1}(t)\boldsymbol{Q}(t)\,dt + \boldsymbol{K}.$$

Then

$$\begin{aligned}
\boldsymbol{Z}(x) &= \boldsymbol{Y}(x)\boldsymbol{V}(x) \\
&= \boldsymbol{Y}(x)\left(\int_{x_0}^{x} \boldsymbol{Y}^{-1}(t)\boldsymbol{Q}(t)\,dt + \boldsymbol{K}\right) \\
&= \boldsymbol{Y}(x)\int_{x_0}^{x} \boldsymbol{Y}^{-1}(t)\boldsymbol{Q}(t)\,dt + \boldsymbol{Y}(x)\boldsymbol{K}
\end{aligned}$$

is the solution to the nonhomogeneous differential equation. Two things are worth noting:

1. The solution exhibits the two parts that the solution of a nonhomogeneous linear equation is supposed to exhibit: the kernel plus a particular solution.
2. This procedure is just the method of **variation of parameters** in its natural setting!

This technique is equally applicable to vector systems—just take $\boldsymbol{V}$ to be a vector to match the shape of the vector $\boldsymbol{Q}$. Verify that in either the vector or matrix case

$$\boldsymbol{Y}_0(x) = \boldsymbol{Y}(x)\int_{x_0}^{x} \boldsymbol{Y}^{-1}(t)\boldsymbol{Q}(t)\,dt$$

is a particular solution of the nonhomogeneous equation $\boldsymbol{L}(\boldsymbol{Y}) = \boldsymbol{Q}$. Observe that if $\boldsymbol{Q}(x)$ is a vector, then $\boldsymbol{Y}_0(x)$ is a vector. It is also true that $\boldsymbol{Y}_0(x_0) = \boldsymbol{0}$, because x_0 is used as the lower limit on the integral.

The relationships between vectors and matrices are important here. We obtained a matrix solution only after solving a vector system and collecting the (column) vector solutions into a matrix. However, after having done this, we see that the matrix solutions play a central role. The ability to compute inverses gives us a powerful notational tool for keeping track of complicated calculations. We can also express the vector solution $\boldsymbol{z} = \boldsymbol{Y}\boldsymbol{k}$ in terms of the fundamental matrix of solutions by using a *column vector* $\boldsymbol{k}$ rather than a matrix.

Example 8.6 (Continuation of Example 8.4) Find a fundamental matrix solution for the homogeneous 2×2 matrix differential equation

$$\boldsymbol{Y}' = \begin{pmatrix} 0 & \omega \\ -\omega & 0 \end{pmatrix} \boldsymbol{Y}$$

that satisfies the initial condition $\boldsymbol{Y}(0) = \boldsymbol{I}_2$.

Solution. From Example 8.4 we know that

$$\boldsymbol{Y}(x) = \omega \begin{pmatrix} \cos\omega x & \sin\omega x \\ -\sin\omega x & \cos\omega x \end{pmatrix}$$

is a fundamental solution. This is easily verified:

$$\boldsymbol{Y}'(x) = \left(\omega \begin{pmatrix} \cos\omega x & \sin\omega x \\ -\sin\omega x & \cos\omega x \end{pmatrix}\right)'$$

$$= \omega \begin{pmatrix} -\omega \sin \omega x & \omega \cos \omega x \\ -\omega \cos \omega x & -\omega \sin \omega x \end{pmatrix}$$

$$= \begin{pmatrix} 0 & \omega \\ -\omega & 0 \end{pmatrix} \begin{pmatrix} \omega \cos \omega x & \omega \sin \omega x \\ -\omega \sin \omega x & \omega \cos \omega x \end{pmatrix}$$

$$= \begin{pmatrix} 0 & \omega \\ -\omega & 0 \end{pmatrix} \boldsymbol{Y}(x).$$

Since

$$\boldsymbol{Y}(0) = \omega \begin{pmatrix} 1 & 0 \\ 0 & 1 \end{pmatrix}, \quad \text{we have} \quad \boldsymbol{Y}(0)^{-1} = \frac{1}{\omega} \begin{pmatrix} 1 & 0 \\ 0 & 1 \end{pmatrix},$$

and the solution we want can be obtained simply by dividing through by ω. Therefore the solution we want is

$$\boldsymbol{U}(x) = \frac{1}{\omega} \left(\omega \begin{pmatrix} \cos \omega x & \sin \omega x \\ -\sin \omega x & \cos \omega x \end{pmatrix} \right) = \begin{pmatrix} \cos \omega x & \sin \omega x \\ -\sin \omega x & \cos \omega x \end{pmatrix}.$$

You should verify that $\boldsymbol{U}(x)$ does indeed satisfy the required conditions. ◇

Exercises 8.2

Part I. Verify that the given matrix-valued function $\boldsymbol{W}(x)$ is a fundamental solution of the first-order linear system $\boldsymbol{Y}' = \boldsymbol{P}(x)\boldsymbol{Y}$. The coefficient matrix $\boldsymbol{P}(x)$ is given in the first column. Let *Mathematica* do the work for you.

1a. $\boldsymbol{P}_1(x) = \begin{pmatrix} 0 & 1 \\ -9 & 0 \end{pmatrix}; \quad \boldsymbol{W}_1(x) = \begin{pmatrix} \cos(3x) & \sin(3x) \\ -3\sin(3x) & 3\cos(3x) \end{pmatrix}.$

2a. $\boldsymbol{P}_2(x) = \begin{pmatrix} 0 & 1 \\ -25 & 0 \end{pmatrix}; \quad \boldsymbol{W}_2(x) = \begin{pmatrix} \cos(5x) & \sin(5x) \\ -5\sin(5x) & 5\cos(5x) \end{pmatrix}.$

3a. $\boldsymbol{P}_3(x) = \begin{pmatrix} 0 & 1 \\ -25 & 10 \end{pmatrix}; \quad \boldsymbol{W}_3(x) = \begin{pmatrix} e^{5x} & xe^{5x} \\ 5e^{5x} & e^{5x} + 5xe^{5x} \end{pmatrix}.$

4a. $\boldsymbol{P}_4(x) = \begin{pmatrix} 0 & 1 \\ \dfrac{-3(7+10x)}{1+2x} & \dfrac{2(5+8x)}{1+2x} \end{pmatrix}; \quad \boldsymbol{W}_4(x) = \begin{pmatrix} e^{3x} & xe^{5x} \\ 3e^{3x} & e^{5x} + 5xe^{5x} \end{pmatrix}.$

5a. $\boldsymbol{P}_5(x) = \begin{pmatrix} \dfrac{-2 - 20e^{5x} + 15x}{-1 - 2e^{5x} + 3x} & \dfrac{5(1 - 14e^{5x} + 6x)}{-1 - 2e^{5x} + 3x} \\ -1 & -2 \end{pmatrix}$

$\boldsymbol{W}_5(x) = \begin{pmatrix} 5e^{3x} & 1 + 7e^{5x} + 2x \\ -e^{3x} & -e^{5x} - x \end{pmatrix}.$

6a. $\boldsymbol{P}_6(x) = \begin{pmatrix} 10 & 35 \\ -1 & -2 \end{pmatrix}; \quad \boldsymbol{W}_6(x) = \begin{pmatrix} 5e^{3x} & 7e^{5x} \\ -e^{3x} & -e^{5x} \end{pmatrix}.$

7a. $\boldsymbol{P}_7(x) = \begin{pmatrix} -4 & 1 \\ -1 & -2 \end{pmatrix}; \quad \boldsymbol{W}_7(x) = \begin{pmatrix} (1 - x/e^{-3x}) & -e^{-3x} \\ -(x/e^{3x}) & -e^{-3x} \end{pmatrix}.$

8a. $\boldsymbol{P}_8(x) = \begin{pmatrix} 2+(7/x) & 4+(15/x^2)+(14/x) \\ -1 & -2 \end{pmatrix}$;

$\boldsymbol{W}_8(x) = \begin{pmatrix} x^2(3+2x) & x^4(5+2x) \\ -x^3 & -x^5 \end{pmatrix}$.

9a. $\boldsymbol{P}_9(x) = \begin{pmatrix} 6 & 25 \\ -1 & -2 \end{pmatrix}$;

$\boldsymbol{W}_9(x) = \begin{pmatrix} e^{2x}(4\cos(3x) - 3\sin(3x) & e^{2x}(3\cos(3x)+4\sin(3x)) \\ -e^{2x}\cos(3x) & -e^{2x}\sin(3x) \end{pmatrix}$.

10a. $\boldsymbol{P}_{10}(x) = \begin{pmatrix} 0 & 5 \\ -1 & -2 \end{pmatrix}$;

$\boldsymbol{W}_{10}(x) = \begin{pmatrix} (\cos(2x) - 2\sin(2x))/e^x & (2\cos(2x)+\sin(2x))/e^x \\ -(\cos(2x))/e^x & -(\sin(2x)/e^x) \end{pmatrix}$.

PART II. When possible, for each system in Part I find a fundamental solution such that at $x = 0$ the solution is an appropriate-sized identity matrix. When necessary, let *Mathematica* do the work for you. These are problem 1b–10b.

PART III. For each nonhomogeneous system $\boldsymbol{Y}' = \boldsymbol{P}(x) + \boldsymbol{Q}(x)$ below, use the fundamental solution given in Part I and the method of variation of parameters to find a particular solution. Then state the complete solution. Let *Mathematica* do the work for you. The operators in these problems correspond to those in Part I.

1c. $\boldsymbol{Y}' = \boldsymbol{P}_1(x)\boldsymbol{Y}(x) + \begin{pmatrix} 0 & 5 \\ -1 & -2 \end{pmatrix}$.

2c. $\boldsymbol{Y}' = \boldsymbol{P}_2(x)\boldsymbol{Y}(x) + \begin{pmatrix} 1 & x \\ 1 & 0 \end{pmatrix}$.

3c. $\boldsymbol{Y}' = \boldsymbol{P}_3(x)\boldsymbol{Y}(x) + \begin{pmatrix} 1 & e^x \\ 1 & 0 \end{pmatrix}$.

4c. $\boldsymbol{Y}' = \boldsymbol{P}_4(x)\boldsymbol{Y}(x) + \begin{pmatrix} 1 & x^2 \\ 1 & 0 \end{pmatrix}$.

5c. $\boldsymbol{Y}' = \boldsymbol{P}_5(x)\boldsymbol{Y}(x) + \begin{pmatrix} 2 & x \\ 1 & 4 \end{pmatrix}$.

6c. $\boldsymbol{Y}' = \boldsymbol{P}_6(x)\boldsymbol{Y}(x) + \begin{pmatrix} 1 & x \\ -1 & 0 \end{pmatrix}$.

7c. $\boldsymbol{Y}' = \boldsymbol{P}_7(x)\boldsymbol{Y}(x) + \begin{pmatrix} 1 & x \\ e^x & 0 \end{pmatrix}$.

8c. $\boldsymbol{Y}' = \boldsymbol{P}_8(x)\boldsymbol{Y}(x) + \begin{pmatrix} 1 & x \\ e^x & -1 \end{pmatrix}$.

9c. $\boldsymbol{Y}' = \boldsymbol{P}_9(x)\boldsymbol{Y}(x) + \begin{pmatrix} 1 & 2x \\ 1 & 3 \end{pmatrix}$.

10c. $\boldsymbol{Y}' = \boldsymbol{P}_{10}(x)\boldsymbol{Y}(x) + \begin{pmatrix} 1 & 2x \\ 3 & 0 \end{pmatrix}$.

PART IV. Manufacture similar problems for yourself. Given $\boldsymbol{W}(x)$ which is invertible, the matrix $\boldsymbol{P}(x) = \boldsymbol{W}'(x)\boldsymbol{W}(x)^{-1}$. Then almost anything simple can be used for $\boldsymbol{Q}(x)$. You may occasionally find an integral that *Mathematica* cannot do. Do not expect `DSolve` to be able to solve your problems. Someday it may be able to do all of them. If the coefficient matrix is constant, and you are willing to wait long enough, *Mathematica* may return a solution. We are about to learn how to manually direct the solution process.

PART V. Theory.

11. Verify that in either the vector or matrix case

$$\boldsymbol{Y}_0(x) = \boldsymbol{Y}(x) \int_{x_0}^{x} \boldsymbol{Y}^{-1}(t)\boldsymbol{Q}(t)\, dt$$

is a particular solution of the nonhomogeneous equation $\boldsymbol{L}(\boldsymbol{Y}) = \boldsymbol{Y}' - \boldsymbol{P}(x)\boldsymbol{Y} = \boldsymbol{Q}$, and that $\boldsymbol{Y}(x_0) = \boldsymbol{0}$.

12. Verify that

$$\boldsymbol{U}(x) = \begin{pmatrix} \cos\omega x & \sin\omega x \\ -\sin\omega x & \cos\omega x \end{pmatrix}$$

does indeed satisfy

$$\boldsymbol{Y}' = \begin{pmatrix} 0 & \omega \\ -\omega & 0 \end{pmatrix} \boldsymbol{Y}$$

and the initial condition $\boldsymbol{Y}(0) = \boldsymbol{I}_2$.

13. Show that the matrix function $\boldsymbol{U} = \boldsymbol{Y}(\boldsymbol{Y}(x_0))^{-1}$ satisfies

$$\boldsymbol{L}(\boldsymbol{U}) = \boldsymbol{U}' - \boldsymbol{P}(x)\boldsymbol{U} = \boldsymbol{0}$$

and

$$\boldsymbol{U}(x_0) = \boldsymbol{I}_n,$$

where $\boldsymbol{I}_n$ is the $n \times n$ identity matrix.

14. Show that if $\boldsymbol{Z}$ is a solution of $\boldsymbol{L}(\boldsymbol{Y}) = \boldsymbol{Y}' - \boldsymbol{P}(x)\boldsymbol{Y} = \boldsymbol{0}$ and $\boldsymbol{Y}$ is a fundamental solution, then $\boldsymbol{Z} = \boldsymbol{Y}(\boldsymbol{Y}(x_0))^{-1}\boldsymbol{Z}(x_0)$.

8.3 First-Order Constant Coefficients Systems

As we found in Section 4.2 the solution of constant coefficients systems leads immediately to auxiliary problems in algebra. This was certainly true for homogeneous nth-order differential equations with constant coefficients where we needed to find the roots of polynomials. We will find a general representation for fundamental solutions of first-order systems, and then explore the algebraic approaches that we must take in order to actually solve a given system. The multidimensional setting in which we are now operating permits some strange things to happen that do not occur in the one-dimensional setting. We have to completely examine the structure of matrices in order to describe every solution, even though the general solution can be described theoretically as an exponential without the need to know about internal structure.

Picard's Method and the Matrix Exponential

Suppose that we wish to solve the homogeneous $n \times n$ constant coefficient matrix differential equation

$$\boldsymbol{Y}' = \boldsymbol{A}\boldsymbol{Y},$$

along with the initial condition

$$\boldsymbol{Y}(0) = \boldsymbol{I}_n.$$

The Picard[1] method, says that to solve $\boldsymbol{Y}' = \boldsymbol{A}\boldsymbol{Y}$ we should produce a sequence of (matrix-valued) functions from the recursion relation

$$\boldsymbol{Y}_{m+1}(x) = \boldsymbol{I}_n + \int_0^x \boldsymbol{A}\boldsymbol{Y}_m(t)\, dt.$$

We start the sequence with $\boldsymbol{Y}_0(x) = \boldsymbol{I}_n$. Then

$$\begin{aligned}
\boldsymbol{Y}_1(x) &= \boldsymbol{I}_n + \int_0^x \boldsymbol{A}\boldsymbol{Y}_0(t)dt = \boldsymbol{I}_n + \boldsymbol{A}x, \\
\boldsymbol{Y}_2(x) &= \boldsymbol{I}_n + \int_0^x \boldsymbol{A}\boldsymbol{Y}_1(t)dt = \boldsymbol{I}_n + \int_0^x \boldsymbol{A}\,(\boldsymbol{I}_n + \boldsymbol{A}t)dt \\
&= \boldsymbol{I}_n + \boldsymbol{A}x + \frac{\boldsymbol{A}^2x^2}{2!}, \\
\boldsymbol{Y}_3(x) &= \boldsymbol{I}_n + \int_0^x \boldsymbol{A}\boldsymbol{Y}_2(t)dt = \boldsymbol{I}_n + \int_0^x \boldsymbol{A}\left(\boldsymbol{I}_n + \boldsymbol{A}t + \frac{\boldsymbol{A}^2t^2}{2!}\right)dt \\
&= \boldsymbol{I}_n + \boldsymbol{A}x + \frac{\boldsymbol{A}^2x^2}{2!} + \frac{\boldsymbol{A}^3x^3}{3!}.
\end{aligned}$$

In general,

$$\boldsymbol{Y}_m(x) = \boldsymbol{I}_n + \sum_{i=1}^{m} \frac{\boldsymbol{A}x^i}{i!},$$

which may be verified by mathematical induction.

[1] Émile Picard (1856–1941), an eminent French mathematician.

This sequence (of matrix-valued functions) has a limit because each of the component sequences has a limit. (This is an exercise.) But we want to know what the limit is. Here is a hint: consider the Maclaurin[2] series expansion for the ordinary exponential function, $e^{ax} = \sum_{i=0}^{\infty}((ax)^i/i!)$. The partial sums of this series look exactly like the terms of the sequence that the Picard method was defining for us. Compare the partial sum

$$\sum_{i=0}^{m} \frac{(ax)^i}{i!} = 1 + \sum_{i=1}^{m} \frac{(ax)^i}{i!}$$

with the expression

$$\boldsymbol{I}_n + \sum_{i=1}^{m} \frac{(\boldsymbol{A}x)^i}{i!}$$

for $\boldsymbol{Y}_n(x)$ above. Isn't the similarity striking? We have the multiplicative identity matrix $\boldsymbol{I}_n$ in place of the number 1 and the matrix $\boldsymbol{A}$ in place of the number a. Since the Picard iterants we are producing look so much like the partial sums for an exponential, we define the limit of the sequence of Picard iterants to be

$$\boldsymbol{U}(x) = \lim_{m\to\infty} \boldsymbol{Y}_m(x) = e^{\boldsymbol{A}x}.$$

This is called the matrix exponential of $\boldsymbol{A}x$. When we observe that $\boldsymbol{Y}'_m(x) = \boldsymbol{A}\boldsymbol{Y}_{m-1}(x)$, we find by taking limits that

$$\frac{d}{dx} e^{\boldsymbol{A}x} = \boldsymbol{A}e^{\boldsymbol{A}x}.$$

In addition, since $\boldsymbol{Y}_m(0) = \boldsymbol{I}_n$ for all m,

$$\boldsymbol{U}(0) = e^{\boldsymbol{A}x}|_{x=0} = \boldsymbol{I}_n.$$

Therefore, $\boldsymbol{U}(x) = e^{\boldsymbol{A}x}$ is the fundamental matrix solution that satisfies $\boldsymbol{U}' = \boldsymbol{A}\boldsymbol{U}$ and $\boldsymbol{U}(0) = \boldsymbol{I}_n$. This means that every solution of $\boldsymbol{Y}' = \boldsymbol{A}\boldsymbol{Y}$ can be expressed as $\boldsymbol{Y} = \boldsymbol{U}\boldsymbol{K}$ for some choice of the constant matrix $\boldsymbol{K}$. It is often convenient to use the notation $\exp(\boldsymbol{A}x)$ for $e^{\boldsymbol{A}x}$.

Mathematica distinguishes between the exponential of a numeric-valued expression and that of a matrix-valued expression. It uses `Exp` for the former, and `MatrixExp` for the latter.

Example 8.7 Find the matrix exponential

$$\exp\left[\begin{pmatrix} 0 & 1 \\ -1 & 0 \end{pmatrix} x\right].$$

Solution. The matrix exponential $\exp\left[\begin{pmatrix} 0 & 1 \\ -1 & 0 \end{pmatrix} x\right]$ is the solution of the differential equation $\boldsymbol{Y}' = \begin{pmatrix} 0 & 1 \\ -1 & 0 \end{pmatrix} \boldsymbol{Y}$ satisfying the initial condition $\boldsymbol{Y}(0) =$

[2]Colin Maclaurin (1698–1746), Scottish mathematician who was a skilled geometer. His work *Treatise on Fluxions* attempted to apply Greek rigor to the new calculus.

I_2. It is left as an exercise for you to apply Picard's method to obtain enough terms for you to recognize the four component series. We will take advantage of Examples 8.4 and 8.5, and observe that since $\boldsymbol{A} = \begin{pmatrix} 0 & 1 \\ -1 & 0 \end{pmatrix}$ is just the system we studied when $\omega = 1$, we know that

$$\exp\left[\begin{pmatrix} 0 & 1 \\ -1 & 0 \end{pmatrix} x\right] = \begin{pmatrix} \cos x & \sin x \\ -\sin x & \cos x \end{pmatrix}.$$

This is verified by the calculations

$$\begin{aligned}\begin{pmatrix} \cos x & \sin x \\ -\sin x & \cos x \end{pmatrix}' &= \begin{pmatrix} -\sin x & \cos x \\ -\cos x & -\sin x \end{pmatrix} \\ &= \begin{pmatrix} 0 & 1 \\ -1 & 0 \end{pmatrix}\begin{pmatrix} \cos x & \sin x \\ -\sin x & \cos x \end{pmatrix}\end{aligned}$$

and

$$\begin{pmatrix} \cos 0 & \sin 0 \\ -\sin 0 & \cos 0 \end{pmatrix} = \begin{pmatrix} 1 & 0 \\ 0 & 1 \end{pmatrix} = \boldsymbol{I}_2.$$

◇

Example 8.7M Find the matrix exponential $\exp\left[\begin{pmatrix} 0 & 1 \\ -1 & 0 \end{pmatrix} x\right]$ by *Mathematica*.

Solution.

```
In[1]:=  m = MatrixExp[{{0,1},{-1,0}}x]

Out[1]=      -I x      I x
            E         E       -I  -I x          2 I x
         {{------- + ------, --- E     (-1 + E     )},
              2         2     2

                                           -I x     I x
           I  -I x           2 I x      E         E
         {- E      (-1 + E      ),   ------- + ------}}
           2                              2         2
```

Note that the answer is expressed in terms of complex numbers. (In *Mathematica*, `I = Complex[0,1] = 0 + 1 i`.) The exponential of a real matrix is real. Use `ComplexExpand` to get the real form.

```
In[2]:=  Simplify[ComplexExpand[m]]

Out[2]=  {{Cos[x], Sin[x]}, {-Sin[x], Cos[x]}}
```

◇

Example 8.8 **(a)** Find the matrix exponential of the diagonal matrix

$$\boldsymbol{D} = \operatorname{diag}(d_1, d_2, \ldots, d_n)x = \operatorname{diag}(d_1 x, d_2 x, \ldots, d_n x).$$

Solution. The solution is simple once you realize that the powers of $\boldsymbol{D}$ are also diagonal matrices having a particularly nice form:

$$\boldsymbol{D}^k = \operatorname{diag}((d_1 x)^k, (d_2 x)^k, \ldots, (d_n x)^k).$$

Therefore

$$\begin{aligned}
&\exp(\operatorname{diag}(d_1, d_2, \ldots, d_n)x) \\
&\quad = \lim_{m\to\infty} \sum_{k=0}^{m} \operatorname{diag}((d_1x)^k, (d_2x)^k, \ldots, (d_nx)^k)/k! \\
&\quad = \operatorname{diag}\left(\lim_{m\to\infty} \sum_{k=0}^{m} (d_1x)^k/k!, \lim_{m\to\infty} \sum_{k=0}^{m} (d_2x)^k/k!, \ldots, \lim_{m\to\infty} \sum_{k=0}^{m} (d_nx)^k/k! \right) \\
&\quad = \operatorname{diag}(\exp(d_1x), \exp(d_2x), \ldots, \exp(d_nx)) \qquad \diamond
\end{aligned}$$

Example 8.8 (b) Apply the result of Example 8.8 to find the matrix exponential of the diagonal matrix $\boldsymbol{D} = \operatorname{diag}(-1, 2, 3)x = \operatorname{diag}(-x, 2x, 3x)$.

Solution. The desired exponential is

$$e^{\boldsymbol{D}} = \operatorname{diag}(e^{-x}, e^{2x}, e^{3x}) = \begin{pmatrix} e^{-x} & 0 & 0 \\ 0 & e^{2x} & 0 \\ 0 & 0 & e^{3x} \end{pmatrix}.$$

Note that

$$\begin{aligned}
\frac{d}{dx} \begin{pmatrix} e^{-x} & 0 & 0 \\ 0 & e^{2x} & 0 \\ 0 & 0 & e^{3x} \end{pmatrix} &= \begin{pmatrix} -e^{-x} & 0 & 0 \\ 0 & 2e^{2x} & 0 \\ 0 & 0 & 3e^{3x} \end{pmatrix} \\
&= \begin{pmatrix} -1 & 0 & 0 \\ 0 & 2 & 0 \\ 0 & 0 & 3 \end{pmatrix} \begin{pmatrix} e^{-x} & 0 & 0 \\ 0 & e^{2x} & 0 \\ 0 & 0 & e^{3x} \end{pmatrix}
\end{aligned}$$

and

$$\begin{pmatrix} e^0 & 0 & 0 \\ 0 & e^0 & 0 \\ 0 & 0 & e^0 \end{pmatrix} = \begin{pmatrix} 1 & 0 & 0 \\ 0 & 1 & 0 \\ 0 & 0 & 1 \end{pmatrix} = \boldsymbol{I}_3. \qquad \diamond$$

Example 8.8M (b) Find the matrix exponential of the diagonal matrix

$$\boldsymbol{D} = \operatorname{diag}(-1, 2, 3)x = \operatorname{diag}(-x, 2x, 3x)$$

by *Mathematica*.

Solution.

```
In[3]:=  diag = DiagonalMatrix[-1,2,3]

Out[3]=  {{-1, 0, 0}, {0, 2, 0}, {0, 0, 3}}

In[4]:=  Y[x_] = MatrixExp[diag x]

Out[4]=     -x               2 x               3 x
         {{E  , 0, 0}, {0, E   , 0}, {0, 0, E   }}

In[5]:=  Y[0]

Out[5]=  {{1, 0, 0}, {0, 1, 0}, {0, 0, 1}}
```

Check that `Y[x]` is a solution of the differential equation.

```
In[6]:= Y'[x]-diag.Y[x]

Out[6]= {{0, 0, 0}, {0, 0, 0}, {0, 0, 0}}
```

◇

A matrix $\boldsymbol{N} \neq \boldsymbol{0}$ is called **nilpotent** of order p if $\boldsymbol{N}^p = \boldsymbol{0}$, but $\boldsymbol{N}^{p-1} \neq \boldsymbol{0}$. Of course, if $\boldsymbol{N}^p = \boldsymbol{0}$, then $\boldsymbol{N}^k = \boldsymbol{0}$ for $k \geq p$. The exponential of a nilpotent matrix is particularly simple to compute, and the result is interesting.

Example 8.9 Show that

$$\boldsymbol{N} = \begin{pmatrix} 0 & 1 & -3 \\ 0 & 0 & 2 \\ 0 & 0 & 0 \end{pmatrix}$$

is nilpotent of order 3 and calculate $\exp(\boldsymbol{N}x)$.

Solution. We have that $\boldsymbol{N}^1 \neq 0$,

$$\boldsymbol{N}^2 = \begin{pmatrix} 0 & 0 & 2 \\ 0 & 0 & 0 \\ 0 & 0 & 0 \end{pmatrix} \neq 0,$$

and $\boldsymbol{N}^k = 0$ for $k \geq 3$. Therefore, $\boldsymbol{N}$ is nilpotent of order 3. It follows that $\boldsymbol{Y}(x)$, the exponential of $\boldsymbol{N}x$, is simply

$$\begin{aligned} \exp(\boldsymbol{N}x) &= \boldsymbol{I}_3 + \frac{\boldsymbol{N}x}{1} + \frac{(\boldsymbol{N}x)^2}{2!} \\ &= \begin{pmatrix} 1 & 0 & 0 \\ 0 & 1 & 0 \\ 0 & 0 & 1 \end{pmatrix} + \begin{pmatrix} 0 & 1 & -3 \\ 0 & 0 & 2 \\ 0 & 0 & 0 \end{pmatrix} x + \begin{pmatrix} 0 & 0 & 2 \\ 0 & 0 & 0 \\ 0 & 0 & 0 \end{pmatrix} \frac{x^2}{2} \\ &= \begin{pmatrix} 1 & x & x^2 - 3x \\ 0 & 1 & 2x \\ 0 & 0 & 1 \end{pmatrix}. \end{aligned}$$

Verify that $\boldsymbol{Y}'(x) = \boldsymbol{N}\boldsymbol{Y}(x)$ and that $\boldsymbol{Y}(0) = \boldsymbol{I}_3$. This solution matrix is a matrix polynomial. ◇

Example 8.9M By *Mathematica*, show that

$$\boldsymbol{N} = \begin{pmatrix} 0 & 1 & -3 \\ 0 & 0 & 2 \\ 0 & 0 & 0 \end{pmatrix}$$

is nilpotent of order 3 and calculate $\exp(\boldsymbol{N}x)$.

Solution. The square of $\boldsymbol{N}$ is nonzero. Use the function `MatrixPower[A, n]` to find the nth power of the square matrix A.

```
In[7]:= MatrixPower[{{0,1,-3},{0,0,2},{0,0,0}},2]

Out[7]= {{0, 0, 2}, {0, 0, 0}, {0, 0, 0}}
```

The third power is 0, so n is nilpotent of order 3.

```
In[8]:= MatrixPower[{{0,1,-3},{0,0,2},{0,0,0}},3]

Out[8]= {{0, 0, 0}, {0, 0, 0}, {0, 0, 0}}

In[9]:= MatrixExp[{{0,1,-3},{0,0,2},{0,0,0}}x]//MatrixForm

Out[9]= 1          x          (-3 + x) x

        0          1          2 x

        0          0          1
```

◇

We now introduce the idea of a characteristic value problem and show how to use *Mathematica* to solve such problems. These ideas are needed immediately.

Characteristic Value Problems

Given an $n \times n$ matrix $\boldsymbol{A}$, a characteristic value problem for $\boldsymbol{A}$ has the form

$$(\boldsymbol{A} - r\boldsymbol{I})\boldsymbol{v} = \boldsymbol{0},$$

where r is a (possibly complex) number, $\boldsymbol{v}$ is a nonzero vector and $\boldsymbol{I}$ is the $n \times n$ identity matrix; it has 1's on the main diagonal and 0's elsewhere. The characteristic equation of the problem is $\det(\boldsymbol{A} - r\boldsymbol{I}) = \boldsymbol{0}$. This is a polynomial equation of degree n, which means that there are n numbers $r_1, r_2, \ldots, r_n$ (counting multiplicities) that satisfy it. These numbers are the characteristic roots of $\boldsymbol{A}$. To each such characteristic root r_i, there corresponds at least one nonzero vector $\boldsymbol{v}_i$ such that $(\boldsymbol{A} - r_i\boldsymbol{I})\boldsymbol{v}_i = \boldsymbol{0}$. Any such nonzero vector corresponding to r_i is called a characteristic vector of $\boldsymbol{A}$. The matrix $\boldsymbol{A}$ is said to be diagonalizable if it has is a set of n linearly independent characteristic vectors. Otherwise $\boldsymbol{A}$ possesses generalized characteristic vectors in addition to characteristic vectors. We will meet generalized characteristic vectors only indirectly, so their definition is left to a linear algebra course.

Here is a list of properties of characteristic value problems, roots, and vectors. The list is for your reference; we will not prove these assertions.

- Any nonzero multiple of a characteristic vector is a characteristic vector.
- The multiplicity of a characteristic root r is called the algebraic multiplicity of the root: the number of linearly independent characteristic vectors that correspond to r is called the geometric multiplicity of the root.
- The geometric multiplicity of a root never exceeds the algebraic multiplicity of that root.
- A matrix possesses generalized characteristic vectors when the algebraic and geometric multiplicities of some root differ.

- Characteristic vectors corresponding to different characteristic roots are linearly independent, so a matrix that has distinct characteristic roots is diagonalizable.
- Some matrices that have repeated characteristic roots are diagonalizable.
- Given an $n \times n$ matrix $\boldsymbol{A}$, there exists an invertible matrix $\boldsymbol{P}$ such that

$$\boldsymbol{P}^{-1}\boldsymbol{A}\boldsymbol{P} = \boldsymbol{J},$$

  the **Jordan canonical form** of $\boldsymbol{A}$. The columns of $\boldsymbol{P}$ can be chosen to be n linearly independent characteristic vectors and generalized characteristic vectors of $\boldsymbol{A}$.
- When $\boldsymbol{A}$ is diagonalizable, $\boldsymbol{J}$ is a diagonal matrix having the characteristic roots of $\boldsymbol{A}$ along the diagonal. In this case, the columns of $\boldsymbol{P}$ can be taken to be n linearly independent characteristic vectors of $\boldsymbol{A}$.

Characteristic Value Problems by *Mathematica*

Example 8.10M Solve the characteristic value problem

$$\begin{cases} 6x + 2y = rx \\ 3x + y = ry. \end{cases}$$

Solution. This can be written in the standard form $(\boldsymbol{A} - r\boldsymbol{I})\boldsymbol{v} = \boldsymbol{0}$ where

$$\boldsymbol{A} = \begin{pmatrix} 6 & 2 \\ 3 & 1 \end{pmatrix}, \boldsymbol{I} = \begin{pmatrix} 1 & 0 \\ 0 & 1 \end{pmatrix}, \quad \text{and} \quad \boldsymbol{v} = \begin{pmatrix} x \\ y \end{pmatrix}.$$

We solve the problem in two ways: (1) using the theory, and (2) using built-in functions from *Mathematica*.

Using the theory:

```
In[10]:= A={{6,2},{3,1}}

Out[10]= {{6, 2}, {3, 1}}
```

This matrix is the coefficient matrix.

```
In[11]:= sys[r_]=A-r*IdentityMatrix[2]

Out[11]= {{6 - r, 2}, {3, 1 - r}}
```

Here is its determinant.

```
In[12]:= p[r_]=Det[sys[r]]

Out[12]=           2
         -7 r + r
```

Find the numbers r1 and r2 that make the determinant zero. These are the characteristic roots: they make the system $(\boldsymbol{A} - r\boldsymbol{I})\boldsymbol{v} = \boldsymbol{0}$ have nonzero solutions.

```
In[13]:= Solve[p[r]==0]

Out[13]= {{r -> 0}, {r -> 7}}
```

Capture the two characteristic roots.

```
In[14]:= {r1,r2}=r/.Solve[p[r]==0]

Out[14]= {0, 7}
```

The first solution. Here we find `v1` corresponding to `r1` by choosing the first item in `NullSpace[sys[r1]]`.

```
In[15]:= v1=NullSpace[sys[r1]][[1]]

Out[15]= {-1, 3}
```

The first solution checks.

```
In[16]:= A.v1==r1*v1

Out[16]= True
```

The second solution. Here we find `v2` corresponding to `r2` by choosing the first item in `NullSpace[sys[r2]`.

```
In[17]:= v2=NullSpace[sys[r2]][[1]]

Out[17]= {2, 1}
```

The second solution checks.

```
In[18]:= A.v2==r2*v2

Out[18]= True
```

Native *Mathematica*. [Here *Mathematica* does the work, but we don't know how. (It does exactly what we did.)]

```
In[19]:= a = {{6,2},{3,1}}

Out[19]= {{6, 2}, {3, 1}}

In[20]:= {vals,vecs} = Eigensystem[a]

Out[20]=               1
         {{0, 7}, {{-(-), 1}, {2, 1}}}
                       3
```

The characteristic roots are `vals == {0,7}` and the corresponding characteristic vectors (written as rows) are `vecs == {{-1/3,1},{2,1}}`. This agrees with the previous results if we multiply the first vector by 3. This is permitted since any nonzero multiple of a characteristic vector is a characteristic vector. ◇

The Solution Process

We are in a situation where the theory says that solutions to a differential system exist, but we do not know how to find them. We present a method in theory and then give examples done both manually and by *Mathematica* following the manual steps. It is worth knowing how to guide *Mathematica* to a solution because you may be able to direct it to do something that it cannot (yet) do on its own. The key to this method is the **characteristic value problem.**

To solve the vector system $\boldsymbol{y}' = \boldsymbol{A}\boldsymbol{y}$, guided by our solution of equation 8.4 in the system setting, we seek at least one solution having the form $\boldsymbol{Y}(x) = \boldsymbol{K}e^{rx}$, where we are to find r and a corresponding nonzero constant vector $\boldsymbol{K}$ such that

$$\boldsymbol{y}' - \boldsymbol{A}\boldsymbol{y} = r\boldsymbol{K}e^{rx} - \boldsymbol{A}\boldsymbol{K}e^{rx} = (r\boldsymbol{K} - \boldsymbol{A}\boldsymbol{K})e^{rx} = -(\boldsymbol{A}\boldsymbol{K} - r\boldsymbol{K})e^{rx} = \mathbf{0}.$$

We have a solution if can solve the **characteristic value problem**

$$\boldsymbol{A}\boldsymbol{K} = r\boldsymbol{K}.$$

We must find numbers r and corresponding *nonzero* vectors $\boldsymbol{K}$ so that

$$\boldsymbol{A}\boldsymbol{K} - r\boldsymbol{K} = (\boldsymbol{A} - r\boldsymbol{I}_n)\boldsymbol{K} = \mathbf{0}.$$

This is an algebraic system of n linear equations in the n unknown components of $\boldsymbol{K}$. The $n \times n$ coefficient matrix is $\boldsymbol{A} - r\boldsymbol{I}_n$. If r happens to be such that this coefficient matrix has an inverse, then $\boldsymbol{K} = \mathbf{0}$ is the only solution. We do not want this to happen. Can we choose r so that the matrix $\boldsymbol{A} - r\boldsymbol{I}_n$ has no inverse? We learned how to accomplish this in Chapter 1: if we can find r so that $\det(\boldsymbol{A} - r\boldsymbol{I}_n) = 0$, then there is a nonzero vector $\boldsymbol{K}$ such that

$$(\boldsymbol{A} - r\boldsymbol{I}_n)\boldsymbol{K} = \mathbf{0}.$$

The pair $\{r, \boldsymbol{K}\}$, which is a characteristic value and corresponding characteristic vector, gives us a solution to the differential system as we verify next.

Consider $\boldsymbol{y}(x) = \boldsymbol{K}e^{rx}$ with r chosen so that $\det(\boldsymbol{A} - r\boldsymbol{I}_n) = 0$ and $\boldsymbol{K} \neq 0$ chosen so that $(\boldsymbol{A} - r\boldsymbol{I}_n)\boldsymbol{K} = \mathbf{0}$, then

$$\begin{aligned}
\boldsymbol{y}'(x) - \boldsymbol{A}\boldsymbol{y}(x) &= r\boldsymbol{K}e^{rx} - \boldsymbol{A}\boldsymbol{K}e^{rx} \\
&= (r\boldsymbol{K} - \boldsymbol{A}\boldsymbol{K})e^{rx} \\
&= (r\boldsymbol{I}_n - \boldsymbol{A})\boldsymbol{K}e^{rx} \\
&= -(\boldsymbol{A} - r\boldsymbol{I}_n)\boldsymbol{K}e^{rx} \\
&= \mathbf{0}e^{rx} \\
&= \mathbf{0}.
\end{aligned}$$

So $\boldsymbol{y}$ does satisfy the differential equation.

We now know what has to be done. How do we do it? An example will illustrate the theory.

Example 8.11 Let

$$\boldsymbol{A} = \begin{pmatrix} -1 & 2 & 2 \\ 2 & 2 & 2 \\ -3 & -6 & -6 \end{pmatrix}.$$

Find a fundamental set of solutions for the first-order vector system $\boldsymbol{y}' = \boldsymbol{A}\boldsymbol{y}$. Then express the kernel as $\boldsymbol{y} = \boldsymbol{W}(x)\boldsymbol{C}$ where $\boldsymbol{W}(x)$ is a fundamental matrix and $\boldsymbol{C}$ is a constant vector.

Solution. The differential operator is $\boldsymbol{L}(\boldsymbol{y}) = \boldsymbol{y}' - \boldsymbol{A}\boldsymbol{y}$. Its kernel is the space of all functions such that $\boldsymbol{y}' - \boldsymbol{A}\boldsymbol{y} = \boldsymbol{0}$ which is those functions where $\boldsymbol{y}' = \boldsymbol{A}\boldsymbol{y}$. We saw above that we would have a solution of the form $\boldsymbol{Y}(x) = \boldsymbol{K}e^{rx}$ if we could choose r so that $\det(\boldsymbol{A} - r\boldsymbol{I}_3) = 0$. Then, after choosing r, choose $\boldsymbol{K}$ to be a nonzero solution of the linear algebraic system $(\boldsymbol{A} - r\boldsymbol{I}_3)\boldsymbol{K} = \boldsymbol{0}$.

Let's see how to make $\det(\boldsymbol{A} - r\boldsymbol{I}_3) = 0$.

$$\begin{aligned} \boldsymbol{A} - r\boldsymbol{I}_3 &= \begin{pmatrix} -1 & 2 & 2 \\ 2 & 2 & 2 \\ -3 & -6 & -6 \end{pmatrix} - r\begin{pmatrix} 1 & 0 & 0 \\ 0 & 1 & 0 \\ 0 & 0 & 1 \end{pmatrix} \\ &= \begin{pmatrix} -1-r & 2 & 2 \\ 2 & 2-r & 2 \\ -3 & -6 & -6-r \end{pmatrix}, \end{aligned}$$

so we want

$$\begin{aligned} \det(\boldsymbol{A} - r\boldsymbol{I}_3) &= \det\begin{pmatrix} -1-r & 2 & 2 \\ 2 & 2-r & 2 \\ -3 & -6 & -6-r \end{pmatrix} \\ &= r^3 + 5r^2 + 6r \\ &= r(r+2)(r+3) = 0. \end{aligned}$$

Notice that we are once again back to finding roots of polynomials, just like we were in Chapter 4. As before, distinct real roots are going to present the simplest case. Repeated roots will again require special attention, and we will have to see how to convert the conjugate complex case into two real solutions. This much is familiar ground.

Now that we have three choices for $r: r_1 = 0, r_2 = -2$, and $r_3 = -3$, what do we do to get the corresponding vectors $\boldsymbol{K}_1, \boldsymbol{K}_2$, and $\boldsymbol{K}_3$? To find $\boldsymbol{K}_1$ we need to solve

$$(\boldsymbol{A} - r_1\boldsymbol{I}_3)\boldsymbol{K}_1 = (\boldsymbol{A} - 0\boldsymbol{I}_3)\boldsymbol{K}_1 = \boldsymbol{A}\boldsymbol{K}_1 = \begin{pmatrix} -1 & 2 & 2 \\ 2 & 2 & 2 \\ -3 & -6 & -6 \end{pmatrix}\boldsymbol{K}_1 = 0.$$

This problem is solved if we have a basis for the kernel of $\boldsymbol{A}$. Row reduce $\boldsymbol{A}$ to get

$$\begin{pmatrix} -1 & 2 & 2 \\ 2 & 2 & 2 \\ -3 & -6 & -6 \end{pmatrix} \sim \begin{pmatrix} 1 & 0 & 0 \\ 0 & 1 & 1 \\ 0 & 0 & 0 \end{pmatrix}.$$

Here a basis for the kernel is $\boldsymbol{K}_1 = \begin{pmatrix} 0 \\ -1 \\ 1 \end{pmatrix}$. (Verify that this is a solution of $\boldsymbol{AK}_1 = 0\boldsymbol{K}_1$.)

To find $\boldsymbol{K}_2$ we solve

$$(\boldsymbol{A} - r_2\boldsymbol{I}_3)\boldsymbol{K}_2 = (\boldsymbol{A} + 2\boldsymbol{I}_3)\boldsymbol{K}_2 = \begin{pmatrix} 1 & 2 & 2 \\ 2 & 4 & 2 \\ -3 & -6 & -4 \end{pmatrix} \boldsymbol{K}_2 = 0.$$

We again want a basis for the kernel of $\boldsymbol{A} + 2\boldsymbol{I}_3$. Row reduce $\boldsymbol{A} + 2\boldsymbol{I}_3$ to get

$$\boldsymbol{A} + 2\boldsymbol{I}_3 = \begin{pmatrix} 1 & 2 & 2 \\ 2 & 4 & 2 \\ -3 & -6 & -4 \end{pmatrix} \sim \begin{pmatrix} 1 & 2 & 0 \\ 0 & 0 & 1 \\ 0 & 0 & 0 \end{pmatrix}.$$

A basis for the kernel is $\boldsymbol{K}_2 = \begin{pmatrix} -2 \\ 1 \\ 0 \end{pmatrix}$. (Again, verify.)

To find $\boldsymbol{K}_3$ solve

$$(\boldsymbol{A} - r_3\boldsymbol{I}_3)\boldsymbol{K}_3 = (\boldsymbol{A} + 3\boldsymbol{I}_3)\boldsymbol{K}_3 = \begin{pmatrix} 2 & 2 & 2 \\ 2 & 5 & 2 \\ -3 & -6 & -3 \end{pmatrix} \boldsymbol{K}_3 = 0.$$

To obtain a basis for the kernel of $\boldsymbol{A} + 3\boldsymbol{I}_3$, row reduce $\boldsymbol{A} + 3\boldsymbol{I}_3$ to get

$$\boldsymbol{A} + 3\boldsymbol{I}_3 = \begin{pmatrix} 2 & 2 & 2 \\ 2 & 5 & 2 \\ -3 & -6 & -3 \end{pmatrix} \sim \begin{pmatrix} 1 & 0 & 1 \\ 0 & 1 & 0 \\ 0 & 0 & 0 \end{pmatrix}.$$

A basis for the kernel is $\boldsymbol{K}_3 = \begin{pmatrix} -1 \\ 0 \\ 1 \end{pmatrix}$. (Again, verify.)

Now that we have the three pairs $\{r_1, \boldsymbol{K}_1\}$, $\{r_2, \boldsymbol{K}_2\}$, and $\{r_3, \boldsymbol{K}_3\}$ we can state the three solution vectors to our differential equation to be

$$\boldsymbol{y}_1(x) = \boldsymbol{K}_1 e^{r_1 x} = \begin{pmatrix} 0 \\ -1 \\ 1 \end{pmatrix} e^0 = \begin{pmatrix} 0 \\ -1 \\ 1 \end{pmatrix}.$$

(This is a constant solution. It is analogous to the constant solutions we found for certain differential equations in Chapter 2.) Similarly,

$$\boldsymbol{y}_2(x) = \boldsymbol{K}_2 e^{r_2 x} = \begin{pmatrix} -2 \\ 1 \\ 0 \end{pmatrix} e^{-2x} = \begin{pmatrix} -2e^{-2x} \\ e^{-2x} \\ 0 \end{pmatrix},$$

and

$$\boldsymbol{y}_3(x) = \boldsymbol{K}_3 e^{r_3 x} = \begin{pmatrix} -1 \\ 0 \\ 1 \end{pmatrix} e^{-3x} = \begin{pmatrix} -e^{-3x} \\ 0 \\ e^{-3x} \end{pmatrix}.$$

A fundamental matrix is

$$\boldsymbol{W}(x) = \begin{pmatrix} 0 & -2e^{-2x} & -e^{-3x} \\ -1 & e^{-2x} & 0 \\ 1 & 0 & e^{-3x} \end{pmatrix},$$

and a general vector solution is

$$\boldsymbol{Y}(x) = \boldsymbol{W}(x) \begin{pmatrix} c_1 \\ c_2 \\ c_3 \end{pmatrix} = \begin{pmatrix} -2c_2e^{-2x} - c_3e^{-3x} \\ -c_1 + c_2e^{-2x} \\ c_1 + c_3e^{-3x} \end{pmatrix}. \qquad \diamond$$

As you can see, even for this simple system where things worked out well, the solution process can be long and complicated. This is why it is imperative for you to fully understand the steps in the solution process. That way, when the calculations become very long and drawn out, you can think to yourself about where you are in the process, what it is that you just did, and what is to come next. You even need to be able to do this when *Mathematica* is doing the calculations for you. This is *especially* important, because you will attempt to have *Mathematica* solve problems that are so complicated that you would never attempt to solve them by hand.

Example 8.11M We repeat the last example, letting *Mathematica* do the calculations for us.

Solution.

```
In[21]:= a = {{-1,2,2},{2,2,2},{-3,-6,-6}}

Out[21]= {{-1, 2, 2}, {2, 2, 2}, {-3, -6, -6}}
```

The matrix `a-rI`.

```
In[22]:= sys[r_] = a-r*IdentityMatrix[3]

Out[22]= {{-1 - r, 2, 2}, {2, 2 - r, 2}, {-3, -6, -6 - r}}
```

The characteristic polynomial.

```
In[23]:= CharacteristicPoly[r_] = Factor[Det[sys[r]]]

Out[23]= -(r (2 + r) (3 + r))
```

Get and name the characteristic roots `r1`, `r2`, and `r3`. Note the parallel assignment.

```
In[24]:= {r1,r2,r3} = r/.Solve[CharacteristicPoly[r] == 0]

Out[24]= {0, -2, -3}
```

Get the characteristic vector and solution corresponding to `r1` $= 0$

```
In[25]:= k1 = NullSpace[sys[r1]][[1]]

Out[25]= {0, -1, 1}
```

The solution corresponding to r1.

```
In[26]:= y1[x_] = Exp[r1 x]*k1

Out[26]= {0, -1, 1}
```

Get the characteristic vector and solution corresponding to $r2 = -2$

```
In[27]:= k2 = NullSpace[sys[r2]][[1]]

Out[27]= {-2, 1, 0}
```

The solution corresponding to r2.

```
In[28]:= y2[x_] = Exp[r2 x]*k2

Out[28]=   -2      -2 x
          {----, E     , 0}
            2 x
           E
```

Get the characteristic vector and solution corresponding to $r3 = -3$

```
In[29]:= k3 = NullSpace[sys[r3]][[1]]

Out[29]= {-1, 0, 1}
```

The solution corresponding to r3.

```
In[30]:= y3[x_] = Exp[r3 x]*k3

Out[30]=     -3 x        -3 x
          {-E     , 0, E    }
```

The complete solution.

```
In[31]:= y[x_] = c1*y1[x]+c2*y2[x]+c3*y3[x]

Out[31]=  -2 c2     c3              c2            c3
         {----- - ----, -c1 + ----, c1 + ----}
           2 x     3 x             2 x           3 x
          E       E               E             E
```

Check:

```
In[32]:= Simplify[y'[x]-a.y[x] == {0,0,0}]

Out[32]= True
```

W[x] is fundamental matrix of the system $Y' = a.y$. Note that we have to transpose to turn the (row) vectors into columns.

```
In[33]:= W[x_] =Transpose[{y1[x],y2[x],y3[x]}]

Out[33]=
              -2         -3 x           -2 x                  -3 x
         {{0, ----, -E       }, {-1, E      , 0}, {1, 0, E       }}
               2 x
              E
```

This matrix is always invertible because it has a nonzero determinant:

```
In[34]:= Det[W[x]]

Out[34]=   -5 x
         -E
```

W[x] solves the matrix equation $Y' = a.Y (Y' - a.Y = 0)$.

```
In[35]:= Simplify[W'[x]-a.W[x] == 0, Trig->False]

Out[35]= True
```

The matrix exponential of the coefficient matrix a.

```
In[36]:= (U[x_] = W[x].Inverse[W[0]])//MatrixForm

Out[36]=   -3 x     2          -2      2           -2      2
         -E     + ----        ---- + ----         ---- + ----
                   2 x         3 x    2 x          3 x    2 x
                  E           E      E            E      E

               -2 x                -2 x                -2 x
         1 - E              2 - E               1 - E

                -3 x                 2                   2
         -1 + E             -2 + ----           -1 + ----
                                  3 x                 3 x
                                 E                   E

In[37]:= U[0]

Out[37]= {{1, 0, 0}, {0, 1, 0}, {0, 0, 1}}

In[38]:= Simplify[U'[x]-a.U[x] == 0, Trig->False]

Out[38]= True
```

◇

You might think from this example that doing these problems by *Mathematica* is longer than by hand. *This would be a mistaken impression.* In this latter example we did everything that was illustrated in the previous example, and, in addition, produced and checked both the natural fundamental matrix and the matrix exponential of the system. These latter two computations lengthened the example considerably, but they illustrate how easily we can make complicated calculations using *Mathematica*. To prove the point, try obtaining $U[x]$ by hand from the definition of $W[x]$. Then manually check that $U[x]$ is a solution of the matrix system. This should convince you that *Mathematica* has simplified our work.

Mathematica provides us with two shortcuts: the function `Eigensystem`, and the function `MatrixExp`. Here is how these would work for our example:

```
In[39]:= {roots, vectors} = Eigensystem[a]

Out[39]= {{-3, -2, 0}, {{-1, 0, 1}, {-2, 1, 0}, {0, -1, 1}}}
```

This output has the form `{roots, vectors}` where `roots` is a list of the characteristic roots of the matrix a, and `vectors` is a list of characteristic vectors in the proper order so that they correspond to the appropriate characteristic root. In this case, `roots ={-3,-2,0}` and `vectors = {{-1, 0, 1}, {-2, 1, 0}, {0, -1, 1}}`. Note that this is not the same ordering that we used before, but that *the correspondence of vectors to roots is the same.* This means that the appropriate correspondence is:

Roots	Vectors
-3	$\{-1,0,1\}$
-2	$\{-2,1,0\}$
0	$\{0,-1,1\}$

Here is a particularly simple fundamental matrix `W[x]`.

```
In[40]:= W[x_] = Transpose[Exp[roots x]*vectors]
         (* Note the use of '*' *)

Out[40]=      -3 x    -2             -2 x          -3 x
         {{-E     ,  ----, 0}, {0, E    , -1}, {E     , 0, 1}}
                      2 x
                     E
```

Mathematica provides `MatrixExp` to find the solution `U[x]` we found above.

```
In[41]:= U1[x_] = MatrixExp[a x]

Out[41]=
                  x            x             x
          -1 + 2 E     2 (-1 + E )   2 (-1 + E )
         {{---------,  -----------,  -----------},
             3 x           3 x           3 x
            E             E             E

               -2 x       -2 x       -2 x
          {1 - E    , 2 - E    , 1 - E    },

                -3 x         2           2
          {-1 + E    , -2 + ----, -1 + ----}}
                             3 x         3 x
                            E           E
```

The form is slightly different, but this is the fundamental solution matrix $U(x)$ that we found before. Simplify the difference `U1[x]-U[x]` to see that it is the zero matrix. The function `MatrixExp` quickly gives us a fundamental solution, but we have also lost something: from `W[x]` we know that in each of the three directions $\{-1,0,1\},\{-2,1,0\}$, and $\{0,-1,1\}$ the solution acts like a simple exponential, and that in all other directions, its behavior is that of a mixture of exponentials. Now we have only such a mixture of exponentials; there is no easy way to obtain from this matrix those directions in which the action of the solution is "pure."

The example suggests (correctly) that it is easy to summarize the distinct real roots case as a theorem.

Theorem 8.4 *If the characteristic equation for the $n \times n$ matrix $\boldsymbol{A}$ has only real and distinct characteristic roots $r_1, r_2, \ldots, r_n$, then every solution of the first-order differential system $\boldsymbol{y}' = \boldsymbol{A}\boldsymbol{y}$ can be written uniquely in the form*

$$\boldsymbol{y}(x) = \sum_{i=1}^{n} c_i e^{r_i x} \boldsymbol{k}_i,$$

where $\boldsymbol{k}_i$ is the real nonzero characteristic vector corresponding to the characteristic root $r_i, 1 \leq i \leq n$.

The individual solutions $\boldsymbol{y}_i(x) = e^{r_i x}\boldsymbol{k}_i, 1 \leq i \leq n$ are linearly independent and form the columns of a fundamental matrix solution

$$\boldsymbol{W}(x) = (\boldsymbol{y}_1(x) \mid \boldsymbol{y}_2(x) \mid \ldots \mid \boldsymbol{y}_n(x))$$

of the $n \times n$ matrix system

$$\boldsymbol{Y}'(x) = \boldsymbol{A}\boldsymbol{Y}(x).$$

Every solution of the matrix system $\boldsymbol{Y}'(x) = \boldsymbol{A}\boldsymbol{Y}(x)$ can be written uniquely as

$$\boldsymbol{Y}(x) = \boldsymbol{W}(x)\boldsymbol{K},$$

where $\boldsymbol{K}$ is an $n \times n$ constant matrix, and every solution of the vector system $\boldsymbol{y}' = \boldsymbol{A}\boldsymbol{y}$ can be written uniquely in the form

$$\boldsymbol{y}(x) = \boldsymbol{W}(x)\boldsymbol{k},$$

where $\boldsymbol{k}$ is an n-vector.

Exercises 8.3

Part I. Set up the recursion suggested by Picard's method for each of these first-order systems. The recursion for the initial-value problem

$$\boldsymbol{Y}'(x) = \boldsymbol{P}(x)\boldsymbol{Y}(x) + \boldsymbol{Q}(x), \quad \boldsymbol{Y}(0) = \boldsymbol{K}$$

is

$$\boldsymbol{Y}_{n+1}(x) = \boldsymbol{K} + \int_0^x \boldsymbol{P}(t)\boldsymbol{Y}_n(t) + \boldsymbol{Q}(t)\, dt,$$

$$\boldsymbol{Y}_0(x) \equiv \boldsymbol{K}.$$

Find the first five nonzero terms of a power series expansion for the solution. It will be centered on $x_0 = 0$. Let *Mathematica* do the work for you.

1. $\boldsymbol{Y}'(x) = \begin{pmatrix} 1 & 2 \\ x & 1 \end{pmatrix} \boldsymbol{Y}(x) + \begin{pmatrix} x & 2 \\ x & 1 \end{pmatrix}, \quad \boldsymbol{Y}(0) = \begin{pmatrix} -1 \\ 2 \end{pmatrix}.$

2. $\boldsymbol{Y}'(x) = \begin{pmatrix} 0 & 2 \\ x & x \end{pmatrix} \boldsymbol{Y}(x) + \begin{pmatrix} x & 2x \\ 3 & 1 \end{pmatrix}, \quad \boldsymbol{Y}(0) = \begin{pmatrix} -1 \\ 3 \end{pmatrix}.$

3. $\boldsymbol{Y}'(x) = \begin{pmatrix} 1 & x \\ x & -1 \end{pmatrix} \boldsymbol{Y}(x) + \begin{pmatrix} 0 & 2 \\ x & 0 \end{pmatrix}, \quad \boldsymbol{Y}(0) = \begin{pmatrix} 2 \\ -3 \end{pmatrix}.$

4. $\boldsymbol{Y}'(x) = \begin{pmatrix} 3 & 2x \\ 1+x & 1-x \end{pmatrix} \boldsymbol{Y}(x) + \begin{pmatrix} 3x & 2-x \\ x & 1 \end{pmatrix}, \quad \boldsymbol{Y}(0) = \begin{pmatrix} 1 \\ 1 \end{pmatrix}.$

PART II. Manually solve each of these diagonal systems.

5. $\boldsymbol{Y}'(x) = \begin{pmatrix} 3 & 0 \\ 0 & -5 \end{pmatrix} \boldsymbol{Y}(x), \quad \boldsymbol{Y}(0) = \begin{pmatrix} 1 & 1 \\ -2 & 3 \end{pmatrix}.$

6. $\boldsymbol{Y}'(x) = \begin{pmatrix} 3 & 0 & 0 \\ 0 & -5 & 0 \\ 0 & 0 & 4 \end{pmatrix} \boldsymbol{Y}(x), \quad \boldsymbol{Y}(0) = \begin{pmatrix} 1 & 1 & 0 \\ 0 & -2 & 3 \\ 1 & 0 & 2 \end{pmatrix}.$

7. $\boldsymbol{Y}'(x) = \begin{pmatrix} 2 & 0 & 0 \\ 0 & -7 & 0 \\ 0 & 0 & 3 \end{pmatrix} \boldsymbol{Y}(x), \quad \boldsymbol{Y}(0) = \begin{pmatrix} 1 & -1 & 0 \\ 0 & -4 & 3 \\ 0 & 0 & -2 \end{pmatrix}.$

8. $\boldsymbol{Y}'(x) = \begin{pmatrix} -3 & 0 & 0 & 0 \\ 0 & -5 & 0 & 0 \\ 0 & 0 & 2 & 0 \\ 0 & 0 & 0 & -4 \end{pmatrix} \boldsymbol{Y}(x), \quad \boldsymbol{Y}(0) = \begin{pmatrix} 1 & 1 & 0 & 0 \\ 0 & 0 & -2 & 3 \\ 0 & 4 & 0 & 1 \\ 0 & 0 & 2 & 0 \end{pmatrix}.$

PART III. Show that each of these matrices is nilpotent. State the order of nilpotency of each matrix. Find the matrix exponential $e^{\boldsymbol{A}x}$ of each matrix. Verify that each exponential is a matrix polynomial. Write each matrix polynomial in the form $\sum_{k=0}^{n} \boldsymbol{C}_k x^k$.

9. $\boldsymbol{N}_1 = \begin{pmatrix} 0 & 1 & 0 & 0 \\ 0 & 0 & -2 & 3 \\ 0 & 0 & 0 & 1 \\ 0 & 0 & 0 & 0 \end{pmatrix}.$

10. $\boldsymbol{N}_2 = \begin{pmatrix} 0 & 1 & 0 & 4 \\ 0 & 0 & -2 & 3 \\ 0 & 0 & 0 & 2 \\ 0 & 0 & 0 & 0 \end{pmatrix}.$

PART IV. Using the techniques of Examples 8.11 and 8.11 M, find a fundamental solution for each of these constant coefficients systems. Even when doing the problems manually, let *Mathematica* solve the characteristic value problems that arise.

11. $\boldsymbol{Y}'(x) = \begin{pmatrix} 8 & -5 \\ 10 & -7 \end{pmatrix} \boldsymbol{Y}(x)$.

12. $\boldsymbol{Y}'(x) = \begin{pmatrix} 3 & 0 & 0 \\ -10 & -2 & 0 \\ 10 & 6 & 1 \end{pmatrix} \boldsymbol{Y}(x)$.

13. $\boldsymbol{Y}'(x) = \begin{pmatrix} 133 & 54 & 22 \\ -300 & -122 & -50 \\ -60 & -24 & -9 \end{pmatrix} \boldsymbol{Y}(x)$.

14. $\boldsymbol{Y}'(x) = \begin{pmatrix} 1 & 0 & 0 \\ -4 & -1 & 0 \\ 8 & 4 & 1 \end{pmatrix} \boldsymbol{Y}(x)$.

15. $\boldsymbol{Y}'(x) = \begin{pmatrix} -2 & 2 & -8 & -7 \\ 0 & -1 & -8 & -8 \\ 0 & 0 & -5 & -4 \\ 0 & 0 & 8 & 7 \end{pmatrix} \boldsymbol{Y}(x)$.

PART V. Theory.

16. Show that if v_1 and v_2 are characteristic vectors of the matrix $\boldsymbol{A}$ corresponding to the characteristic root r, then $v = c_1 v_1 + c_2 v_2$ is also a characteristic vector. This means that there corresponds to r a subspace of characteristic vectors.

17. Verify by induction that

$$\boldsymbol{Y}_n(x) = \boldsymbol{I}_n + \sum_{i=1}^{n} \frac{(\boldsymbol{A}x)^i}{i!}$$

is the unique solution of

$$\boldsymbol{Y}_{n+1}(x) = \boldsymbol{I}_n + \int_0^x \boldsymbol{A}\boldsymbol{Y}_n(t)\,dt, \quad Y_0(x) = \boldsymbol{I}_n$$

for $n \geq 0$.

18. Show that each component $(\boldsymbol{Y}_m)_{ij}$ of $\boldsymbol{Y}_m(x) = \boldsymbol{I}_n + \sum_{i=1}^{m} (\boldsymbol{A}x)^i/i!$ is bounded by

$$|(\boldsymbol{Y}_m)_{ij}| \leq 1 + \sum_{i=1}^{m} \frac{(nM)^i}{i!}$$

when M is chosen such that $\max_{i,j}|\boldsymbol{A}_{i,j}| \leq M$. Hence conclude that $\lim_{m\to\infty} \boldsymbol{Y}_m(x)$ exists for each real number x.

19. Make enough iterations of Picard's method on the equation of Example 8.7 to obtain recognizable series.

8.4 Repeated and Complex Roots

We have just seen an example of how to solve first-order vector or matrix systems when the characteristic roots of the coefficient matrix are real and distinct. The theory for the next case is not always so neat. We begin where the characteristic roots are not distinct, but where there is a complete linearly independent set of characteristic vectors. In the study of linear algebra, the case we are considering is identified by saying that the coefficient matrix $\boldsymbol{A}$ is **diagonalizable**. This means that there is an invertible matrix $\boldsymbol{P}$ such that $\boldsymbol{D} = \boldsymbol{P}^{-1}\boldsymbol{A}\boldsymbol{P}$ has only zeros off of the main diagonal. That is, $\boldsymbol{D}$ is a diagonal matrix. The entries on the diagonal of $\boldsymbol{D}$ are the characteristic roots of $\boldsymbol{A}$, and the columns of $\boldsymbol{P}$ are the corresponding characteristic vectors. This means that we have already learned how to use *Mathematica* to do all of the calculations that we need.

Repeated Roots (Diagonalizable)

In Section 4.2 we learned that the multiplicity of a characteristic root r of the matrix $\boldsymbol{A}$ is called the **algebraic multiplicity** of the root, and the dimension of the kernel of $\boldsymbol{A} - r\boldsymbol{I}$ is called the **geometric multiplicity** of the root r. The geometric multiplicity is the number of linearly independent characteristic vectors that correspond to r. The geometric multiplicity is always at least 1 and is never greater than the algebraic multiplicity. The matrix $\boldsymbol{A}$ is **diagonalizable** provided that the algebraic multiplicity of each characteristic root equals the geometric multiplicity of that root. This follows from the fact that characteristic vectors corresponding to distinct characteristic roots are linearly independent. Equivalently, the $n \times n$ matrix $\boldsymbol{A}$ is diagonalizable provided that $\boldsymbol{A}$ has n linearly independent characteristic vectors.

Consider the matrix

$$\boldsymbol{A} = \begin{pmatrix} 2 & 6 & 3 & 12 & 6 \\ 0 & 26 & 6 & 54 & 6 \\ 0 & 6 & 5 & 12 & 6 \\ 0 & -12 & -3 & -25 & -3 \\ 0 & -6 & -3 & -12 & -4 \end{pmatrix}$$

and the matrix

$$\boldsymbol{P} = \begin{pmatrix} 3 & 2 & 2 & -1 & -1 \\ 2 & 9 & 2 & -4 & -2 \\ 2 & 2 & 2 & -1 & -1 \\ -1 & -4 & -1 & 2 & 1 \\ -1 & -2 & -1 & 1 & 1 \end{pmatrix}.$$

Then we have this matrix equality:

$$\boldsymbol{A}\boldsymbol{P} = \begin{pmatrix} 6 & 4 & 4 & 1 & 1 \\ 4 & 18 & 4 & 4 & 2 \\ 4 & 4 & 4 & 1 & 1 \\ -2 & -8 & -2 & -2 & -1 \\ -2 & -4 & -2 & -1 & -1 \end{pmatrix}$$

$$= \begin{pmatrix} 3 & 2 & 2 & -1 & -1 \\ 2 & 9 & 2 & -4 & -2 \\ 2 & 2 & 2 & -1 & -1 \\ -1 & -4 & -1 & 2 & 1 \\ -1 & -2 & -1 & 1 & 1 \end{pmatrix} \begin{pmatrix} 2 & 0 & 0 & 0 & 0 \\ 0 & 2 & 0 & 0 & 0 \\ 0 & 0 & 2 & 0 & 0 \\ 0 & 0 & 0 & -1 & 0 \\ 0 & 0 & 0 & 0 & -1 \end{pmatrix}$$
$$= \boldsymbol{PD},$$

where

$$\boldsymbol{D} = \begin{pmatrix} 2 & 0 & 0 & 0 & 0 \\ 0 & 2 & 0 & 0 & 0 \\ 0 & 0 & 2 & 0 & 0 \\ 0 & 0 & 0 & -1 & 0 \\ 0 & 0 & 0 & 0 & -1 \end{pmatrix}.$$

We note for later use that

$$\boldsymbol{P}^{-1} = \begin{pmatrix} 1 & 0 & -1 & 0 & 0 \\ 0 & 1 & 0 & 2 & 0 \\ -1 & 0 & 2 & 0 & 1 \\ 0 & 2 & 0 & 5 & -1 \\ 0 & 0 & 1 & -1 & 3 \end{pmatrix}.$$

From $\boldsymbol{AP} = \boldsymbol{PD}$ we can see that $\boldsymbol{D} = \boldsymbol{P}^{-1}\boldsymbol{AP}$, which was to be illustrated. From these results, we can read off the characteristic roots from the matrix $\boldsymbol{D}$ and see that $r_1 = 2$ is a triple root and that $r_2 = -1$ is a double root. The first three columns of $\boldsymbol{P}$ are the characteristic vectors corresponding to r_1 and the last two columns of $\boldsymbol{P}$ are the characteristic vectors corresponding to r_2. So if we can solve the characteristic value problem completely we can easily diagonalize $\boldsymbol{A}$, and if we have diagonalized $\boldsymbol{A}$, we can read off a complete solution of the characteristic value problem, given that we know $\boldsymbol{P}$ and $\boldsymbol{D}$.

Convince yourself that the matrix exponential of the matrix $\boldsymbol{D}x$ is

$$e^{\boldsymbol{D}x} = \operatorname{diag}(e^{r_1 x}, e^{r_2 x}, \ldots, e^{r_n x})$$

and that

$$\begin{aligned} \exp(\boldsymbol{P}^{-1}(\boldsymbol{A}x)\boldsymbol{P}) &= \boldsymbol{P}^{-1}\exp(\boldsymbol{A}x)\boldsymbol{P} \\ &= \boldsymbol{P}^{-1}e^{\boldsymbol{A}x}\boldsymbol{P}, \end{aligned}$$

so that from

$$\begin{aligned} \exp(\boldsymbol{P}^{-1}(\boldsymbol{A}x)\boldsymbol{P}) &= \boldsymbol{P}^{-1}\exp(\boldsymbol{A}x)\boldsymbol{P} \\ &= \boldsymbol{P}^{-1}e^{\boldsymbol{A}x}\boldsymbol{P} \\ &= e^{\boldsymbol{D}x} \\ &= \operatorname{diag}(e^{r_1 x}, e^{r_2 x}, \ldots, e^{r_n x}), \end{aligned}$$

we have

$$e^{\boldsymbol{A}x} = \boldsymbol{P}\operatorname{diag}(e^{r_1 x}, e^{r_2 x}, \ldots, e^{r_n x})\boldsymbol{P}^{-1}.$$

Example 8.12 The observation of the last paragraph leads to the calculation:

$$e^{\boldsymbol{A}x} = \boldsymbol{P} \begin{pmatrix} e^{2x} & 0 & 0 & 0 & 0 \\ 0 & e^{2x} & 0 & 0 & 0 \\ 0 & 0 & e^{2x} & 0 & 0 \\ 0 & 0 & 0 & e^{-x} & 0 \\ 0 & 0 & 0 & 0 & e^{-x} \end{pmatrix} \boldsymbol{P}^{-1}.$$

You should use *Mathematica* to multiply these matrices out and then check that the resulting matrix is a solution of the original problem, $\boldsymbol{Y}' = \boldsymbol{A}\boldsymbol{Y}$.

The diagonalizable case is summarized in this theorem.

Theorem 8.5 *If the $n \times n$ matrix $\boldsymbol{A}$ is diagonalizable, that is, there is an invertible $n \times n$ matrix $\boldsymbol{P}$ such that $\boldsymbol{P}^{-1}\boldsymbol{A}\boldsymbol{P} = \boldsymbol{D}$, a diagonal matrix, then a fundamental solution of the matrix differential system $\boldsymbol{Y}' = \boldsymbol{A}\boldsymbol{Y}$ is $\boldsymbol{W}(x) = \boldsymbol{P}\exp(\boldsymbol{D}x)\boldsymbol{P}^{-1}$.*

Every solution of the matrix differential system $\boldsymbol{Y}' = \boldsymbol{A}\boldsymbol{Y}$ has the form

$$\boldsymbol{Y}(x) = \boldsymbol{W}(x)\boldsymbol{K},$$

where $\boldsymbol{K}$ is a constant $n \times n$ matrix.

Every solution of the vector differential system $\boldsymbol{y}' = \boldsymbol{A}\boldsymbol{y}$ has the form $\boldsymbol{y}(x) = \boldsymbol{W}(x)\boldsymbol{k}$, where $\boldsymbol{k}$ is a constant n-vector.

The theorem is deliberately stated so as to not prohibit the characteristic roots from being complex, even though the case of complex roots has yet to be discussed.

Also, not every matrix differential system has a coefficient matrix that is diagonalizable, so the situation is not always as simple as the case described by Theorem 8.5. The next paragraphs describe this more complicated case.

Repeated Roots (Not Diagonalizable)

When the characteristic root r of $\boldsymbol{A}$ has algebraic multiplicity m, but geometric multiplicity $m_1 < m$, then the root r of $\boldsymbol{A}$ is said to be **deficient**. In theory, if one were to write down $n \times n$ matrices with random rational numbers as entries, the probability is zero of encountering a deficient matrix. This does not mean that they cannot happen, but that they are rare. Just as one has to carefully construct polynomials of degree greater than 4 in order for the polynomials to be solvable exactly, so one has to construct deficient matrices artificially. When doing so, for each example the starting point is a matrix $\boldsymbol{J}$ that is in **Jordan canonical form**. Then $\boldsymbol{J}$ is transformed into $\boldsymbol{A} = \boldsymbol{P}^{-1}\boldsymbol{J}\boldsymbol{P}$ by the choice of some suitable matrix $\boldsymbol{P}$. The problem $\boldsymbol{Y}' = \boldsymbol{A}\boldsymbol{Y}$ is then presented.

Jordan canonical forms of matrices are an important topic of study in linear algebra. We will not elaborate on the theory, except to say that every square matrix $\boldsymbol{A}$ is similar to a matrix of the form $\boldsymbol{D} + \boldsymbol{N}$, where $\boldsymbol{D}\boldsymbol{N} = \boldsymbol{N}\boldsymbol{D}, \boldsymbol{D}$ is diagonal, and $\boldsymbol{N}$ is nilpotent. In summary, $\boldsymbol{P}^{-1}\boldsymbol{A}\boldsymbol{P} = \boldsymbol{D}+\boldsymbol{N}$. The matrix $\boldsymbol{D}+\boldsymbol{N}$

is (essentially) the Jordan canonical form of $\boldsymbol{A}$. Furthermore, $\boldsymbol{A}$ is diagonalizable if and only if $\boldsymbol{N} = \boldsymbol{0}$.

It is true that $\exp((\boldsymbol{D}+\boldsymbol{N})x) = \exp(\boldsymbol{D}x)\exp(\boldsymbol{N}x)$, and, as we saw in Example 8.9, $\exp(\boldsymbol{N}x)$ is a matrix of polynomials. The orders of the polynomials depend on the last power of $\boldsymbol{N}$ that is nonzero.

Knowing this, the solution of $\boldsymbol{Y}' = \boldsymbol{A}\boldsymbol{Y}$ proceeds as follows. Let $\boldsymbol{P}$ be such that $\boldsymbol{P}^{-1}\boldsymbol{A}\boldsymbol{P} = \boldsymbol{D}+\boldsymbol{N}$, where $\boldsymbol{D}$ is diagonal, $\boldsymbol{N}$ is nilpotent and $\boldsymbol{D}\boldsymbol{N} = \boldsymbol{N}\boldsymbol{D}$. Define $\boldsymbol{Z} = \boldsymbol{P}^{-1}\boldsymbol{Y}$. Then $\boldsymbol{Z}' = \boldsymbol{P}^{-1}\boldsymbol{Y}'$, or $\boldsymbol{Y}' = \boldsymbol{P}\boldsymbol{Z}'$. Thus, from $\boldsymbol{P}\boldsymbol{Z}' = \boldsymbol{Y}' = \boldsymbol{A}\boldsymbol{Y} = \boldsymbol{A}\boldsymbol{P}\boldsymbol{Z}$ we obtain

$$\boldsymbol{Z}' = \boldsymbol{P}^{-1}\boldsymbol{A}\boldsymbol{P}\boldsymbol{Z} = (\boldsymbol{D}+\boldsymbol{N})\boldsymbol{Z}.$$

Hence

$$\boldsymbol{Z}(x) = \exp((\boldsymbol{D}+\boldsymbol{N})x) = \exp(\boldsymbol{D}x)\exp(\boldsymbol{N}x) = \boldsymbol{P}^{-1}\boldsymbol{Y}(x)$$

and we have

$$\boldsymbol{Y}(x) = \boldsymbol{P}\boldsymbol{Z}(x) = \boldsymbol{P}\exp(\boldsymbol{D}x)\exp(\boldsymbol{N}x).$$

You are encouraged to verify that this is indeed a solution of the matrix system $\boldsymbol{Y}' = \boldsymbol{A}\boldsymbol{Y}$. In order to do this, you will need to be aware that $\boldsymbol{N}$ and $\boldsymbol{D}$ not only commute with one another, but each commutes with $\exp(\boldsymbol{D}x)$ and $\exp(\boldsymbol{N}x)$.

One obvious question is: *where did the matrix* $\boldsymbol{P}$ *come from*? Clearly, once $\boldsymbol{P}$ is known things proceed smoothly, but $\boldsymbol{P}$ is not very easy to obtain. Once again, the theory is fairly straightforward, but it is difficult to carry out. A linear algebra course would help your understanding of how to obtain $\boldsymbol{P}$. We will give a partial explanation by way of examples.

If the root r of $\boldsymbol{A}$ has multiplicity m, then there will be m linearly independent solutions of the form

$$\begin{cases} \boldsymbol{y}_1(x) = \boldsymbol{K}_{m-1}e^{rx} \\ \boldsymbol{y}_2(x) = (\boldsymbol{K}_{m-2} + \boldsymbol{K}_{m-1}x)e^{rx} \\ \boldsymbol{y}_3(x) = \left(\boldsymbol{K}_{m-3} + \boldsymbol{K}_{m-2}x + \boldsymbol{K}_{m-1}\dfrac{x^2}{2!}\right)e^{rx} \\ \quad\vdots \\ \boldsymbol{y}_m(x) = \left(\boldsymbol{K}_0 + \boldsymbol{K}_1 x + \cdots + \boldsymbol{K}_{m-1}\dfrac{x^{m-1}}{(m-1)!}\right)e^{rx}. \end{cases} \tag{8.10}$$

In these expressions the $\boldsymbol{K}_i$ are constant n-vectors. There are exactly m linearly independent $\boldsymbol{K}_i$ that appear in the solution, even though not all of these forms need actually occur. A sequence of $\boldsymbol{K}_i$ in one of these forms is called a **chain**. $\boldsymbol{K}_0, \boldsymbol{K}_1, \ldots, \boldsymbol{K}_{m-2}$ are **generalized characteristic vectors**; only $\boldsymbol{K}_{m-1}$ is a characteristic vector.

As we have seen, when r has a complete system of characteristic vectors, every one of the solutions corresponding to r has the form of $\boldsymbol{y}_1(x)$, and there are m linearly independent choices for $\boldsymbol{K}_{m-1}$. If r has deficiency 1, then there are $m-1$ linearly independent solutions having the form of $\boldsymbol{y}_1(x)$ and one other having the form of $\boldsymbol{y}_2(x)$. The situation can get exceedingly complicated, and there is no way to know in advance which of the forms will actually occur. It does not hurt to assume a solution of the form of $\boldsymbol{y}_m(x)$, since the solution procedure will zero out any unneeded $\boldsymbol{K}_i$. The list 8.10 contains the possible forms for the solutions. *It is not a formula for the solutions.*

Here is an example presented from the standpoint of differential equations rather than linear algebra. It will suggest to us the steps that are required to obtain a complete solution of a differential equation where the coefficient matrix has a deficient root. The discussion applies to every characteristic root of $\boldsymbol{A}$ whose algebraic and geometric multiplicities differ.

Suppose that r is a characteristic root of $\boldsymbol{A}$ that has deficiency 3. This means that r is at least a quadruple root of the characteristic equation. That is, r has algebraic multiplicity $m \geq 4$. Since we are assuming that the deficiency is 3, there are $m-3$ linearly independent characteristic vectors, and we can use them as before to provide us with $m-3$ linearly independent solutions. To find the three missing solutions we can assume a solution of the form $\boldsymbol{y}(x) = (\boldsymbol{K}_0 + \boldsymbol{K}_1 x + \boldsymbol{K}_2 x^2/2!)e^{rx}$ and try to determine values for the three constant vectors $\boldsymbol{K}_0, \boldsymbol{K}_1$, and $\boldsymbol{K}_2$. Our attempt to find the three missing solutions will actually give us all m linearly independent solutions that correspond to the characteristic root r.

For this example there will be $m-3$ linearly independent choices for $\boldsymbol{K}_0$. We cannot say in advance which form(s) from 8.10 will occur, but some will involve $\boldsymbol{K}_1$ and some may involve $\boldsymbol{K}_2$. What will actually occur is determined by the Jordan canonical form of the coefficient matrix $\boldsymbol{A}$ which we will know only after our solution process is complete.

Suppose that $\boldsymbol{y}(x) = (\boldsymbol{K}_0 + \boldsymbol{K}_1 x + \boldsymbol{K}_2 x^2/2!)e^{rx}$ is a solution of $\boldsymbol{y}' = \boldsymbol{A}\boldsymbol{y}$. Then

$$(\boldsymbol{K}_1 + \boldsymbol{K}_2 x)e^{rx} + r\left(\boldsymbol{K}_0 + \boldsymbol{K}_1 x + \boldsymbol{K}_2 \frac{x^2}{2!}\right)e^{rx} = \boldsymbol{A}\left(\boldsymbol{K}_0 + \boldsymbol{K}_1 x + \boldsymbol{K}_2 \frac{x^2}{2!}\right)e^{rx}$$

should hold for all x. The equation simplifies to

$$[(\boldsymbol{A}\boldsymbol{K}_0 - r\boldsymbol{K}_0 - \boldsymbol{K}_1) + (\boldsymbol{A}\boldsymbol{K}_1 - r\boldsymbol{K}_1 - \boldsymbol{K}_2)x + (\boldsymbol{A}\boldsymbol{K}_2 - r\boldsymbol{K}_2)\frac{x^2}{2!} = \boldsymbol{0}.$$

This equation holds identically if these equations are satisfied simultaneously:

$$\begin{cases} (\boldsymbol{A} - r\boldsymbol{I})\boldsymbol{K}_2 = \boldsymbol{0} \\ (\boldsymbol{A} - r\boldsymbol{I})\boldsymbol{K}_1 = \boldsymbol{K}_2 \\ (\boldsymbol{A} - r\boldsymbol{I})\boldsymbol{K}_0 = \boldsymbol{K}_1 \end{cases}. \tag{8.11}$$

There will be exactly m linearly independent solutions of these equations since the root r has multiplicity m. It is important to note that not every choice of characteristic vector $\boldsymbol{K}_2$ permits a solution for $\boldsymbol{K}_1$. Likewise, not every choice of $\boldsymbol{K}_1$ permits a solution for $\boldsymbol{K}_0$. We let *Mathematica* solve the system simultaneously, which takes care of these considerations.

When we solve the system

$$\begin{cases} (\boldsymbol{A} - r\boldsymbol{I})\boldsymbol{K}_2 = \boldsymbol{0} \\ (\boldsymbol{A} - r\boldsymbol{I})\boldsymbol{K}_1 = \boldsymbol{K}_2 \\ (\boldsymbol{A} - r\boldsymbol{I})\boldsymbol{K}_0 = \boldsymbol{K}_1 \end{cases}.$$

seeking three linearly independent solutions we get all m solutions. These solutions are substituted into the form

$$\boldsymbol{y}(x) = \left(\boldsymbol{K}_0 + \boldsymbol{K}_1 x + \boldsymbol{K}_2 \frac{x^2}{2!}\right) e^{rx}$$

to obtain all of the linearly independent solutions.

A complete description of the solution when the coefficient matrix has deficient roots is very elaborate. Though all of the mechanisms necessary to find all solution in any particular instance are now in place, we will not give the general description, but will be satisfied with the ability to solve each individual problem that is encountered. We illustrate these ideas with an example. It appears in the *Mathematica* supplement as *Repeated Root-Example 1.*

Example 8.13 Let

$$\boldsymbol{A} = \begin{pmatrix} 2 & 1 & 0 & 0 & 0 & 0 \\ 0 & 2 & 1 & 0 & 0 & 0 \\ 0 & 0 & 2 & 0 & 0 & 0 \\ 0 & 0 & 0 & 2 & 0 & 0 \\ 0 & 0 & 0 & 0 & 2 & 1 \\ 0 & 0 & 0 & 0 & 0 & 2 \end{pmatrix}.$$

Solve the vector system $\boldsymbol{y}' = \boldsymbol{A}\boldsymbol{y}$.

Solution. We first form the matrix

$$\boldsymbol{A} - r\boldsymbol{I} = \begin{pmatrix} 2-r & 1 & 0 & 0 & 0 & 0 \\ 0 & 2-r & 1 & 0 & 0 & 0 \\ 0 & 0 & 2-r & 0 & 0 & 0 \\ 0 & 0 & 0 & 2-r & 0 & 0 \\ 0 & 0 & 0 & 0 & 2-r & 1 \\ 0 & 0 & 0 & 0 & 0 & 2-r \end{pmatrix},$$

from which it is seen immediately that the characteristic polynomial is $p(r) = (r-2)^6$. This means that $r = 2$ is repeated 6 times as a characteristic root. The corresponding characteristic vectors are found from row-reducing the matrix

$$\boldsymbol{A} - 2\boldsymbol{I} = \begin{pmatrix} 0 & 1 & 0 & 0 & 0 & 0 \\ 0 & 0 & 1 & 0 & 0 & 0 \\ 0 & 0 & 0 & 0 & 0 & 0 \\ 0 & 0 & 0 & 0 & 0 & 0 \\ 0 & 0 & 0 & 0 & 0 & 1 \\ 0 & 0 & 0 & 0 & 0 & 0 \end{pmatrix}.$$

This produces the three characteristic vectors,

$$\boldsymbol{v}_1 = \begin{pmatrix} 1 \\ 0 \\ 0 \\ 0 \\ 0 \\ 0 \end{pmatrix}, \quad \boldsymbol{v}_2 = \begin{pmatrix} 0 \\ 0 \\ 0 \\ 1 \\ 0 \\ 0 \end{pmatrix}, \quad \text{and} \quad \boldsymbol{v}_3 = \begin{pmatrix} 0 \\ 0 \\ 0 \\ 0 \\ 1 \\ 0 \end{pmatrix}.$$

We needed six characteristic vectors and found only three. The matrix $\boldsymbol{A}$ therefore has a deficiency of three. We thus seek vectors $\boldsymbol{K}_0, \boldsymbol{K}_1$, and $\boldsymbol{K}_2$ so that

$$\boldsymbol{y}(x) = \left(\boldsymbol{K}_0 + \boldsymbol{K}_1 x + \boldsymbol{K}_2 \frac{x^2}{2!} \right) e^{2x}$$

is a solution. Let the jth component of $\boldsymbol{K}_i$, $(\boldsymbol{K}_i)_j = c_{6i+j}$, for $i = 0, 1, 2$, and $1 \leq j \leq 6$. Then, when we solve

$$\begin{cases} (\boldsymbol{A} - r\boldsymbol{I})\boldsymbol{K}_2 = \boldsymbol{0} \\ (\boldsymbol{A} - r\boldsymbol{I})\boldsymbol{K}_1 = \boldsymbol{K}_2 \\ (\boldsymbol{A} - r\boldsymbol{I})\boldsymbol{K}_0 = \boldsymbol{K}_1 \end{cases}.$$

a system of equations in the 18 $(= 3 \times 6)$ unknowns, $c_1, \ldots, c_{18}$, we find that

$$\boldsymbol{K}_0 = \begin{pmatrix} c_1 \\ c_7 \\ c_{13} \\ c_4 \\ c_5 \\ c_{11} \end{pmatrix}, \quad \boldsymbol{K}_1 = \begin{pmatrix} c_7 \\ c_{13} \\ 0 \\ 0 \\ c_{11} \\ 0 \end{pmatrix}, \quad \text{and} \quad \boldsymbol{K}_2 = \begin{pmatrix} c_{13} \\ 0 \\ 0 \\ 0 \\ 0 \\ 0 \end{pmatrix}. \tag{8.12}$$

Three vectors are expressed in terms of the six arbitrary constants c_1, c_4, c_5, c_7, c_{11}, and c_{13}. It is important to observe that c_1, c_4, and c_5 occur only in $\boldsymbol{K}_0$, c_7, and c_{11} occur in both $\boldsymbol{K}_0$ and $\boldsymbol{K}_1$, while c_{13} occurs in all three. We will produce each of the six solutions from the form

$$\boldsymbol{y}(x) = \left(\boldsymbol{K}_0 + \boldsymbol{K}_1 x + \boldsymbol{K}_2 \frac{x^2}{2!} \right) e^{2x}$$

by successively setting one of the $c_1, c_4, c_5, c_7, c_{11}$, and c_{13} to 1 and the others to 0. This results in these solutions:

When $c_1 = 1, c_4 = 0, c_5 = 0, c_7 = 0, c_{11} = 0$ and $c_{13} = 0$, we get

$$\boldsymbol{y}_1(x) = \begin{pmatrix} 1 \\ 0 \\ 0 \\ 0 \\ 0 \\ 0 \end{pmatrix} e^{2x} = \begin{pmatrix} e^{2x} \\ 0 \\ 0 \\ 0 \\ 0 \\ 0 \end{pmatrix}.$$

When $c_1 = 0, c_4 = 1, c_5 = 0, c_7 = 0, c_{11} = 0$ and $c_{13} = 0$, we get

$$\boldsymbol{y}_2(x) = \begin{pmatrix} 0 \\ 0 \\ 0 \\ 1 \\ 0 \\ 0 \end{pmatrix} e^{2x} = \begin{pmatrix} 0 \\ 0 \\ 0 \\ e^{2x} \\ 0 \\ 0 \end{pmatrix}.$$

When $c_1 = 0$, $c_4 = 0$, $c_5 = 1$, $c_7 = 0$, $c_{11} = 0$, and $c_{13} = 0$, we get

$$\boldsymbol{y}_3(x) = \begin{pmatrix} 0 \\ 0 \\ 0 \\ 0 \\ 1 \\ 0 \end{pmatrix} e^{2x} = \begin{pmatrix} 0 \\ 0 \\ 0 \\ 0 \\ e^{2x} \\ 0 \end{pmatrix}.$$

The three solutions $\boldsymbol{y}_1(x), \boldsymbol{y}_2(x)$, and $\boldsymbol{y}_3(x)$ that involve c_1, c_4, and c_5 correspond respectively to the three characteristic vectors $\boldsymbol{v}_1$, $\boldsymbol{v}_2$, and $\boldsymbol{v}_3$ that were found originally. The remaining solutions involve chains of two or more vectors. Note that c_7 and c_{11} each occur twice. Observe how they produce chains of two vectors. Finally, c_{13} occurs three times and produces a chain of three vectors in the solution it generates.

When $c_1 = 0$, $c_4 = 0$, $c_5 = 0$, $c_7 = 1$, $c_{11} = 0$, and $c_{13} = 0$, we get

$$\boldsymbol{y}_4(x) = \left[\begin{pmatrix} 0 \\ 1 \\ 0 \\ 0 \\ 0 \\ 0 \end{pmatrix} + \begin{pmatrix} 1 \\ 0 \\ 0 \\ 0 \\ 0 \\ 0 \end{pmatrix} x \right] e^{2x} = \begin{pmatrix} xe^{2x} \\ e^{2x} \\ 0 \\ 0 \\ 0 \\ 0 \end{pmatrix}.$$

The coefficient of x is the characteristic vector $\boldsymbol{v}_1$, and the constant term is new.

When $c_1 = 0$, $c_4 = 0$, $c_5 = 0$, $c_7 = 0$, $c_{11} = 1$, and $c_{13} = 0$, we get

$$\boldsymbol{y}_5(x) = \left[\begin{pmatrix} 0 \\ 0 \\ 0 \\ 0 \\ 0 \\ 1 \end{pmatrix} + \begin{pmatrix} 0 \\ 0 \\ 0 \\ 0 \\ 1 \\ 0 \end{pmatrix} x \right] e^{2x} = \begin{pmatrix} 0 \\ 0 \\ 0 \\ 0 \\ xe^{2x} \\ e^{2x} \end{pmatrix}.$$

The coefficient of x is the characteristic vector $\boldsymbol{v}_3$, and the constant term is new.

And when $c_1 = 0$, $c_4 = 0$, $c_5 = 0$, $c_7 = 0$, $c_{11} = 0$, and $c_{13} = 1$, we get

$$\boldsymbol{y}_6(x) = \left[\begin{pmatrix} 0 \\ 0 \\ 1 \\ 0 \\ 0 \\ 0 \end{pmatrix} + \begin{pmatrix} 0 \\ 1 \\ 0 \\ 0 \\ 0 \\ 0 \end{pmatrix} x + \begin{pmatrix} 1 \\ 0 \\ 0 \\ 0 \\ 0 \\ 0 \end{pmatrix} \frac{x^2}{2!} \right] e^{2x} = \begin{pmatrix} (x^2/2)e^{2x} \\ xe^{2x} \\ e^{2x} \\ 0 \\ 0 \\ 0 \end{pmatrix}.$$

The coefficient of x^2 is the characteristic vector $\boldsymbol{v}_1$. The constant coefficient in the chain is new.

These are the six linearly independent solutions that we sought. They can be used as the columns of a fundamental matrix, and every solution can be produced as a linear combination of them. Note that only the first three of the possible six forms the solution might contain actually occurred.

All six solutions are contained in the single expression

$$\boldsymbol{y}(x) = \left[\begin{pmatrix} c_1 \\ c_7 \\ c_{13} \\ c_4 \\ c_5 \\ c_{11} \end{pmatrix} + \begin{pmatrix} c_7 \\ c_{13} \\ 0 \\ 0 \\ c_{11} \\ 0 \end{pmatrix} x + \begin{pmatrix} c_{13} \\ 0 \\ 0 \\ 0 \\ 0 \\ 0 \end{pmatrix} \frac{x^2}{2!} \right] e^{2x},$$

which comes from equation 8.12. This representation is somewhat obscure, in that it is relatively hard to visualize any specific one of the six solutions listed above. In general, the entries in the vectors are very complicated linear expressions in the arbitrary constants; this example was especially simple since the coefficient matrix was in Jordan form. ◇

The solution to this example was produced with the help of *Mathematica*. Without it the calculations would have been exceedingly difficult. With it, the main problem is keeping in mind what steps are required to produce the solution(s). In case the matrix had had other roots that were deficient, then steps similar to the ones just executed would have to be executed for each such deficient root. The collection of all n vector solutions would then constitute the complete solution of the differential system, and the matrix with them as columns would be a fundamental matrix.

Complex Roots

When a real polynomial has a complex root, the complex conjugate of the root is also a root. Since we are studying real differential equations, the same is true here. In the present context, not only is the conjugate of the root also a root, but the conjugate of the corresponding characteristic vector is the characteristic vector corresponding to the conjugate root. Suppose that r is a complex characteristic root of the real matrix $\boldsymbol{A}$ with characteristic vector $\boldsymbol{v}$, then

$$\overline{(\boldsymbol{A} - r\boldsymbol{I})\boldsymbol{v}} = (\boldsymbol{A} - \overline{r}\boldsymbol{I})\overline{\boldsymbol{v}},$$

so $\overline{\boldsymbol{v}}$ is a characteristic vector corresponding to the conjugate root $\overline{r}$.

The characteristic vector corresponding to a complex root is complex, so its conjugate is a different vector. This is important to know. Also, as we saw before in our earlier studies of linear operators, if a complex object is a solution of a real homogeneous linear equation, then both the real and imaginary parts of this object are (real) solutions of the homogeneous linear equation. We will use these ideas to obtain real solutions that correspond to complex characteristic roots.

The complex conjugate characteristic roots are found along with the real characteristic roots when the characteristic equation is solved. Given a pair $(r, \boldsymbol{v})$ and $(\overline{r}, \overline{\boldsymbol{v}})$, of complex solutions to $(\boldsymbol{A} - r\boldsymbol{I})\boldsymbol{v} = \boldsymbol{0}$, we find two complex solutions of the differential system $y' = \boldsymbol{A}y$ to be $\boldsymbol{y}_{c_1}(x) = \boldsymbol{v}e^{rx}$, and $\boldsymbol{y}_{c_2}(x) = \overline{\boldsymbol{v}}e^{\overline{r}x}$. Write $r = r_1 = a + bi$ and $\boldsymbol{v} = \boldsymbol{K}_1 + \boldsymbol{K}_2 i$, where $\boldsymbol{K}_1$ and $\boldsymbol{K}_2$ are real constant vectors. Using these two complex solutions we get the two linearly independent real solutions $\boldsymbol{y}_1(x)$ and $\boldsymbol{y}_2(x)$ from this expansion:

$$\begin{aligned}
\boldsymbol{y}_{c_1}(x) &= \boldsymbol{v}e^{rx} \\
&= (\boldsymbol{K}_1 + \boldsymbol{K}_2 i)e^{(a+bi)x} \\
&= (\boldsymbol{K}_1 + \boldsymbol{K}_2 i)e^{ax}(\cos(bx) + i\sin(bx)) \\
&= e^{ax}(\boldsymbol{K}_1\cos(bx) - \boldsymbol{K}_2\sin(bx)) + ie^{ax}(\boldsymbol{K}_1\sin(bx) + \boldsymbol{K}_2\cos(bx)) \\
&= \boldsymbol{y}_1(x) + i\boldsymbol{y}_2(x),
\end{aligned}$$

where

$$\boldsymbol{y}_1(x) = e^{ax}(\boldsymbol{K}_1\cos(bx) - \boldsymbol{K}_2\sin(bx)) = e^{ax}(\mathrm{Re}(\boldsymbol{v})\cos(bx) - \mathrm{Im}(\boldsymbol{v})\sin(bx))$$

and

$$\boldsymbol{y}_2(x) = e^{ax}(\boldsymbol{K}_1\sin(bx) + \boldsymbol{K}_2\cos(bx)) = e^{ax}(\mathrm{Re}(\boldsymbol{v})\sin(bx) + \mathrm{Im}(\boldsymbol{v})\cos(bx)).$$

We would have obtained the same result had we used $\boldsymbol{y}_{c_2}(x)$.

Example 8.14 Find a real fundamental solution for the homogeneous second-order differential system

$$\boldsymbol{y}'(x) = \begin{pmatrix} a & b \\ -b & a \end{pmatrix} \boldsymbol{y}(x).$$

Solution. The characteristic equation is

$$\det\begin{pmatrix} a-r & b \\ -b & a-r \end{pmatrix} = (r-a)^2 + b^2 = 0,$$

which has as solutions $r_{1,2} = a \pm bi$. The characteristic vector $\boldsymbol{v}$ corresponding to $r_1 = a + bi$ is found from

$$(\boldsymbol{A} - r_1\boldsymbol{I})\boldsymbol{v} = \begin{pmatrix} a-r_1 & b \\ -b & a-r_1 \end{pmatrix}\boldsymbol{v} = \begin{pmatrix} -bi & b \\ -b & -bi \end{pmatrix}\boldsymbol{v} = \boldsymbol{0},$$

which has as basis for its kernel $\boldsymbol{v} = \begin{pmatrix} 1 \\ i \end{pmatrix}$. Corresponding to the second root $r_2 = a - bi$ is the vector $\bar{\boldsymbol{v}} = \begin{pmatrix} 1 \\ -i \end{pmatrix}$. Since

$$\boldsymbol{v} = \begin{pmatrix} 1 \\ 0 \end{pmatrix} + i\begin{pmatrix} 0 \\ 1 \end{pmatrix},$$

we have that

$$\boldsymbol{K}_1 = \operatorname{Re}(\boldsymbol{v}) = \begin{pmatrix} 1 \\ 0 \end{pmatrix}$$

and

$$\boldsymbol{K}_2 = \operatorname{Im}(\boldsymbol{v}) = \begin{pmatrix} 0 \\ 1 \end{pmatrix}.$$

The two complex solutions are therefore

$$\boldsymbol{y}_{c_1}(x) = \begin{pmatrix} 1 \\ i \end{pmatrix} e^{(a+bi)x}$$

and

$$\boldsymbol{y}_{c_2}(x) = \begin{pmatrix} 1 \\ -i \end{pmatrix} e^{(a-bi)x}$$

To these correspond the real solutions

$$\begin{aligned} \boldsymbol{y}_1(x) &= e^{ax}(\boldsymbol{K}_1 \cos(bx) - \boldsymbol{K}_2 \sin(bx)) \\ &= e^{ax}\left(\begin{pmatrix} 1 \\ 0 \end{pmatrix} \cos(bx) - \begin{pmatrix} 0 \\ 1 \end{pmatrix} \sin(bx)\right) \\ &= e^{ax}\begin{pmatrix} \cos(bx) \\ -\sin(bx) \end{pmatrix}, \end{aligned}$$

and

$$\begin{aligned} \boldsymbol{y}_2(x) &= e^{ax}(\boldsymbol{K}_1 \sin(bx) + \boldsymbol{K}_2 \cos(bx)) \\ &= e^{ax}\left(\begin{pmatrix} 1 \\ 0 \end{pmatrix} \sin(bx) + \begin{pmatrix} 0 \\ 1 \end{pmatrix} \cos(bx)\right) \\ &= e^{ax}\begin{pmatrix} \sin(bx) \\ \cos(bx) \end{pmatrix}. \end{aligned}$$

A real fundamental matrix is therefore

$$\boldsymbol{Y}(x) = e^{ax}\begin{pmatrix} \cos(bx) & \sin(bx) \\ -\sin(bx) & \cos(bx) \end{pmatrix}.$$

Note that if $a = 0$, then this is just the solution we found in Example 8.4. Observe also that if $b = 0$, then the solution is just

$$\boldsymbol{Y}(x) = e^{ax}\begin{pmatrix} 1 & 0 \\ 0 & 1 \end{pmatrix} = \begin{pmatrix} e^{ax} & 0 \\ 0 & e^{ax} \end{pmatrix},$$

as was to be expected when the coefficient matrix is diagonal. ◇

The matrix $\begin{pmatrix} a & b \\ -b & a \end{pmatrix}$ plays a fundamental role in the theory of complex characteristic roots of real matrices. You are encouraged to explore further.

Exercises 8.4

Part I. Each of these systems has a diagonalizable coefficient matrix. In each case find a fundamental solution. Let *Mathematica* do the work for you. The methods of Section 8.3 still work.

1. $\boldsymbol{Y}'(x) = \begin{pmatrix} 7 & 2 & 1 \\ 0 & 2 & 0 \\ -30 & -12 & -4 \end{pmatrix} \boldsymbol{Y}(x).$

2. $\boldsymbol{Y}'(x) = \begin{pmatrix} 32 & 12 & 5 \\ -60 & -22 & -10 \\ -30 & -12 & -3 \end{pmatrix} \boldsymbol{Y}(x).$

3. $\boldsymbol{Y}'(x) = \begin{pmatrix} -2 & 15 & -10 & 5 \\ 0 & 8 & -10 & 0 \\ 0 & 5 & -7 & 0 \\ 0 & -5 & 10 & 3 \end{pmatrix} \boldsymbol{Y}(x).$

4. $\boldsymbol{Y}'(x) = \begin{pmatrix} -2 & 15 & -10 & 5 \\ 0 & 8 & -10 & 0 \\ 0 & 5 & -7 & 0 \\ 0 & -5 & 10 & 3 \end{pmatrix} \boldsymbol{Y}(x).$

Part II. None of these systems has a coefficient matrix that is diagonalizable. Nevertheless, in each case find a fundamental solution. Let *Mathematica* do the work for you. Model your investigations after Example 8.13 and the discussion that precedes it.

5. $\boldsymbol{Y}'(x) = \begin{pmatrix} 3 & 0 & 0 \\ 3 & 4 & 1 \\ -11 & -4 & 0 \end{pmatrix} \boldsymbol{Y}(x).$

6. $\boldsymbol{Y}'(x) = \begin{pmatrix} -3 & 0 & 0 \\ 13 & 3 & 1 \\ -6 & -4 & -1 \end{pmatrix} \boldsymbol{Y}(x).$

7. $\boldsymbol{Y}'(x) = \begin{pmatrix} -2 & 14 & -12 & 3 \\ 0 & 6 & -10 & -2 \\ 0 & 4 & -7 & -1 \\ 0 & -3 & 10 & 5 \end{pmatrix} \boldsymbol{Y}(x).$

8. $\boldsymbol{Y}'(x) = \begin{pmatrix} -2 & 16 & -2 & 15 \\ 0 & 8 & 0 & 10 \\ 0 & 5 & -2 & 5 \\ 0 & -5 & 0 & -7 \end{pmatrix} \boldsymbol{Y}(x).$

Part III. Each of these systems has at least one pair of complex characteristic roots, and consequently a corresponding pair of complex characteristic vectors. Find a fundamental matrix solution expressed in real (not complex) terms.

9. $$\boldsymbol{Y}'(x) = \begin{pmatrix} -3 & 0 & 0 \\ 13 & 3 & 1 \\ -8 & -5 & -1 \end{pmatrix} \boldsymbol{Y}(x).$$

10. $$\boldsymbol{Y}'(x) = \begin{pmatrix} 6 & 2 & 0 \\ -10 & -2 & 0 \\ -15 & -8 & -1 \end{pmatrix} \boldsymbol{Y}(x).$$

11. $$\boldsymbol{Y}'(x) = \begin{pmatrix} -1 & 14 & -2 & 14 \\ 1 & 6 & 0 & 9 \\ 1 & 3 & -2 & 4 \\ -1 & -3 & 0 & -6 \end{pmatrix} \boldsymbol{Y}(x).$$

12. $$\boldsymbol{Y}'(x) = \begin{pmatrix} -1 & 10 & -18 & -6 \\ 1 & 2 & -14 & -9 \\ 1 & 1 & -9 & -5 \\ -1 & 1 & 12 & 10 \end{pmatrix} \boldsymbol{Y}(x).$$

PART IV. Theory.

13. Show that an $n \times n$ matrix $\boldsymbol{A}$ is diagonalizable provided that $\boldsymbol{A}$ has n linearly independent characteristic vectors.

14. Use *Mathematica* to multiply out the matrices that occur in Example 8.12 and then check that the resulting matrix is a solution of the original problem, $\boldsymbol{Y}' = \boldsymbol{A}\boldsymbol{Y}$.

15. Show that if $\boldsymbol{N}$ is nilpotent, then $\exp(\boldsymbol{N}x)$ is a polynomial.

16. Verify that the matrix function

$$\boldsymbol{Y}(x) = \boldsymbol{P}\boldsymbol{Z}(x) = \boldsymbol{P}\exp(\boldsymbol{D}x)\exp(\boldsymbol{N}x)$$

is a solution of the matrix system $\boldsymbol{Y}' = \boldsymbol{A}\boldsymbol{Y}$. You will need to be aware that $\boldsymbol{N}$ and $\boldsymbol{D}$ not only commute with one another, but each commutes with $\exp(\boldsymbol{D}x)$ and $\exp(\boldsymbol{N}x)$.

17. Show that if $\boldsymbol{D} = \operatorname{diag}(d_1, \ldots, d_n)$ is diagonal, then

$$\exp(\boldsymbol{D}x) = \operatorname{diag}(e^{d_1 x}, \ldots, e^{d_n x}).$$

18. If $\boldsymbol{B} = \boldsymbol{P}^{-1}\boldsymbol{A}\boldsymbol{P}$, show that $\boldsymbol{B}^k = \boldsymbol{P}^{-1}\boldsymbol{A}^k\boldsymbol{P}$. Hence show that

$$\exp(\boldsymbol{B}x) = \boldsymbol{P}^{-1}\exp(\boldsymbol{A}x)\boldsymbol{P}.$$

19. Let

$$\boldsymbol{A} = \boldsymbol{D} + \boldsymbol{N} = \begin{pmatrix} 2 & 0 & 0 \\ 0 & 2 & 0 \\ 0 & 0 & 1 \end{pmatrix} + \begin{pmatrix} 0 & 1 & 0 \\ 0 & 0 & 0 \\ 0 & 0 & 0 \end{pmatrix}.$$

Show that $\boldsymbol{DN} = \boldsymbol{ND}$ and that $\boldsymbol{N}$ is nilpotent. Use these two properties to write out

$$\exp(\boldsymbol{A}x) = \exp(\boldsymbol{D}x)\exp(\boldsymbol{N}x).$$

Verify that both $\boldsymbol{D}$ and $\boldsymbol{N}$ commute with $\exp(\boldsymbol{A}x)$.

8.5 Nonhomogeneous Equations and Boundary-Value Problems

We saw in Section 8.2 how the method of variation of parameters works. Now that we can obtain fundamental solutions to constant coefficients differential systems we can apply the method of variation of parameters to real problems. It is still true that the method of undetermined coefficients remains available to us. There are times when undetermined coefficients can get a solution more easily than variation of parameters because of the number and complexity of the integrals involved, but when the systems are of large order, the number of unknowns can be very large. In cases such as this the *Mathematica* `Solve` function can be quite slow. It is worth knowing that there are alternative methods available.

Nonhomogeneous Systems

We will reconsider nonhomogeneous versions of some of the homogeneous systems that we have solved in this chapter. We will primarily illustrate the use of variation of parameters which, for the vector differential system $\boldsymbol{y}' = \boldsymbol{A}\boldsymbol{y} + \boldsymbol{q}(x)$ having fundamental matrix $\boldsymbol{W}(x)$, says that a particular solution is

$$\boldsymbol{y}_p(x) = \boldsymbol{W}(x)\int \boldsymbol{W}(x)^{-1}\boldsymbol{q}(x)\,dx.$$

So a complete solution is

$$\boldsymbol{y}(x) = \boldsymbol{W}(x)\boldsymbol{K} + \boldsymbol{W}(x)\int \boldsymbol{W}(x)^{-1}\boldsymbol{q}(x)\,dx.$$

If we are solving the initial-value problem $\boldsymbol{y}' = \boldsymbol{A}\boldsymbol{y} + \boldsymbol{q}(x), \boldsymbol{y}(x_0) = \boldsymbol{\alpha}$, then the definition

$$\boldsymbol{y}_p(x) = \boldsymbol{W}(x)\int_{x_0}^{x} \boldsymbol{W}(t)^{-1}\boldsymbol{q}(t)\,dt$$

makes $\boldsymbol{y}_p(x_0) = \boldsymbol{0}$. The solution to the initial-value problem in closed form is therefore

$$\begin{aligned}\boldsymbol{y}(x) &= \boldsymbol{W}(x)(\boldsymbol{W}(x_0))^{-1}\boldsymbol{\alpha} + \boldsymbol{W}(x)\int_{x_0}^{x} (\boldsymbol{W}(t))^{-1}\boldsymbol{q}(t)\,dt \\ &= \exp(\boldsymbol{A}x)\boldsymbol{\alpha} + \boldsymbol{W}(x)\int_{x_0}^{x} (\boldsymbol{W}(t))^{-1}\boldsymbol{q}(t)\,dt.\end{aligned}$$

Example 8.15 Find a complete solution of the nonhomogeneous vector differential system

$$\boldsymbol{y}'(x) = \begin{pmatrix} 0 & 1 & 0 \\ 0 & 0 & 1 \\ -3 & 11/2 & 3/2 \end{pmatrix} \boldsymbol{y}(x) + \begin{pmatrix} x-1 \\ 2 \\ e^x \end{pmatrix}.$$

Solution. We found in Example 8.5 that a fundamental matrix for the homogeneous system is the matrix function

$$\boldsymbol{W}(x) = \begin{pmatrix} e^{x/2} & e^{-2x} & e^{3x} \\ e^{x/2}/2 & -2e^{-2x} & 3e^{3x} \\ e^{x/2}/4 & 4e^{-2x} & 9e^{3x} \end{pmatrix},$$

which has an inverse

$$(\boldsymbol{W}(x))^{-1} = \begin{pmatrix} 24/(25e^{x/2}) & 4/(25e^{x/2}) & -4/(25e^{x/2}) \\ 3e^{2x}/25 & -7e^{2x}/25 & 2e^{2x}/25 \\ -2/(25e^{3x}) & 3/(25e^{3x}) & 2/(25e^{3x}) \end{pmatrix}.$$

The vector $\boldsymbol{v}(x)$ we need is

$$\begin{aligned} \boldsymbol{v}(x) &= \int (\boldsymbol{W}(x))^{-1}\boldsymbol{q}(x)\,dx \\ &= \int \begin{pmatrix} 24/(25e^{x/2}) & 4/(25e^{x/2}) & -4/(25e^{x/2}) \\ 3e^{2x}/25 & -7e^{2x}/25 & 2e^{2x}/25 \\ -2/(25e^{3x}) & 3/(25e^{3x}) & 2/(25e^{3x}) \end{pmatrix} \begin{pmatrix} x-1 \\ 2 \\ e^x \end{pmatrix} dx \\ &= \int \begin{pmatrix} -4(4+e^x-6x)/(25e^{x/2}) \\ e^{2x}(-17+2e^x+3x)/25 \\ 2(4+e^x-x)/(25e^{3x}) \end{pmatrix} dx \\ &= \begin{pmatrix} -8(8+e^x+6x)/(25e^{x/2}) \\ e^{2x}(-111+8e^x+18x)/300 \\ -22-9e^x+6x/(225e^{3x}) \end{pmatrix}. \end{aligned}$$

Thus

$$\begin{aligned} \boldsymbol{y}_p(x) &= \begin{pmatrix} e^{x/2} & e^{-2x} & e^{3x} \\ e^{x/2}/2 & -2e^{-2x} & 3e^{3x} \\ e^{x/2}/4 & 4e^{-2x} & 9e^{3x} \end{pmatrix} \begin{pmatrix} -8(8+e^x+6x)/(25e^{x/2}) \\ e^{2x}(-111+8e^x+18x)/300 \\ (-22-9e^x+6x)/(225e^{3x}) \end{pmatrix} \\ &= \begin{pmatrix} (-109-12e^x-66x)/36 \\ -(5/6)-(e^x/3)-x \\ -3-(e^x/3) \end{pmatrix}. \\ &= \begin{pmatrix} -109/36 \\ -5/6 \\ -3 \end{pmatrix} - \begin{pmatrix} 1/3 \\ 1/3 \\ 1/3 \end{pmatrix} e^x - \begin{pmatrix} 11/6 \\ 1 \\ 0 \end{pmatrix} x. \end{aligned}$$

Observe that we have three kinds of vector objects here: a constant, a multiple of e^x, and a multiple of x. These are precisely the three kinds of objects that we would have expected to obtain from undetermined coefficients. This suggests the form to assume to solve the problem by undetermined coefficients: three terms, a constant vector, a vector multiple of e^x, and a vector multiple of x. The three unknown vectors are to be determined. This means that nine equations in the nine unknown coefficients are required. ◇

Here is the solution of Example 8.15 by *Mathematica*.

Example 8.15M Given the fundamental matrix $\boldsymbol{W}(x)$ defined above, find a particular solution of the nonhomogeneous vector differential system

$$\boldsymbol{y}'(x) = \begin{pmatrix} 0 & 1 & 0 \\ 0 & 0 & 1 \\ -3 & 11/2 & 3/2 \end{pmatrix} \boldsymbol{y}(x) + \begin{pmatrix} x-1 \\ 2 \\ e^x \end{pmatrix}.$$

Solution.

```
In[42]:= W[x_] = {{E^(x/2),E^(-2x),E^(3x)},
                  {E^(x/2)/2,-2E^(-2x),3E^(3x)},
                  {E^(x/2)/4,4E^(-2x),9E^(3x)}}

Out[42]=
                                   x/2
          x/2    -2 x    3 x      E        -2         3 x
       {{E   , E    , E   }, {----, ----, 3 E   },
                                  2       2 x
                                         E

           x/2
          E        4       3 x
        {----, ----, 9 E   }}
           4      2 x
                 E

In[43]:= q[x_] = {x-1,2,E^x}

Out[43]=                  x
         {-1 + x, 2, E }

In[44]:= y[x_] = Simplify[W[x].Integrate[Simplify[
                 Inverse[W[x]].q[x]],x]]

Out[44]=
                    x                        x                x
         -109 - 12 E  - 66 x       5     E                   E
        {-------------------, -(-) - -- - x, -3 - --}
                36                 6     3                   3
```

We should check our solution.

```
In[45]:= Simplify[y'[x]-A.y[x] == q[x], Trig->False]

Out[45]= True
```

◇

Except for the formalities of defining `W[x_]` and `q[x_]`, and checking the result, the particular solution was obtained in one step at *In[3]*. You need not really be concerned about the work that *Mathematica* expended in that one statement unless an exceedingly difficult integra was encountered. In which case the problem would possibly have been very difficult by hand. In that somewhat unlikely event you should consider an alternative method.

Example 8.16 Solve the problem of Example 8.15 by undetermined coefficients.

Solution. We do this with the help of *Mathematica*. Assume $\boldsymbol{y}_p(x) = \boldsymbol{K}_0 + \boldsymbol{K}_1 x + \boldsymbol{K}_2 e^x$, where $\boldsymbol{K}_0$, $\boldsymbol{K}_1$, and $\boldsymbol{K}_2$ are constant vectors. This form is suggested by the functions that appear in the nonhomogeneous term $\boldsymbol{q}(x) = \begin{pmatrix} x-1 \\ 2 \\ e^x \end{pmatrix}$.

Since $\boldsymbol{y}_p(x)$ is to be a solution of $\boldsymbol{y}' = \boldsymbol{A}\boldsymbol{y} + \boldsymbol{q}(x)$, we must have

$$\boldsymbol{K}_1 + \boldsymbol{K}_2 e^x = \boldsymbol{A}\boldsymbol{K}_0 + \boldsymbol{A}\boldsymbol{K}_1 x + \boldsymbol{A}\boldsymbol{K}_2 e^x + \begin{pmatrix} -1 \\ 2 \\ 0 \end{pmatrix} + \begin{pmatrix} 1 \\ 0 \\ 0 \end{pmatrix} x + \begin{pmatrix} 0 \\ 0 \\ 1 \end{pmatrix} e^x.$$

Equate coefficients to obtain the equations that we must solve.

$$\boldsymbol{K}_1 = \boldsymbol{A}\boldsymbol{K}_0 + \begin{pmatrix} -1 \\ 2 \\ 0 \end{pmatrix}$$

$$\boldsymbol{K}_2 = \boldsymbol{A}\boldsymbol{K}_2 + \begin{pmatrix} 0 \\ 0 \\ 1 \end{pmatrix}$$

$$0 = \boldsymbol{A}\boldsymbol{K}_1 + \begin{pmatrix} 1 \\ 0 \\ 0 \end{pmatrix}.$$

Here we have

$$\boldsymbol{A} = \begin{pmatrix} 0 & 1 & 0 \\ 0 & 0 & 1 \\ -3 & 11/2 & 3/2 \end{pmatrix}, \quad \boldsymbol{K}_0 = \begin{pmatrix} k_1 \\ k_2 \\ k_3 \end{pmatrix},$$

$$\boldsymbol{K}_1 = \begin{pmatrix} k_4 \\ k_5 \\ k_6 \end{pmatrix}, \quad \boldsymbol{K}_2 = \begin{pmatrix} k_7 \\ k_8 \\ k_9 \end{pmatrix}.$$

In terms of the k_i, the equations are expressed in *Mathematica* as

```
k4 == -1 + k2
k5 == 2 + k3
                11 k2   3 k3
k6 == -3 k1 + ----- + ----
                  2      2
k7 == k8
k8 == k9
                   11 k8   3 k9
k9 == 1 - 3 k7 + ----- + ----
                     2      2
0 == 1 + k5
0 == k6
               11 k5   3 k6
0 == -3 k4 + ----- + ---- .
                 2      2
```

The solution process continues in *Mathematica*. Note the conjunction of the equations (&&).

```
In[46]:= Solve[k4 == -1+k2 && k5 == 2+k3 &&
         k6 == -3*k1+(11*k2)/2+(3*k3)/2 &&
         k7 == k8 && k8 == k9 &&
         k9 == 1-3*k7+(11*k8)/2+(3*k9)/2 && 0 == 1+k5 &&
         0 == k6 && 0 == -3*k4+(11*k5)/2+(3*k6)/2]

Out[46]=          5             11                                    109
         {{k2 ->-(-), k4 ->-(--), k3 ->-3, k5 ->-1, k1 ->-(---),
                  6             6                                     36

                      1            1            1
           k6 -> 0, k7 -> -(-), k8 -> -(-), k9 -> -(-)}}
                      3            3            3
```

Now capture the coefficient vectors.

```
In[47]:= {K0,K1,K2} = {K0,K1,K2}/.%[[1]]

Out[47]=    109      5             11              1      1      1
         {{-(---), -(-), -3}, {-(--), -1, 0}, {-(-), -(-), -(-)}}
             36      6             6               3      3      3
```

Combine the terms into the desired particular solution.

```
In[48]:= {1,x,E^x}.{K0,K1,K2}

Out[48]=              x                  x              x
           109      E     11 x     5    E               E
         {-(---) - -- - ----, -(-) - -- - x, -3 - --}
            36      3      6       6    3               3
```

◇

This agrees with the solution that we had before. Note that undetermined coefficients are somewhat awkward even in this low-order setting. The awkwardness increases as the number of different kinds of objects in $\boldsymbol{q}(x)$ increases. There soon arise very many equations to solve. Also it is not easy to select and equate coefficients automatically, as you will see if you try it. Undetermined coefficients does save having to compute some very complicated integrals, but you have to know the form the solution will have, and the situation can become complicated, as it did in Chapter 4, when objects that appear on the right-hand side are in the kernel of the differential operator.

Boundary-Value Problems

Now that we can completely solve initial value problems for homogeneous and nonhomogeneous linear differential systems, it is reasonable to consider **boundary-value problems**. We will study these problems only for matrix systems. Such a problem can be represented for matrix systems as

$$\begin{cases} \boldsymbol{Y}' = \boldsymbol{A}\boldsymbol{Y} + \boldsymbol{Q}(x) \\ \boldsymbol{H}(\boldsymbol{Y}) = \boldsymbol{C}, \end{cases}$$

where the **boundary condition operator** $\boldsymbol{H}$ has the **homogeneity property**

$$\boldsymbol{H}(\boldsymbol{U}(x)\boldsymbol{K}) = \boldsymbol{H}(\boldsymbol{U}(x))\boldsymbol{K}$$

for all $n \times n$ constant matrices $\boldsymbol{K}$. Under this definition, the initial-value problem

$$\begin{cases} \boldsymbol{Y}' = \boldsymbol{A}\boldsymbol{Y} + \boldsymbol{Q}(x) \\ \boldsymbol{Y}(x_0) = \boldsymbol{C}, \end{cases}$$

is a boundary-value problem with $\boldsymbol{H}(\boldsymbol{Y}(x)) = \boldsymbol{Y}(x_0)$. The boundary condition operator $\boldsymbol{H}$ always performs some sort of evaluation on the matrix function it receives as its argument. This means that the domain of $\boldsymbol{H}$ is the $n \times n$ continuously differentiable matrix-valued functions, and the range is the $n \times n$ (constant) matrices. The value of $\boldsymbol{H}$ at any matrix function is a constant matrix.

Examples of typical kinds of boundary condition operators $\boldsymbol{H}$ are:

- $\boldsymbol{H}(\boldsymbol{Y}(x)) = \boldsymbol{Y}(x_0)$;
- $\boldsymbol{H}(\boldsymbol{Y}(x)) = \boldsymbol{B}\boldsymbol{Y}(x_0)$, $\boldsymbol{B}$ an $n \times n$ (constant) matrix;
- $\boldsymbol{H}(\boldsymbol{Y}(x)) = \boldsymbol{B}_1\boldsymbol{Y}(x_0) + \boldsymbol{B}_2\boldsymbol{Y}(x_1)$, $\boldsymbol{B}_1$ and $\boldsymbol{B}_2$ are $n \times n$ (constant) matrices;
- $\boldsymbol{H}(\boldsymbol{Y}(x)) = \int_a^b \boldsymbol{Y}(x)\,dx$;
- $\boldsymbol{H}(\boldsymbol{Y}(x)) = \int_a^b \boldsymbol{F}(x)\boldsymbol{Y}(x)\,dx$, $\boldsymbol{F}(x)$ is $n \times n$ and continuous on $[a, b]$;
- $\boldsymbol{H}(\boldsymbol{Y}(x)) = \boldsymbol{B}_1\boldsymbol{Y}(x_0) + \boldsymbol{B}_2\boldsymbol{Y}(x_1) + \int_a^b \boldsymbol{F}(x)\boldsymbol{Y}(x)\,dx$; $\boldsymbol{B}_1$, $\boldsymbol{B}_2$, and $\boldsymbol{F}(x)$ as before.
- $\boldsymbol{H}(\boldsymbol{Y}(x)) = \sum_{i=1}^m \boldsymbol{B}_i\boldsymbol{Y}(x_i)$.

There are others, but by now you get the idea. You should verify that each of these forms for the boundary condition operator $\boldsymbol{H}$ has both the ordinary linearity property and the homogeneity property for constant right-multiplier matrices.

We will not study boundary value problems exhaustively, but will rely on this theorem for our activities.

Theorem 8.6 *The nonhomogeneous boundary-value problem*

$$\begin{cases} \boldsymbol{Y}' = \boldsymbol{A}\boldsymbol{Y} + \boldsymbol{Q}(x) \\ \boldsymbol{H}(\boldsymbol{Y}) - \boldsymbol{C} \end{cases}$$

has a unique solution if and only if the homogeneous boundary-value problem

$$\begin{cases} \boldsymbol{Y}' = \boldsymbol{A}\boldsymbol{Y} \\ \boldsymbol{H}(\boldsymbol{Y}) = \boldsymbol{0} \end{cases}$$

has the zero matrix as its unique solution. These both happen if and only if $\boldsymbol{H}(\boldsymbol{W})$ *is invertible whenever* $\boldsymbol{W}(x)$ *is a fundamental solution of* $\boldsymbol{Y}' = \boldsymbol{A}\boldsymbol{Y}$.

Proof. As was shown above, the nonhomogeneous system

$$\boldsymbol{Y}' = \boldsymbol{A}\boldsymbol{Y} + \boldsymbol{Q}(x)$$

has as its complete set of solutions

$$\boldsymbol{Y}(x) = \boldsymbol{W}(x)\boldsymbol{K} + \boldsymbol{W}(x)\int_a^x (\boldsymbol{W}(t))^{-1}\boldsymbol{Q}(t)\,dt = \boldsymbol{W}(x)\boldsymbol{K} + \boldsymbol{Y}_p(x).$$

We must find out when a unique value for $\boldsymbol{K}$ can be found so that $\boldsymbol{H}(\boldsymbol{Y}) = \boldsymbol{C}$. Make the calculation

$$\begin{aligned} \boldsymbol{H}(\boldsymbol{Y}) &= \boldsymbol{H}(\boldsymbol{W}(x)\boldsymbol{K} + \boldsymbol{Y}p(x)) \\ &= \boldsymbol{H}(\boldsymbol{W})\boldsymbol{K} + \boldsymbol{H}(\boldsymbol{Y}_p). \end{aligned}$$

This has the value $\boldsymbol{C}$ provided that

$$\boldsymbol{H}(\boldsymbol{W})\boldsymbol{K} + \boldsymbol{H}(\boldsymbol{Y}_p) = \boldsymbol{C},$$

which is equivalent to

$$\boldsymbol{H}(\boldsymbol{W})\boldsymbol{K} = \boldsymbol{C} - \boldsymbol{H}(\boldsymbol{Y}_p).$$

Now, since $\boldsymbol{C}$ is completely arbitrary, so is $\boldsymbol{C} - \boldsymbol{H}(\boldsymbol{Y}_p)$. From linear algebra, we learn that

$$\boldsymbol{A}x = \boldsymbol{b}$$

has a solution x for every choice of $\boldsymbol{b}$ if and only if $\boldsymbol{A}$ is invertible. It follows that

$$\boldsymbol{H}(\boldsymbol{W})\boldsymbol{K} = \boldsymbol{C} - \boldsymbol{H}(\boldsymbol{Y}_p)$$

has a solution for every choice of $\boldsymbol{C}$ if and only if $\boldsymbol{H}(\boldsymbol{W})$ is invertible. The solution $\boldsymbol{Y}(x) = \boldsymbol{W}(x)\boldsymbol{K}$ of $\boldsymbol{Y}' = \boldsymbol{A}\boldsymbol{Y}$ must also satisfy $\boldsymbol{H}(\boldsymbol{W})\boldsymbol{K} = \boldsymbol{0}$, and $\boldsymbol{H}(\boldsymbol{W})\boldsymbol{K} = \boldsymbol{0}$ has only the zero solution $\boldsymbol{K} = \boldsymbol{0}$. This makes $\boldsymbol{Y}(x) = \boldsymbol{0}$ if and only if $\boldsymbol{H}(\boldsymbol{W})$ is invertible, which proves the theorem. □

That is all there is to the theory of boundary value problems as we will study them. Some interesting special cases of boundary condition operators will be discussed in the exercises. What remains is to learn how to solve the problems that we encounter. It will only take a few examples for us to understand clearly what to do. If you attempt to do any one of these problems without the aid of *Mathematica* you will see why they are rarely covered in an introductory text: the manipulations are gigantic. By using *Mathematica* we will never see these huge calculations, and we will proceed simply and directly to the answer and check it. The answers themselves may be huge, however.

Example 8.17 Does the boundary-value problem

$$\boldsymbol{Y}'(x) = \begin{pmatrix} 0 & 1 & 0 \\ 0 & 0 & 1 \\ -3 & 11/2 & 3/2 \end{pmatrix} \boldsymbol{Y}(x) + \begin{pmatrix} x-1 & x-1 & x-1 \\ 2 & 2 & 2 \\ e^x & e^x & e^x \end{pmatrix},$$

$$\boldsymbol{H}(\boldsymbol{Y}) = \boldsymbol{Y}(0) - \boldsymbol{Y}(1) = \begin{pmatrix} 1 & 0 & 0 \\ 0 & 1 & 0 \\ 0 & 0 & 1 \end{pmatrix}.$$

have a unique solution?

Solution. We begin by calculating $\boldsymbol{H}(\boldsymbol{W}(x)) = \boldsymbol{W}(0) - \boldsymbol{W}(1)$. We can solve the problem uniquely if $\boldsymbol{H}(\boldsymbol{W}(x))$ is invertible. `W[x_]` was defined in Example 8.15. We have these calculations

```
In[49]:= W[x_] = {{E^(x/2),E^(-2x),E^(3x)},
                  {E^(x/2)/2,-2E^(-2x),3E^(3x)},
                  {E^(x/2)/4,4E^(-2x),9E^(3x)}}

Out[49]=
                                   x/2
          x/2    -2 x    3 x      E         -2         3 x
       {{E    , E     , E   }, {----, ------, 3 E    },
                                   2       2 x
                                          E

                x/2
         E     4        3 x
        {--, ----, 9 E   }}
         4     2 x
              E

In[50]:= Det[W[0]-W[1]]                    (* H[W] = W[0]-W[1] *)

Out[50]=
                                                                  3/2
       125     125      125      125 Sqrt[E]    125 E     125 E
     -(---) + ---- - ------- + ----------- - ------- + -------- +
        4        2       3/2         4            4         4
              4 E    4 E

              3        7/2
        125 E     125 E
        ------ - --------
          4          4
```

This is nonzero, but that fact is obscure. Look at the numeric value, which is nonzero.

```
In[51]:= N[%]

Out[51]= -334.549
```

It follows that $\boldsymbol{H}(\boldsymbol{W})$ is invertible, so our problem has a unique solution. The calculation of the actual solution is done in the notebook *Boundary Value Examples*, but is far too long to include here. The solution is also checked. ◇

Example 8.18 Solve the boundary value problem

$$\boldsymbol{Y}' = \begin{pmatrix} 0 & 1 \\ -1 & 0 \end{pmatrix} \boldsymbol{Y} + \begin{pmatrix} x & 0 \\ 0 & 1 \end{pmatrix}$$

$$\boldsymbol{H}(\boldsymbol{Y}) = \boldsymbol{Y}(0) + 2\boldsymbol{Y}(\pi) = \begin{pmatrix} 0 & 1 \\ 1 & -1 \end{pmatrix}.$$

Solution. Here we have $\boldsymbol{W}(x) = \begin{pmatrix} \cos x & \sin x \\ -\sin x & \cos x \end{pmatrix}$, so that

$$\begin{aligned} \boldsymbol{H}(\boldsymbol{W}) &= \boldsymbol{W}(0) + 2\boldsymbol{W}(\pi) \\ &= \begin{pmatrix} 1 & 0 \\ 0 & 1 \end{pmatrix} + 2\begin{pmatrix} -1 & 0 \\ 0 & -1 \end{pmatrix} \end{aligned}$$

$$= \begin{pmatrix} -1 & 0 \\ 0 & -1 \end{pmatrix},$$

which is invertible. So there is a unique solution. All we need to do is find it.

By variation of parameters or undetermined coefficients we find that

$$\boldsymbol{Y}_p(x) = \begin{pmatrix} 1 & 1 \\ -x & 0 \end{pmatrix}.$$

The system that the solution $\boldsymbol{Y}(x) = \boldsymbol{W}(x)\boldsymbol{K} + \boldsymbol{Y}_p(x)$ must satisfy is

$$\boldsymbol{H}(\boldsymbol{Y}) = \boldsymbol{H}(\boldsymbol{W}\boldsymbol{K} + \boldsymbol{Y}_p) = \boldsymbol{H}(\boldsymbol{W})\boldsymbol{K} + \boldsymbol{H}(\boldsymbol{Y}_p) = \begin{pmatrix} 0 & 1 \\ 1 & -1 \end{pmatrix}$$

which is equivalent to

$$\boldsymbol{H}(\boldsymbol{W})\boldsymbol{K} = \begin{pmatrix} 0 & 1 \\ 1 & -1 \end{pmatrix} - \boldsymbol{H}(\boldsymbol{Y}_p),$$

or

$$\begin{pmatrix} -1 & 0 \\ 0 & -1 \end{pmatrix} \boldsymbol{K} = \begin{pmatrix} 0 & 1 \\ 1 & -1 \end{pmatrix} - \begin{pmatrix} 1 & 1 \\ 0 & 0 \end{pmatrix} - 2\begin{pmatrix} 1 & 1 \\ -\pi & 0 \end{pmatrix}$$
$$= \begin{pmatrix} -3 & -2 \\ 1+2\pi & -1 \end{pmatrix}.$$

Thus

$$\boldsymbol{K} = -\begin{pmatrix} -3 & -2 \\ 1+2\pi & -1 \end{pmatrix} = \begin{pmatrix} 3 & 2 \\ -1-2\pi & 1 \end{pmatrix}.$$

This means that the solution is

$$\begin{aligned} \boldsymbol{Y}(x) &= \boldsymbol{W}(x)\boldsymbol{K} + \boldsymbol{Y}_p(x) \\ &= \begin{pmatrix} \cos x & \sin x \\ -\sin x & \cos x \end{pmatrix} \begin{pmatrix} 3 & 2 \\ -1-2\pi & 1 \end{pmatrix} + \begin{pmatrix} 1 & 1 \\ -x & 0 \end{pmatrix} \\ &= \begin{pmatrix} 1 + 3\cos x + (-1-2\pi)\sin x & 1 + 2\cos x + \sin x \\ -(x + \cos x + 2\pi\cos x + 3\sin x) & \cos x - 2\sin x \end{pmatrix}. \end{aligned}$$

That this is the desired solution can be verified without difficulty. ◇

Example 8.18M The entire solution process in *Mathematica* follows.

Solution. Define the system

```
In[52]:= A = {{0,1},{-1,0}}

Out[52]= {{0, 1}, {-1, 0}}

In[53]:= Q[x_] = {{x,0},{0,1}}

Out[53]= {{x, 0}, {0, 1}}

In[54]:= BvpRhs = {{0,1},{1,-1}}

Out[54]= {{0, 1}, {1, -1}}
```

Define the boundary value operator

```
In[55]:= H[U_] = U[0]+2U[Pi]

Out[55]= U[0] + 2 U[Pi]
```

Define solution parts, first the fundamental solution, and then a particular solution by variation of parameters.

```
In[56]:= W[x_] =
         Simplify[ComplexExpand[MatrixExp[A x]]]

Out[56]= {{Cos[x], Sin[x]}, {-Sin[x], Cos[x]}}

In[57]:= Yp[x_] = Simplify[W[x].Integrate[Inverse[W[x]].Q[x],x]]

Out[57]= {{1, 1}, {-x, 0}}
```

Is there a unique solution? Yes:

```
In[58]:= Det[H[W]]

Out[58]= 1
```

Find the constant matrix multiplier, K=KMat.

```
In[59]:= KMat = Table[k[i,j],{i,1,2},{j,1,2}]

Out[59]= {{k[1, 1], k[1, 2]}, {k[2, 1], k[2, 2]}}

In[60]:= gg
         sys = H[W].KMat == BvpRhs-H[Yp]

Out[60]= {{-k[1, 1], -k[1, 2]}, {-k[2, 1], -k[2, 2]}} ==

           {{-3, -2}, {1 + 2 Pi, -1}}
```

Solve for K=KMat. (One can actually compute this using Inverse[H[W]]).

```
In[61]:= Simplify[Solve[sys,Flatten[KMat]]]

Out[61]= {{k[1, 1] -> 3, k[1, 2] -> 2, k[2, 1] -> -1 - 2 Pi,

            k[2, 2] -> 1}}
```

Capture the solution.

```
In[62]:= SolnKMat = KMat/.%[[1]]

Out[62]= {{3, 2}, {-1 - 2 Pi, 1}}
```

Determine the unique solution.

```
In[63]:= Y[x_] = Simplify[W[x].SolnKMat+Yp[x]]

Out[63]= {{1 + 3 Cos[x] + (-1 - 2 Pi) Sin[x], 1 + 2 Cos[x] + Sin[x]},
          {-(x + Cos[x] + 2 Pi Cos[x] + 3 Sin[x]), Cos[x] - 2 Sin[x]}}
```

Check the solution in the differential equation and the boundary condition.

```
In[64]:= Simplify[Y'[x]-A.Y[x] == Q[x]  &&  H[Y] == BvpRhs]

Out[64]= True
```

◇

Once again, to solve a problem by *Mathematica* the steps we follow are essentially identical to those we follow to solve the problem manually. The differences are usually definitions of auxiliary quantities and functions. We were easily able to check the results. This is important when the problems get huge, as in Example 8.17.

Two-Point Boundary-Value Problems

Some types of boundary-value problems that are familiar to users of differential equations can be put into the context we are studying, but it is not immediately obvious how it should be done.

Example 8.19 Convert this two-point boundary-value problem

$$\begin{cases} y_1' = y_2 + x \\ y_2' = -y_1 \\ y_1(0) = 0 \\ y_2(\pi) = 0 \end{cases}$$

to a 2×2 system.

Solution. The first observation to make is that the problem is a vector problem:

$$\begin{pmatrix} y_1 \\ y_2 \end{pmatrix}' = \begin{pmatrix} 0 & 1 \\ -1 & 0 \end{pmatrix} \begin{pmatrix} y_1 \\ y_2 \end{pmatrix} + \begin{pmatrix} x \\ 0 \end{pmatrix}.$$

But how should the boundary conditions be written? The secret is to separately cut off the top and bottom of the vector y, evaluate the parts separately at 0 and at π, and "glue" the pieces back into a single vector. Here is how:

$$\begin{pmatrix} 1 & 0 \\ 0 & 0 \end{pmatrix} \begin{pmatrix} y_1 \\ y_2 \end{pmatrix} (0) + \begin{pmatrix} 0 & 0 \\ 0 & 1 \end{pmatrix} \begin{pmatrix} y_1 \\ y_2 \end{pmatrix} (\pi) = \begin{pmatrix} y_1(0) \\ y_2(\pi) \end{pmatrix}.$$

Now that the two pieces are separately evaluated, all that remains to be done to completely pose the problem in the context of this section is to state the problem twice: once as the first column of a matrix system and again, unchanged, as the second column. That's all there is to it. The resulting problem appears as:

$$\boldsymbol{Y}(x) = \begin{pmatrix} 0 & 1 \\ -1 & 0 \end{pmatrix} \boldsymbol{Y}(x) + \begin{pmatrix} x & x \\ 0 & 0 \end{pmatrix}$$

$$\boldsymbol{H}(\boldsymbol{Y}) = \begin{pmatrix} 1 & 0 \\ 0 & 0 \end{pmatrix} \boldsymbol{Y}(0) + \begin{pmatrix} 0 & 0 \\ 0 & 1 \end{pmatrix} \boldsymbol{Y}(\pi) = \begin{pmatrix} 0 & 0 \\ 1 & 1 \end{pmatrix}.$$

When you solve this system, you will note that the two columns of the solution are identical, since we merely have a single vector problem duplicated. The solution of this matrix system appears in the notebook *Boundary Value Examples.* In case you are interested, the solution turns out to be

$$\boldsymbol{Y}(x) = \begin{pmatrix} -(-1 + \cos x + \sin x + \pi \sin x) & -(-1 + \cos x + \sin x + \pi \sin x) \\ -x + (-1 - \pi) \cos x + \sin x & -x + (-1 - \pi) \cos x + \sin x \end{pmatrix}.$$

If you evaluate the appropriate portion of the cited notebook, you will also be able to check that this answer is correct. ◇

It is clear that this technique for evaluating the various component functions of a vector differential system at different points extends to higher dimensions: one just has to slice the vector into more pieces using matrices such as

$$\begin{pmatrix} 1 & 0 & 0 \\ 0 & 0 & 0 \\ 0 & 0 & 0 \end{pmatrix}, \quad \begin{pmatrix} 0 & 0 & 0 \\ 0 & 1 & 0 \\ 0 & 0 & 0 \end{pmatrix}, \quad \text{and} \quad \begin{pmatrix} 0 & 0 & 0 \\ 0 & 0 & 0 \\ 0 & 0 & 1 \end{pmatrix}$$

in three dimensions. These vectors select the first, second, and third components of the vector, respectively. For example,

$$\begin{pmatrix} 1 & 0 & 0 \\ 0 & 0 & 0 \\ 0 & 0 & 0 \end{pmatrix} \begin{pmatrix} y_1 \\ y_2 \\ y_3 \end{pmatrix} (x_1) + \begin{pmatrix} 0 & 0 & 0 \\ 0 & 1 & 0 \\ 0 & 0 & 0 \end{pmatrix} \begin{pmatrix} y_1 \\ y_2 \\ y_3 \end{pmatrix} (x_2)$$

$$+ \begin{pmatrix} 0 & 0 & 0 \\ 0 & 0 & 0 \\ 0 & 0 & 1 \end{pmatrix} \begin{pmatrix} y_1 \\ y_2 \\ y_3 \end{pmatrix} (x_3) = \begin{pmatrix} y_1(x_1) \\ y_2(x_2) \\ y_3(x_3) \end{pmatrix}$$

evaluates the three components of y at three possibly different points and combines the results. This permits the expression of a three-point boundary condition as a single operator.

Conclusions

With the help of *Mathematica*, some substantial problems involving systems of differential equations having constant coefficients can be solved with relative ease. For the most part the theory is simple and direct, even though the practice may be messy. But with the power of *Mathematica* available to you, hard problems should be within your ability to solve. There will still be systems that are outside the scope of even the power of *Mathematica*. But their number should decrease with each new release of the software.

In the next chapter, we turn our attention to systems of differential equations where the coefficient matrix is not constant.

Exercises 8.5

Part I. For each matrix that follows, verify that it is differentiable and invertible for all x, and then find a linear system for which it is a fundamental solution. Use the fact that if $\boldsymbol{V}(x)$ is differentiable and everywhere invertible, then $\boldsymbol{P}(x) = \boldsymbol{V}'(x)\boldsymbol{V}(x)^{-1}$ is such that $\boldsymbol{V}'(x) = \boldsymbol{P}(x)\boldsymbol{V}(x)$. Finally, find a fundamental solution such that $\boldsymbol{Y}(0) = \boldsymbol{I}$. You should use the function `Inverse` to find $\boldsymbol{P}(x)$ and to determine the appropriate right-multiplier.

1. $\boldsymbol{V}_1(x) = \begin{pmatrix} x/e^{2x} & e^{-2x} & e^{3x} \\ e^{-2x} - (2x/e^{2x}) & -2/e^{2x} & 3e^{3x} \\ (-4/e^{2x}) + (4x/e^{2x}) & 4/e^{2x} & 9e^{3x} \end{pmatrix}$. Find $\boldsymbol{P}_1(x)$.

2. $\boldsymbol{V}_2(x) = \begin{pmatrix} -\sin(2x) & \cos(2x) & e^{2x} \\ -2\cos(2x) & -2\sin(2x) & 2e^{2x} \\ 4\sin(2x) & -4\cos(2x) & 4e^{2x} \end{pmatrix}$. Find $\boldsymbol{P}_2(x)$.

3. $\boldsymbol{V}_3(x) = \begin{pmatrix} e^{-x} & -\sin(2x) & \cos(2x) & e^{2x} \\ -e^{-x} & -2\cos(2x) & -2\sin(2x) & 2e^{2x} \\ e^{-x} & 4\sin(2x) & -4\cos(2x) & 4e^{2x} \\ -e^{-x} & 8\cos(2x) & 8\sin(2x) & 8e^{2x} \end{pmatrix}$. Find $\boldsymbol{P}_3(x)$.

4. $\boldsymbol{V}_4(x) = \begin{pmatrix} -e^{-x} & -2/e^{2x} & 3e^{3x} & 2e^{2x} \\ -e^{-x} & -8/e^{2x} & 27e^{3x} & 8e^{2x} \\ e^{-x} & e^{-2x} & e^{3x} & e^{2x} \\ e^{-x} & 4/e^{2x} & 9e^{3x} & 4e^{2x} \end{pmatrix}$. Find $\boldsymbol{P}_4(x)$.

Part II. Consider the nonhomogeneous form of each of the systems derived in Part I: $\boldsymbol{Y}_i'(x) = \boldsymbol{P}_i(x)\boldsymbol{Y}_i(x) + \boldsymbol{Q}_i(x)$ where $\boldsymbol{Q}_i(x)$ is given below. Use the fundamental solution you were given and the method of variation of parameters to find a particular solution and hence the complete solution of the nonhomogeneous system. For each problem, determine which one of the members of the complete solution has the property that $\boldsymbol{Y}_i(0) = \boldsymbol{0}$.

5. $\boldsymbol{Q}_1(x) = \begin{pmatrix} 1 & 0 & x \\ 0 & 0 & 1 \\ -x & 1 & 0 \end{pmatrix}$.

6. $\boldsymbol{Q}_2(x) = \begin{pmatrix} 1 & 0 & e^{x} \\ 0 & 0 & -1 \\ x & 0 & 1 \end{pmatrix}$.

7. $\boldsymbol{Q}_3(x) = \begin{pmatrix} 0 & 1 & 0 & 0 \\ 1 & 0 & x & 0 \\ x & 0 & 0 & 1 \\ 0 & 0 & 1 & 0 \end{pmatrix}$.

8. $\boldsymbol{Q}_4(x) = \begin{pmatrix} 0 & 1 & 0 & 0 \\ 1 & 0 & x & 0 \\ x & 0 & 0 & 1 \\ 0 & 0 & 1 & 0 \end{pmatrix}$.

PART III. Given the nonhomogeneous differential system

$$\boldsymbol{Y}'(x) = \begin{pmatrix} 0 & 1 \\ -1 & 0 \end{pmatrix} \boldsymbol{Y}(x) + \begin{pmatrix} 1 & x \\ 0 & 1 \end{pmatrix},$$

for each boundary value operator $\boldsymbol{H}_i$ below. Verify that each $\boldsymbol{H}_i$ in problems 9–15 is right homogeneous for constant matrices and by Theorem 8.6 the system

$$\begin{cases} \boldsymbol{Y}'(x) = \begin{pmatrix} 0 & 1 \\ -1 & 0 \end{pmatrix} \boldsymbol{Y}(x) + \begin{pmatrix} 1 & x \\ 0 & 1 \end{pmatrix}, \\ \boldsymbol{H}_i(\boldsymbol{Y}) = \begin{pmatrix} 1 & 2 \\ 0 & -1 \end{pmatrix} \end{cases}$$

has a unique solution. Find that solution.

9. $\boldsymbol{H}_1(\boldsymbol{Y}) = \boldsymbol{Y}(0)$.

10. $\boldsymbol{H}_2(\boldsymbol{Y}) = \boldsymbol{Y}(0) - \boldsymbol{Y}(p)$.

11. $\boldsymbol{H}_3(\boldsymbol{Y}) = \begin{pmatrix} 1 & 0 \\ 0 & 0 \end{pmatrix} \boldsymbol{Y}(0) + \begin{pmatrix} 0 & 0 \\ 0 & 1 \end{pmatrix} \boldsymbol{Y}(\pi)$.

12. $\boldsymbol{H}_4(\boldsymbol{Y}) = \begin{pmatrix} 1 & 0 \\ 0 & 0 \end{pmatrix} \boldsymbol{Y}(0) + 2\boldsymbol{Y}\left(\frac{\pi}{2}\right) + \begin{pmatrix} 0 & 0 \\ 0 & 1 \end{pmatrix} \boldsymbol{Y}(\pi)$.

13. $\boldsymbol{H}_5(\boldsymbol{Y}) = \boldsymbol{Y}(0) + \int_0^p \boldsymbol{Y}(t)\, dt$.

14. $\boldsymbol{H}_6(\boldsymbol{Y}) = \boldsymbol{Y}(0) + \int_0^\pi \boldsymbol{Y}(t) dt + 2\int_\pi^{2\pi} \boldsymbol{Y}(t)\, dt$.

15. $\boldsymbol{H}_7(\boldsymbol{Y}) = \begin{pmatrix} 1 & 0 \\ 0 & 0 \end{pmatrix} \boldsymbol{Y}(0) + 2\boldsymbol{Y}\left(\frac{\pi}{2}\right) + \begin{pmatrix} 0 & 0 \\ 0 & 1 \end{pmatrix} \boldsymbol{Y}(\pi) + \int_0^\pi \boldsymbol{Y}(t)\, dt$.

PART IV.

16. Solve the system of Example 8.19 in *Mathematica*.

17. Repeat the solution of Example 8.19 in *Mathematica* with the boundary conditions changed to $\boldsymbol{y}_1(0) = 1$, $\boldsymbol{y}_2(\pi) = 1$.

18. Repeat the solution of Example 8.19 in *Mathematica* with the boundary conditions changed to $\boldsymbol{y}_1(0) = 1$, $\boldsymbol{y}_2(\pi) = 0$.

8.6 Applications of Systems of Differential Equations

There are many processes in the world that can be modeled by systems of differential equations. In biology, modeling competing species requires coupled systems of nonlinear equations. Chemical mixing problems involving a single solute dissolved in solutions throughout several interconnected containers whose contents are being intermixed, require one differential equation for the amount of solute in each container. In physics, mechanical systems involving multiple springs and multiple masses in various configurations introduce linear (and sometimes nonlinear) systems of differential equations. In electrical engineering, passive L-R-C circuits having multiple loops and hence multiple currents require a system of linear differential equations, one equation for the current in each loop. In fact, passive L-R-C circuits are modeled by a hybrid set of equations consisting of several differential equations and several algebraic equations all to be satisfied simultaneously. In international relations, a linear system of equations proposed by Richardson[3] effectively models the forces that govern the armament and disarmament of nations that do not trust one another.

Passive L-R-C Circuits Having Multiple Loops

An example should suffice to illustrate the kinds of reasoning that apply to problems of this kind. The example is set up manually and solved by *Mathematica*, given numerical values for the various components. You will use **Kirchhoff's voltage law** in each loop: *The applied voltage equals the sum of the voltage drops across the components in the loop.* You will use **Kirchhoff's current law** at each junction of two or more branches: *The sum of the currents directed toward a junction equals the sum of the currents directed out of that junction.* Differential equations arise from Kirchhoff's voltage law and the algebraic equations from Kirchhoff's current law.

We outline an algorithm that produces a system of equations that models the currents in a circuit having N loops that have branches in common. Observe carefully how the first two steps have been followed in the labeling of Figure 8.1.

First. Identify and number the loops that occur in the circuit. Then propose (arbitrarily) a direction for positive current in each separate branch of the circuit. Draw an arrow in each branch to indicate (assumed) positive current. Do not be concerned with signs. The solution will supply the correct signs.

[3]Lewis Frye Richardson, 1881–1953. British physicist. Richardson applied his education in physics to forecasting the weather during the 1920's. He can be called the 'father' of the modern science of meterology. During the depression of the 1930's, he turned his attention to economics and established several of the basic principles of that field. Then, as World War II began to spread throughout Europe, he turned his research to armament and disarmament. The Richardson Models of warfare are still widely studied.

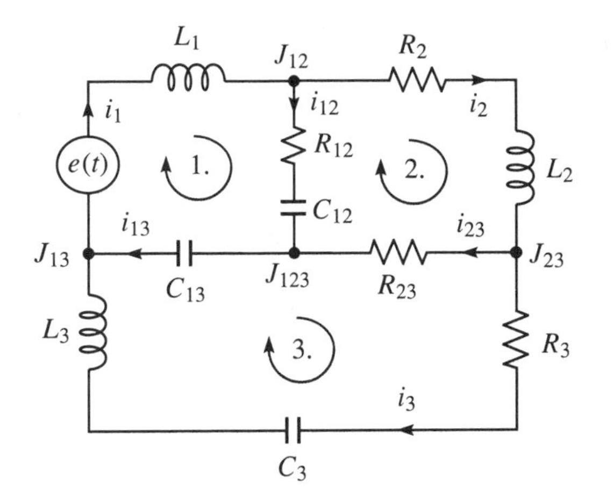

Figure 8.1: A passive L-R-C circuit having three loops.

Second. Name each component and each current. Using a consistent naming convention such as that illustrated below will help you keep track of the meaning of the symbols you introduce. Briefly, a useful naming convention is:

- Use L_m, R_m, C_m, and i_m for objects that occur only in loop m.
- Use L_{mk}, R_{mk}, C_{mk}, and i_{mk} for objects occurring both in loop m and in loop k (always have $m < k$). Internal branch currents typically have the form i_{mk}.
- Name every junction of two or more circuit branches so that J_{mk} means the junction is in common to two loops m and k and J_{mkl} means the junction is in common with the three loops m, k, and l. Similar conventions apply to junctions common to more than three loops. Maintain $m < k < l <$ etc.

Third. Apply Kirchhoff's voltage law (in derivative form to give each term as a current) to each loop of the circuit. Process the components in order around each loop, honoring the assumed current direction through each component: opposite direction needs opposite sign. Use de/dt as the applied current.

Fourth. Apply Kirchhoff's current law to each junction.

Fifth. Solve the system of equations obtained in the fourth step for all of the branch currents having two subscripts in terms of the loop currents that have a single subscript. Substitute these values into the differential system obtained in the third step to obtain a system of N second-order linear differential equations in the primary loop currents $i_1, i_2, \ldots, i_N$.

Sixth. Solve this system for $i_1, i_2, \ldots, i_N$. Substitute $i_1, i_2, \ldots, i_N$ into the formulas obtained for the various branch currents if values for the branch currents are needed.

This may seem like a complicated process, but each step is simple, and with the help of *Mathematica*, a solution is easy to obtain. In systems such as these, the initial conditions soon have very little effect (transient solution), and all lasting effects (steady-state solution) come from the particular solution determined by the applied voltage.

Example 8.20 Set up a system of differential equations that models the current in each of the three loops of Figure 8.1.

Solution. The components and currents have been named according to the stated conventions. The results of applying Kirchhoff's voltage law in derivative form to each of the three loops are:

$$\textbf{(loop 1)}\ L_1\frac{d^2 i_1}{dt^2} + R_{12}\frac{di_{12}}{dt} + \frac{1}{C_{12}}i_{12} + \frac{1}{C_{13}}i_{13} = \frac{de}{dt}$$

$$\textbf{(loop 2)}\ L_2\frac{d^2 i_2}{dt^2} + R_{23}\frac{di_{23}}{dt} - R_{12}\frac{di_{12}}{dt} - \frac{1}{C_{12}}i_{12} + R_2\frac{di_2}{dt} = 0$$

$$\textbf{(loop 3)}\ L_3\frac{d^2 i_3}{dt^2} - \frac{1}{C_{13}}i_{13} - R_{23}\frac{di_{23}}{dt} + R_3\frac{di_3}{dt} + \frac{1}{C_3}i_3 = 0.$$

The results of applying Kirchhoff's current law to each of the four junctions are:

$$\begin{aligned}
\textbf{(J12)}\ & i_1 = i_2 + i_{12}\\
\textbf{(J13)}\ & i_1 = i_{13} + i_3\\
\textbf{(J23)}\ & i_2 = i_{23} + i_3\\
\textbf{(J123)}\ & i_{13} = i_{12} + i_{23}.
\end{aligned}$$

Solve these algebraic equations for i_{12}, i_{13}, and i_{23} in terms of i_1, i_2, and i_3 to get:

$$\begin{aligned}
\textbf{(branch)}\ & i_{12} = i_1 - i_2\\
& i_{13} = i_1 - i_3\\
& i_{23} = i_2 - i_3.
\end{aligned}$$

Substitute these branch currents into the three loop equations to get:

$$\textbf{(loop 1)}\ L_1\frac{d^2 i_1}{dt^2} + R_{12}\frac{d(i_1 - i_2)}{dt} + \frac{1}{C_{12}}(i_1 - i_2) + \frac{1}{C_{13}}(i_1 - i_3) = \frac{de}{dt}$$

$$\begin{aligned}
\textbf{(loop 2)}\ L_2\frac{d^2 i_2}{dt^2} + R_{23}\frac{d(i_2 - i_3)}{dt} - R_{12}\frac{d(i_1 - i_2)}{dt}&\\
- \frac{1}{C_{12}}(i_1 - i_2) + R_2\frac{di_2}{dt} = 0&
\end{aligned}$$

$$\textbf{(loop 3)}\ L_3\frac{d^2 i_3}{dt^2} - \frac{1}{C_{13}}(i_1 - i_3) - R_{23}\frac{d(i_2 - i_3)}{dt} + R_3\frac{di_3}{dt} + \frac{1}{C_3}i_3 = 0.$$

These equations simplify to this second-order system:

$$\text{(loop 1)}\ L_1\frac{d^2i_1}{dt^2}+R_{12}\frac{di_1}{dt}-R_{12}\frac{di_2}{dt}$$
$$+\frac{1}{C_{12}}i_1-\frac{1}{C_{12}}i_2+\frac{1}{C_{13}}i_1-\frac{1}{C_{13}}i_3=\frac{de}{dt}$$

$$\text{(loop 2)}\ L_2\frac{d^2i_2}{dt^2}+R_{23}\frac{di_2}{dt}-R_{23}\frac{di_3}{dt}-R_{12}\frac{di_1}{dt}-R_{12}\frac{di_2}{dt}$$
$$-\frac{1}{C_{12}}i_1-\frac{1}{C_{12}}i_2+R_2\frac{di_2}{dt}=0$$

$$\text{(loop 3)}\ L_3\frac{d^2i_3}{dt^2}-\frac{1}{C_{13}}i_1-\frac{1}{C_{13}}i_3-R_{23}\frac{di_2}{dt}-R_{23}\frac{di_3}{dt}+R_3\frac{di_3}{dt}+\frac{1}{C_3}i_3=0.$$

One can collect terms involving like currents or simplify in other ways, but these equations serve as a system of second-order equations that describes the behavior of the three primary currents, i_1, i_2, and i_3. The three branch currents i_{12}, i_{13}, and i_{23} can be obtained from the branch equations once i_1, i_2, and i_3 are known. We have accomplished what the example required: we have set up a system of differential equations that models the current in each branch of the three loops of Figure 8.1. The substitutions introduced in Section 8.1 show that the second-order system is equivalent to six first-order differential equations, so the order of the system defining the loop currents is six. ◇

The Spread of AIDS

Tudor (1992) gave a model for the spread of AIDS in a population in which an attempt is being made to educate the population about the disease and its modes of propagation. His model was of the SI type (see Hethcote (1976)). Models of SI type assume that every individual in the population is **susceptible** (S) to the disease and that after adequate contact becomes **infected** (I) and remains so thereafter. Tudor assumed that in the presence of efforts at education, a portion, S_c, of the entire S population would change their behavior so as to avoid contact with the infected population and that the remainder of the S population, S_u, would be unwilling to change. The dynamics of the model assumes that everyone entering the sexually active population is susceptible, with the rate of entry into the population S being α. One should actually write $\alpha = \alpha_c + \alpha_u$ where α_c denotes the rate of entry into the population S_c that is willing to change and α_u denoting the rate of entry into the population S_u that is unwilling to change. Assume further that the death rate $\mu = \mu_c + \mu_u$ with the designations being analogous to those for α. Denote the population size by N, and assume that $\alpha = \mu$ so that the population is of constant size. An education campaign is undertaken, which is designed to increase α_c and decrease λ. The resulting model has the form

$$S_u' = -\lambda I S_u + \alpha_u - \mu_u S_u,$$
$$S_c' = \alpha_c - \mu_c S_c,$$
$$I' = \lambda I S_u - \mu_u I,$$

where λ contacts per unit time are made by the $I(t)$ infectious individuals. This is a nonlinear system of three equations in three unknown functions. The second equation can be explicitly solved for S_c which reduces the system to two equations in S_u and I. Note that the rate of *decrease* in the S_u population due to contact with the infected population ($= \lambda I S_u$) is exactly the same as the rate of *increase* of the infected population I due to this same contact. This describes those susceptibles who become infected and thus become part of the infected population. Note also that the rate of change of the S_c population is independent of the size of the infected population because they have modified their behavior to avoid contact with infected individuals.

Tudor goes on to study the effects of increased levels of educational campaigns at specific later times. He concludes with the statement: "Although the models presented here are meant primarily for use in the classroom, we have seen that, based on our analysis here, to effectively combat AIDS, we must either be assured that behavioral changes are permanent, which is doubtful, or we may never cease in our [educational] campaign efforts."

Chemical Mixing

Chemical mixing problems are basically transport problems where a substance is moved through a collection of containers or compartments. The following example discusses the mixing of salt in an aqueous solution that is in motion through two containers. Another example would be that of the circulation throughout the body of a substance that has been injected directly into the blood stream and is dissipated throughout the body. An interesting problem is when a short-lived radioactive tracer is injected into the body and its progress monitored as it is dissipated (differentially) throughout the body, is simultaneously decaying, and the remnant is eventually excreted.

Example 8.21 Suppose that fluid is able to pass through an interconnected collection of containers as depicted in Figure 8.2. The upper container holds at least 100 gallons and the lower at least 200 gallons. Initially, 25 pounds of salt (NaCl) is dissolved in 100 gallons of water in the upper container and the lower container contains 200 gallons of pure water (no salt). After the start of the process, the incoming fluid is brine containing a concentration of 1 pound per gallon of salt that is entering at the rate of 3 gallons per minute and exiting at a rate of 5 gallons per minute. Both containers are kept well mixed. The solution in the lower container is recycled back into the upper container at the rate of 2 gallons per minute. Solution also flows out of the second container at the rate of 3 gallons per minute and is lost to the system. Set up and solve a system of equations that describes the amounts $x_1(t)$ and $x_2(t)$ of salt in the respective

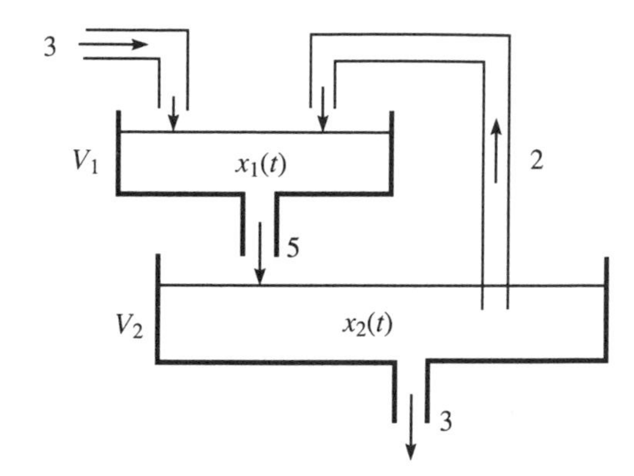

Figure 8.2: Two containers of constant volume with recycling.

containers after the process starts. What is the limiting concentration of salt in each container?

Solution. Let $x_1(t)$ be the amount of salt in the first container and $x_2(t)$ be the amount in the second. Denote the volume of the first container by $V_1(t)$ and that of the second by $V_2(t)$. Both the volumes involved and the amounts of salt obey the general rule

$$\text{Rate of change} = \text{rate}_{\text{in}} - \text{rate}_{\text{out}}.$$

This permits us to express the four related quantities of salt and volume as:

$$\frac{dV_1}{dt} = \text{rate}_{\text{in}} - \text{rate}_{\text{out}} = \left(3\frac{\text{gallon}}{\text{minute}} + 2\frac{\text{gallon}}{\text{minute}}\right) - 5\frac{\text{gallon}}{\text{minute}} = 0\frac{\text{gallon}}{\text{minute}}.$$

This means that $V_1(t) = \text{constant} = V_1(0) = 100$ gallons.

$$\frac{dV_2}{dt} = \text{rate}_{\text{in}} - \text{rate}_{\text{out}} = 5\frac{\text{gallon}}{\text{minute}} - \left(2\frac{\text{gallon}}{\text{minute}} + 3\frac{\text{gallon}}{\text{minute}}\right) = 0\frac{\text{gallon}}{\text{minute}}.$$

This means that $V_2(t) = \text{constant} = V_2(0) = 200$ gallons. Furthermore,

$$\begin{aligned}\frac{dx_1}{dt} &= \text{rate}_{\text{in}} - \text{rate}_{\text{out}} \\ &= \left(3\frac{\text{gallon}}{\text{minute}}\right)\left(1\frac{\text{pound}}{\text{gallon}}\right) + \left(2\frac{\text{gallon}}{\text{minute}}\right)\left(\frac{x_2}{V_2}\frac{\text{pound}}{\text{gallon}}\right) \\ &\quad - \left(5\frac{\text{gallon}}{\text{minute}}\right)\left(\frac{x_1}{V_1}\frac{\text{pound}}{\text{gallon}}\right) \\ &= 3 + \frac{2x_2}{V_2} - \frac{5x_1}{V_1},\end{aligned}$$

and

$$\begin{aligned}\frac{dx_2}{dt} &= \text{rate}_{\text{in}} - \text{rate}_{\text{out}} \\ &= \left(5\frac{\text{gallon}}{\text{minute}}\right)\left(\frac{x_1}{V_1}\frac{\text{pound}}{\text{gallon}}\right) - \left(2\frac{\text{gallon}}{\text{minute}}\right)\left(\frac{x_2}{V_2}\frac{\text{pound}}{\text{gallon}}\right) \\ &\quad - \left(3\frac{\text{gallon}}{\text{minute}}\right)\left(\frac{x_2}{V_2}\frac{\text{pound}}{\text{gallon}}\right) \\ &= \frac{5x_1}{V_1} - \frac{5x_2}{V_2}.\end{aligned}$$

The system of equations is

$$\begin{aligned}\frac{dV_1}{dt} &= 0 \\ \frac{dV_2}{dt} &= 0 \\ \frac{dx_1}{dt} &= 3 + \frac{2x_2}{V_2} - \frac{5x_1}{V_1} \\ \frac{dx_2}{dt} &= \frac{5x_1}{V_1} - \frac{5x_2}{V_2}.\end{aligned}$$

Since V_1 and V_2 are constant, these equations simplify to this initial value problem for the amounts of salt in the two containers:

$$\begin{aligned}\frac{dx_1}{dt} &= 3 - \frac{5x_1}{100} + \frac{2x_2}{200}, \quad x_1(0) = 25 \\ \frac{dx_2}{dt} &= \frac{5x_1}{100} - \frac{5x_2}{200}, \quad x_2(0) = 0.\end{aligned}$$

The solution of these two equations by the standard methods we discussed earlier in this chapter is rather complicated. (You can do it as an exercise.) Numerically evaluating the various coefficients yields the (approximate) solution:

$$\begin{aligned}x_1(t) &= 100 - 16.7621e^{-0.0631174t} - 58.2379e^{-0.0118826t} \\ x_2(t) &= 200 + 21.9875e^{-0.0631174t} - 221.988e^{-0.0118826t}.\end{aligned}$$

In *Mathematica* these can be plotted this way

```
In[65]:= p = Plot[Evaluate[N[{100,200,s1[t],s2[t]}]],{t,0,400}];
```

-Omitted-

```
In[66]:= Show[p,Graphics[{Text["x1[t]",{250,80}],
                          Text["x2[t]",{250,170}]}],
                AxesLabel->{"t","NaCl"}]
```

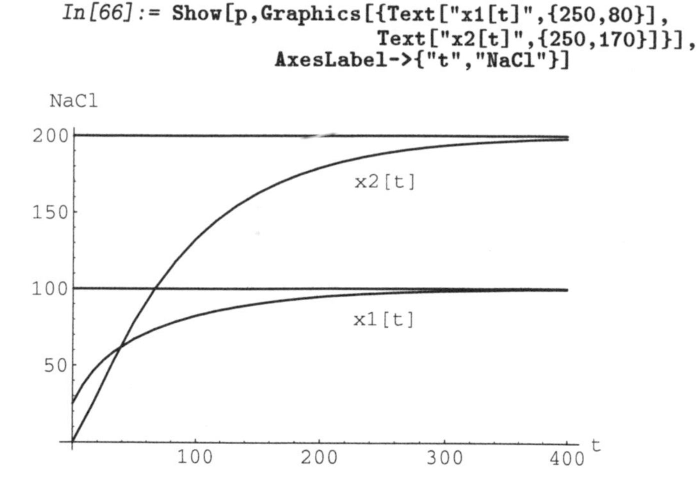

It is clear both graphically and analytically that

$$\lim_{t\to\infty} x_1(t) = 100$$

and

$$\lim_{t\to\infty} x_2(t) = 200. \qquad \diamond$$

Suppose that for the same two containers as in Example 8.21 the inflow and outflow of the containers is not in balance so that the volume(s) vary with time. We illustrate the phenomenon. The system turns out to not have constant coefficients, and so technically belongs in the next chapter, but we solve it numerically using `NDSolve`. The example actually consists of three separate parts, of which we indicate the solution only of the first.

Example 8.22 Suppose that fluid is able to pass through an interconnected collection of containers as depicted in Figure 8.3. The upper container holds at least 150 gallons and the lower at least 300 gallons. Initially, 25 pounds of salt (NaCl) is dissolved in 100 gallons of water in the upper container, and the lower container contains 200 gallons of pure water (no salt). After the start of the process, the incoming fluid is brine containing a concentration of 1 pound per gallon of salt that is entering at the rate of 3 gallons per minute and exiting at a rate of 6 gallons per minute. Both containers are kept well mixed. The solution in the lower container is recycled back into the upper container at the rate of 2 gallons per minute. Solution also flows out of the second container at the rate of 4 gallons per minute and is lost to the system. Set up and solve a system of equations that describes the amounts $x_1(t)$ and $x_2(t)$ of salt in the respective containers after the process starts. What is the limiting concentration of salt in each container?

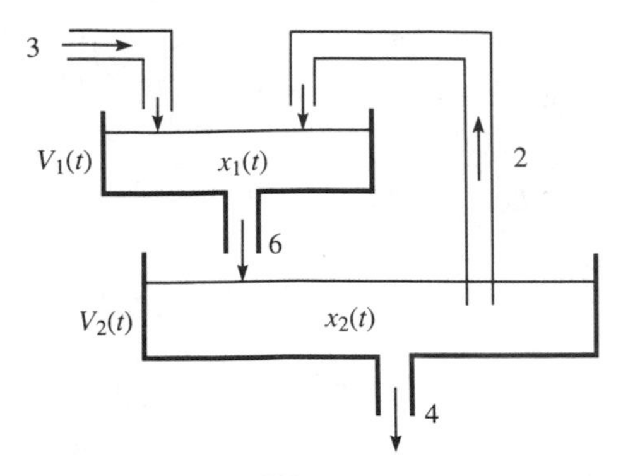

Figure 8.3: Two containers of varying volume with recycling.

Solution. Let $x_1(t), x_2(t), V_1(t)$, and $V_2(t)$ be as in Example 8.21. Initially express the four related quantities of salt and volume as:

$$\frac{dV_1}{dt} = \text{rate}_{\text{in}} - \text{rate}_{\text{out}} = \left(3\frac{\text{gallon}}{\text{minute}} + 2\frac{\text{gallon}}{\text{minute}}\right) - 6\frac{\text{gallon}}{\text{minute}} = -1\frac{\text{gallon}}{\text{minute}}.$$

This means that $V_1(t) = V_1(0) - t = (100 - t)$ gallons. Therefore, after 100 minutes, the first container empties. At this point the system enters a new state where the salt solution that enters the first container flows directly on into the second container. Therefore, for the first 100 minutes $V_2(t)$ obeys this rule:

$$\frac{dV_2}{dt} = \text{rate}_{\text{in}} - \text{rate}_{\text{out}} = 6\frac{\text{gallon}}{\text{minute}} - \left(2\frac{\text{gallon}}{\text{minute}} + 4\frac{\text{gallon}}{\text{minute}}\right) = 0\frac{\text{gallon}}{\text{minute}}.$$

This means that for the first 100 minutes, $V_2(t) = \text{constant} = V_2(0) = 200$ gallons. This holds until the first container empties. (What is the new equation for how V_2 changes? Over what interval is it in effect?) During this first 100 minutes,

$$\begin{aligned}
\frac{dx_1}{dt} &= \text{rate}_{\text{in}} - \text{rate}_{\text{out}} \\
&= \left(3\frac{\text{gallon}}{\text{minute}}\right)\left(1\frac{\text{pound}}{\text{gallon}}\right) + \left(2\frac{\text{gallon}}{\text{minute}}\right)\left(\frac{x_2}{V_2(t)}\right)\frac{\text{pound}}{\text{gallon}} \\
&\quad - \left(6\frac{\text{gallon}}{\text{minute}}\right)\left(\frac{x_1}{V_1(t)}\right)\frac{\text{pound}}{\text{gallon}} \\
&= 3 + \frac{2x_2}{V_2(t)} - \frac{6x_1}{V_1(t)},
\end{aligned}$$

and

$$\begin{aligned}\frac{dx_2}{dt} &= \text{rate}_{\text{in}} - \text{rate}_{\text{out}} \\ &= \left(6\frac{\text{gallon}}{\text{minute}}\right)\left(\frac{x_1}{V_1(t)}\right)\frac{\text{pound}}{\text{gallon}} - \left(2\frac{\text{gallon}}{\text{minute}}\right)\left(\frac{x_2}{V_2(t)}\right)\frac{\text{pound}}{\text{gallon}} \\ &\quad - \left(4\frac{\text{gallon}}{\text{minute}}\right)\left(\frac{x_2}{V_2(t)}\right)\frac{\text{pound}}{\text{gallon}} \\ &= \frac{6x_1}{V_1(t)} - \frac{6x_2}{V_2(t)}.\end{aligned}$$

The system of equations in effect for the first 100 minutes is

$$\begin{aligned}\frac{dV_1}{dt} &= -1, V_1(0) = 100, \\ \frac{dV_2}{dt} &= 0, V_2(0) = 200, \\ \frac{dx_1}{dt} &= 3 + \frac{2x_2}{V_2(t)} - \frac{6x_1}{V_1(t)}, \quad x_1(0) = 25, \\ \frac{dx_2}{dt} &= \frac{6x_1}{V_1(t)} - \frac{6x_1}{V_2(t)}, \quad x_2(0) = 0.\end{aligned}$$

Since $V_1(t) = 100 - t$ and $V_2(t) = 200$ during the first 100 minutes, these equations simplify to this initial value problem for the amounts of salt in the two containers:

$$\begin{aligned}\frac{dx_1}{dt} &= 3 - \frac{6x_1}{100-t} + \frac{2x_2}{200}, \quad x_1(0) = 25 \\ \frac{dx_2}{dt} &= \frac{6x_1}{100-t} - \frac{6x_2}{200}, \quad x_2(0) = 0, \quad \text{for} \quad 0 \le t \le 100.\end{aligned}$$

As yet, we have no formal technique for solving this problem, but we can numerically solve it using `NDSolve`. That is left as an exercise. The system is a singular system of the type discussed in Chapter 9. ◇

Radioactive Decay Series

In Chapter 3 we considered the radioactive decay of a single element. For some elements—notably uranium, thorium, and actinium—as an isotope of one of these elements decays it decays into an isotope of another element. This new isotope may itself be radioactive and decay into another isotope, which may be radioactive and decay further. The identification of the isotopes in these decay series took years of careful analysis by numerous chemists and physicists, starting with Henri Becquerel in 1896, Pierre and Marie Curie during 1898–1902, and Willard Libby during 1947–1950, whose method of radiocarbon dating was also considered in Chapter 3.

The differential equation for the amount x_1 of the initial element in a decay series is $dx_1/dt = -\lambda_1 x_1$, which we saw in Chapter 3. For decay products later in the series the differential equation for the amount x_k of product k is $dx_k/dt = \lambda_{k-1}x_{k-1} - \lambda_k x_k$, because the amount of product k increases due to the decay of product $k-1$, but decreases through radioactive decay with decay rate λ_k. The sizes of the decay rates within a single decay series may vary by several orders of magnitude. Ultimately, after several products have been produced, a final stable product is produced that decays no further. The differential equation for the amount x_n of the final stable product is $dx_n/dt = \lambda_{n-1}x_{n-1}$ because product n just increases through the decay of product $n-1$. This final product is normally some isotope of the element lead.

For purposes of illustration, suppose that e_1, e_2, e_3, and e_4 are consecutive products in a decay series and that $x_1(t)$, $x_2(t)$, $x_3(t)$, and $x_4(t)$ are the amounts present at time t. Let λ_1, λ_2, λ_3, and $\lambda_4 = 0$ be the respective decay rates. The statement $\lambda_4 = 0$ says that e_4 is the stable end product of the series. The differential equations of the decay series can be written as

$$\begin{cases} \dfrac{dx_1}{dt} = -\lambda_1 x_1 \\ \dfrac{dx_2}{dt} = \lambda_1 x_1 - \lambda_2 x_2 \\ \dfrac{dx_3}{dt} = \lambda_2 x_2 - \lambda_3 x_3 \\ \dfrac{dx_4}{dt} = \lambda_3 x_3. \end{cases}$$

The choice of initial conditions depends on the experiment that is being observed, but a common set of initial conditions is $x_1(0) = \alpha_0, x_2(0) = 0, x_3(0) = 0, x_4(0) = 0$. This assumes that an amount α_0 of substance e_1 is isolated and begins to decay. For a while, both dx_2/dt and dx_3/dt are positive, and hence the amounts of substances e_2 and e_3 are increasing. As the amount of substance e_1 decreases, dx_2/dt eventually becomes negative, which eventually causes dx_3/dt to also become negative. The rate dx_4/dt remains positive throughout. After some (possible very long) period of time, the masses of all three radioactive products decrease, with their total mass having been converted into the final stable product e_4. The total mass of the four isotopes is constant since

$$\frac{d(x_1 + x_2 + x_3 + x_4)}{dt} = (-\lambda_1 x_1) + (\lambda_1 x_1 - \lambda_2 x_2) + (\lambda_2 x_2 - \lambda_3 x_3) + (\lambda_3 x_3) = 0.$$

The phenomenon of radioactive decay series has naturally introduced a system of first-order linear differential equations. See Figure 8.4. The differential equations for the decay series of uranium, thorium, and actinium have respectively 14, 10, and 11 equations. Though the same elements appear in several of these series, different isotopes of the elements occur in the different series. These different isotopes have markedly different half-lives and hence different decay rates. In order to set up the differential equations for any one of these series, it is necessary to look up the half-lives of the various isotopes that occur

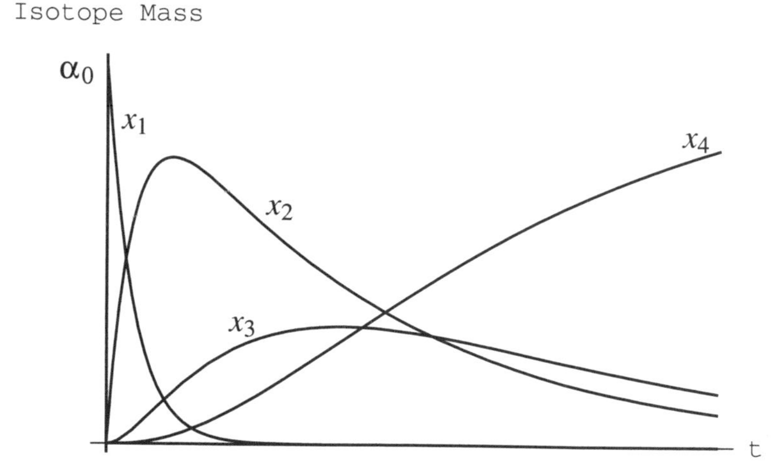

Figure 8.4: Solutions of a typical decay series showing the rise and ultimate fall of the radioactive products and the rise of the stable product.

in the series and convert each half-life into a decay rate. Tables of half-lives of the various isotopes are readily available. Since the half-life $t_{1/2}$ of a substance is related to the decay rate λ by $t_{1/2} = \ln 2/\lambda$, one finds the necessary value of λ from $\lambda = \ln 2/t_{1/2}$. Remember that all of the decay rates should be based on the second as standard unit of time, even though the half-lives in a series can range from microseconds to millions of years.

An interesting laboratory exercise using the thorium series consists of determining the date of manufacture of Coleman lantern mantles. Thorium is put in Coleman lantern mantles because it glows with a particularly bright white light. The manufacture date of a particular mantle can be determined with some certainty by measuring the relative quantities of the components of the isotopes in the decay series and comparing with the quantities predicted by the differential equations.

It is not difficult to write down the complete solution for the system that describes a decay series because each differential equation can easily be solved explicitly in terms of the solution to the differential equation that precedes it in the system. These explicit solution functions have the form $x_k(t) = \sum_{j=1}^{k} c_k \exp(\lambda_j t)$ where the c_k are determined by the initial conditions. This description of decay series does not address the fact that occasionally an isotope has two decay modes that produce different isotopes. These two product isotopes each head decay series, and produce a common product after one or several steps. One accounts for these separate paths by observing the probabilities with which the two modes of decay occur and using these probabilities to state how later recombinations take place. The probabilities of various modes of decay are available in standard tables.

International Armament and Disarmament

Beginning shortly after the First World War, Lewis F. Richardson, who almost single-handedly established the science of meteorology, studied the causes and effects of deadly quarrels. This was his terminology for anything from a murder through global warfare. The distillations of his efforts appeared posthumously in two books, *Statistics of Deadly Quarrels* and *Arms and Insecurity*. One of the ideas that he proposed was that it was possible to model the circumstances under which an arms race might be undertaken.

Here is a summary of Richardson's model of international armament. Suppose that there are two nations that mistrust each another. Suppose that one nation feels the need to defend itself against the other. Let x and y denote the amount of armament of the two nations. The first nation begins to arm itself proportionally to the perceived threat from the other nation: $dx/dt = ky$. The second nation does the same: $dy/dt = kx$. But these are not the only factors, for if they were, both nations would never be able to stop the arms race. For each nation there is a cost of armament that tends to oppose the increase. Call these terms αx and βy. In addition, there are grievances g_1 and g_2 present that are more or less constant with respect to armament considerations. Thus the final pair of equations is

$$\begin{cases} \dfrac{dx}{dt} = ky - \alpha x + g_1 \\ \dfrac{dy}{dt} = kx - \beta y + g_2. \end{cases}$$

Unilateral disarmament on the part of the second nation corresponds to putting $y = 0$. If this is done, note that y does not remain zero! It causes only a temporary decrease in the rate of growth of armament of the first nation, the armament of both nations will again rise unless something else changes. Since mathematically it is possible for either x or y to be negative, Richardson concluded after some study that this situation is not "peace," but rather "co-operation" which is primarily manifested through commerce as trade. "Peace" then became something that happens only briefly in transition between two other states. If two nations trust one another, it is possible to assign negative values to the "grievances" g_1 and g_2. In this case control of the arms race is dependent on the relative sizes of the costs α and β compared to k.

Richardson generalized the model to three nations and then to n nations. In the case of three nations he showed that the above equations for any pair of nations could predict a basically neutral arms buildup, but that the third nation could destabilize any one of the pairs, and hence the entire triple of nations. He concluded on the basis of applying the model to real data that the world of 1938 was unstable (!), but that when the instability was resolved, a "just barely stable" world situation would result. In fact, he suggested that any "peaceful" portion of our century was only barely stable, with renewed arms races always ready to break out.

EXERCISES 8.6

1. The three loop equations

$$\textbf{(loop 1)}\ L_1\frac{d^2 i_1}{dt^2}+R_{12}\frac{di_1}{dt}-R_{12}\frac{di_2}{dt}$$
$$+\frac{1}{C_{12}}i_1-\frac{1}{C_{12}}i_2+\frac{1}{C_{13}}i_1-\frac{1}{C_{13}}i_3=\frac{de}{dt}$$

$$\textbf{(loop 2)}\ L_2\frac{d^2 i_2}{dt^2}+R_{23}\frac{di_2}{dt}-R_{23}\frac{di_3}{dt}-R_{12}\frac{di_1}{dt}-R_{12}\frac{di_2}{dt}$$
$$-\frac{1}{C_{12}}i_1-\frac{1}{C_{12}}i_2+R_2\frac{di_2}{dt}=0$$

$$\textbf{(loop 3)}\ L_3\frac{d^2 i_3}{dt^2}-\frac{1}{C_{13}}i_1-\frac{1}{C_{13}}i_3$$
$$-R_{23}\frac{di_2}{dt}-R_{23}\frac{di_3}{dt}+R_3\frac{di_3}{dt}+\frac{1}{C_3}i_3=0.$$

describe the currents in the three loops in this circuit.

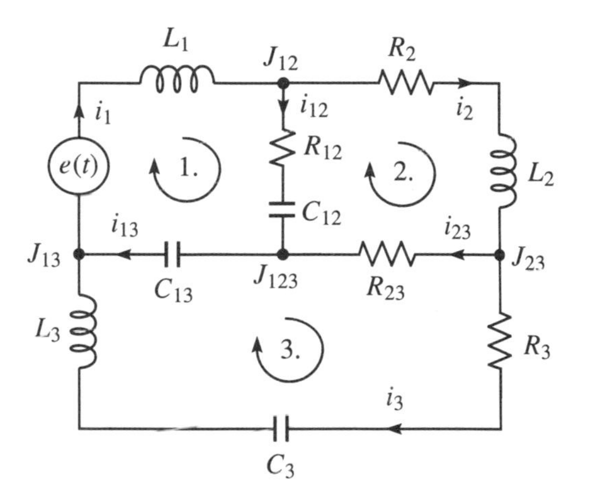

Transform the system into a system of six first-order equations in the loop currents i_1, i_2, and i_3. Use `NDSolve` to obtain an approximate solution to the system when $L_1=0.2h$, $R_2=10\Omega$, $L_2=1h$, $R_3=100\Omega$, $L_3=0.5h$, $C_3=0.22\mu f$, $R_{12}=220\Omega$, $C_{12}=0.1\mu f$, $C_{13}=1.0\mu f$, $R_{23}=330\Omega$, $e(t)=3\cos(20\pi t)$.

2. Use `NDSolve` to obtain an approximate solution to the system that models the spread of AIDS. Make individual plots of the three functions and combine the plots on a single set of axes. Use these (fictional) values: $\alpha_c=0.15, \mu_c=$

0.05, $\alpha_u = 0.15, \mu_u = 0.25, \lambda = 0.20$. It would be worth your while to obtain good estimates for these parameters based on reliable data and then repeat the problem.

3. The definitions for this problem are given in Example 8.22.

   (a) Solve the differential system numerically using `NDSolve`. `Plot` $V_1(t)$, $V_2(t)$, $x_1(t)$, and $x_2(t)$ on the same axes for $0 \le t \le 100$.
   (b) What system of differential equations governs the next phase of the process? Pay particular attention to the initial conditions that are in effect at the beginning of this second phase.
   (c) Determine the time at which the second phase ends, and plot the same four functions over the time interval during which this second phase is in effect.
   (d) Determine the equations that are in effect thereafter. Solve them.
   (e) `Plot` a composite graph of the solution throughout the three phases studied.

4. Here is a drawing that represents part of the uranium decay series.

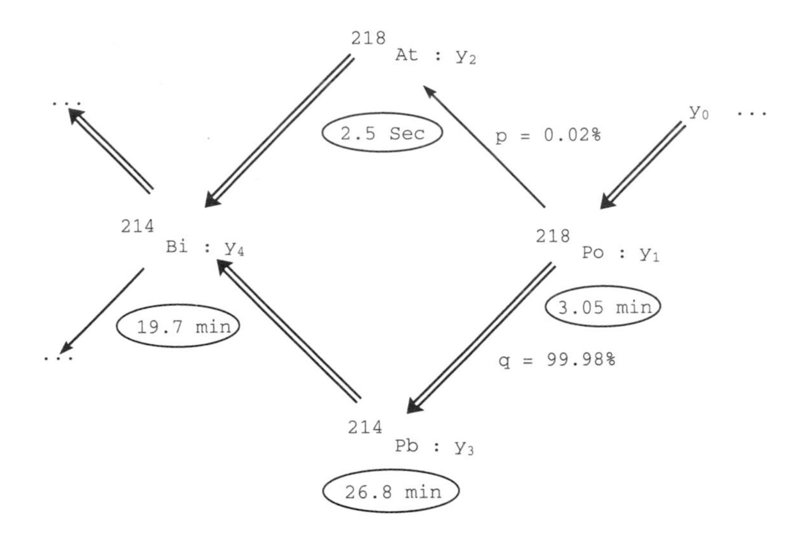

The double-line arrows indicate the main modes of decay. The percentages $p = 0.02\%$ and $q = 99.98\%$ represent the relative occurrences of the indicated decay modes of ^{218}Po into ^{218}At and ^{214}Pb, respectively. The encircled numbers are the half-lives of the nearest isotopes. We will include y_0 in the differential equations, but will let $y_0(t) = 0$ in order to isolate this portion of the series. For brevity of notation we use $y_i(t)$ for the amount of isotope i that is present at time t. The association of the y-names and isotopes is on the picture.

   (a) Determine each of the four decay constants λ_i that are associated with the appropriate y_i.

(b) Explain why these differential equations describe the rates of change of the amounts of the isotopes:

$$\frac{dy_1}{dt} = \lambda_0 y_0 - \lambda_1 y_1$$

$$\frac{dy_2}{dt} = p\lambda_1 y_1 - \lambda_2 y_2$$

$$\frac{dy_3}{dt} = q\lambda_1 y_1 - \lambda_3 y_3$$

$$\frac{dy_4}{dt} = \lambda_2 y_2 + \lambda_3 y_3 - \lambda_4 y_4.$$

Note how the probabilities of the two decay modes of ^{218}Po are included in the second and third equations.

(c) Show that under the assumption $y_0 = 0$, the change of mass in the system consisting of these four isotopes is $d(y_1 + y_2 + y_3 + y_4)/dt = -\lambda_4 y_4$. Recall that $p + q = 1$.

(d) Solve the system of exercise 2 above using the values for the λ_i that you obtained in part (a) letting the initial conditions be $y_1(0) = 1, y_2(0) = 0, y_3(0) = 0$, and $y_4(0) = 0$. Take $y_0 = 0$.

(e) `Plot` the solution functions on a common pair of axes over an appropriate time interval. Choose your time interval large enough to reveal the maxima for y_2, y_3, and y_3. If you cannot seem to find all of the curves you expect, plot each curve individually. It is sometimes helpful (and useful to investigators) to plot `Log[curves]` on an interval $[small, large]$, with $small > 0$.

8.7 Phase Portraits

In Chapter 4 we studied differential equations that could be put into the form

$$\frac{dy}{dx} = \frac{f(x,y)}{g(x,y)}. \tag{8.13}$$

In this chapter we have studied linear systems that have the form

$$\begin{cases} \dfrac{dy}{dt} = f(x,y) \\ \dfrac{dx}{dt} = g(x,y) \end{cases} \tag{8.14}$$

Equation 8.13 and system 8.14 look too similar to be unrelated, and, indeed, they are related. Recall from the chain rule that $(dy/dx)(dx/dt) = dy/dt$. It follows that

$$\frac{dy}{dx} = \frac{dy/dt}{dx/dt} = \frac{f(x,y)}{g(x,y)}.$$

In system 8.14 we have disconnected the numerator and denominator of equation 8.13 and have asked that x and y be simultaneously parameterized (by t). Thus in 8.14 we have made the static situation of 8.13 into a dynamic situation: both $x(t)$ and $y(t)$ can change with t.

The parameterized solutions of 8.14 trace out the (static) solutions of 8.13. For this reason we refer to the solutions of 8.13, where y is given as a function of x or x is given as a function of y, as **trajectories** or **orbits** of the solutions of 8.14. System 8.14 is often called a **dynamical system**.

Note also that system 8.14 is **autonomous**—it does not explicitly involve the independent variable t. This means that the system has the property that any t-translate of a solution is a solution. Indeed, let $(x(t), y(t))$ be a solution of 8.14 for $t_1 < t < t_2$, $a \neq 0$ be a number, and consider the functions $u(t) = x(t+a)$, and $v(t) = y(t+a)$. Then

$$v'(t) = y'(t+a) = f(x(t+a), y(t+a)) = f(u(t), v(t))$$

and

$$u'(t) = x'(t+a) = g(x(t+a), y(t+a)) = g(u(t), v(t)),$$

so the pair $(u(t), v(t))$ is also a solution of 8.14 for $t_1 - a < t < t_2 - a$. The solution $(u(t), v(t))$ is a different solution from the solution $(x(t), y(t))$, but the two solutions trace out the same trajectory, or solution, of equation 8.13. Of great interest is the limiting behavior of these solutions as $t \to \pm\infty$ if the domain is infinite.

In Chapter 4 we discovered that differential equations such as 8.13 can have solutions where either x is constant, y is constant, or both are constant. The case where both x and y are constant is of special importance in the study of system 8.14. If this happens, both derivatives are 0, so we have

$$f(x, y) = g(x, y) = 0.$$

Such a point is called a **stationary point** or an **equilibrium point** of the system. This idea is related to that of a critical point of a vector field. There is a unique solution that is stationary at each equilibrium point, and uniqueness says that no nonequilibrium solution ever passes through an equilibrium point. So at any point in the plane, either there is a solution that "moves through" the point, or else there is a solution that is stationary at the point.

How do the "moving" solutions behave in the neighborhood of an equilibrium point? This is a very interesting topic for study, and we will look at it briefly.

The Linear Approximation to a System

We first concentrate on what happens in a neighborhood of a point. Expand both $f(x, y)$ and $g(x, y)$ in a Taylor series (in two variables) about the point (a, b). This expresses values of the functions at (x, y) in terms of information that is concentrated at the point (a, b). We then have

$$f(x, y) = f(a, b) + f_x(a, b)(x - a) + f_y(a, b)(y - b) + \text{higher-order terms}$$

and

$$g(x,y) = g(a,b) + g_x(a,b)(x-a) + g_y(a,b)(y-b) + \text{higher-order terms},$$

where f_x denotes $\partial f/\partial x$ and similarly for the other partials, f_y, g_x, and g_y. Since higher-order terms approach 0 faster than linear terms, we can estimate local behavior by simply ignoring all higher-order terms. When we do so, the system

$$\begin{cases} \dfrac{d(y-b)}{dt} = \dfrac{dy}{dt} = f(a,b) + f_x(a,b)(x-a) + f_y(a,b)(y-b) \\ \dfrac{d(x-a)}{dt} = \dfrac{dx}{dt} = g(a,b) + g_x(a,b)(x-a) + g_y(a,b)(y-b) \end{cases} \tag{8.15}$$

results. This says that near any point, system 8.14 is essentially a linear system. System 8.15 is called the **linear approximation** to system 8.14. An especially interesting thing happens at an equilibrium point: $f(a,b) = g(a,b) = 0$, so the linear system that gives clues to the behavior in the vicinity of an equilibrium point is actually homogeneous.

$$\begin{cases} \dfrac{d(y-b)}{dt} = f_x(a,b)(x-a) + f_y(a,b)(y-b) \\ \dfrac{d(x-a)}{dt} = g_x(a,b)(x-a) + g_y(a,b)(y-b) \end{cases}. \tag{8.16}$$

If we write $Y = y - b$, $X = x - a$, $g_x(a,b) = \alpha$, $g_y(a,b) = \beta$, $f_x(a,b) = \gamma$, $f_y(a,b) = \delta$, and reverse the order of the equations, then system 8.16 becomes even simpler.

$$\begin{cases} \dfrac{dX}{dt} = \alpha X + \beta Y \\ \dfrac{dY}{dt} = \gamma X + \delta Y \end{cases},$$

or

$$\frac{d}{dt}\begin{pmatrix} X \\ Y \end{pmatrix} = \begin{pmatrix} \alpha X + \beta Y \\ \gamma X + \delta Y \end{pmatrix} = \begin{pmatrix} \alpha & \beta \\ \gamma & \delta \end{pmatrix}\begin{pmatrix} X \\ Y \end{pmatrix} = \boldsymbol{A}\begin{pmatrix} X \\ Y \end{pmatrix} \tag{8.17}$$

where $\boldsymbol{A} = \begin{pmatrix} \alpha & \beta \\ \gamma & \delta \end{pmatrix}$ is a constant matrix. Constant coefficients systems of this type are the topic under discussion in this chapter, and this is the form we will study in this section. We now classify the ways an autonomous first-order system such as 8.14 can behave in a neighborhood of an equilibrium point.

Behavior in the Neighborhood of an Equilibrium

The coefficient matrix $\boldsymbol{A} = \begin{pmatrix} \alpha & \beta \\ \gamma & \delta \end{pmatrix}$ can have one of a quite small number of canonical forms that describe every possibility. We can immediately identify the solution in each of these cases. The basic classification is into real or complex characteristic roots. What then matters is distinct or repeated roots. Here are three real cases that arise from real matrices. They are basically all of the possibilities.

$$\begin{pmatrix} \alpha & 0 \\ 0 & \delta \end{pmatrix}, \quad \begin{pmatrix} \alpha & 1 \\ 0 & \alpha \end{pmatrix}, \quad \begin{pmatrix} \alpha & \beta \\ -\beta & \alpha \end{pmatrix},$$

where α, β, γ, and δ are all real. We will encounter one special case of these. A related case that is not truly canonical also often arises. These are:

$$\begin{pmatrix} 0 & \beta \\ -\beta & 0 \end{pmatrix}, \quad \begin{pmatrix} 0 & \beta \\ \gamma & 0 \end{pmatrix}.$$

In each of the above cases, we want the nonzero entries that appear to truly be nonzero. This means that we want $\det \boldsymbol{A} \neq 0$. Situations in which $\det \boldsymbol{A} = 0$ are called **degenerate**. For the matrix $\boldsymbol{A}$ taken to be any of these cases, we determine all solutions (as functions of t) and the trajectories to which they correspond. As we go along further interesting special cases will arise. Note that a linear differential system $\boldsymbol{y}' = \boldsymbol{A}\boldsymbol{y}$ always has the origin as an equilibrium point, and that the requirement that $\det \boldsymbol{A} \neq 0$ means that there are no other equilibrium points. In what follows, the origin is the equilibrium point under discussion.

Consider the special case where $\alpha = \delta$ and $\beta = \gamma = 0$. We know the solution of

$$\begin{pmatrix} x \\ y \end{pmatrix}' = \begin{pmatrix} \alpha & 0 \\ 0 & \alpha \end{pmatrix} \begin{pmatrix} x \\ y \end{pmatrix}$$

to be $x(t) = c_1 e^{\alpha t}$ and $y(t) = c_2 e^{\alpha t}$. This means that

$$c_1 y(t) = c_2 x(t).$$

For each choice of c_1 and c_2 this is a parameterization of a straight line through the origin (our equilibrium point). The static form of this differential system is

$$\frac{dy}{dx} = \frac{\alpha\, y}{\alpha\, x} = \frac{y}{x},$$

a separable differential equation whose solutions are just $c_1 y = c_2 x$. So the trajectories are straight lines that radiate from the equilibrium point, and the parameterized solutions lie on those trajectories. Dynamically, we find that if $\alpha > 0$, then both $x(t)$ and $y(t)$ become infinite as $t \to \infty$, and $x(t)$ and $y(t)$ both approach 0 as $t \to -\infty$. Thus the parameterized solution radiates outward from the origin as t increases. When this happens to all solutions, the equilibrium point is called a **source**. If $\alpha < 0$, the static solution is unchanged, but the parameterization becomes $x(t) = c_1 e^{-\alpha t}$ and $y(t) = c_2 e^{-\alpha t}$. This parameterized solution approaches the origin radially as t increases. When this happens to all solutions, the equilibrium point is called a **sink**. A sink and a source are illustrated in Figures 8.5 and 8.6.

An equilibrium point that has trajectories approaching it from all directions is called a **node** or **proper node**. An equilibrium that has trajectories approaching it from only two directions is an **improper node**.

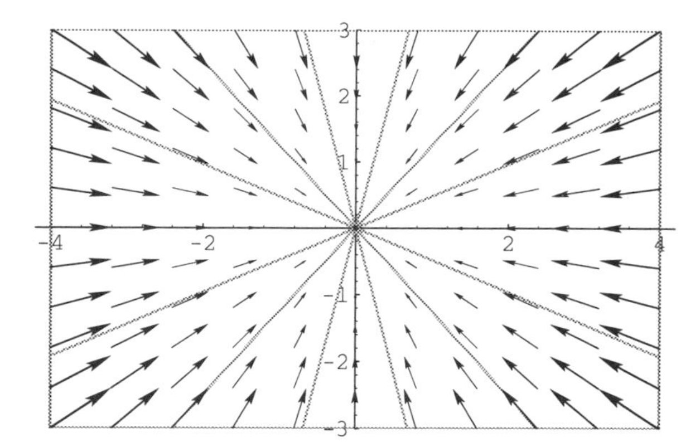

Figure 8.5: A proper node. Sink.

Let $\boldsymbol{A} = \begin{pmatrix} \alpha & 0 \\ 0 & \delta \end{pmatrix}$, where $\alpha \neq \delta$. The degenerate special case when $\delta = 0$ is illustrated in Figures 8.7 and 8.8. There, every point on the y-axis is an equilibrium point. This is an example of the type of degeneracy mentioned above: there are equilibrium points other than the origin, and the equilibrium points are not isolated from one another. When $\boldsymbol{A}$ is not degenerate, other very interesting things happen.

For example, let $\alpha > 0$, and consider

$$\begin{pmatrix} x \\ y \end{pmatrix}' = \begin{pmatrix} \alpha & 0 \\ 0 & 2\alpha \end{pmatrix} \begin{pmatrix} x \\ y \end{pmatrix}.$$

The characteristic roots are α and 2α, so this system has as solution $x(t) = c_1 e^{\alpha t}, y(t) = c_2 e^{2\alpha t}$. It follows that both $x(t)/c_1$ and $y(t)/c_2$ are positive when c_1 and c_2 are nonzero. In addition,

$$\frac{y(t)}{c_2} = \left(\frac{x(t)}{c_1}\right)^2.$$

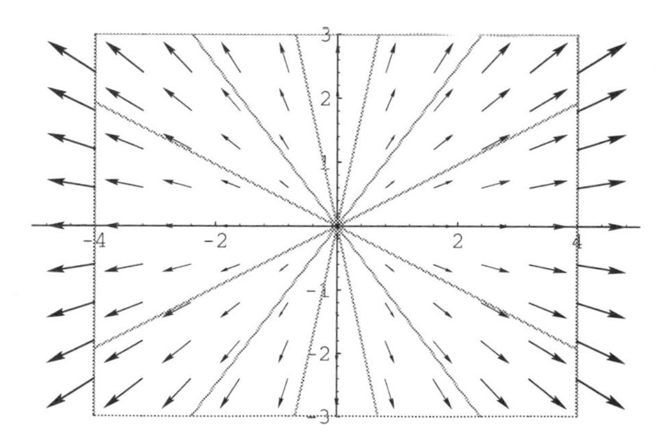

Figure 8.6: A proper node. Source.

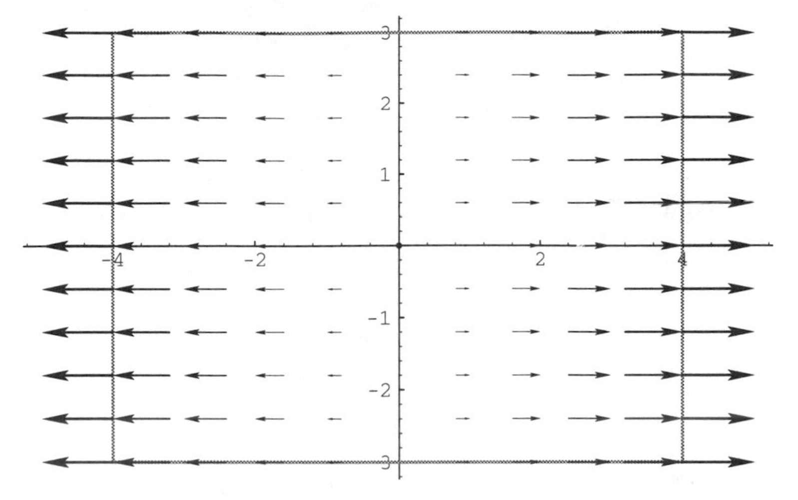

Figure 8.7: Degenerate–every point of the y-axis is an equilibrium.

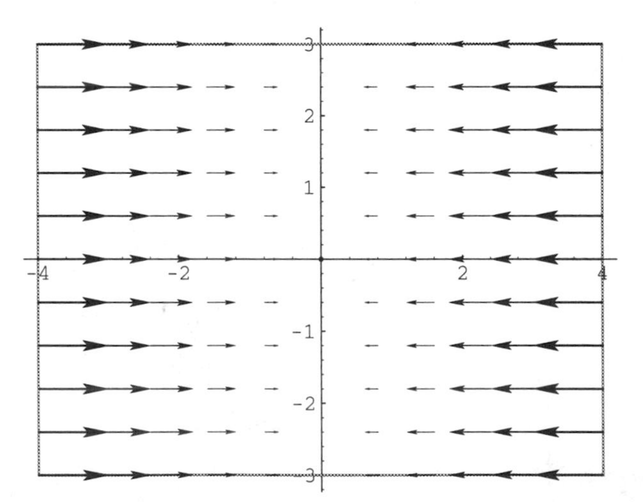

Figure 8.8: Also degenerate–every point of the y-axis is an equilibrium.

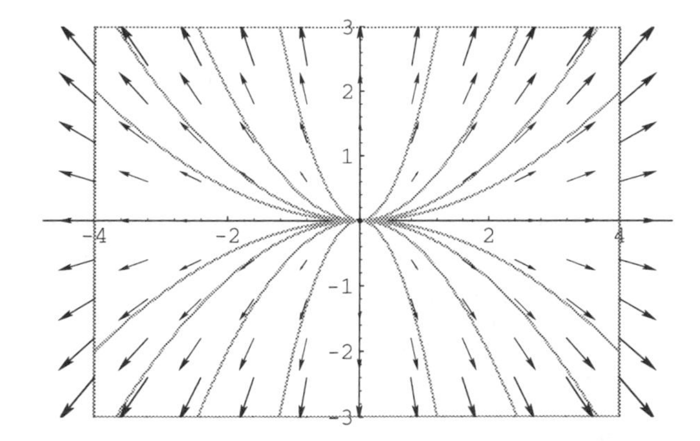

Figure 8.9: Improper node. Source. Trajectories are parabolas.

Thus $(x(t), y(t))$ parameterizes a portion of the parabola

$$y = c_2 \left(\frac{x}{c_1}\right)^2, \quad \text{or} \quad c_1^2 y = c_2 x^2.$$

The differential equation of the trajectories is $dy/dx = (2\alpha y)/(\alpha x) = 2y/x$. The trajectories, the solutions of this separable equation, are $\ln|y| = \ln x^2 + \ln c = \ln x^2 + \ln c_2 - \ln c_1^2$. This latter is equivalent to $c_1^2 y = c_2 x^2$, so each parametric solution parameterizes a member of this family of parabolas (including the degenerate cases when one of c_1 or c_2 is zero). Since $\alpha > 0$, when c_1 and c_2 are not both zero, $(x(t), y(t))$ moves away from $(0,0)$ as $t \to \infty$ (so $(0,0)$ is a source) and $(x(t), y(t))$ approaches $(0,0)$ as $t \to -\infty$. If $c_2 = 0$ the solution stays on the x-axis, whereas, if $c_1 = 0$, the solution stays on the y- axis. This is illustrated in Figures 8.9 and 8.10.

In the general case where $\boldsymbol{A} = \begin{pmatrix} \alpha & 0 \\ 0 & \delta \end{pmatrix}$, the characteristic roots are α and δ, and the parametric solution is $x(t) = c_1 e^{\alpha t}$ and $y(t) = c_2 e^{\delta t}$. From this it

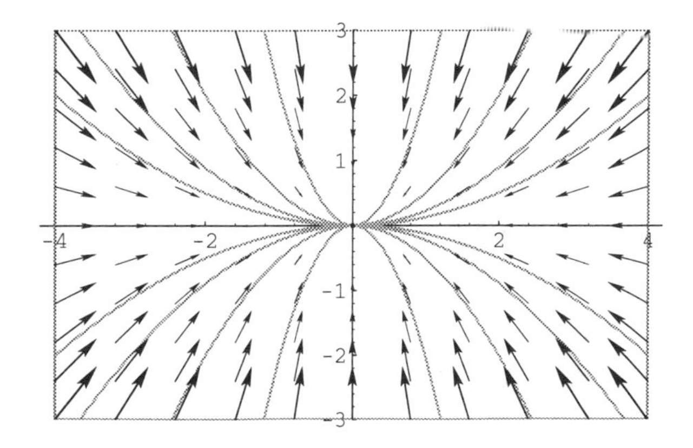

Figure 8.10: Improper node. Sink. Trajectories are parabolas.

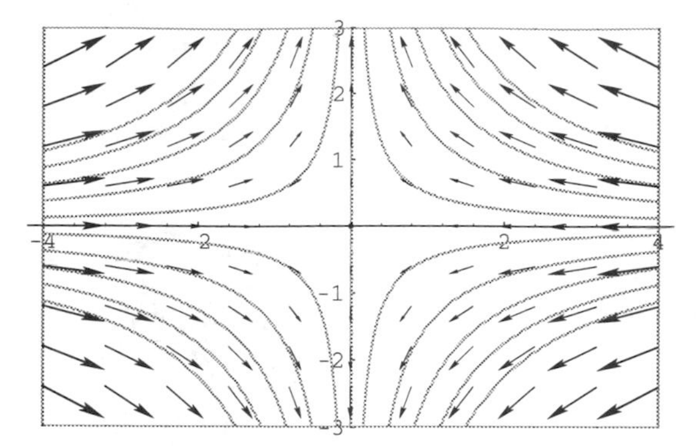

Figure 8.11: Saddle point.

follows that when neither c_1 nor c_2 is zero,

$$\left(\frac{y}{c_2}\right)^\delta = \left(\frac{x}{c_1}\right)^\alpha.$$

Because both y/c_2 and x/c_1 are positive, this equation holds for any real α and δ. Solutions can exist in any quadrant, and no solution leaves a quadrant in which it starts. A solution that starts away from zero on either axis stays on that axis on the same side of the origin. When α and δ are both positive, any parameterized solution that starts away from the origin moves away from the origin, and the origin is a source. When α and δ are both negative, any parameterized solution that starts away from the origin stays away from the origin, but approaches the origin as t increases. Hence the origin is a sink. These statements deserve careful thought.

A new kind of behavior occurs when α and δ have different signs. Then, as $t \to \infty$, one of $x(t)$ or $y(t)$ approaches the origin and the other approaches either positive or negative infinity, depending on the sign of the appropriate c_i. This produces the behavior illustrated in Figures 8.11 and 8.12. The trajectories act

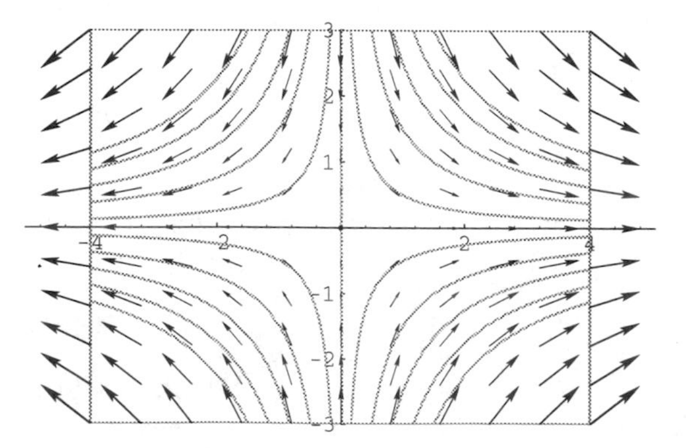

Figure 8.12: Another saddle point.

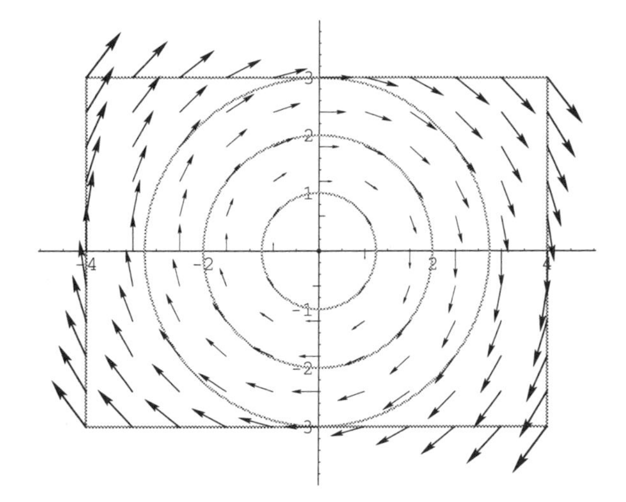

Figure 8.13: Center. Trajectories are circles.

rather like hyperbolas in the neighborhood of the origin, first approaching, then turning, and finally receding from the origin. Such an equilibrium point is called a **saddle point**.

A still different kind of behavior occurs when $\boldsymbol{A} = \begin{pmatrix} \alpha & \beta \\ -\beta & \alpha \end{pmatrix}$. Here the characteristic roots are $\alpha \pm \beta i$, so

$$\begin{pmatrix} x(t) \\ y(t) \end{pmatrix} = e^{\alpha t} \begin{pmatrix} \cos(\beta t) & \sin(\beta t) \\ -\sin(\beta t) & \cos(\beta t) \end{pmatrix} \begin{pmatrix} c_1 \\ c_2 \end{pmatrix}.$$

In the special case when $\alpha = 0$, every solution is clearly periodic, and the equilibrium point is called a **center**. See Figures 8.13 and 8.14.

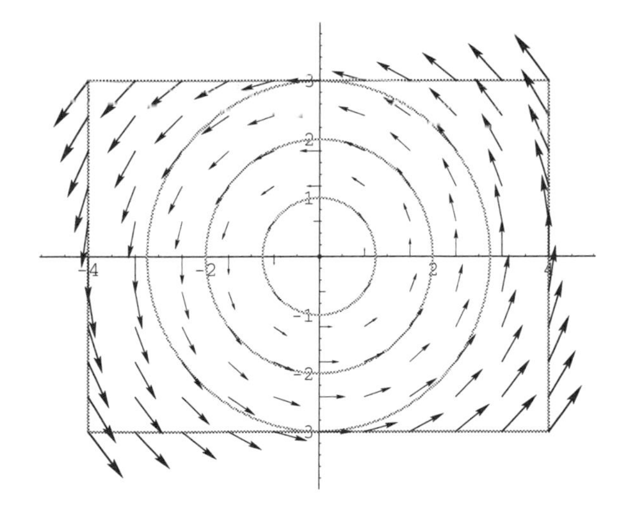

Figure 8.14: Center. Trajectories are circles. Opposite orientation.

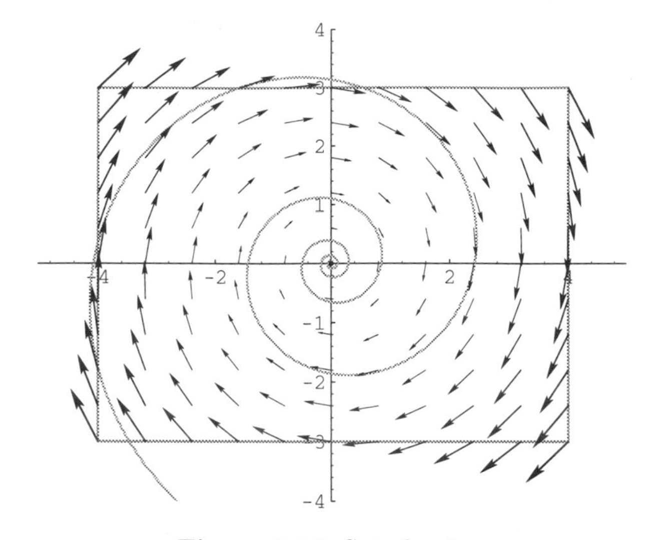

Figure 8.15: Spiral point.

When $\alpha < 0$, each solution approaches 0 along a spiral path as $t \to \infty$. This follows from the calculation that the distance of the solution from the origin is a constant multiple of $e^{\alpha t}$, and when $\alpha < 0$ this approaches 0 as $t \to \infty$. In this case, the origin is called a **spiral point**. This case is illustrated in Figures 8.15 and 8.16.

When $\alpha > 0$, each solution spirals outward away from the origin. The direction of the spiral depends on the sign of β. In this case also the origin is called a **spiral point**.

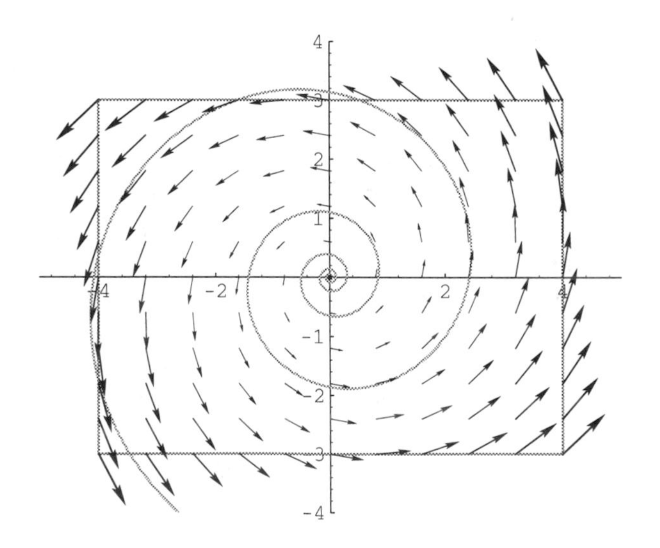

Figure 8.16: Spiral point. Opposite orientation.

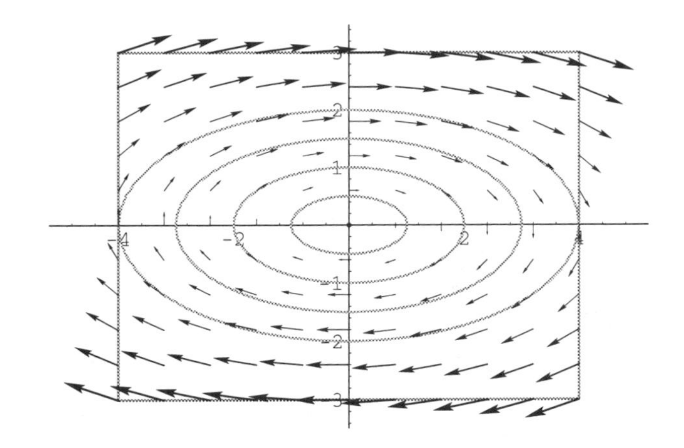

Figure 8.17: Center. Elliptical trajectories.

In the special case $A = \begin{pmatrix} 0 & \beta \\ \gamma & 0 \end{pmatrix}$ with β and γ having opposite signs, the characteristic roots are pure imaginary, the solutions are periodic, follow ellipses, and the origin is again a **center**. See Figures 8.17 and 8.18.

If β and γ have the same sign, then the characteristic roots are real and distinct and this is an earlier diagonalizable case. Note that you must be alert to signs as a part of the form that you are examining.

The remaining case is that of repeated real roots, where $A = \begin{pmatrix} \alpha & 1 \\ 0 & \alpha \end{pmatrix}$. Here the matrix A is not diagonalizable. In this case, the parametric solution swirls around the origin, but does not spiral. The solution is $x(t) = (c_1 + c_2 t)e^{\alpha t}$, $y(t) = c_1 e^{\alpha t}$. Note that, depending on the sign of α, $x(t)$ has exactly one maximum (or minimum), and that $y(t)$ has none. See Figures 8.19 and 8.20 for illustrations

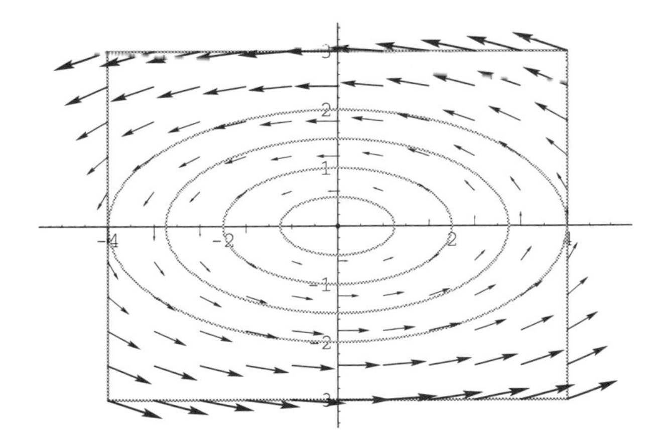

Figure 8.18: Center. Elliptical trajectories. Opposite orientation.

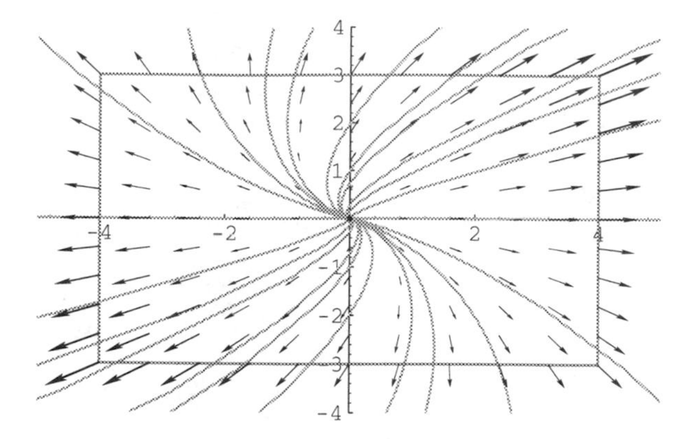

Figure 8.19: Improper node. Repeated root.

of this behavior. In this case the origin is called a **node** or **improper node**. Depending on the sign of α, the parameterized solution can either approach or depart from the origin as $t \to \infty$.

This briefly summarizes types of behavior that can occur in the neighborhood of an equilibrium point.

First-Order Systems with More than Two Equations

It is reasonable to expect, based on the above discussion, that systems such as

$$\boldsymbol{y}' = \boldsymbol{A}\boldsymbol{y}$$

having more than two equations should behave much like systems of two equations, except that there is the possibility of somewhat greater complication, and, indeed, this is the case. Basically, it matters whether or not the characteristic

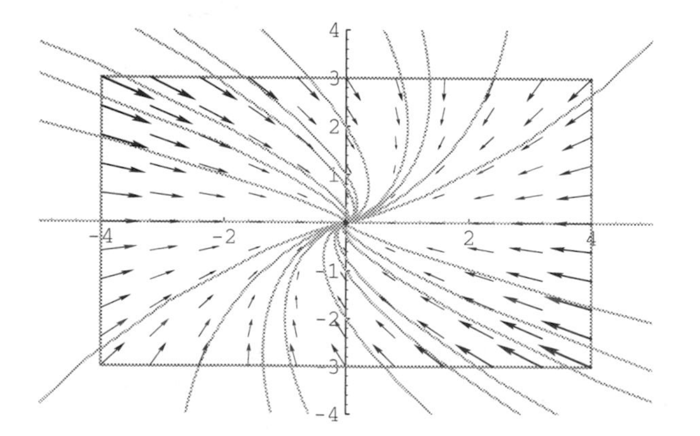

Figure 8.20: Another improper node. Repeated root. Opposite orientation.

roots are repeated or distinct, and what the sign of the real part of each root is. The real part of the characteristic root determines how the exponential factor behaves, and this is the factor that ultimately determines how far the solution gets from the origin. Repeated roots cause polynomial factors to appear, but these polynomial factors are dominated by exponentials. Periodic factors, which arise from complex roots, are well-behaved; here is where the impact of the sign of the real part can be seen clearly. If the real part of a complex root is 0, then a periodic solution occurs. If the real part of a complex root is negative, the exponential goes to 0 as $t \to \infty$, and the solution spirals to the origin. If the real part of a complex root is positive, then the solution spirals away from the origin.

If a characteristic root is real, its sign says whether solutions built from it approach or depart from 0.

Since solutions that result from real characteristic roots typically look like

$$\boldsymbol{y}(t) = c_1 \boldsymbol{k_1} e^{r_1 t} + c_2 \boldsymbol{k_2} e^{r_2 t} + \cdots + c_m \boldsymbol{k_m} e^{r_m t}.$$

If all but one c_i are 0, then the solution lies on a line through the origin. (Why?) If exactly two c_i are nonzero, then the solution lies in a plane, and is essentially like the two by two case discussed earlier. The pattern continues, with the complexity increasing as the dimension increases. But in each two-dimensional subspace, we know the behavior.

Nonlinear Dynamical Systems: Some Examples

The previous discussion was reasonably complete for linear systems. It is therefore reasonably complete for neighborhoods of equilibria of nonlinear systems. There is another fascinating phenomenon that can occur for nonlinear systems which does not occur for linear systems. For this reason alone, it is not sufficient to concentrate entirely on local behavior of a system about a point of equilibrium.

To illustrate this additional phenomenon, consider the nonlinear differential system

$$\begin{cases} \dfrac{dx}{dt} = 4y + x(4 - x^2 - y^2) \\ \dfrac{dy}{dt} = -4x + y(4 - x^2 - y^2) \end{cases}. \tag{8.18}$$

The origin is the only equilibrium point of the system. At the origin, the system behaves like

$$\begin{cases} \dfrac{dx}{dt} = 4x + 4y \\ \dfrac{dy}{dt} = -4x + 4y \end{cases},$$

which has characteristic roots $4 \pm 4i$. The origin is therefore a spiral point with the solution receding with increasing t.

In order to get global information about the solution, let $x = r\cos\theta$ and $y = r\sin\theta$, where x, y, r, and θ are all functions of t. This expresses the problem in polar coordinates. We determine $r(t)$ and $\theta(t)$ and then construct $x(t)$ and $y(t)$ from these. From $r^2 = x^2 + y^2$, differentiation with respect to t reveals that

$$r\frac{dr}{dt} = x\frac{dx}{dt} + y\frac{dy}{dt}.$$

Using the values of dx/dt and dy/dt from system 8.18, and the polar definitions of x and y reduces this further to

$$r\frac{dr}{dt} = r^2(4 - r^2).$$

Thus, either $r = 0$, which gives us the equilibrium solution, or

$$\frac{dr}{dt} = r(4 - r^2).$$

This separable equation has constant solutions $r = 0$, $r = 2$ and $r = -2$. We can then determine $r(t)$ when $r(t)$ is not constant from

$$\int \frac{dr}{r(4 - r^2)} = \int dt = t + k.$$

Integrating gives

$$\frac{1}{4}\ln|r| - \frac{1}{8}\ln|4 - r^2| = t + k.$$

After multiplying through by 8 and combining logarithms, this finally reduces to

$$\frac{r^2}{|4 - r^2|} = e^{8(t+k)}.$$

In order to proceed, it is important to know whether we want a solution where $4 - r^2 > 0$ or one where $4 - r^2 < 0$. That is, will our solution start inside or outside the circle $x^2 + y^2 = 4$ which is discreetly hiding in system 8.18?

Let $c^2 = e^{8k}$. When $4 - r^2 > 0$, the solution is

$$r(t) = \frac{2c}{\sqrt{c^2 + e^{-8t}}},$$

for all real t, and when $4 - r^2 < 0$, the solution is

$$r(t) = \frac{2c}{\sqrt{c^2 - e^{-8t}}}$$

for $t > (-\ln c^2)/8$.

Now determine $\theta(t)$. Differentiate $\tan\theta = y/x$ with respect to t to find that

$$\sec^2\theta\frac{d\theta}{dt} = \frac{x(dy/dt) - y(dx/dt)}{x^2}.$$

Using the differential equations and the polar definitions for $x(t)$ and $y(t)$ further reduces this to

$$\sec^2\theta \frac{d\theta}{dt} = -4\sec^2\theta,$$

or

$$\frac{d\theta}{dt} = -4.$$

Thus $\theta(t) = -4t + c$.

Combining the definitions of $r(t)$ and $\theta(t)$ yields these possibilities:

$$\begin{cases} x(t) = 0 \\ y(t) = 0 \end{cases} \quad \text{for all t. [This is the equilibrium solution.]}$$

$$\begin{cases} x(t) = 2\cos(-4t + c) \\ y(t) = 2\sin(-4t + c) \end{cases}. \quad \text{[This is a parameterized circle.]}$$

$$\begin{cases} x(t) = \dfrac{2c}{\sqrt{c^2 + e^{-8t}}}\cos(-4t + c) \\ y(t) = \dfrac{2c}{\sqrt{c^2 + e^{-8t}}}\sin(-4t + c) \end{cases}, \quad r < 2. \quad \text{[Valid for all } t\text{.]}$$

$$\begin{cases} x(t) = \dfrac{2c}{\sqrt{c^2 - e^{-8t}}}\cos(-4t + c) \\ y(t) = \dfrac{2c}{\sqrt{c^2 - e^{-8t}}}\sin(-4t + c) \end{cases}, \quad r > 2. \quad \text{[Valid for } t > (-\ln c^2)/8.\text{]}$$

The solutions obtained for $r < 0$ do not differ from these. (Replace c by $c + \pi$.)

Figure 8.21 depicts the underlying vector field and one solution heading outward from the origin. The limiting circle is the new object we were seeking. Every

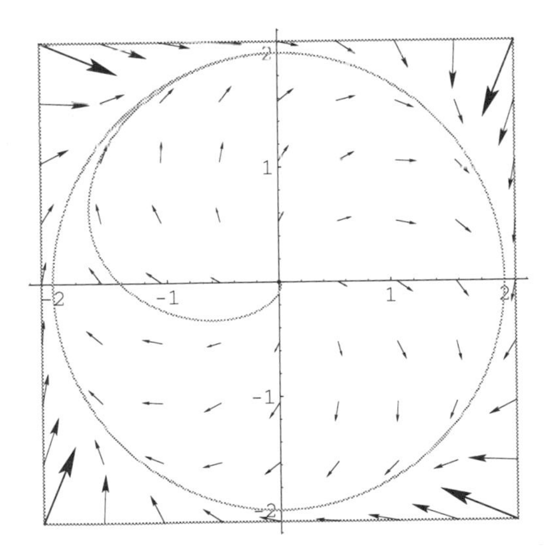

Figure 8.21: Circular limit cycle.

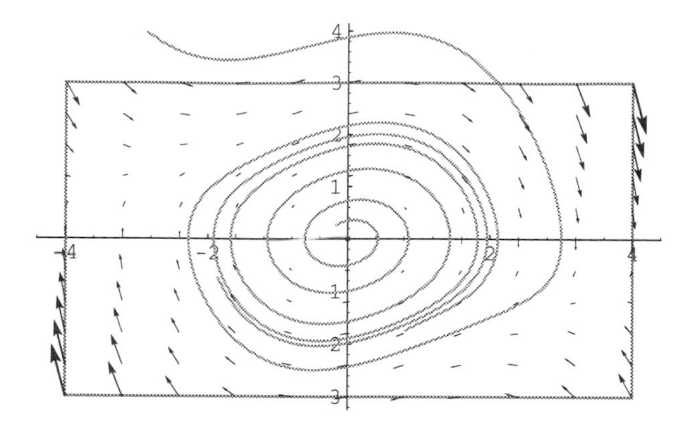

Figure 8.22: Van der Pol. Two trajectories—one going out; one going in.

nonconstant solution approaches it. This circle is an example of a **limit cycle**. Of course, limit cycles can be much more complicated than this.

A famous example of a complicated limit cycle that arose in the study of radio tubes is the van der Pol equation, which in its system form is:

$$\begin{cases} \dfrac{dx}{dt} = y \\ \dfrac{dy}{dt} = (-1/4)(x^2 - 1)y - x. \end{cases}$$

Two solutions and the underlying vector field of the van der Pol equation are shown in Figure 8.22. This system is studied extensively in texts on dynamical systems.

Chaos

This quick look at phase portraits and geometric differential equations cannot do justice to the richness of the field. Soon after beginning such a study, it is apparent that there are systems of nonlinear differential equations that exhibit a startling sensitivity to their initial conditions. No equation that we have yet studied has this property, but the Lorenz[4] equations

$$\begin{cases} \dfrac{dx}{dt} = \sigma(y - x) \\ \dfrac{dy}{dt} = \rho x - y - xz \\ \dfrac{dz}{dt} = -\beta z + xy \end{cases}$$

[4]Edward Lorenz (1917-), American meteorologist. He proposed his famous equations in 1963 while studying models of air flowing as a fluid within the atmosphere. They were the first truly simple differential equations to exhibit chaotic behavior.

are a famous example. Such equations are described as having solutions that exhibit **chaotic** behavior. Chaos and the dynamical systems that exhibit it are an important topic for further study.

EXERCISES 8.7

PART I. For the coefficient matrices below, the differential system $\boldsymbol{y}' = \boldsymbol{A}\boldsymbol{y}$ is in a canonical form or one of the standard special cases. Examine the characteristic roots of each matrix. From them classify the behavior of the solutions in the vicinity of the origin. Write down the general solution of each system. Plot one or more solutions using `ParametricPlot`. Use the package `:Graphics:PlotVectorField.m` to plot the underlying vector field. Optionally, use `Show` to combine plots of a solution and the vector field into a single plot.

1. $\boldsymbol{A} = \begin{pmatrix} -2 & 0 \\ 0 & -2 \end{pmatrix}$
2. $\boldsymbol{A} = \begin{pmatrix} 2 & 0 \\ 0 & 2 \end{pmatrix}$
3. $\boldsymbol{A} = \begin{pmatrix} -2 & 0 \\ 0 & -1 \end{pmatrix}$
4. $\boldsymbol{A} = \begin{pmatrix} 2 & 0 \\ 0 & 1 \end{pmatrix}$
5. $\boldsymbol{A} = \begin{pmatrix} -2 & 0 \\ 0 & 4 \end{pmatrix}$
6. $\boldsymbol{A} = \begin{pmatrix} 2 & 0 \\ 0 & 4 \end{pmatrix}$
7. $\boldsymbol{A} = \begin{pmatrix} 3 & 1 \\ 0 & 3 \end{pmatrix}$
8. $\boldsymbol{A} = \begin{pmatrix} -3 & 1 \\ 0 & -3 \end{pmatrix}$
9. $\boldsymbol{A} = \begin{pmatrix} 0 & 2 \\ -2 & 0 \end{pmatrix}$
10. $\boldsymbol{A} = \begin{pmatrix} 0 & -2 \\ 2 & 0 \end{pmatrix}$
11. $\boldsymbol{A} = \begin{pmatrix} -1 & 2 \\ -2 & -1 \end{pmatrix}$
12. $\boldsymbol{A} = \begin{pmatrix} 1 & 2 \\ -2 & 1 \end{pmatrix}$

PART II. The matrices below are similar to the matrices in Part I that have the same number. The transforming matrix is $\boldsymbol{P} = \begin{pmatrix} 2 & 3 \\ 1 & 2 \end{pmatrix}$. The similarity transform is $\boldsymbol{P}^{-1}\boldsymbol{A}\boldsymbol{P}$, where $\boldsymbol{A}$ is from Part I. Analyze the behavior near the equilibrium point and plot the solutions as was done in Part I. Describe the differences that the transform by $\boldsymbol{P}$ has produced. (Problems 1 and 2 are unchanged under similarity transformations.)

3a. $\boldsymbol{P}^{-1}\boldsymbol{A}\boldsymbol{P} = \begin{pmatrix} -5 & -6 \\ 2 & 2 \end{pmatrix}$

4a. $\boldsymbol{P}^{-1}\boldsymbol{A}\boldsymbol{P} = \begin{pmatrix} 5 & 6 \\ -2 & -2 \end{pmatrix}$

5a. $\boldsymbol{P}^{-1}\boldsymbol{A}\boldsymbol{P} = \begin{pmatrix} -20 & -36 \\ 12 & 22 \end{pmatrix}$

6a. $\boldsymbol{P}^{-1}\boldsymbol{A}\boldsymbol{P} = \begin{pmatrix} 20 & 36 \\ -12 & -22 \end{pmatrix}$

7a. $\boldsymbol{P}^{-1}\boldsymbol{A}\boldsymbol{P} = \begin{pmatrix} 5 & 4 \\ -1 & 1 \end{pmatrix}$

8a. $\boldsymbol{P}^{-1}\boldsymbol{A}\boldsymbol{P} = \begin{pmatrix} -1 & 4 \\ -1 & -5 \end{pmatrix}$

9a. $\boldsymbol{P}^{-1}\boldsymbol{A}\boldsymbol{P} = \begin{pmatrix} 16 & 26 \\ -10 & -16 \end{pmatrix}$

10a. $\boldsymbol{P}^{-1}\boldsymbol{A}\boldsymbol{P} = \begin{pmatrix} -16 & -26 \\ 10 & 16 \end{pmatrix}$

11a. $\boldsymbol{P}^{-1}\boldsymbol{A}\boldsymbol{P} = \begin{pmatrix} 15 & 26 \\ -10 & -17 \end{pmatrix}$

12a. $\boldsymbol{P}^{-1}\boldsymbol{A}\boldsymbol{P} = \begin{pmatrix} 17 & 26 \\ -10 & -15 \end{pmatrix}$

PART III. Nonlinear differential systems.

13. Consider the nonlinear differential system

$$\begin{cases} \dfrac{dx}{dt} = xy \\ \dfrac{dy}{dt} = -y \end{cases}$$

Show that each point of the x-axis is an equilibrium point, and that there are no other equilibrium points. This means that this is a degenerate case.

Show that the set of solutions is

$$\begin{cases} x(t) = ae^{-(ce^{-t})} \\ y(t) = ce^{-t} \end{cases},$$

and that the parameters a and c can be chosen so that any point of the plane is on some solution. Solve the equation $dy/dx = -1/x$ for the trajectories, noting that $x = 0$ is a constant solution. Show that each point on the x-axis is approached by solution curves from exactly two directions. (This is symptomatic of a degeneracy.) Plot the vector field and some representative solutions.

14. Consider the nonlinear differential system

$$\begin{cases} \dfrac{dx}{dt} = -y\cos x \\ \dfrac{dy}{dt} = x\cos y \end{cases}$$

Show that the origin and the rectangular lattice of points

$$\left(\frac{(2m+1)\pi}{2}, \frac{(2n+1)\pi}{2}\right)$$

for m and n integer are all of the equilibrium points of the system. Show that if (p, q) is one of the equilibrium points, then the linear approximation to the system near (p, q) is

$$\begin{pmatrix} X \\ Y \end{pmatrix}' = \begin{pmatrix} q\sin p & -\cos p \\ \cos q & -p\sin q \end{pmatrix} \begin{pmatrix} X \\ Y \end{pmatrix},$$

where $X = x - p$ and $Y = y - q$. Classify the sixteen equilibria nearest the origin. Can you determine a pattern for the distribution of the equilibria? Plot the vector field to graphically support your classification. Find all vertical and horizontal constant solutions of the equation

$$\frac{dy}{dx} = \frac{x\cos y}{-y\cos x}$$

of the trajectories. Note the sign of dy/dx in the regions between the vertical and horizontal solutions. Note how, except near the origin, any solution seems to "join" two equilibrium points.

15. Consider the nonlinear differential system

$$\begin{cases} \dfrac{dx}{dt} = 4y + x(4 - x^2 - y^2)(1 - x^2 - y^2) \\ \dfrac{dy}{dt} = -4x + y(4 - x^2 - y^2)(1 - x^2 - y^2) \end{cases}$$

Transform to polar coordinates. Show that the origin is the only equilibrium. Show that either $r = 0$ or $dr/dt = r(4 - r^2)(1 - r^2)$, so that r has constant solutions $0, \pm 1, \pm 2$. Show that r is increasing between 0 and 1, decreasing between 1 and 2, and is increasing when $r > 2$. Show also that $d\theta/dt = -4$. Identify two limit cycles. Describe the global behavior of the solution. Plot the underlying vector field to obtain support for your description. Note how the behavior near the two limit cycles is different.

16. Consider the nonlinear differential system

$$\begin{cases} \dfrac{dx}{dt} = 36y + x(4 - x^2 - y^2)(1 - x^2 - y^2)(9 - x^2 - y^2)^2 \\ \dfrac{dy}{dt} = -36x + y(4 - x^2 - y^2)(1 - x^2 - y^2)(9 - x^2 - y^2)^2 \end{cases}$$

Perform the analysis that was done on the last problem. Take note of the fact that still another kind of behavior takes place near the outer limit cycle. Limit cycles are sometimes called **stable**, or **unstable**, or **semistable**. Propose a definition for these terms based on the behavior that you see for the solutions of this system. Seek out a textbook on dynamical systems or other reference to compare their accepted definitions with yours.

PART IV. Project. Suppose that you want to visualize the behavior of a solution that lies in some two-dimensional subspace of $\boldsymbol{R}^n$. Let the solution have the form $\boldsymbol{y}(t) = f_1(t)\boldsymbol{v_1} + f_2(t)\boldsymbol{v_2}$ where $\boldsymbol{v_1}$ and $\boldsymbol{v_2}$ are nonzero, nonparallel vectors in $\boldsymbol{R}^n$, and $f_1(t)$ and $f_2(t)$ are differentiable scalar functions defined for $t \in \boldsymbol{R}$. It is easy to make in $\boldsymbol{R}^2$ a faithful copy of the plane in $\boldsymbol{R}^n$ that contains $y(t)$ for all t.

The linear transformation T on $\boldsymbol{R}^n$ that takes $\boldsymbol{v_1}$ to $\boldsymbol{u_1}$ and $\boldsymbol{v_2}$ to $\boldsymbol{u_2}$ and preserves the angle between $\boldsymbol{v_1}$ and $\boldsymbol{v_2}$ is what we want. We know the angle θ from

$$\cos\theta = \frac{\boldsymbol{v_1} \cdot \boldsymbol{v_2}}{|\boldsymbol{v_1}|\,|\boldsymbol{v_2}|}$$

We can therefore take

$$\boldsymbol{u_1} = T(\boldsymbol{v_1}) = |\boldsymbol{v_1}|(1, 0) = (|\boldsymbol{v_1}|, 0),$$

and

$$\boldsymbol{u_2} = T(\boldsymbol{v_2}) = |\boldsymbol{v_2}|(\cos\theta, \sin\theta) = (|\boldsymbol{v_2}|\cos\theta, |\boldsymbol{v_2}|\sin\theta)$$

Then

$$\begin{aligned} T(\boldsymbol{y}(t)) &= T(f_1(t)\boldsymbol{v_1} + f_2(t)\boldsymbol{v_2}) \\ &= f_1(t)T(\boldsymbol{v_1}) + f_2(t)T(\boldsymbol{v_2}) \\ &= f_1(t)\boldsymbol{u_1} + f_2(t)\boldsymbol{u_2} \end{aligned}$$

Check that $|\boldsymbol{v_1}| = |\boldsymbol{u_1}|$, $|\boldsymbol{v_2}| = |\boldsymbol{u_2}|$ and the angle between $\boldsymbol{u_1}$ and $\boldsymbol{u_2}$ is θ. Thus the copy in the plane is faithful to the original in $\boldsymbol{R}^n$, though it is not "identical."

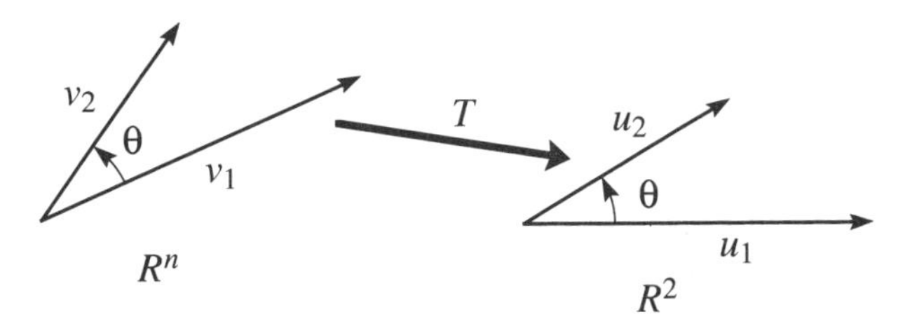

Figure 8.23: The transformation T.

Apply this technique to plot

$$\boldsymbol{y}(t) = \begin{pmatrix} -1 \\ 2 \\ 3 \end{pmatrix} \cos t + \begin{pmatrix} 2 \\ 0 \\ -1 \end{pmatrix} \sin t$$

faithfully in the plane.

8.8 Two Nonlinear Examples (Optional)

The n-body problem governing the idealized motion of a body in a gravitational field produced by $n-1$ larger masses produces a system of n nonlinear second-order differential equations. In this section we look at the two-body problem. The **Volterra-Lotka equations**, also known as the **predator-prey equations,** are also discussed.

Newton's Second Law and Planetary Orbits

We generally follow the exposition of Blake Temple and C. A. Tracy (1992). Suppose that in a three-dimensional coordinate system we have a large object of mass M_s located at $\boldsymbol{r}_s$ and a small object of mass M_p located at $\boldsymbol{r}_p$ and we wish to study the motion of the smaller object about the larger. (Think of the sun and one planet, for instance.) From Newton's second law we can express these two vector equations for the force on each object due to the presence of the other. The force on object a due to object b has magnitude inversely proportional to the square of the distance separating the objects and acts in the direction opposite the unit vector from b to a. Equation 8.19 shows the force equation for the smaller object and the force equation for the larger object.

$$\begin{aligned} M_p \frac{d^2 \boldsymbol{r}_p}{dt^2} &= -\frac{G_0 M_s M_p}{|\boldsymbol{r}_p - \boldsymbol{r}_s|^3} (\boldsymbol{r}_p - \boldsymbol{r}_s) \\ M_s \frac{d^2 \boldsymbol{r}_s}{dt^2} &= -\frac{G_0 M_s M_p}{|\boldsymbol{r}_s - \boldsymbol{r}_p|^3} (\boldsymbol{r}_s - \boldsymbol{r}_p). \end{aligned} \tag{8.19}$$

Let

$$\boldsymbol{r} = \boldsymbol{r}_p - \boldsymbol{r}_s,$$

and

$$\boldsymbol{r}_0 = \frac{M_p \boldsymbol{r}_p + M_s \boldsymbol{r}_s}{M_p + M_s},$$

which is the center of mass of the system. We seek implications of equations 8.19.

Add equations 8.19 to get

$$\begin{aligned}\frac{d^2\boldsymbol{r}_0}{dt^2} &= \frac{M_p(d^2\boldsymbol{r}_p/dt^2) + M_s(d^2\boldsymbol{r}_s/dt^2)}{M_p + M_s} \\ &= -\frac{G_0 M_s M_p}{(M_p + M_s)|\boldsymbol{r}_s - \boldsymbol{r}_p|^3}(\boldsymbol{r}_s - \boldsymbol{r}_s) = \boldsymbol{0}.\end{aligned}$$

This says that $\boldsymbol{r}_0(t) = c_1 t + c_2$, which says that $\boldsymbol{r}_0$ is undergoing unrestricted motion in a straight line. This is the motion to which Newton's first law refers.

Subtract the slightly simplified equations

$$\frac{d^2\boldsymbol{r}_p}{dt^2} = -\frac{G_0 M_s}{|\boldsymbol{r}_p - \boldsymbol{r}_s|^3}(\boldsymbol{r}_p - \boldsymbol{r}_s)$$

and

$$\frac{d^2\boldsymbol{r}_s}{dt^2} = -\frac{G_0 M_p}{|\boldsymbol{r}_s - \boldsymbol{r}_p|^3}(\boldsymbol{r}_s - \boldsymbol{r}_p)$$

to get

$$\begin{aligned}\frac{d^2\boldsymbol{r}}{dt^2} &= \frac{d^2\boldsymbol{r}_p}{dt^2} - \frac{d^2\boldsymbol{r}_s}{dt^2} \\ &= -\frac{G_0}{|\boldsymbol{r}|^3}(M_s\boldsymbol{r}_p - M_s\boldsymbol{r}_s + M_p\boldsymbol{r}_s - M_p\boldsymbol{r}_p) \\ &= -\frac{G_0}{|\boldsymbol{r}|^3}(M_s\boldsymbol{r} - M_p\boldsymbol{r}) \\ &= -\frac{G_0(M_s - M_p)}{|\boldsymbol{r}|^3}\boldsymbol{r},\end{aligned}$$

Now

$$G_0(M_s - M_p) = G_0 M_s\left(1 - \frac{M_p}{M_s}\right) \approx G_0 M_s = G,$$

since M_s is much larger than M_p. Hence

$$\frac{d^2\boldsymbol{r}}{dt^2} = -\frac{G}{|\boldsymbol{r}|^3}\boldsymbol{r}. \tag{8.20}$$

This is the equation of motion that Newtonian mechanics predicts. Initial conditions associated with 8.20 are $\boldsymbol{r}(0) = \boldsymbol{\alpha}_0$ and $(d\boldsymbol{r}/dt)(0) = \boldsymbol{\alpha}_1$, where $\boldsymbol{\alpha}_0$ and $\boldsymbol{\alpha}_1$

are nonzero and nonparallel. If $\boldsymbol{\alpha}_0$ and $\boldsymbol{\alpha}_1$ are parallel, then the smaller object is moving toward or away from the larger object and the discussion of escape velocity applies if the objects are separating. In astronomy it is often convenient to determine the initial conditions at perihelion, the moment of minimum separation, because then the form of the solution is very simple.

The point $\boldsymbol{r}_s$ and the vectors $\boldsymbol{\alpha}_0$ and $\boldsymbol{\alpha}_1$ determine a plane π_s. The vector $\boldsymbol{r}_p(t) = \boldsymbol{r}(t) + \boldsymbol{r}_s$ lies in the plane π_s as we now show. The normal to π_s is the vector cross product $\boldsymbol{\alpha}_0 \times \boldsymbol{\alpha}_1$. We show that the vector $\boldsymbol{v}(t) = \boldsymbol{r}(t) \times (d\boldsymbol{r}/dt)$, which is perpendicular to both $\boldsymbol{r}(t)$ and $(d\boldsymbol{r}/dt)$, is constant. This constant value is $\boldsymbol{\alpha}_0 \times \boldsymbol{\alpha}_1$. Recall from equation 8.20 that $(d^2\boldsymbol{r}/dt^2)$ is parallel to $\boldsymbol{r}$ so that

$$\frac{d\boldsymbol{v}}{dt} = \frac{d\boldsymbol{r}}{dt} \times \frac{d\boldsymbol{r}}{dt} + \boldsymbol{r}(t) \times \frac{d^2\boldsymbol{r}}{dt^2} = \mathbf{0} + \mathbf{0} = \mathbf{0}.$$

Consider a rectangular coordinate system in π_s with origin at $\boldsymbol{r}_s$. Let $\boldsymbol{r}(t) = (x(t), y(t))$ in this coordinate system. In terms of x and y equation 8.20 becomes

$$\begin{cases} \dfrac{d^2x}{dt^2} = -\dfrac{G}{|\boldsymbol{r}|^3 x} \\ \dfrac{d^2y}{dt^2} = -\dfrac{G}{|\boldsymbol{r}|^3 y} \end{cases} \tag{8.21}$$

It is convenient to express $(x(t), y(t))$ in polar coordinates as $x(t) = r(t)\cos(\theta(t))$, $y(t) = r(t)\sin(\theta(t))$. Then in the polar coordinates $(r(t), \theta(t))$ we find that

$$\begin{aligned} \frac{d^2x}{dt^2} &= \frac{d^2r}{dt^2}\cos(\theta) - 2\frac{dr}{dt}\sin(\theta)\frac{d\theta}{dt} - r\cos(\theta)\left(\frac{d\theta}{dt}\right)^2 - r\sin(\theta)\frac{d^2\theta}{dt^2} \\ &= -\frac{G}{r^2}\cos(\theta) \end{aligned}$$

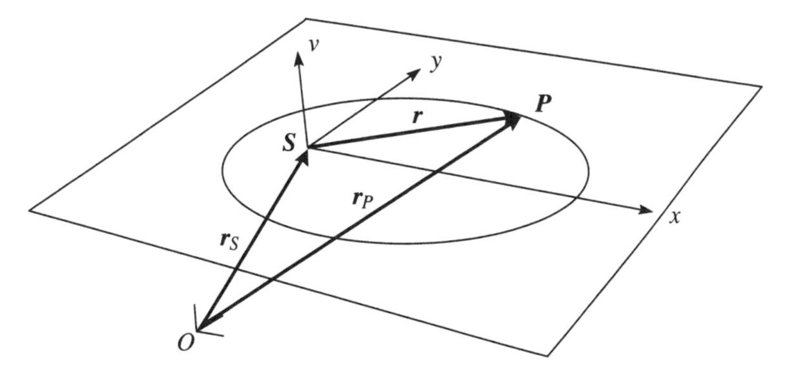

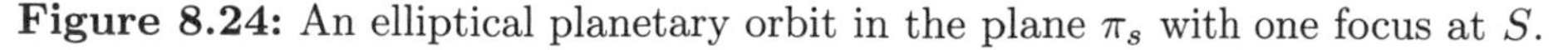
Figure 8.24: An elliptical planetary orbit in the plane π_s with one focus at S.

and

$$\frac{d^2y}{dt^2} = \frac{d^2r}{dt^2}\sin(\theta) + 2\frac{dr}{dt}\cos(\theta)\frac{d\theta}{dt} - r\sin(\theta)\left(\frac{d\theta}{dt}\right)^2 + r\cos(\theta)\frac{d^2\theta}{dt^2}$$
$$= -\frac{G}{r^2}\sin(\theta).$$

This special form permits us to manipulate further. Calculate

$$\sin(\theta)\frac{d^2x}{dt^2} - \cos(\theta)\frac{d^2y}{dt^2} = -2\frac{dr}{dt}\frac{d\theta}{dt} - r\frac{d^2\theta}{dt^2} = \frac{1}{r}\frac{d}{dt}\left(r^2\frac{d\theta}{dt}\right) = 0.$$

This means that $r^2(dq/dt) = H$, a nonzero constant. Thus, from $(d\theta/dt) = (H/r^2) \neq 0$, we learn that θ is monotone. [Note that if $H = 0$, then either $r = 0$ or $(d\theta/dt) = 0$. If the former, then the smaller object has fallen into the larger, and if the latter, then the motion is in a straight line.] Now calculate

$$\cos(\theta)\frac{d^2x}{dt^2} + \sin(\theta)\frac{d^2y}{dt^2} = \frac{d^2r}{dt^2} - r\left(\frac{d\theta}{dt}\right)^2 = -\frac{G}{r^2}$$

or

$$\frac{d^2r}{dt^2} - \frac{H^2}{r^3} = -\frac{G}{r^2}.$$

By the chain rule we find that

$$\frac{dr}{dt} = \frac{dr}{d\theta}\frac{d\theta}{dt}$$

and

$$\frac{d^2r}{dt^2} = \frac{d^2r}{d\theta^2}\left(\frac{d\theta}{dt}\right)^2 + \frac{dr}{d\theta}\frac{d^2\theta}{dt^2}.$$

We can express

$$\frac{d^2\theta}{dt^2} = \frac{-2(dr/dt)(d\theta/dt)}{r}$$
$$= \frac{-2(dr/dt)H}{r^3}$$
$$= \frac{-2H}{r^3}\frac{dr}{d\theta}\frac{d\theta}{dt}$$
$$= \frac{-2H^2}{r^5}\frac{dr}{d\theta}.$$

Therefore we have

$$\frac{d^2r}{d\theta^2}\left(\frac{H}{r^2}\right)^2 + \frac{dr}{d\theta}\left(\frac{-2H^2}{r^5}\frac{dr}{d\theta}\right) - \frac{H^2}{r^3} = -\frac{G}{r^2}.$$

Multiply by r^2 to get

$$\frac{d^2r}{d\theta^2}\frac{H^2}{r^2} - 2\left(\frac{dr}{d\theta}\right)^2\frac{H^2}{r^3} = \frac{H^2}{r} - G.$$

This is the equation that was studied in Chapter 5 by reducing it to a second-order linear problem. Its solutions describe the usual elliptical orbits of planets, and the more unusual paths taken by some comets.

Predator-Prey Equations

In 1926, Lotka[5] (1926) built on foundations established by Volterra[6] to propose a model of two competing species. Let $x(t)$ denote the number of some prey species such as rabbits that are present at time t, and let $y(t)$ denote the number of some predator species such as foxes that feeds exclusively on the prey species. The prey species feeds on an external source of food such as grass or other vegetation that is assumed to be plentifully in supply. The prey propagate in accordance with the model of unrestricted growth that we studied earlier, but they die due to encounters with the predator population. The predators, on the other hand, would tend to die off were they not sustained by the feeding that results from successful encounters with the prey.

Here are the **Volterra-Lotka** equations that were proposed to model this situation.

$$\begin{cases} \dfrac{dx}{dt} = ax - bxy \\ \dfrac{dy}{dt} = -cy + dxy \end{cases}, \tag{8.22}$$

where a, b, c, and d are positive constants of proportionality. Observe that there is an equilibrium point at $x = y = 0$ and at $x = c/d, y = a/b$. The point $(0, 0)$ is a saddle point and the point $(c/d, a/b)$ is a center. (See Section 8.7.) This system is another where it is possible to reduce the system to a single differential equation and it is productive to do so. For equations 8.22 dividing yields

$$\frac{dy}{dx} = \frac{-cy + dxy}{ax - bxy} = \left(\frac{dx - c}{x}\right)\left(\frac{y}{a - by}\right),$$

which is an equation with variables separable. We observe that two new constant solutions have appeared: $x = 0$ and $y = 0$. These correspond to the extinction of one or the other of the two species. There are two other equations of interest: $dx - c = 0$ and $a - by = 0$. At these places the derivative has an especially

[5]Alfred J. Lotka (1880–1949), American physicist-turned-biologist. He was born in Austria of American parents, built on foundations established by Volterra to propose a model of two competing species.

[6]Vito Volterra (1860–1940), Italian mathematician. As a student, he published a paper of examples that prompted Lebesgue to propose his own integral that was more powerful than the Riemann integral.

interesting behavior. When $x = (c/d)$, then $(dy/dx) = 0$, and when $y = (a/b)$ then (dy/dx) is vertical $((dx/dy) = 0)$. We will see that along each solution curve each of these happens twice during each "cycle." After solving the separable equation we will use each of equations 8.22 to study the behavior of solutions to this model.

Separate variables over intervals where $x \neq 0$, $y \neq 0$, $x \neq (c/d)$, and $y \neq (a/b)$. This leads to

$$\int \frac{a - by}{y} dy = \int \frac{dx - c}{x} dx,$$

or

$$a \ln y - by = dx - c \ln x + k.$$

The constant k can be estimated by a count of populations in the field. This says that the solutions to the Volterra-Lotka predator-prey equations are the level curves of the function $f(x, y) = a \ln y - by - dx + c \ln x$. Figure 8.25 is how some of these level curves might look.

These curves, properly parameterized, are the solution curves we seek. As you can see in Figure 8.25, the solution curves are closed. Why this should be is often examined in a second course in differential equations. We would like to determine just how a typical solution curve is traversed. Does the solution go around clockwise or counterclockwise?

Suppose that $(x(t_0), y(t_0)) = (x_0, y_0)$ is on a solution curve and that $x_0 > (c/d)$ and $y_0 > (a/b)$. Then $x'(t_0) = ax_0 - bx_0y_0 = x_0(a - by_0) < 0$, which says that $x(t)$ is decreasing at t_0. Also $y'(t_0) = -cy_0 + dx_0y_0 = y_0(-c + dx_0) > 0$, which says that $y(t)$ is increasing at t_0. This says that $y(x)$ is moving left and up, which means that the solution traverses the closed path in a counterclockwise

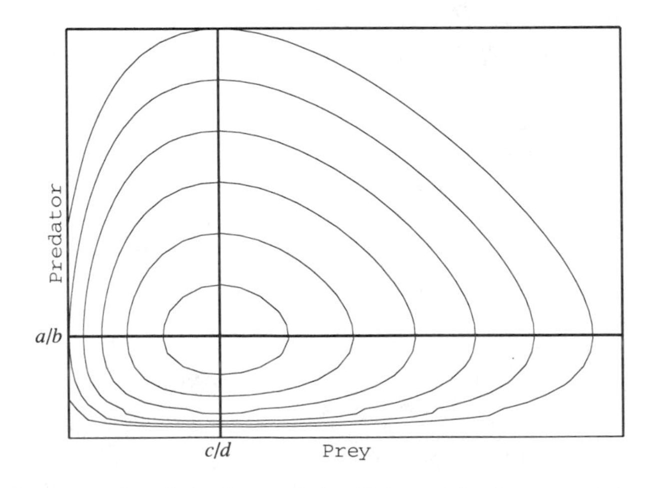

Figure 8.25: Solutions of the Volterra-Lotka equations.

direction. Since $(dy/dt) = 0$ when $x = (c/d)$, both the top and bottom of each closed path occurs when $x = (c/d)$. Also, $(dx/dt) = 0$ when $y = (a/b)$, so the left and right extremes of each path occur when $y = (a/b)$. The minimum of the prey population occurs when $x = (c/d)$ and $y < (a/b)$ because

$$\begin{aligned}\left.\frac{d^2y}{dt^2}\right|_{x=c/d} &= -c\frac{dy}{dt} + d\frac{dx}{dt}y + dx\frac{dy}{dt}\\ &= (dx - c)\frac{dy}{dt} + dy\frac{dx}{dt}\\ &= dxy(a - by) > 0.\end{aligned}$$

Similar remarks and calculations can be made about the other extremes of $x(t)$ and $y(t)$.

During the years 1847 to 1903 the Hudson Bay Company kept data on the populations of lynx and hares in Canada. The variations conform quite closely to what this model would predict, even though there were more than two species competing within the ecosystem under study. This and other studies are reported in Leigh (1968).

8.9 Defective Systems of First-Order Differential Equations (Optional)

Consider the vector constant coefficients linear differential operator

$$\boldsymbol{L}(\boldsymbol{y}) = \boldsymbol{A}\boldsymbol{y}' - \boldsymbol{P}\boldsymbol{y} = \begin{pmatrix} \boldsymbol{I}_r & \boldsymbol{0} \\ \boldsymbol{0} & \boldsymbol{0}_{n-r} \end{pmatrix}\boldsymbol{y}' - \boldsymbol{P}\boldsymbol{y},$$

where the matrices $\boldsymbol{A}$ and $\boldsymbol{P}$ are $n \times n$ and $\boldsymbol{A}$ has only $r < n$ linearly independent rows. Regularly partition $\boldsymbol{P}$ like $\boldsymbol{A}$ as follows:

$$\boldsymbol{P} = \begin{pmatrix} \boldsymbol{P}_{11} & \boldsymbol{P}_{12} \\ \boldsymbol{P}_{21} & \boldsymbol{P}_{22} \end{pmatrix},$$

so that $\boldsymbol{P}_{11}$ is $r \times r$ and $\boldsymbol{P}_{22}$ is $(n - r) \times (n - r)$. Assume that $\boldsymbol{P}_{22}$ is invertible. The differential equation is then

$$\boldsymbol{L}(\boldsymbol{y}) = \boldsymbol{A}\boldsymbol{y}' - \boldsymbol{P}\boldsymbol{y} = \begin{pmatrix} \boldsymbol{I}_r & \boldsymbol{0} \\ \boldsymbol{0} & \boldsymbol{0}_{n-r} \end{pmatrix}\boldsymbol{y}' - \begin{pmatrix} \boldsymbol{P}_{11} & \boldsymbol{P}_{12} \\ \boldsymbol{P}_{21} & \boldsymbol{P}_{22} \end{pmatrix}\boldsymbol{y} = \boldsymbol{0}.$$

Let $\boldsymbol{y} = \begin{pmatrix} \boldsymbol{u} \\ \boldsymbol{v} \end{pmatrix}$, where $\boldsymbol{u}$ has r rows and $\boldsymbol{v}$ has $n - r$ rows. Our equation looks like

$$\begin{pmatrix} \boldsymbol{I}_r & \boldsymbol{0} \\ \boldsymbol{0} & \boldsymbol{0}_{n-r} \end{pmatrix}\begin{pmatrix} \boldsymbol{u} \\ \boldsymbol{v} \end{pmatrix}' = \begin{pmatrix} \boldsymbol{P}_{11} & \boldsymbol{P}_{12} \\ \boldsymbol{P}_{21} & \boldsymbol{P}_{22} \end{pmatrix}\begin{pmatrix} \boldsymbol{u} \\ \boldsymbol{v} \end{pmatrix},$$

which can be multiplied out to give

$$\begin{cases} \boldsymbol{u}' = \boldsymbol{P}_{11}\boldsymbol{u} + \boldsymbol{P}_{12}\boldsymbol{v} \\ \boldsymbol{0} = \boldsymbol{P}_{21}\boldsymbol{u} + \boldsymbol{P}_{22}\boldsymbol{v} \end{cases}. \tag{8.23}$$

Since $\boldsymbol{P}_{22}$ is invertible, $\boldsymbol{v} = -\boldsymbol{P}_{22}^{-1}\boldsymbol{P}_{21}\boldsymbol{u}$, and

$$\begin{aligned} \boldsymbol{u}' &= \boldsymbol{P}_{11}\boldsymbol{u} + \boldsymbol{P}_{12}\boldsymbol{v} \\ &= \boldsymbol{P}_{11}\boldsymbol{u} - \boldsymbol{P}_{12}\boldsymbol{P}_{22}^{-1}\boldsymbol{P}_{21}\boldsymbol{u} \\ &= (\boldsymbol{P}_{11} - \boldsymbol{P}_{12}\boldsymbol{P}_{22}^{-1}\boldsymbol{P}_{21})\boldsymbol{u} \end{aligned}$$

or

$$\boldsymbol{u}' = \boldsymbol{B}\boldsymbol{u},$$

where $\boldsymbol{B} = \boldsymbol{P}_{11} - \boldsymbol{P}_{12}\boldsymbol{P}_{22}^{-1}\boldsymbol{P}_{21}$. To solve the original problem $\boldsymbol{L}(\boldsymbol{y}) = \boldsymbol{A}\boldsymbol{y}' - \boldsymbol{P}\boldsymbol{y} = \boldsymbol{0}$ for $\boldsymbol{y}$ we need to solve $\boldsymbol{u}' = \boldsymbol{B}\boldsymbol{u}$ for $\boldsymbol{u}$. This is an $r \times r$ system with constant coefficients having r linearly independent solutions, all representable in the form $\boldsymbol{u}(x) = \exp(\boldsymbol{B}x)\boldsymbol{K}$, where $\boldsymbol{K}$ is an arbitrary constant r-vector. Obtain $\boldsymbol{v}$ from

$$\boldsymbol{v}(x) = -\boldsymbol{P}_{22}^{-1}\boldsymbol{P}_{21}\boldsymbol{u}(x) = -\boldsymbol{P}_{22}^{-1}\boldsymbol{P}_{21}\exp(\boldsymbol{B}x)\boldsymbol{K}.$$

Then

$$\begin{aligned} \boldsymbol{y}(x) &= \begin{pmatrix} \boldsymbol{u}(x) \\ \boldsymbol{v}(x) \end{pmatrix} \\ &= \begin{pmatrix} \exp(\boldsymbol{B}x)\boldsymbol{K} \\ -\boldsymbol{P}_{22^{-1}}\boldsymbol{P}_{21}\exp(\boldsymbol{B}x)\boldsymbol{K} \end{pmatrix} \\ &= \begin{pmatrix} \boldsymbol{I}_r \\ -\boldsymbol{P}_{22}^{-1}\boldsymbol{P}_{21} \end{pmatrix} \exp(\boldsymbol{B}x)\boldsymbol{K}. \end{aligned}$$

A basis for the kernel of equations 8.23, which has dimension r, is the r columns of the matrix

$$\begin{pmatrix} \boldsymbol{I}_r \\ -\boldsymbol{P}_{22}^{-1}\boldsymbol{P}_{21} \end{pmatrix} \exp(\boldsymbol{B}x).$$

These vectors span the set of solutions of $\boldsymbol{L}(\boldsymbol{y}) = \boldsymbol{0}$, the original problem. We see that when the leading coefficient is not of full rank, then the differential system is effectively of lower order than it appears to be, and consequently there are fewer linearly independent solutions than we would have expected. The second equation in 8.23 is a linear system, so it is proper to think of $\boldsymbol{L}(\boldsymbol{y}) = \boldsymbol{0}$ as a differential system with algebraic side conditions. Because $\boldsymbol{P}_{22}$ is invertible, the algebraic system $\boldsymbol{0} = \boldsymbol{P}_{21}\boldsymbol{u} + \boldsymbol{P}_{22}\boldsymbol{v}$ has a kernel of dimension r (solve for $\boldsymbol{v}$ in terms of the r-vector $\boldsymbol{u}$). This means that not every possible point in $\boldsymbol{R}^n$ is a candidate to be an initial value for $\boldsymbol{L}(\boldsymbol{y}) = \boldsymbol{0}$. We saw that the subspace of possible initial values has dimension r. We found a subspace of solutions of dimension r. So the problem does actually only determine an r-dimensional subspace in $\boldsymbol{R}^n$. If $\boldsymbol{P}_{22}$ is not invertible, then the actual order of the system may be less than r. [See Ross, 1964.]

Example 8.23 Describe all of the solutions of the defective vector system

$$\begin{pmatrix} 1 & 0 & 0 & 0 & 0 \\ 0 & 1 & 0 & 0 & 0 \\ 0 & 0 & 0 & 0 & 0 \\ 0 & 0 & 0 & 0 & 0 \\ 0 & 0 & 0 & 0 & 0 \end{pmatrix} \boldsymbol{y}'(x) = \begin{pmatrix} -2 & 6 & 6 & 0 & 1 \\ -1 & -3 & 2 & 1 & 1 \\ -1 & 0 & 2 & 1 & 1 \\ 1 & 1 & -2 & 1 & 0 \\ 1 & -2 & -1 & 2 & 1 \end{pmatrix} \boldsymbol{y}(x).$$

Solution. Here there are two linearly independent rows in $\boldsymbol{A}$, so $r = 2$. Since there are five rows, we expect to solve the lower three rows for three variables in terms of two variables to get a system of two differential equations from which to produce the complete set of solutions, a linear space of dimension two. Partition $\boldsymbol{P}$ into four submatrices

$$\boldsymbol{P} = \left(\begin{array}{cc|ccc} -2 & 6 & 6 & 0 & 1 \\ -1 & -3 & 2 & 1 & 1 \\ \hline -1 & 0 & 2 & 1 & 1 \\ 1 & 1 & -2 & 1 & 0 \\ 1 & -2 & -1 & 2 & 1 \end{array}\right),$$

where

$$\boldsymbol{P}_{11} = \begin{pmatrix} -2 & 6 \\ -1 & -3 \end{pmatrix}, \quad \boldsymbol{P}_{12} = \begin{pmatrix} 6 & 0 & 1 \\ 2 & 1 & 1 \end{pmatrix}, \quad \boldsymbol{P}_{21} = \begin{pmatrix} -1 & 0 \\ 1 & 1 \\ 1 & -2 \end{pmatrix},$$

and

$$\boldsymbol{P}_{22} = \begin{pmatrix} 2 & 1 & 1 \\ -2 & 1 & 0 \\ -1 & 2 & 1 \end{pmatrix}.$$

Then $\det \boldsymbol{P}_{22} = 1$, and

$$\boldsymbol{v} = -\boldsymbol{P}_{22}^{-1}\boldsymbol{P}_{21}\boldsymbol{u} = \begin{pmatrix} 1 & -3 \\ 1 & -7 \\ -2 & 13 \end{pmatrix} \boldsymbol{u},$$

so the system we need to solve for $\boldsymbol{u}$ is

$$\boldsymbol{u}' = (\boldsymbol{P}_{11} - \boldsymbol{P}_{12}\boldsymbol{P}_{22}^{-1}\boldsymbol{P}_{21})\boldsymbol{u} = \begin{pmatrix} 2 & 1 \\ 0 & -3 \end{pmatrix} \boldsymbol{u},$$

which has as fundamental solution matrix

$$\boldsymbol{u}(x) = \begin{pmatrix} -e^{-3x} & 5e^{2x} \\ 5e^{-3x} & 0 \end{pmatrix}.$$

From this we find that

$$\begin{aligned} \boldsymbol{v}(x) &= -\boldsymbol{P}_{22}^{-1}\boldsymbol{P}_{21}\boldsymbol{u}(x) \\ &= \begin{pmatrix} 1 & -3 \\ 1 & -7 \\ -2 & 13 \end{pmatrix} \boldsymbol{u}(x) \end{aligned}$$

$$= \begin{pmatrix} 1 & -3 \\ 1 & -7 \\ -2 & 13 \end{pmatrix} \begin{pmatrix} -e^{-3x} & 5e^{2x} \\ 5e^{-3x} & 0 \end{pmatrix}$$

$$= \begin{pmatrix} -16e^{-3x} & 5e^{2x} \\ -36e^{-3x} & 5e^{2x} \\ 67e^{-3x} & -10e^{2x} \end{pmatrix}.$$

Then two vector solutions to the original problem are the columns of the matrix

$$\boldsymbol{y}(x) = \begin{pmatrix} \boldsymbol{u}(x) \\ \boldsymbol{v}(x) \end{pmatrix} = \begin{pmatrix} -e^{-3x} & 5e^{2x} \\ 5e^{-3x} & 0 \\ -16e^{-3x} & 5e^{2x} \\ -36e^{-3x} & 5e^{2x} \\ 67e^{-3x} & -10e^{2x} \end{pmatrix}.$$

These are easily checked by substitution into the original problem. Note that the system was actually of order two and not order five as it first appeared. When the form of the matrix $\boldsymbol{A}$ is not so clear, then it is more difficult to determine the order, because $\boldsymbol{A}$ must be row- and column-reduced to put it into the form used here before the order of the system can be determined. When performing the row- and column-reductions, the matrices that actually do the reduction must be captured for use to simultaneously reduce $\boldsymbol{P}$. ◇

When the submatrix $\boldsymbol{P}_{22}$ is singular but the $(n-r) \times n$ matrix $(\boldsymbol{P}_{21} \mid \boldsymbol{P}_{22})$ has full rank $n-r$, then the effective order of the defective system

$$\boldsymbol{L}(\boldsymbol{y}) = \boldsymbol{A}\boldsymbol{y}' - \boldsymbol{P}\boldsymbol{y} = \begin{pmatrix} \boldsymbol{I}_r & \boldsymbol{0} \\ \boldsymbol{0} & \boldsymbol{0} \end{pmatrix} \boldsymbol{y}' - \begin{pmatrix} \boldsymbol{P}_{11} & \boldsymbol{P}_{12} \\ \boldsymbol{P}_{21} & \boldsymbol{P}_{22} \end{pmatrix} \boldsymbol{y} = \boldsymbol{0}$$

is actually less than r, as the following example illustrates.

Example 8.24 Consider the defective system

$$\begin{pmatrix} 1 & 0 & 0 \\ 0 & 0 & 0 \\ 0 & 0 & 0 \end{pmatrix} \boldsymbol{y}' = \begin{pmatrix} -1 & 2 & c \\ 2 & 2 & -3 \\ 5 & 4 & -6 \end{pmatrix} \boldsymbol{y}$$

in which $\boldsymbol{P}_{22} = \begin{pmatrix} 2 & -3 \\ 4 & -6 \end{pmatrix}$ is not invertible. Show that the character of the solutions changes depending on whether or not the parameter $c = -3$.

Solution. The submatrix $(\boldsymbol{P}_{21} \mid \boldsymbol{P}_{22}) = \begin{pmatrix} 2 & 2 & -3 \\ 5 & 4 & -6 \end{pmatrix}$ has rank 2. We make the change of dependent variable

$$\boldsymbol{z} = \begin{pmatrix} 0 & 0 & 1 \\ 0 & 1 & 0 \\ 1 & 0 & 0 \end{pmatrix} \boldsymbol{y},$$

which interchanges the first and third rows of $\boldsymbol{y}$ and acts to interchange the first and third columns of $\boldsymbol{A}$ and of $\boldsymbol{P}$ when the substitution is made. This results in the system

$$\boldsymbol{B}\boldsymbol{z}' - \boldsymbol{Q}\boldsymbol{z} = \begin{pmatrix} 0 & 0 & 1 \\ 0 & 0 & 0 \\ 0 & 0 & 0 \end{pmatrix} \boldsymbol{z}' - \begin{pmatrix} c & 2 & -1 \\ -3 & 2 & 2 \\ -6 & 4 & 5 \end{pmatrix} \boldsymbol{z} = \boldsymbol{0}.$$

Solve the algebraic system in the lower two rows:

$$(\boldsymbol{Q}_{21} \mid \boldsymbol{Q}_{22})\boldsymbol{z} = \begin{pmatrix} -3 & 2 & 2 \\ -6 & 4 & 5 \end{pmatrix} \boldsymbol{z} = \boldsymbol{0}$$

in which $\boldsymbol{Q}_{22}$ is invertible. Then

$$\begin{pmatrix} z_2 \\ z_3 \end{pmatrix} = -\boldsymbol{Q}_{22}^{-1}\boldsymbol{Q}_{21} z_1 = \begin{pmatrix} 3/2 \\ 0 \end{pmatrix} z_1.$$

The substitution

$$\boldsymbol{z} = \begin{pmatrix} z_1 \\ 3z_1/2 \\ 0 \end{pmatrix}$$

leads to the "differential" equation $0 = cz_1 + 3z_1 = (c+3)z_1 = 0$. This is not a differential equation at all, but another algebraic equation: the fact that $\boldsymbol{P}_{22}$ was singular has reduced the order of the differential equation from 1 to 0. Observe that from the equation $(c+3)z_1 = 0$ we learn that if $c = -3$, then the original system has the one parameter family of solutions

$$\boldsymbol{y} = \begin{pmatrix} 0 & 0 & 1 \\ 0 & 1 & 0 \\ 1 & 0 & 0 \end{pmatrix}^{-1} \boldsymbol{z} = \begin{pmatrix} 0 & 0 & 1 \\ 0 & 1 & 0 \\ 1 & 0 & 0 \end{pmatrix} \begin{pmatrix} z_1 \\ 3z_1/2 \\ 0 \end{pmatrix} = \begin{pmatrix} 0 \\ 3z_1/2 \\ z_1 \end{pmatrix},$$

whereas, if $c \neq -3$, then the only solution is $\boldsymbol{y} = \boldsymbol{0}$ since if $c \neq -3$, then $z_1 = 0$. ◇

Exercises 8.9

1. Consider the deficient differential system $\boldsymbol{L}(\boldsymbol{y}) = \boldsymbol{A}\boldsymbol{y}' - \boldsymbol{P}\boldsymbol{y} = \boldsymbol{0}$. Suppose that the $n \times n$ constant matrix $\boldsymbol{A}$ has rank $r < n$ and that there are invertible matrices $\boldsymbol{M}$ and $\boldsymbol{Q}$ such that

$$\boldsymbol{M}\boldsymbol{A}\boldsymbol{Q} = \begin{pmatrix} \boldsymbol{I}_r & \boldsymbol{0} \\ \boldsymbol{0} & \boldsymbol{0} \end{pmatrix}.$$

   Let $\boldsymbol{y} = \boldsymbol{Q}w$.

   (a) Multiply the system $\boldsymbol{L}(\boldsymbol{Q}w) = \boldsymbol{0}$ on the left by $\boldsymbol{M}$.
   (b) Show that the resulting system is

$$\boldsymbol{M}\boldsymbol{A}\boldsymbol{Q}w' = \boldsymbol{M}\boldsymbol{P}\boldsymbol{Q}w = \boldsymbol{P}_1 w,$$

or

$$\begin{pmatrix} \boldsymbol{I}_r & \mathbf{0} \\ \mathbf{0} & \mathbf{0} \end{pmatrix} w' = \boldsymbol{P}_1 w,$$

which has the same form as the system we discussed above.

(c) Show further that if w is a solution of

$$\begin{pmatrix} \boldsymbol{I}_r & \mathbf{0} \\ \mathbf{0} & \mathbf{0} \end{pmatrix} w' = \boldsymbol{P}_1 w,$$

then $\boldsymbol{y} = \boldsymbol{Q}w$ is a solution of $\boldsymbol{L}(\boldsymbol{y}) = \mathbf{0}$.

2. Apply the results of exercise 1. Let

$$\boldsymbol{A} = \begin{pmatrix} 6 & -8 & 9 \\ 0 & 0 & 1 \\ -15 & 20 & -23 \end{pmatrix},$$

then

$$\boldsymbol{M} = \begin{pmatrix} 3 & 0 & 1 \\ 0 & 1 & 0 \\ 5 & 1 & 2 \end{pmatrix} \quad \text{and} \quad \boldsymbol{Q} = \begin{pmatrix} 3 & 0 & 4 \\ 2 & 1 & 3 \\ 0 & 1 & 0 \end{pmatrix}.$$

(a) Show that $\boldsymbol{M}$ and $\boldsymbol{Q}$ are invertible matrices such that

$$\boldsymbol{MAQ} = \begin{pmatrix} \boldsymbol{I}_2 & \mathbf{0} \\ \mathbf{0} & \mathbf{0} \end{pmatrix}.$$

(b) Use this fact to solve the defective differential system

$$\boldsymbol{L}(\boldsymbol{y}) = \begin{pmatrix} 6 & -8 & 9 \\ 0 & 0 & 1 \\ -15 & 20 & -23 \end{pmatrix} \boldsymbol{y}' - \begin{pmatrix} 20 & -28 & 31 \\ 1 & -1 & -1 \\ -50 & 70 & -77 \end{pmatrix} \boldsymbol{y} = \mathbf{0}.$$

[*Hint:* Let $\boldsymbol{z} = \boldsymbol{Qy}$. Substitute into the differential equation and premultiply by $\boldsymbol{M}$.]

3. Let $\boldsymbol{A} = \begin{pmatrix} -1 & -2 & 4 \\ -2 & -3 & 6 \\ -2 & -3 & 6 \end{pmatrix}$ and $\boldsymbol{P} = \begin{pmatrix} -2 & 0 & 0 \\ -4 & -3 & 6 \\ -4 & -3 & 6 \end{pmatrix}$.

(a) Find invertible matrices $\boldsymbol{M}$ and $\boldsymbol{Q}$ such that $\boldsymbol{MAQ} = \begin{pmatrix} \boldsymbol{I}_2 & \mathbf{0} \\ \mathbf{0} & \mathbf{0} \end{pmatrix}$.

(b) Solve the defective differential system $\boldsymbol{Ay}' - \boldsymbol{Py} = \mathbf{0}$. The notebook *Rank Canonical Form* may help you find $\boldsymbol{M}$ and $\boldsymbol{Q}$.

8.10 Solution of Linear Systems by Laplace Transforms (Optional)

The theorems and formulas about Laplace transforms developed in Chapter 6 are applicable without change to initial value problems for linear systems of differential equations. The simplified package **LPT.m** and the full-featured package

LaplaceTransform.m both work for systems as they did for a single differential equation initial value problem. Our approach thus far in this chapter has been to convert a mixed-order system into a first-order system and then use linear algebra to find the solution. When using Laplace transforms to solve initial value problems for systems, that conversion is unnecessary. In addition, once a fundamental solution of the homogeneous system has been obtained, one can proceed to solve boundary value problems of the kind covered in Section 8.5.

As discussed in Section 6.4, the nonhomogeneous parts of systems may be piecewise continuous as long as they are of exponential order. The Laplace transform still handles them easily, but the solutions may be very complicated. We still must piece together the parts of a piecewise continuous function by using $\mathcal{U}(t-c)$ and $\mathcal{P}(t,a,b)$. The steps in the solution process are otherwise unchanged from those introduced in Chapter 6. To apply a Laplace transform or inverse Laplace transform to a vector or matrix of expressions, simply apply the transform to each entry of the vector or matrix.

Here is a sequence of examples that progresses from simple to rather complicated.

Example 8.25 Use Laplace transforms to solve the initial value problem

$$y'(t) - z(t) = 0,$$
$$z'(t) + y(t) = 0; \quad y(0) = 1, \; z(0) = 0.$$

Solution. Transform the differential system.

$$s\,\mathcal{L}(y(t)) - y(0) - \mathcal{L}(z(t)) = 0,$$
$$s\,\mathcal{L}(z(t)) - z(0) + \mathcal{L}(y(t)) = 0.$$

Substitute the initial conditions $y(0) = 1, z(0) = 0$.

$$s\,\mathcal{L}(y(t)) - 1 - \mathcal{L}(z(t)) = 0,$$
$$s\,\mathcal{L}(z(t)) + \mathcal{L}(y(t)) = 0.$$

In its matrix form this system is

$$\begin{pmatrix} s & -1 \\ 1 & s \end{pmatrix} \begin{pmatrix} \mathcal{L}(y(t)) \\ \mathcal{L}(z(t)) \end{pmatrix} = \begin{pmatrix} 1 \\ 0 \end{pmatrix}.$$

Solve this system for $\mathcal{L}(y(t))$ and $\mathcal{L}(z(t))$.

$$\begin{pmatrix} \mathcal{L}(y(t)) \\ \mathcal{L}(z(t)) \end{pmatrix} = \begin{pmatrix} s & -1 \\ 1 & s \end{pmatrix}^{-1} \begin{pmatrix} 1 \\ 0 \end{pmatrix} = \begin{pmatrix} s/1+s^2 \\ -1/1+s^2 \end{pmatrix}.$$

Finally, apply inverse Laplace transforms.

$$\begin{pmatrix} y(t) \\ z(t) \end{pmatrix} = \begin{pmatrix} \mathcal{L}^{-1}(s/1+s^2) \\ \mathcal{L}^{-1}(-1/1+s^2) \end{pmatrix} = \begin{pmatrix} \cos t \\ -\sin t \end{pmatrix}.$$

This is the solution, as is easily checked. ◇

Example 8.25M Solve the system of Example 8.25 by using the function `LPTSolve` found in the package **LPT.m**.

Solution. Load the package `LPT.m`.

```
In[67]:= <<:RossDE:LPT.m
```

LPTSolve displays intermediate results as it finds them.

```
In[68]:= LPTSolve[{y'[t]-z[t] == 0,
                   z'[t]+y[t] == 0,
                   y[0] == 1,
                   z[0] == 0},
                   {y[t], z[t]}, t, s]
```

The transformed system

```
{s LPT[y[t], t, s] - LPT[z[t], t, s] - y[0] == 0,
 LPT[y[t], t, s] + s LPT[z[t], t, s] - z[0] == 0,
 y[0] == 1, z[0] == 0}
```

The unknown(s) isolated.

```
                        s                                  1
{LPT[y[t], t, s] -> ------, LPT[z[t], t, s] -> -(------)}
                         2                                  2
                    1 + s                              1 + s
```

```
Out[68]= {y[t] -> Cos[t], z[t] -> -Sin[t]}
```

This solution can now be captured and verified if desired. ◇

Example 8.26 Use Laplace transforms to solve the initial-value problem

$$y''(t) - z(t) = 0,$$
$$z'(t) + y(t) = 0; \quad y(0) = 1,\ y'(0) = 0; \quad z(0) = 0.$$

Solution. Transform the differential system.

$$s^2 \mathscr{L}(y(t)) - y'(0) - sy(0) - \mathscr{L}(z(t)) = 0,$$
$$s\mathscr{L}(z(t)) - z(0) + \mathscr{L}(y(t)) = 0.$$

Substitute the initial conditions $y(0) = 1$, $y'(0) = 0$, $z(0) = 0$.

$$s^2 \mathscr{L}(y(t)) - s - \mathscr{L}(z(t)) = 0,$$
$$s\mathscr{L}(z(t)) + \mathscr{L}(y(t)) = 0.$$

In its matrix form this system is

$$\begin{pmatrix} s^2 & -1 \\ 1 & s \end{pmatrix} \begin{pmatrix} \mathscr{L}(y(t)) \\ \mathscr{L}(z(t)) \end{pmatrix} = \begin{pmatrix} s \\ 0 \end{pmatrix}.$$

Solve this system for $\mathscr{L}(y(t))$ and $\mathscr{L}(z(t))$.

$$\begin{pmatrix} \mathscr{L}(y(t)) \\ \mathscr{L}(z(t)) \end{pmatrix} = \begin{pmatrix} s^2 & -1 \\ 1 & s \end{pmatrix}^{-1} \begin{pmatrix} s \\ 0 \end{pmatrix} = \begin{pmatrix} s^2/1+s^3 \\ -s/1+s^3 \end{pmatrix}.$$

Finally, apply inverse Laplace transforms.

$$\begin{pmatrix} y(t) \\ z(t) \end{pmatrix} = \begin{pmatrix} \mathscr{L}^{-1}\left(\dfrac{s^2}{1+s^3}\right) \\ \mathscr{L}^{-1}\left(\dfrac{-s}{1+s^3}\right) \end{pmatrix}$$

$$= \begin{pmatrix} \mathscr{L}^{-1}\left(\dfrac{1}{3(1+s)} + \dfrac{-1+2s}{3(1-s+s^2)}\right) \\ \mathscr{L}^{-1}\left(\dfrac{1}{3(1+s)} + \dfrac{1+s}{3(1-s+s^2)}\right) \end{pmatrix}$$

$$= \frac{1}{3}\begin{pmatrix} e^{-t} + 2e^{t/2}\cos(\sqrt{3}\,t/2) \\ e^{-t} + 2e^{t/2}\left(\cos(\sqrt{3}\,t/2) + \sqrt{3}\sin(\sqrt{3}\,t/2)\right) \end{pmatrix}.$$

This is the solution, as you should check. ◇

Example 8.26M Use the function `LPTSolve` provided in the package **LPT.m** to solve the initial value problem stated in Example 8.26.

Solution.

```
In[69]:= LPTSolve[{y''[t]-z[t] == 0,
                    z'[t]+y[t] == 0,
                    y[0] == 1,
                    y'[0] == 0,
                    z[0] == 0},
                   {y[t], z[t]}, t, s]
```

The transformed system

```
  2
{s  LPT[y[t], t, s] - LPT[z[t], t, s] - s y[0] - y'[0] ==

   0, LPT[y[t], t, s] + s LPT[z[t], t, s] - z[0] == 0,

  y[0] == 1, y'[0] == 0, z[0] == 0}
```

The unknown(s) isolated.

```
                         1             -1 + 2 s
{LPT[y[t], t, s] -> ---------- + ----------------,
                     3 (1 + s)                 2
                                   3 (1 - s + s )

                          1            1 + s
  LPT[z[t], t, s] -> ---------- - ----------------}
                      3 (1 + s)                 2
                                    3 (1 - s + s )
```

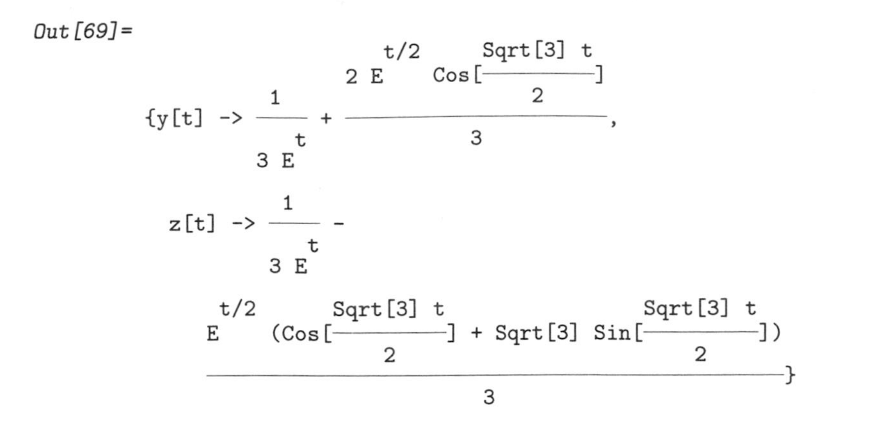

Example 8.27M Define

$$f(t) = \begin{cases} 1, & 0 \le t \le 2 \\ 0, & 2 < t \end{cases}.$$

Use the function `LPTSolve` provided in the package **LPT.m** to solve this initial value problem that has a piecewise continuous right-hand side:

$$y''(t) - z(t) = 0,$$
$$z'(t) + y(t) = f(t);$$
$$y(0) = 1$$
$$y'(0) = 0$$
$$z(0) = 0.$$

Solution.

```
In[70]:= LPTSolve[{y''[t]-z[t] == 0,
                   z'[t]+y[t] == 1-UnitStep[t-2],
                   y[0] == 1,
                   y'[0] == 0,
                   z[0] == 0},
                   {y[t], z[t]}, t, s]
```

The transformed system

```
  2
{s  LPT[y[t], t, s] - LPT[z[t], t, s] - s y[0] - y'[0] ==

   0, LPT[y[t], t, s] + s LPT[z[t], t, s] - z[0] ==

   1    1
   - - ------, y[0] == 1, y'[0] == 0, z[0] == 0}
   s    2 s
       E    s
```

The unknown(s) isolated.

```
{LPT[y[t], t, s] ->

        2 s
  -1 + E              1                    -1 + 2 s
  ----------- + ------------------ + -----------------------,
    2 s            2 s                  2 s            2
   E    s       3 E    (1 + s)       3 E    (1 - s + s )

                          1                   1 + s
  LPT[z[t], t, s] -> ------------------ - ----------------------}
                        2 s                  2 s           2
                     3 E    (1 + s)       3 E    (1 - s + s )

Out[70]=
                                                  2 - t
                                                 E      UnitStep[-2 + t]
          {y[t] -> 1 - UnitStep[-2 + t] + ------------------------- +
                                                         3

                 -1 + t/2      Sqrt[3] (-2 + t)
              2 E          Cos[----------------] UnitStep[-2 + t]
                                       2
              ------------------------------------------------------,
                                    3

                      -1 - t   3     (3 t)/2      Sqrt[3] (-2 + t)
            z[t] -> (E       (E  - E         Cos[----------------] -
                                                          2

                            (3 t)/2      Sqrt[3] (-2 + t)
                   Sqrt[3] E         Sin[----------------])
                                                 2

              UnitStep[-2 + t]) / 3}
```

◇

A close look at this solution shows that it has a two-part definition with the break at $t = 2$. If you run this example for yourself, you will find that it takes quite a bit of time to run to completion. But imagine doing it by hand!

This brief introduction to the use of Laplace transforms to solve nonhomogeneous initial value problems should help convince you that the Laplace transform is a powerful tool. As it has been introduced, the Laplace transform is applicable only to initial value problems for ordinary differential equations and systems, but they are very common problems. The Laplace transform has other applications, as well.

EXERCISES 8.10

Use the function `LPTSolve` from the package **LPT.m** to solve these initial value problems for systems.

1. $y_1'(t) = -y_1(t) + 2y_2(t) + 2y_3(t),$
   $y_2'(t) = 2y_1(t) + 2y_2(t) + 2y_3(t),$
   $y_3'(t) = -3y_1(t) - 6y_2(t) - 6y_3(t);$
   $y_1(0) = 1, y_2(0) = -3, y_3(0) = 2.$

2. $y_1'(t) = 2y_2(t)$
   $y_2'(t) = 2y_1(t)$
   $y_1(0) = 1, y_2(0) = 0.$

3. $y_1'(t) = -y_1(t) + 2y_2(t) + 2y_3(t),$
   $y_2'(t) = 2y_1(t) + 2y_2(t) + 2y_3(t) + 1,$
   $y_3'(t) = -3y_1(t) - 6y_2(t) - 6y_3(t);$
   $y_1(0) = 1, y_2(0) = -3, y_3(0) = 2.$

4. $y_1'(t) = 2y_2(t),$
   $y_2'(t) = 2y_1(t);$
   $y_1(0) = 0, y_2(0) = 1.$

5. $y_1'(t) = -y_1(t) + 2y_2(t) + 2y_3(t),$
   $y_2'(t) = 2y_1(t) + 2y_2(t) + 2y_3(t) + 1,$
   $y_3'(t) = -3y_1(t) - 6y_2(t) - 6y_3(t) + \sin 5t;$
   $y_1(0) = 1, y_2(0) = 0, y_3(0) = 0.$

6. $y_1'(t) = 2y_2(t) + e^{3t},$
   $y_2'(t) = 2y_1(t) - e^{t};$
   $y_1(0) = 1, y_2(0) = 0.$

Let $f(t) = \begin{cases} e^t, & 0 \le t \le 2 \\ 0, & 2 < t \end{cases}$, and $g(t) = \begin{cases} 0, & 0 \le t \le 2 \\ 4, & 2 < t \end{cases}$. Solve these nonhomogeneous initial-value problems.

7. $y_1'(t) = 2y_2(t)$
   $y_2'(t) = 2y_1(t) + f(t);$
   $y_1(0) = 1, y_2(0) = 0.$

8. $y_1'(t) = 2y_2(t) + f(t)$
   $y_2'(t) = 2y_1(t) + g(t);$
   $y_1(0) = 0, y_2(0) = 0.$

9. $y_1'(t) = 2y_2(t) + f(t)$
   $y_2'(t) = 2y_1(t) + g(t);$
   $y_1(0) = 0, y_2(0) = 0.$

10. $y_1'(t) = 2y_2(t) + f(t)$
    $y_2'(t) = 2y_1(t) + g(t);$
    $y_1(0) = 0, y_2(0) = 0.$

11. $y_1'(t) = 2y_2(t) + g(t)$
$y_2'(t) = 2y_1(t) + f(t);$
$y_1(0) = 0, y_2(0) = 0.$

12. $y_1'(t) = y_1(t) + 2y_2(t) + g(t)$
$y_2'(t) = -2y_1(t) + y_2(t) + f(t);$
$y_1(0) = 0, y_2(0) = 0.$

9

Differential Systems with Variable Coefficients

9.0 Introduction

The introduction to Chapter 8 also serves to introduce this chapter. The idea of a fundamental solution is still of special importance to us. Even though we will occasionally find just the vector solution that is requested, at other times we will search for a fundamental solution and then supply the desired vector solution. Except in relatively rare instances we will find that it is difficult to give a closed-form expression for the solutions that we obtain. The solutions that we find will usually be in the form of an infinite series. In many instances, we will have to be satisfied with knowing only the first few terms. Even though we can produce more terms on request, we will not have "solved" the differential equation. When we check such a result, we can only verify that the terms that we have are correct. Sometimes we can say that our series are computationally useful, but rarely will we be able to recognize the series as representing some function that we know. Indeed, there is a course of study in mathematics known as **special functions** where many of the series are identified and their properties demonstrated. Other series that we will encounter will remain unidentified, because they have limited general interest. Of course, if you can show that one of the series you find deserves special interest, give it all of the publicity it deserves.

The topic of Cauchy-Euler systems is included because it is such an accessible generalization of the Cauchy-Euler differential equations that were encountered in Chapter 6. These are differential systems in which the origin is a singular point. The solution techniques for Cauchy-Euler differential equations are actually an interesting application of the methods of Chapter 8.

Other than systems of **Cauchy-Euler** type, the systems that we will study can usually only be solved by infinite series centered on either an ordinary point or a regular singular point. Homogeneous differential systems centered on an ordinary point x_0 can be written in the form

$$\boldsymbol{y}'(x) = \boldsymbol{P}(x)\boldsymbol{y}(x), \tag{9.1}$$

where $\boldsymbol{y}(x)$ is a differentiable n-vector function and the $n \times n$ matrix $\boldsymbol{P}(x)$ is analytic at x_0. Here again, to say that $\boldsymbol{P}(x)$ is **analytic** at x_0 means that

$$\boldsymbol{P}(x) = \sum_{i=0}^{\infty} \boldsymbol{B}_i (x - x_0)^i, \tag{9.2}$$

where the coefficients of the (possibly finite) series are $n \times n$ matrices. The convergence is component-by-component. The matrix $\boldsymbol{P}(x)$ can be a polynomial of degree m, which occurs when $\boldsymbol{B}_m \neq \boldsymbol{0}$, but $\boldsymbol{B}_i = \boldsymbol{0}$, for $i > m$. In most of the examples that we consider this will be the case because the process for finding $\boldsymbol{y}(x)$ is much simpler when $\boldsymbol{P}(x)$ is a polynomial. This is due to the fact that the recursion relation that defines all but the first few coefficients will have a definite number of terms when $\boldsymbol{P}(x)$ is a polynomial. The recursion relation will not have a fixed number of terms when $\boldsymbol{P}(x)$ is defined by a series.

The idea of a **determining set** is still important here, and will tell us the same information that it did in Chapter 7. The determining set is only useful when $\boldsymbol{P}(x)$ is a polynomial. As before, the determining set reveals what happens to each power of x in the solution series.

When the differential system can be written in the form

$$(x - x_0)\boldsymbol{y}'(x) = \boldsymbol{P}(x)\boldsymbol{y}(x), \tag{9.3}$$

where $\boldsymbol{y}(x)$ is a differentiable n-vector function and the $n \times n$ matrix $\boldsymbol{P}(x)$ is **analytic** at x_0, then x_0 is said to be a **regular singular point** of the system. We can always make the change of variables $t = x - x_0$ to change the system into

$$t\boldsymbol{y}'(t) = \boldsymbol{P}(t + x_0)\boldsymbol{y}(t) = \boldsymbol{Q}(t)\boldsymbol{y}(t),$$

where $\boldsymbol{Q}(t)$ is analytic at 0. For this reason, we will let the singular point be $t = 0$. Once the solution $\boldsymbol{y} = \boldsymbol{u}(t)$ is found, the desired solution can be obtained as $\boldsymbol{y}(x) = \boldsymbol{u}(x - x_0)$. There do exist differential systems where x_0 is an **irregular singular point**, but we will not cover these.

Even in the study of systems where a solution is to be centered on a regular singular point, the idea of characteristic value problems arises. Only for those systems having a particularly simple characteristic value problem is the solution process relatively easy.

9.1 Cauchy-Euler Systems

A Reduction Process for Higher-Order Cauchy-Euler Differential Equations

To motivate the definition of Cauchy-Euler systems, we will transform the fourth-order Cauchy-Euler differential equation

$$a_4 x^4 y^{(4)} + a_3 x^3 y^{(3)} + a_2 x^2 y^{(2)} + a_1 x y' + a_0 y = 0, \quad a_4 \neq 0$$

into a singular system of first-order differential equations. To accomplish this, let

$$y_1 = y, y_2 = xy', y_3 = x^2y'', \quad \text{and} \quad y_4 = x^3y'''.$$

Notice the extra factors: the powers of x. These factors are necessary whenever we transform a singular differential equation into a singular system. The transformation proceeds as follows:

$$\begin{aligned}
xy_1' &= xy' = y_2 \\
xy_2' &= x(y' + xy'') = xy' + x^2y'' = y_2 + y_3 \\
xy_3' &= x(2xy'' + x^2y''') = 2x^2y'' + x^3y''' = 2y_3 + y_4 \\
xy_4' &= x(3x^2y''' + x^3y^{(4)}) = 3x^3y''' + x^4y^{(4)} \\
&= 3y_4 - \frac{1}{a_4}[a_3x^3y^{(3)} + a_2x^2y^{(2)} + a_1xy' + a_0y] \\
&= 3y_4 - \frac{a_3}{a_4}x^3y^{(3)} - \frac{a_2}{a_4}x^2y^{(2)} - \frac{a_1}{a_4}xy' - \frac{a_0}{a_4}y \\
&= 3y_4 - \frac{a_3}{a_4}y_4 - \frac{a_2}{a_4}y_3 - \frac{a_1}{a_4}y_2 - \frac{a_0}{a_4}y_1.
\end{aligned}$$

This gives us the vector system

$$x\begin{pmatrix} y_1 \\ y_2 \\ y_3 \\ y_4 \end{pmatrix}' = \begin{pmatrix} 0 & 1 & 0 & 0 \\ 0 & 1 & 1 & 0 \\ 0 & 0 & 2 & 1 \\ -a_0/a_4 & -a_1/a_4 & -a_2/a_4 & 3 - a_3/a_4 \end{pmatrix}\begin{pmatrix} y_1 \\ y_2 \\ y_3 \\ y_4 \end{pmatrix}.$$

This is a system having the form

$$x\boldsymbol{y}' = \boldsymbol{A}\boldsymbol{y},$$

where $\boldsymbol{A}$ is a constant 4×4 matrix. We make the following definition.

Definition 9.1 *Suppose that $\boldsymbol{A}$ is a constant $n \times n$ matrix. A first-order vector (or matrix) differential system consisting of n equations in the form*

$$x\boldsymbol{y}' = \boldsymbol{A}\boldsymbol{y}$$

is called a **Cauchy-Euler system**.

The solution of Cauchy-Euler systems and their relationship with constant coefficients systems is directly analogous to the solution of Cauchy-Euler equations and their relationship with constant coefficient equations that we studied in Chapter 7.

Example 9.1 Convert the second-order Cauchy-Euler differential equation

$$x^2y'' - 2xy' - 4y = 0$$

to a singular system of first-order differential equations in Cauchy-Euler form.

Solution. Let $y_1 = y$, and $y_2 = xy'$. Then

$$\begin{aligned} xy_1' &= xy' = y_2 \\ xy_2' &= x(y' + xy'') \\ &= xy' + x^2y'' \\ &= y_2 + (2xy' + 4y) \\ &= y_2 + (2y_2 + 4y_1) \\ &= 3y_2 + 4y_1. \end{aligned}$$

The resulting system is

$$x\begin{pmatrix} y_1 \\ y_2 \end{pmatrix}' = \begin{pmatrix} 0 & 1 \\ 4 & 3 \end{pmatrix}\begin{pmatrix} y_1 \\ y_2 \end{pmatrix}. \qquad \diamond$$

How do we solve a system such as this? In $x\boldsymbol{y}' = \boldsymbol{A}\boldsymbol{y}$ let $t = \ln x$, which is the same as letting $x = e^t$. Define $\boldsymbol{w}(t) = \boldsymbol{y}(e^t)$. Then

$$\begin{aligned} \frac{d}{dt}\boldsymbol{w}(t) &= \frac{d}{dt}(\boldsymbol{y}(e^t)) = e^t\boldsymbol{y}'(e^t) \\ &= x\boldsymbol{y}'(x) = \boldsymbol{A}\boldsymbol{y}(x) = \boldsymbol{A}\boldsymbol{y}(e^t) = \boldsymbol{A}\boldsymbol{w}(t). \end{aligned}$$

From

$$\boldsymbol{w}' = \boldsymbol{A}\boldsymbol{w} \tag{9.4}$$

we obtain $\boldsymbol{w}(t) = \Phi(t)\boldsymbol{k}$, by the methods of Chapter 7, where $\Phi(t)$ is fundamental and $\boldsymbol{k}$ is an arbitrary constant vector. Then from $\boldsymbol{w}(t) = \boldsymbol{y}(e^t) = \Phi(t)\boldsymbol{k}$, we find that $\boldsymbol{y}(x) = \Phi(\ln x)\boldsymbol{k}$. The matrix-valued function $\Phi(\ln x)$ is still fundamental and is a solution of the system $x\boldsymbol{y}'(x) = \boldsymbol{A}\boldsymbol{y}(x)$, as this calculation shows:

$$\begin{aligned} x\boldsymbol{y}'(x) &= x\Phi'(\ln x)\boldsymbol{k}\frac{d}{dx}\ln x \\ &= x\Phi'(\ln x)\boldsymbol{k}\frac{1}{x} \\ &= \boldsymbol{A}\Phi(\ln x)\boldsymbol{k} \\ &= \boldsymbol{A}\boldsymbol{y}(x), \end{aligned}$$

since $\Phi' = \boldsymbol{A}\Phi$. The Cauchy-Euler problem is thus solved.

We can also solve $x\boldsymbol{y}' = \boldsymbol{A}\boldsymbol{y}$ directly by assuming that $\boldsymbol{y} = \boldsymbol{k}x^r$, where r is a number and $\boldsymbol{k}$ is a nonzero constant vector. With this approach,

$$x\boldsymbol{y}' = \boldsymbol{A}\boldsymbol{y} \quad \text{becomes} \quad x(rx^{r-1})\boldsymbol{k} = \boldsymbol{A}x^r\boldsymbol{k},$$

which says that

$$x^r(\boldsymbol{A} - r\boldsymbol{I})\boldsymbol{k} = \boldsymbol{0}$$

must hold for all x. Thus the characteristic value problem

$$(\boldsymbol{A} - r\boldsymbol{I})\boldsymbol{k} = \boldsymbol{0}$$

must be solved for pairs $(r, \boldsymbol{k})$ with $\boldsymbol{k}$ nonzero. This is precisely the characteristic value problem that must be solved to solve 9.4. So the methods are equivalent (as they were in Chapter 7).

Example 9.2 Solve the singular system that resulted from the transformation in Example 9.1.

Solution. The singular system

$$x\begin{pmatrix} y_1 \\ y_2 \end{pmatrix}' = \begin{pmatrix} 0 & 1 \\ 4 & 3 \end{pmatrix}\begin{pmatrix} y_1 \\ y_2 \end{pmatrix}$$

may be solved by letting $\begin{pmatrix} y_1 \\ y_2 \end{pmatrix} = x^r\boldsymbol{k}$. Then

$$x\begin{pmatrix} y_1 \\ y_2 \end{pmatrix}' = x(rx^{r-1}\boldsymbol{k}) = rx^r\boldsymbol{k} = \begin{pmatrix} 0 & 1 \\ 4 & 3 \end{pmatrix} x^r\boldsymbol{k}.$$

We therefore need to solve

$$\begin{pmatrix} 0 & 1 \\ 4 & 3 \end{pmatrix} x^r\boldsymbol{k} = rx^r\boldsymbol{k},$$

for all x, or

$$\begin{pmatrix} 0 & 1 \\ 4 & 3 \end{pmatrix} \boldsymbol{k} = r\boldsymbol{k}.$$

This characteristic value problem may be expressed as

$$\begin{pmatrix} -r & 1 \\ 4 & 3-r \end{pmatrix} \boldsymbol{k} = \boldsymbol{0},$$

which has the characteristic equation

$$(-r)(3-r) - 4 = r^2 - 3r - 4 = (r-4)(r+1) = 0.$$

We have the characteristic roots $r_1 = -1$ and $r_2 = 4$, to which correspond the characteristic vectors $\boldsymbol{k}_1 = \begin{pmatrix} 1 \\ -1 \end{pmatrix}$ and $\boldsymbol{k}_2 = \begin{pmatrix} 1 \\ 4 \end{pmatrix}$, respectively. From these, we obtain the linearly independent solutions $\boldsymbol{y}_1(x) = \begin{pmatrix} 1 \\ -1 \end{pmatrix} x^{-1}$ and $\boldsymbol{y}_2(x) = \begin{pmatrix} 1 \\ 4 \end{pmatrix} x^4$. This finally gives us the fundamental matrix solution $\Phi(x) = \begin{pmatrix} x^{-1} & x^4 \\ -x^{-1} & 4x^4 \end{pmatrix}$, from which we obtain the most general solution to the vector system as

$$\begin{aligned} \boldsymbol{y}(x) &= \Phi(x)\boldsymbol{k} \\ &= \begin{pmatrix} x^{-1} & x^4 \\ -x^{-1} & 4x^4 \end{pmatrix} \begin{pmatrix} c_1 \\ c_2 \end{pmatrix} \\ &= \begin{pmatrix} x^{-1}c_1 + x^4 c_2 \\ -x^{-1}c_1 + 4x^4 c_2 \end{pmatrix}. \end{aligned}$$

The first component of this vector is the solution $\boldsymbol{y}(x) = x^{-1}c_1 + x^4 c_2$ of the original second-order problem $x^2y'' - 2xy' - 4y = 0$. ◇

It is worth noting that the solution process followed in Chapter 4 is more direct for this problem, but there are systems, such as that illustrated in the next example, where the system may not have come from a single higher-order equation, and the methods of this chapter are the only ones we have available to us.

Example 9.3 Find a fundamental solution matrix for the singular differential system

$$x\boldsymbol{y}'(x) = \begin{pmatrix} 1 & 2 & 1 \\ 6 & -1 & 0 \\ -1 & -2 & -1 \end{pmatrix} \boldsymbol{y}(x),$$

which has the Cauchy-Euler form.

Solution. Assume a solution of the form $\boldsymbol{y}(x) = x^r\boldsymbol{k}$. Then

$$x\boldsymbol{y}'(x) = x(rx^{r-1}\boldsymbol{k}) = \begin{pmatrix} 1 & 2 & 1 \\ 6 & -1 & 0 \\ -1 & -2 & -1 \end{pmatrix} x^r\boldsymbol{k},$$

from which we see that we must solve the characteristic value problem

$$\begin{pmatrix} 1 & 2 & 1 \\ 6 & -1 & 0 \\ -1 & -2 & -1 \end{pmatrix} \boldsymbol{k} = r\boldsymbol{k}.$$

The characteristic equation is

$$\begin{aligned} &\det \begin{pmatrix} 1-r & 2 & 1 \\ 6 & -1-r & 0 \\ -1 & -2 & -1-r \end{pmatrix} \\ &= -r^3 - r^2 + 12r = r(r+4)(r-3) = 0. \end{aligned}$$

To the three characteristic roots $r_1 = 0$, $r_2 = -4$, and $r_3 = 3$, there correspond characteristic vectors that are multiples of

$$\boldsymbol{k}_1 = \begin{pmatrix} 1 \\ 6 \\ -13 \end{pmatrix}, \quad \boldsymbol{k}_2 = \begin{pmatrix} -1 \\ 2 \\ 1 \end{pmatrix}, \quad \text{and} \quad \boldsymbol{k}_3 = \begin{pmatrix} 2 \\ 3 \\ -2 \end{pmatrix}, \quad \text{respectively,}$$

as one may easily verify by a manual or *Mathematica* calculation. From these one obtains the three linearly independent solution vectors

$$\boldsymbol{y}_1(x) = \boldsymbol{k}_1 x^{r_1} = \begin{pmatrix} 1 \\ 6 \\ -13 \end{pmatrix} x^0 = \begin{pmatrix} 1 \\ 6 \\ -13 \end{pmatrix},$$

$$\boldsymbol{y}_2(x) = \boldsymbol{k}_2 x^{r_2} = \begin{pmatrix} -1 \\ 2 \\ 1 \end{pmatrix} x^{-4} = \begin{pmatrix} -x^{-4} \\ 2x^{-4} \\ x^{-4} \end{pmatrix},$$

and

$$\boldsymbol{y}_3(x) = \boldsymbol{k}_3 x^{r_3} = \begin{pmatrix} 2 \\ 3 \\ -2 \end{pmatrix} x^3 = \begin{pmatrix} 2x^3 \\ 3x^3 \\ -2x^3 \end{pmatrix}.$$

As requested, a fundamental matrix for $x > 0$ is therefore

$$\Phi(x) = \begin{pmatrix} 1 & -x^{-4} & 2x^3 \\ 6 & 2x^{-4} & 3x^3 \\ -13 & x^{-4} & -2x^3 \end{pmatrix},$$

so that a complete vector solution to the given equation is

$$\boldsymbol{y}(x) = \Phi(x)\boldsymbol{k} = \begin{pmatrix} 1 & -x^{-4} & 2x^3 \\ 6 & 2x^{-4} & 3x^3 \\ -13 & x^{-4} & -2x^3 \end{pmatrix} \begin{pmatrix} k_1 \\ k_2 \\ k_3 \end{pmatrix}.$$

This should be checked by substitution. Do it by hand and by *Mathematica*. ◇

In the case of diagonalizable coefficient matrices, it is very easy to obtain solutions with *Mathematica*. Here is an example.

Example 9.2M Solve the singular Cauchy-Euler system

$$x \begin{pmatrix} y_1 \\ y_2 \end{pmatrix}' = \begin{pmatrix} 0 & 1 \\ 4 & 3 \end{pmatrix} \begin{pmatrix} y_1 \\ y_2 \end{pmatrix}$$

by *Mathematica*.

Solution. We duplicate the manual steps. Define the coefficient matrix.

```
In[1]:=  a = {{0,1},{4,3}}

Out[1]=  {{0,1}, {4, 3}}

In[2]:=  CharPoly = Factor[Det[a-r*IdentityMatrix[2]]]

Out[2]=  (-4 + r) (1 + r)
```

Since a is diagonalizable, the characteristic value problem can be solved in one step.

```
In[3]:=  {vals,vecs} = Eigensystem[a]

Out[3]=  {{-1, 4}, {{-1, 1}, {1, 4}}}
```

Here are the powers of x.

```
In[4]:=  x^vals

Out[4]=   1    4
         {-, x }
          x
```

The solution vectors as rows.

```
In[5]:=  (x^vals)*vecs          (* Note the use of '*' *)

Out[5]=        1    1       4        4
         {{-(-), -}, {x , 4 x }}
               x    x
```

A fundamental matrix.

```
In[6]:=  (W[x_] = Transpose[(x^vals)*vecs])//MatrixForm

Out[6]=    1      4
         -(-)   x
           x

          1         4
          -     4 x
          x

In[7]:=  Det[W[x]]

Out[7]=       3
         -5 x
```

This is nonzero when $x \neq 0$. W[0] is undefined, so the solution is only defined when $x \neq 0$. Check the solution.

```
In[8]:=  Simplify[x*W'[x]-a.W[x]]

Out[8]=  {{0, 0}, {0, 0}}
```

It is so simple to construct a *Mathematica* function to produce such a solution automatically that it is worth doing so.

This function finds a fundamental solution for diagonalizable Cauchy-Euler systems.

```
In[9]:=  DSolveCauchyEulerSystem[Coeff_,x_?AtomQ] :=
                 Module[{vals, vecs},
                         {vals,vecs} = Eigensystem[Coeff];
                         Transpose[(x^vals)*vecs]
                         ]
```

Here is how it is invoked.

```
In[10]:= DSolveCauchyEulerSystem[a,x]

Out[10]=      1     4      1      4
         {{-(-), x }, {-, 4 x }}
              x            x
```

Change the name of the independent variable.

```
In[11]:= DSolveCauchyEulerSystem[a,t]

Out[11]=      1     4      1      4
         {{-(-), t }, {-, 4 t }}
              t            t
```

◇

Now that we have such a function at our disposal, let's use it to solve Example 9.3.

Example 9.3M Use the function `DSolveCauchyEulerSystem` to find a fundamental solution matrix for the singular differential system

$$x\boldsymbol{y}'(x) = \begin{pmatrix} 1 & 2 & 1 \\ 6 & -1 & 0 \\ -1 & -2 & -1 \end{pmatrix} \boldsymbol{y}(x)$$

which has the Cauchy-Euler form.

Solution. Define the coefficient matrix.

```
In[12]:= a1 = {{1, 2, 1}, {6,-1, 0},{-1,-2,-1}}

Out[12]= {{1, 2, 1}, {6, -1, 0}, {-1, -2, -1}}

In[13]:= W[x_] = DSolveCauchyEulerSystem[a1,x]

Out[13]=     -4            3   2            3      -4           3
         {{-x  , -1, -2 x }, {--, -6, -3 x }, {x  , 13, 2 x }}
                               4
                              x
```

Check the solution.

```
In[14]:= Simplify[x W'[x]-a1.W[x], Trig->False]

Out[14]= {{0, 0, 0}, {0, 0, 0}, {0, 0, 0}}
```

◇

Here is an example where a root is repeated but the matrix is diagonalizable.

Example 9.4M Solve the diagonalizable third-order Cauchy-Euler system

$$x\boldsymbol{y}'(x) = \begin{pmatrix} -23 & 30 & 10 \\ -10 & 12 & 5 \\ -20 & 30 & 7 \end{pmatrix} \boldsymbol{y}(x)$$

which has a double root.

Solution.

```
In[15]:= a2 = {{-23, 30, 10}, {-10, 12, 5}, {-20, 30, 7}}

Out[15]= {{-23, 30, 10}, {-10, 12, 5}, {-20, 30, 7}}

In[16]:= W[x_] = DSolveCauchyEulerSystem[a2,x]

Out[16]=     -3  3      2        2   2   2        2
         {{x  , --, 2 x }, {0, --, x }, {--, 0, 2 x }}
                 3              3         3
                x              x         x
```

The solution is fundamental.

```
In[17]:= Det[W[x]]

Out[17]= 2
         --
          4
         x
```

It checks.

```
In[18]:= Simplify[x W'[x]-a2.W[x], Trig->False]

Out[18]= {{0, 0, 0}, {0, 0, 0}, {0, 0, 0}}
```

◇

Complex Roots

We saw in Section 8.4 that any complex conjugate pairs of characteristic roots are found along with the real characteristic roots when the characteristic equation is solved. Given a pair $(r, \boldsymbol{v})$ and $(\overline{r}, \overline{\boldsymbol{v}})$ of complex solutions to $(\boldsymbol{A} - r\boldsymbol{I})\boldsymbol{v} = 0$, we find two complex solutions of the differential system $x\boldsymbol{y}' = \boldsymbol{A}\boldsymbol{y}$ to be $\boldsymbol{y}_{c_1}(x) = \boldsymbol{v}x^r$, and $\boldsymbol{y}_{c_2}(x) = \overline{\boldsymbol{v}}x^{\overline{r}}$. Write $r = r_1 = a + bi$ and $\boldsymbol{v} = \boldsymbol{k}_1 + \boldsymbol{k}_2 i$, where $\boldsymbol{k}_1$ and $\boldsymbol{k}_2$ are real constant vectors. Using these two complex solutions we get the two linearly independent real solutions $\boldsymbol{y}_1(x)$ and $\boldsymbol{y}_2(x)$ from this expansion:

$$\begin{aligned}
\boldsymbol{y}_{c_1}(x) &= \boldsymbol{v}x^r \\
&= (\boldsymbol{k}_1 + \boldsymbol{k}_2 i)x^{(a+bi)} \\
&= (\boldsymbol{k}_1 + \boldsymbol{k}_2 i)x^a x^b \\
&= (\boldsymbol{k}_1 + \boldsymbol{k}_2 i)x^a e^{bi(\ln x)} \\
&= (\boldsymbol{k}_1 + \boldsymbol{k}_2 i)x^a(\cos(b\ln x) + i\sin(b\ln x)) \\
&= x^a(\boldsymbol{k}_1\cos(b\ln x) - \boldsymbol{k}_2\sin(b\ln x)) \\
&\quad + ix^a(\boldsymbol{k}_1\sin(b\ln x) + \boldsymbol{k}_2\cos(b\ln x)) \\
&= \boldsymbol{y}_1(x) + i\boldsymbol{y}_2(x),
\end{aligned}$$

where

$$\begin{aligned}\boldsymbol{y}_1(x) &= x^a(\boldsymbol{k}_1\cos(b\ln x) - \boldsymbol{k}_2\sin(b\ln x))\\ &= x^a(\mathrm{Re}(\boldsymbol{v})\sin(b\ln x) + \mathrm{Im}(\boldsymbol{v})\cos(b\ln x)).\end{aligned}$$

and

$$\begin{aligned}\boldsymbol{y}_2(x) &= x^a(\boldsymbol{k}_1\sin(b\ln x) + \boldsymbol{k}_2\cos(b\ln x))\\ &= x^a(\mathrm{Re}(\boldsymbol{v})\sin(b\ln x) + \mathrm{Im}(\boldsymbol{v})\cos(b\ln x)).\end{aligned}$$

We would have obtained the same result had we used $\boldsymbol{y}_{c_2}(x)$.

Example 9.5 Find a real fundamental solution for the homogeneous second-order Cauchy-Euler differential system

$$x\frac{d\boldsymbol{y}}{dx} = \begin{pmatrix} a & b \\ -b & a \end{pmatrix}\boldsymbol{y}(x).$$

Solution. The corresponding constant coefficients problem was solved in Example 8.18. There it was found that $r_{1,2} = a \pm bi$, and the characteristic vector $\boldsymbol{K}_1$ corresponding to $r_1 = a+bi$ is $\boldsymbol{K}_1 = \begin{pmatrix} 1 \\ 0 \end{pmatrix} + i\begin{pmatrix} 0 \\ 1 \end{pmatrix}$, so that $\boldsymbol{k}_1 = \mathrm{Re}(\boldsymbol{K}_1) = \begin{pmatrix} 1 \\ 0 \end{pmatrix}$ and $\boldsymbol{k}_2 = \mathrm{Im}(\boldsymbol{K}_1) = \begin{pmatrix} 0 \\ 1 \end{pmatrix}$.

The two complex solutions are therefore

$$\boldsymbol{y}_{c_1}(x) = \begin{pmatrix} 1 \\ i \end{pmatrix} x^{(a+bi)} \quad \text{and} \quad \boldsymbol{y}_{c_2}(x) = \begin{pmatrix} 1 \\ -i \end{pmatrix} x^{(a-bi)}.$$

To these correspond the real solutions

$$\begin{aligned}\boldsymbol{y}_1(x) &= x^a(\boldsymbol{k}_1\cos(b\ln x) - \boldsymbol{k}_2\sin(b\ln x))\\ &= x^a\left(\begin{pmatrix} 1 \\ 0 \end{pmatrix}\cos(b\ln x) - \begin{pmatrix} 0 \\ 1 \end{pmatrix}\sin(b\ln x)\right)\\ &= x^a\begin{pmatrix} \cos(b\ln x) \\ -\sin(b\ln x) \end{pmatrix},\end{aligned}$$

and

$$\begin{aligned}\boldsymbol{y}_2(x) &= x^a(\boldsymbol{k}_1\sin(b\ln x) + \boldsymbol{k}_2\cos(b\ln x))\\ &= x^a\left(\begin{pmatrix} 1 \\ 0 \end{pmatrix}\sin(b\ln x) + \begin{pmatrix} 0 \\ 1 \end{pmatrix}\cos(b\ln x)\right)\\ &= x^a\begin{pmatrix} \sin(b\ln x) \\ \cos(b\ln x) \end{pmatrix}.\end{aligned}$$

We have the real fundamental matrix solution

$$\boldsymbol{Y}(x) = x^a\begin{pmatrix} \cos(b\ln x) & \sin(b\ln x) \\ -\sin(b\ln x) & \cos(b\ln x) \end{pmatrix}.$$

If $b = 0$, then this solution is

$$\boldsymbol{Y}(x) = x^a \begin{pmatrix} 1 & 0 \\ 0 & 1 \end{pmatrix} = \begin{pmatrix} x^a & 0 \\ 0 & x^a \end{pmatrix},$$

which was to be expected if the coefficient matrix is diagonal. ◇

EXERCISES 9.1

PART I. Use the transformation rules for singular systems to transform these Cauchy-Euler equations into Cauchy-Euler systems.

1. $x\dfrac{dy}{dx} + 2y = 0.$

2. $x^2\dfrac{d^2y}{dx^2} - 6y = 0.$

3. $x^3\dfrac{d^3y}{dx^3} + 2x^2\dfrac{d^2y}{dx^2} - 6x\dfrac{dy}{dx} = 0.$

4. $2x^3\dfrac{d^3y}{dx^3} - 5x^2\dfrac{d^2y}{dx^2} + 8x\dfrac{dy}{dx} - 6y = 0.$

5. $6x^3\dfrac{d^3y}{dx^3} - 13x^2\dfrac{d^2y}{dx^2} + 5x\dfrac{dy}{dx} - 8y = 0.$

6. $4x^3\dfrac{d^3y}{dx^3} - 12x^2\dfrac{d^2y}{dx^2} + 15x\dfrac{dy}{dx} - 12y = 0.$

7. $15x^3\dfrac{d^3y}{dx^3} - 2x^2\dfrac{d^2y}{dx^2} + 6x\dfrac{dy}{dx} - 8y = 0.$

8. $9x^3\dfrac{d^3y}{dx^3} + 36x^2\dfrac{d^2y}{dx^2} - 4x\dfrac{dy}{dx} - 8y = 0.$

PART II. Solve these Cauchy-Euler systems. The coefficient matrices are diagonalizable. Solve manually using *Mathematica* to help with the algebra and again using `DSolveCauchyEulerSystem`.

9. $x\boldsymbol{Y}'(x) = \begin{pmatrix} 8 & -5 \\ 10 & -7 \end{pmatrix} \boldsymbol{Y}(x).$

10. $x\boldsymbol{Y}'(x) = \begin{pmatrix} 3 & 0 & 0 \\ -10 & -2 & 0 \\ 10 & 6 & 1 \end{pmatrix} \boldsymbol{Y}(x).$

11. $x\boldsymbol{Y}'(x) = \begin{pmatrix} 133 & 54 & 22 \\ -300 & -122 & -50 \\ -60 & -24 & -9 \end{pmatrix} \boldsymbol{Y}(x).$

12. $xY'(x) = \begin{pmatrix} 1 & 0 & 0 \\ -4 & -1 & 0 \\ 8 & 4 & 1 \end{pmatrix} Y(x).$

13. $xY'(x) = \begin{pmatrix} -2 & 2 & -8 & -7 \\ 0 & -1 & -8 & -8 \\ 0 & 0 & -5 & -4 \\ 0 & 0 & 8 & 7 \end{pmatrix} Y(x).$

PART III. The coefficient matrices of these Cauchy-Euler systems are diagonalizable, but there may be repeated roots. Solve manually using *Mathematica* to help with the algebra. Do not forget the factors of `Log[x]` ($= \ln x$) that are required when there is a repeated root.

14. $xY'(x) = \begin{pmatrix} 7 & 2 & 1 \\ 0 & 2 & 0 \\ -30 & -12 & -4 \end{pmatrix} Y(x).$

15. $xY'(x) = \begin{pmatrix} 32 & 12 & 5 \\ -60 & -22 & -10 \\ -30 & -12 & -3 \end{pmatrix} Y(x).$

16. $xY'(x) = \begin{pmatrix} -2 & 15 & -10 & 5 \\ 0 & 8 & -10 & 0 \\ 0 & 5 & -7 & 0 \\ 0 & -5 & 10 & 3 \end{pmatrix} Y(x).$

17. $xY'(x) = \begin{pmatrix} -2 & 0 & -10 & -10 \\ 0 & -2 & -10 & -10 \\ 0 & 0 & -7 & -5 \\ 0 & 0 & 10 & 8 \end{pmatrix} Y(x).$

PART IV.

18. Show that the solution found in Example 9.3 checks. Do the calculation both manually and by *Mathematica* .

19. The system $xy'(x) = \begin{pmatrix} 4 & 1 \\ -4 & 0 \end{pmatrix} y(x)$ has a double root $r = 2$ and is defective. Solve it by assuming a solution of the form $y(x) = K_0 x^r + K_1 x^r \ln x$, where the coefficient vectors K_0 and K_1 are to be determined. Show that you need to solve a system of the form

$$(A - rI)K_1 = 0$$
$$(A - rI)K_0 = K_1.$$

Then solve.

20. Generalize the technique of the last problem.

21. Find a fundamental solution of the Cauchy-Euler system

$$xy'(x) = \begin{pmatrix} -25/6 & 22 & -31/6 \\ -7/3 & 9 & -4/3 \\ -14/3 & 22 & -14/3 \end{pmatrix} y(x).$$

Check your answer. Is the fundamental matrix defined on an open interval containing 0?

9.2 Solution About Ordinary Points

The differential systems that we will encounter in this section have the form

$$\boldsymbol{y}'(x) = \boldsymbol{P}(x)\boldsymbol{y}(x), \tag{9.5}$$

where $\boldsymbol{P}(x)$ is analytic at 0. For such a system, the point $x_0 = 0$ is an **ordinary point**. Every homogeneous system that we solved in Chapter 8 was of this form, because in that chapter $\boldsymbol{P}(x) = \boldsymbol{A}$ was a constant matrix, which is certainly analytic. We even used Picard's method to find some of the partial sums of an infinite series expansion for the matrix exponential $e^{\boldsymbol{A}x}$. Picard's method is unnecessarily complicated, because we can find solutions by a series version of undetermined coefficients as was done in Chapter 4 for a single differential equation of higher order.

You will see that the methods we use here are almost exactly the same as those of Chapter 7, except that in Chapter 7 the coefficients were numbers and here they are matrices. We will need to watch out for the noncommutativity of matrix multiplication, but otherwise the process used in this chapter is the same as that used in Chapter 7. Once you see that this generalization is possible, then you can be alert to other possible generalizations. Many generalizations have been made; there are certainly others awaiting pursuit by interested investigators.

The following theorem tells us that it is possible to solve system 9.5 by series methods.

Theorem 9.1 *Let the $n \times n$ matrix-valued function $\boldsymbol{P}(x)$ be analytic and have the expansion*

$$\boldsymbol{P}(x) = \sum_{j=0}^{\infty} \boldsymbol{B}_j x^j \quad \text{for} \quad |x| < R,$$

where $\{\boldsymbol{B}_j\}_{j=0}^{\infty}$ are constant $n \times n$ matrices. Let $\boldsymbol{C}_0$ be any $n \times n$ matrix (or n-vector). Then the expressions

$$\boldsymbol{C}_1 = \boldsymbol{B}_0\boldsymbol{C}_0,$$

$$\boldsymbol{C}_2 = \frac{1}{2}(\boldsymbol{B}_0\boldsymbol{C}_1 + \boldsymbol{B}_1\boldsymbol{C}_0),$$

$$\vdots$$

$$\boldsymbol{C}_{k+1} = \frac{1}{k+1}(\boldsymbol{B}_0\boldsymbol{C}_k + \boldsymbol{B}_1\boldsymbol{C}_{k-1} + \cdots + \boldsymbol{B}_k\boldsymbol{C}_0), \quad k \geq 1, \tag{E}$$

determine $\boldsymbol{C}_1, \boldsymbol{C}_2, \ldots$ *uniquely and the function defined by*

$$\boldsymbol{y}(x) = \sum_{k=0}^{\infty} \boldsymbol{C}_k x^k \quad \text{for} \quad |x| < R$$

is a solution of system 9.5 that is analytic at 0.

One immediate consequence of Theorem 9.1 is that if we choose $\boldsymbol{C}_0$ to be an invertible matrix, then the matrix solution we obtain is a fundamental matrix solution for system 9.5 that can then be used to produce any desired solution of system 9.5.

There are often several alternative forms that a system may have. Choosing an appropriate one may save us much unnecessary complication. Suppose that the system has the special form $a(x)\boldsymbol{y}'(x) = \boldsymbol{Q}(x)\boldsymbol{y}(x)$, where $a(x)$ is a nonconstant scalar polynomial multiplier such that $a(0) \neq 0$, and $\boldsymbol{Q}(x)$ is a matrix polynomial. Then $\boldsymbol{P}(x) = \boldsymbol{Q}(x)/a(x)$, being the quotient of two polynomials with nonzero denominator in a neighborhood of 0, is analytic, but $\boldsymbol{P}(x)$ has an infinite expansion as a series, which means that the recursion relation for the coefficient matrices in the assumed solution $\boldsymbol{y}(x) = \sum_{k=0}^{\infty} \boldsymbol{C}_k x^k$ of the system $\boldsymbol{y}'(x) = (\boldsymbol{Q}(x)/a(x))\boldsymbol{y}(x)$ will have a varying number of terms in its definition as equation E of Theorem 9.1 indicates. On the other hand, if we assume $\boldsymbol{y}(x) = \sum_{k=0}^{\infty} \boldsymbol{C}_k x^k$ and substitute directly into

$$a(x)\boldsymbol{y}'(x) = \boldsymbol{Q}(x)\boldsymbol{y}(x),$$

then the resulting recursion relation will eventually have a fixed number of terms, and therefore offers us a better chance of determining a closed form for the solution $\boldsymbol{y}(x)$ of the system. The next example will let us see how the determining set, which was introduced in Chapter 7, should be defined for matrix systems having polynomial coefficients.

Example 9.6 Compare the procedures for solving the two equivalent systems

$$(x^2+1)\boldsymbol{y}'(x) = \begin{pmatrix} 0 & x^2+1 \\ 1 & -x \end{pmatrix} \boldsymbol{y}(x) \tag{9.6}$$

and

$$\begin{aligned} \boldsymbol{y}'(x) &= \frac{1}{(x^2+1)} \begin{pmatrix} 0 & x^2+1 \\ 1 & -x \end{pmatrix} \boldsymbol{y}(x) \\ &= \begin{pmatrix} 0 & 1 \\ 1/(x^2+1) & -x/(x^2+1) \end{pmatrix} \boldsymbol{y}(x) \end{aligned} \tag{9.7}$$

Solution. System 9.6 can be expanded in powers of x to read

$$(x^2+1)\boldsymbol{y}'(x) = \left(\begin{pmatrix} 0 & 1 \\ 1 & 0 \end{pmatrix} + \begin{pmatrix} 0 & 0 \\ 0 & -1 \end{pmatrix} x + \begin{pmatrix} 0 & 1 \\ 0 & 0 \end{pmatrix} x^2 \right) \boldsymbol{y}(x).$$

When we substitute the assumed form $\boldsymbol{y}(x) = \sum_{k=0}^{\infty} \boldsymbol{C}_k x^k$ for $\boldsymbol{y}(x)$, we see that on the left the degree of each term is either changed by 1 ($= 2 - 1$) or by -1

($= 0 - 1$). On the right, the degrees of some terms stay the same, some are raised by 1, and some are raised by 2. So here our determining set is $\{1, -1\} \cup \{0, 1, 2\} = \{-1, 0, 1, 2\}$, which says that the recursion relation is of third ($3 = 2 - (-1)$) order, and it is in effect for $k \geq 2$. In addition, since there are four entries in the determining set, the recursion relation has four terms. The details will be illustrated in later examples.

On the other hand, when the system 9.7,

$$\boldsymbol{y}'(x) = \begin{pmatrix} 0 & 1 \\ 1/(x^2+1) & -x/(x^2+1) \end{pmatrix} \boldsymbol{y}(x),$$

is expanded in powers of x, there results

$$\begin{aligned} \boldsymbol{y}'(x) &= \begin{pmatrix} 0 & 1 \\ 1 - x^2 + x^4 - x^6 + \cdots & -x + x^3 - x^5 + x^7 - \cdots \end{pmatrix} \boldsymbol{y}(x), \\ &= \left(\begin{pmatrix} 0 & 1 \\ 1 & 0 \end{pmatrix} + \begin{pmatrix} 0 & 0 \\ 0 & -1 \end{pmatrix} x + \begin{pmatrix} 0 & 0 \\ -1 & 0 \end{pmatrix} x^2 \right. \\ &\quad \left. + \begin{pmatrix} 0 & 0 \\ 0 & -1 \end{pmatrix} x^3 + \cdots \right) \boldsymbol{y}(x), \end{aligned}$$

which causes the determining set to be $\{-1, 0, 1, 2, 3, 4, \ldots\}$, which is an infinite set. This means that the recursion relation for determining the coefficient matrices, as Theorem 9.1 says, has more and more terms the farther out the series we go. The terms will be the same as would be obtained from system 9.6, but it will be much harder to see how to express them in closed form, because the recursion relation is not of fixed finite order. Not only that, but even before the coefficients of like powers can be equated, the matrix product on the right must be performed, and this means multiplying two power series, each having matrix coefficients. (This is how equation E arises.)

So there are two difficulties with system 9.7: the recursion relation is not of finite order, and two matrix power series must be multiplied. On the other hand, system 9.7 is in the standard form for the solution of systems about an ordinary point, whereas system E is not. ◇

Example 9.7 Rework Example 7.17 by converting the differential equation

$$y'' + (2x + x^2)y' + (4 + x)y = 0$$

to a system. Solve the resulting system by power series centered on the ordinary point $x_0 = 0$.

Solution. When the point x_0 is an ordinary point, as 0 is in this case, the standard method for reducing an equation to a system can be used. Let $y_1 = y$, and $y_2 = y'$. Then

$$y_1' = y' = y_2$$

and

$$y_2' = -(2x + x^2)y' - (4 + x)y = -(2x + x^2)y_2 - (4 + x)y_1.$$

Thus we get the system

$$\begin{pmatrix} y_1 \\ y_2 \end{pmatrix}'(x) = \begin{pmatrix} 0 & 1 \\ -(4+x) & -(2x+x^2) \end{pmatrix} \begin{pmatrix} y_1 \\ y_2 \end{pmatrix}(x)$$

$$= \left(\begin{pmatrix} 0 & 1 \\ -4 & 0 \end{pmatrix} + \begin{pmatrix} 0 & 0 \\ -1 & -2 \end{pmatrix} x + \begin{pmatrix} 0 & 0 \\ 0 & -1 \end{pmatrix} x^2 \right) \begin{pmatrix} y_1 \\ y_2 \end{pmatrix}(x),$$

or,

$$\boldsymbol{y}'(x) = \left(\boldsymbol{B}_0 + \boldsymbol{B}_1 x + \boldsymbol{B}_2 x^2\right) \boldsymbol{y}(x),$$

where $\boldsymbol{B}_0 = \begin{pmatrix} 0 & 1 \\ -4 & 0 \end{pmatrix}$, $\boldsymbol{B}_1 = \begin{pmatrix} 0 & 0 \\ -1 & -2 \end{pmatrix}$, and $\boldsymbol{B}_2 = \begin{pmatrix} 0 & 0 \\ 0 & -1 \end{pmatrix}$. As the calculation proceeds, the symbolic names will prove more revealing to us than would numeric matrices.

Substitute $\boldsymbol{y}(x) = \sum_{k=0}^{\infty} \boldsymbol{C}_k x^k$. It is worth recalling that the determining set is $\{-1, 0, 1, 2\}$ which says that the recursion relation will be third-order and establish a relationship between four terms: $\boldsymbol{C}_{k+1}$, $\boldsymbol{C}_k$, $\boldsymbol{C}_{k-1}$, and $\boldsymbol{C}_{k-2}$. It will be in effect for $k \geq 2$. The substitution gives

$$\sum_{k=1}^{\infty} k \boldsymbol{C}_k x^{k-1} = \left(\boldsymbol{B}_0 + \boldsymbol{B}_1 x + \boldsymbol{B}_2 x^2\right) \sum_{k=0}^{\infty} \boldsymbol{C}_k x^k.$$

Thus

$$\sum_{k=1}^{\infty} k \boldsymbol{C}_k x^{k-1} = \boldsymbol{B}_0 \sum_{k=0}^{\infty} \boldsymbol{C}_k x^k + \boldsymbol{B}_1 \sum_{k=0}^{\infty} \boldsymbol{C}_k x^{k+1} + \boldsymbol{B}_2 \sum_{k=0}^{\infty} \boldsymbol{C}_k x^{k+2},$$

or

$$\sum_{k=1}^{\infty} k \boldsymbol{C}_k x^{k-1} - \sum_{k=0}^{\infty} \boldsymbol{B}_0 \boldsymbol{C}_k x^k - \sum_{k=0}^{\infty} \boldsymbol{B}_1 \boldsymbol{C}_k x^{k+1} - \sum_{k=0}^{\infty} \boldsymbol{B}_2 \boldsymbol{C}_k x^{k+2} = 0.$$

Adjust the indices so that every summation contains x^k.

$$\sum_{k=0}^{\infty} (k+1) \boldsymbol{C}_{k+1} x^k - \sum_{k=0}^{\infty} \boldsymbol{B}_0 \boldsymbol{C}_k x^k - \sum_{k=1}^{\infty} \boldsymbol{B}_1 \boldsymbol{C}_{k-1} x^k - \sum_{k=2}^{\infty} \boldsymbol{B}_2 \boldsymbol{C}_{k-2} x^k = 0.$$

Separate the terms involving x_0 and x_1 and combine the others to get the expression

$$(\boldsymbol{C}_1 - \boldsymbol{B}_0 \boldsymbol{C}_0) x^0 + (2\boldsymbol{C}_2 - \boldsymbol{B}_0 \boldsymbol{C}_1 - \boldsymbol{B}_1 \boldsymbol{C}_0) x^1$$
$$+ \sum_{k=2}^{\infty} [(k+1)\boldsymbol{C}_{k+1} - \boldsymbol{B}_0 \boldsymbol{C}_k - \boldsymbol{B}_1 \boldsymbol{C}_{k-1} - \boldsymbol{B}_2 \boldsymbol{C}_{k-2}] x^k = 0.$$

Now solve for the coefficient matrices $\boldsymbol{C}_k$, $k \geq 0$, to get

$$\boldsymbol{C}_0 = \boldsymbol{K} \quad \text{(arbitrary)}$$

$$\boldsymbol{C}_1 = \boldsymbol{B}_0 \boldsymbol{C}_0 = \boldsymbol{B}_0 \boldsymbol{K}$$

$$\boldsymbol{C}_2 = \frac{1}{2}(\boldsymbol{B}_0 \boldsymbol{C}_1 + \boldsymbol{B}_1 \boldsymbol{C}_0)$$

$$= \frac{1}{2}(\boldsymbol{B}_0\boldsymbol{B}_0\boldsymbol{K} + \boldsymbol{B}_1\boldsymbol{K}) = \frac{1}{2}(\boldsymbol{B}_0^2 + \boldsymbol{B}_1)\boldsymbol{K}$$

$$\boldsymbol{C}_{k+1} = \frac{1}{(k+1)}[\boldsymbol{B}_0\boldsymbol{C}_k + \boldsymbol{B}_1\boldsymbol{C}_{k-1} + \boldsymbol{B}_2\boldsymbol{C}_{k-2}], \quad k \geq 2.$$

This latter equation has the form suggested by Theorem 9.1: it is the recursion relation, a third-order difference equation having four terms, with variable coefficients. It is in effect for $k \geq 2$. Except for the "variable coefficients" observation, all of these facts were revealed to us by the determining set. The arbitrary constant $\boldsymbol{K}$ can be either a vector or a matrix object depending on whether a vector or a matrix solution is sought. All of the coefficients of $\boldsymbol{K}$ in the expressions for the $\boldsymbol{C}_k$ are matrices. It is clear (think about it as a problem in mathematical induction) that each of the expressions for the $\boldsymbol{C}_k$ are in terms of $\boldsymbol{K}$ and the coefficients $\boldsymbol{B}_0$, $\boldsymbol{B}_1$, and $\boldsymbol{B}_2$ of $\boldsymbol{P}(x)$.

Numerically, the coefficients are

$$\boldsymbol{C}_0 = \begin{pmatrix} 1 & 0 \\ 0 & 1 \end{pmatrix} \boldsymbol{K}$$

$$\boldsymbol{C}_1 = \begin{pmatrix} 0 & 1 \\ -4 & 0 \end{pmatrix} \boldsymbol{K}$$

$$\boldsymbol{C}_2 = \begin{pmatrix} -2 & 0 \\ -1/2 & -3 \end{pmatrix} \boldsymbol{K}$$

$$\boldsymbol{C}_{k+1} = \frac{1}{(k+1)} \left[\begin{pmatrix} 0 & 1 \\ -4 & 0 \end{pmatrix} \boldsymbol{C}_k + \begin{pmatrix} 0 & 0 \\ -1 & -2 \end{pmatrix} \boldsymbol{C}_{k-1} + \begin{pmatrix} 0 & 0 \\ 0 & -1 \end{pmatrix} \boldsymbol{C}_{k-2} \right],$$

for $k \geq 2$. The next several coefficients are

$$\boldsymbol{C}_3 = \begin{pmatrix} -1/6 & -1 \\ 16/3 & -2/3 \end{pmatrix} \boldsymbol{K}$$

$$\boldsymbol{C}_4 = \begin{pmatrix} 4/3 & -1/6 \\ 23/12 & 5/2 \end{pmatrix} \boldsymbol{K}$$

$$\boldsymbol{C}_5 = \begin{pmatrix} 23/60 & 1/2 \\ -46/15 & 6/5 \end{pmatrix} \boldsymbol{K}$$

$$\boldsymbol{C}_6 = \begin{pmatrix} -23/45 & 1/5 \\ -361/180 & -37/36 \end{pmatrix} \boldsymbol{K}.$$

This results in a solution of the form

$$\boldsymbol{Y}(x) = \left[\begin{pmatrix} 1 & 0 \\ 0 & 1 \end{pmatrix} + \begin{pmatrix} 0 & 1 \\ -4 & 0 \end{pmatrix} x + \begin{pmatrix} -2 & 0 \\ -1/2 & -3 \end{pmatrix} x^2 + \begin{pmatrix} -1/6 & -1 \\ 16/3 & -2/3 \end{pmatrix} x^3 \right.$$
$$+ \begin{pmatrix} 4/3 & -1/6 \\ 23/12 & 5/2 \end{pmatrix} x^4 + \begin{pmatrix} 23/60 & 1/2 \\ -46/15 & 6/5 \end{pmatrix} x^5$$
$$\left. + \begin{pmatrix} -23/45 & 1/5 \\ -361/180 & -37/36 \end{pmatrix} x^6 + \cdots \right] \boldsymbol{K}.$$

Observe that the coefficient of $\boldsymbol{K}$ consists of a fundamental solution matrix for the system, where only the first seven terms have been listed. Whether $\boldsymbol{Y}(x)$ is a matrix or a vector depends on the choice of matrix or vector for the constant $\boldsymbol{K}$. Four-term recursion relations are difficult to solve, with the result that it is not clear what patterns might exist within the coefficient series that could suggest to us whether or not there is a closed form for this solution. ◇

It is of interest to complete the solution of the system 9.6, which appeared in Example 9.6, where the scalar coefficient polynomial of the derivative is nonzero, but is not constant.

Example 9.8 Solve the differential system

$$(x^2+1)\boldsymbol{y}'(x) = \begin{pmatrix} 0 & x^2+1 \\ 1 & -x \end{pmatrix} \boldsymbol{y}(x)$$

centered on the ordinary point $x_0 = 0$.

Solution. Here the determining set is $\{1,-1,0,1,2\} = \{-1,0,1,2\}$. We expect a four-term recursion relation of order three $(= 2-(-1))$ that takes effect for $k \geq 2$. The left-hand side contributed the set $\{1,-1\}$ and the right-hand side $\{0,1,2\}$; the determining set is the union of these.

Because $x_0 = 0$ is an ordinary point, we assume a series solution of the form $\boldsymbol{y}(x) = \sum_{k=0}^{\infty} \boldsymbol{C}_k x^k$. Substitution gives

$$(1+x^2)\sum_{k=1}^{\infty} k\boldsymbol{C}_k x^{k-1} = (\boldsymbol{B}_0 + \boldsymbol{B}_1 x + \boldsymbol{B}_2 x^2)\sum_{k=0}^{\infty} \boldsymbol{C}_k x^k,$$

where

$$\boldsymbol{B}_0 = \begin{pmatrix} 0 & 1 \\ 1 & 0 \end{pmatrix}, \quad \boldsymbol{B}_1 = \begin{pmatrix} 0 & 0 \\ 0 & -1 \end{pmatrix}, \quad \boldsymbol{B}_2 = \begin{pmatrix} 0 & 1 \\ 0 & 0 \end{pmatrix},$$

and the $\boldsymbol{C}_k$ are to be determined. Distribute the sums to get

$$\sum_{k=1}^{\infty} k\boldsymbol{C}_k x^{k-1} + x^2 \sum_{k=1}^{\infty} k\boldsymbol{C}_k x^{k-1}$$
$$= \boldsymbol{B}_0 \sum_{k=0}^{\infty} \boldsymbol{C}_k x^k + \boldsymbol{B}_1 \sum_{k=0}^{\infty} \boldsymbol{C}_k x^{k+1} + \boldsymbol{B}_2 \sum_{k=0}^{\infty} \boldsymbol{C}_k x^{k+2}.$$

Put everything on the left and adjust the indices so that every summation contains x^k.

$$\sum_{k=0}^{\infty} (k+1)\boldsymbol{C}_{k+1} x^k + \sum_{k=2}^{\infty} (k-1)\boldsymbol{C}_{k-1} x^k$$
$$- \sum_{k=0}^{\infty} \boldsymbol{B}_0 \boldsymbol{C}_k x^k - \sum_{k=1}^{\infty} \boldsymbol{B}_1 \boldsymbol{C}_{k-1} x^k - \sum_{k=2}^{\infty} \boldsymbol{B}_2 \boldsymbol{C}_{k-2} x^k = 0.$$

Separate the terms involving x_0 and x_1 to get the expression

$$(\boldsymbol{C}_1 - \boldsymbol{B}_0\boldsymbol{C}_0)x^0 + (2\boldsymbol{C}_2 - \boldsymbol{B}_0\boldsymbol{C}_1 - \boldsymbol{B}_1\boldsymbol{C}_0)x^1 \\ + \sum_{k=2}^{\infty}[(k+1)\boldsymbol{C}_{k+1} - \boldsymbol{B}_0\boldsymbol{C}_k - (\boldsymbol{B}_1 - (k-1)I_2)\boldsymbol{C}_{k-1} - \boldsymbol{B}_2\boldsymbol{C}_{k-2}]x^k.$$

Now solve for thc coefficient matrices $\boldsymbol{C}_k$, $k > 0$, to get

$$\begin{aligned}
\boldsymbol{C}_0 &= \boldsymbol{K} \quad \text{(arbitrary)} \\
\boldsymbol{C}_1 &= \boldsymbol{B}_0\boldsymbol{C}_0 = \boldsymbol{B}_0\boldsymbol{K} \\
\boldsymbol{C}_2 &= \frac{1}{2}(\boldsymbol{B}_0\boldsymbol{C}_1 + \boldsymbol{B}_1\boldsymbol{C}_0) \\
&= \frac{1}{2}(\boldsymbol{B}_0\boldsymbol{B}_0\boldsymbol{K} + \boldsymbol{B}_1\boldsymbol{K}) = \frac{1}{2}(\boldsymbol{B}_0^2 + \boldsymbol{B}_1)\boldsymbol{K} \\
\boldsymbol{C}_{k+1} &= \frac{1}{k+1}[\boldsymbol{B}_0\boldsymbol{C}_k + (\boldsymbol{B}_1 - (k-1)I_2)\boldsymbol{C}_{k-1} + \boldsymbol{B}_2\boldsymbol{C}_{k-2}],
\end{aligned}$$

for $k \geq 2$. Numerically, the coefficients are found to be

$$\boldsymbol{C}_0 = \begin{pmatrix} 1 & 0 \\ 0 & 1 \end{pmatrix}\boldsymbol{K},$$

$$\boldsymbol{C}_1 = \begin{pmatrix} 0 & 1 \\ 1 & 0 \end{pmatrix}\boldsymbol{K},$$

$$\boldsymbol{C}_2 = \begin{pmatrix} 1/2 & 0 \\ 0 & 0 \end{pmatrix}\boldsymbol{K},$$

and

$$\boldsymbol{C}_{k+1} = \frac{1}{k+1}\left[\begin{pmatrix} 0 & 1 \\ 1 & 0 \end{pmatrix}\boldsymbol{C}_k + \begin{pmatrix} -k+1 & 0 \\ 0 & -k \end{pmatrix}\boldsymbol{C}_{k-1} + \begin{pmatrix} 0 & 1 \\ 0 & 0 \end{pmatrix}\boldsymbol{C}_{k-2}\right],$$

for $k \geq 2$. The next several coefficients are

$$\boldsymbol{C}_3 = \begin{pmatrix} 0 & 0 \\ -1/2 & 0 \end{pmatrix}\boldsymbol{K},$$

$$\boldsymbol{C}_4 = \begin{pmatrix} -1/8 & 0 \\ 0 & 0 \end{pmatrix}\boldsymbol{K},$$

$$\boldsymbol{C}_5 = \begin{pmatrix} 0 & 0 \\ 3/8 & 0 \end{pmatrix}\boldsymbol{K},$$

$$\boldsymbol{C}_6 = \begin{pmatrix} 1/16 & 0 \\ 0 & 0 \end{pmatrix}\boldsymbol{K}.$$

This results in a solution of the form

$$\boldsymbol{Y}(x) = \left[\begin{pmatrix} 1 & 0 \\ 0 & 1 \end{pmatrix} + \begin{pmatrix} 0 & 1 \\ 1 & 0 \end{pmatrix} x + \begin{pmatrix} 1/2 & 0 \\ 0 & 0 \end{pmatrix} x^2 + \begin{pmatrix} 0 & 0 \\ -1/2 & 0 \end{pmatrix} x^3 \right.$$
$$\left. + \begin{pmatrix} -1/8 & 0 \\ 0 & 0 \end{pmatrix} x^4 + \begin{pmatrix} 0 & 0 \\ 3/8 & 0 \end{pmatrix} x^5 + \begin{pmatrix} 1/16 & 0 \\ 0 & 0 \end{pmatrix} x^6 + \cdots \right] \boldsymbol{K}.$$

The second column of the fundamental matrix solution is

$$\boldsymbol{y}_2(x) = \begin{pmatrix} 0 \\ 1 \end{pmatrix} + \begin{pmatrix} 1 \\ 0 \end{pmatrix} x + \sum_{k=2}^{\infty} \begin{pmatrix} 0 \\ 0 \end{pmatrix} x^k = \begin{pmatrix} x \\ 1 \end{pmatrix},$$

which is a polynomial solution of the differential system. ◇

A package of *Mathematica* function definitions named *SeriesSysLeadingPrint* has been provided in the supplement that automates the production of such series solutions. The function is invoked as

`DSolveOrdinarySeries[Leading, PMatrix, Initial, xName, x0, NTerms]`

where the various parameters have these definitions:

`Leading` the scalar multiplier of $y'(x)$;
`PMatrix` the analytic coefficient matrix $P(x)$;
`Initial` the constant matrix (or vector) K;
`xName` the symbol (usually 'x') that is to be the independent variable;
`x0` the center for the series;
`NTerms` the number of terms desired.

Here is how this function solves Example 9.7.

Example 9.8M Use `DSolveOrdinarySeries` to find the terms through x^6 of a series fundamental solution of the differential system

$$(x^2+1)\boldsymbol{y}'(x) = \begin{pmatrix} 0 & x^2+1 \\ 1 & -x \end{pmatrix} \boldsymbol{y}(x)$$

centered on the ordinary point $x_0 = 0$.

Solution. Here `Leading = (x^2 + 1)`, `PMatrix = {{0,1+x^2},{1,-x}}`, `Initial = {{1,0},{0,1}}`, `xName = x`, `x0 = 0`, and `NTerms = 6`.

```
In[19]:= w[x_] = DSolveOrdinarySeries[(1+x^2),{{0,1+x^2},{1,-x}},
                 {{1,0},{0,1}},x,0,6]

        Given the linear differential system:
        2                     2
(1 + x )*u[1]'[x] == {0, 1 + x }.u[x]
        2
(1 + x )*u[2]'[x] == {1, -x}.u[x]

with initial condition:
     u[1][0] == {1, 0},
     u[2][0] == {0, 1},

find a series solution about the ordinary point x == 0.
```

```
The first few coefficients in the solution series:

                    i
u[x] = Sum[c[i]*x ,{i,0,6}],

are:

c[0] = {{1, 0}, {0, 1}}

c[1] = ({{0, 1}, {1, 0}}.c[0]
       )/1
     = {{0, 1}, {1, 0}}

c[2] = ({{0, 1}, {1, 0}}.c[1]
      + {{0, 0}, {0, -1}}.c[0]
       )/2
         1
     = {{-, 0}, {0, 0}}
         2

and the coefficients propagate according to the rule
c[k] = ({{0, 1}, {1, 0}}.c[k-1]
      + {{2 - k, 0}, {0, 1 - k}}.c[k-2]
      + {{0, 1}, {0, 0}}.c[k-3]
       )/(k)
for k >= 3.
```

```
Out[19]=         2      4      6
                x      x      x              7             7
        {{1 + -- - -- + -- + O[x] , x + O[x] },
               2      8     16

                  3       5
                 x     3 x          7
          {x - -- + ---- + O[x] , 1}}
                 2       8
```

Check

```
In[20]:= Simplify[(1+x^2)*w'[x]-{{0,1+x^2},{1,-x}}.w[x], Trig->False]

Out[20]=        6        6          6        7
         {{O[x] , O[x] }, {O[x] , O[x] }}
```

This last output says that the portion of the series that we produced satisfies the differential equation through terms in x^5. The lines between `In[19]` and `Out[19]` are printed outputs from various `Print` statements; they are not part of the result that is returned at `Out[19]`. ◇

As a second example of using *Mathematica* to obtain a series solution, we obtain a vector solution for the system that defines the sine and cosine functions.

Example 9.9M Use `DSolveOrdinarySeries` to find the terms through x^8 of the vector solution of

$$\boldsymbol{y}'(x) = \begin{pmatrix} 0 & 1 \\ -1 & 0 \end{pmatrix} \boldsymbol{y}(x)$$

which satisfies $\boldsymbol{y}(0) = \begin{pmatrix} 1 \\ 0 \end{pmatrix}$.

Solution. Here `Leading = 1`, `PMatrix = {{0,1},{-1,0}}`, `Initial = {1,0}`, `xName = x`, `x0 = 0`, and `NTerms = 8`. Since Initial is a vector, the solution is a vector.

```
In[21]:= w[x_] = DSolveOrdinarySeries[1,{{0,1},{-1,0}},{1,0},x,0,8]

        Given the linear differential system:
(1)*u[1]'[x] == {0, 1}.u[x]
(1)*u[2]'[x] == {-1, 0}.u[x]
with initial condition:
    u[1][0] == 1,
    u[2][0] == 0,
find a series solution about the ordinary point x == 0.

The first few coefficients in the solution series:
                    i
u[x] = Sum[c[i]*x  ,{i,0,8}],

        are: c[0] = {1, 0}
and the coefficients propagate according to the rule
c[k] = ({{0, 1}, {-1, 0}}.c[k-1]
       )/(k)
for k >= 1.

Out[21]=        2     4     6      8
               x     x     x      x          9
         {1 - -- + -- - --- + ----- + O[x] ,
               2    24   720   40320

                3     5      7
               x     x      x          9
          -x + -- - --- + ---- + O[x]  }
               6    120   5040
```

Check

```
In[22]:= Simplify[w'[x]-{{0,1},{-1,0}}.w[x]]

Out[22]=       8      8
         {O[x] , O[x] }
```

So the two solutions, which you immediately should recognize as representing the cosine function and the negative of the sine function, satisfy the differential equation through terms in x^7.

It is of interest to show that a closed-form solution of the recursion relation

$$\boldsymbol{C}_k = \begin{pmatrix} 0 & 1/k \\ -1/k & 0 \end{pmatrix} \boldsymbol{C}_{k-1}, k \geq 1; \quad \boldsymbol{C}_0 = \begin{pmatrix} 1 \\ 0 \end{pmatrix}$$

can be given. Apply the relation several times to produce

$$\boldsymbol{C}_1 = \begin{pmatrix} 0 & 1/1 \\ -1/1 & 0 \end{pmatrix} \begin{pmatrix} 1 \\ 0 \end{pmatrix} = \begin{pmatrix} 0 \\ -1 \end{pmatrix}$$

$$\boldsymbol{C}_2 = \begin{pmatrix} 0 & 1/2 \\ 1/2 & 0 \end{pmatrix} \begin{pmatrix} 0 \\ -1 \end{pmatrix} = \begin{pmatrix} -1/2 \\ 0 \end{pmatrix} = \begin{pmatrix} -1/2! \\ 0 \end{pmatrix}$$

$$C_3 = \begin{pmatrix} 0 & 1/3 \\ -1/3 & 0 \end{pmatrix} \begin{pmatrix} -1/2! \\ 0 \end{pmatrix} = \begin{pmatrix} 0 \\ 1/3! \end{pmatrix},$$

etc. These suggest that we attempt an inductive proof that

$$C_{2m} = \begin{pmatrix} (-1)^m/(2m)! \\ 0 \end{pmatrix}, \quad \text{and} \quad C_{2m+1} = \begin{pmatrix} 0 \\ (-1)^{m+1}/(2m+1)! \end{pmatrix}.$$

The statements are true for $m = 0$ (and 1). Assume them true for m. Then calculate

$$\begin{aligned}
&\begin{pmatrix} 0 & 1/2m+1 \\ -1/2m+1 & 0 \end{pmatrix} C_{2m} \\
&= \begin{pmatrix} 0 & 1/2m+1 \\ -1/2m+1 & 0 \end{pmatrix} \begin{pmatrix} (-1)^m/(2m)! \\ 0 \end{pmatrix} \\
&= \begin{pmatrix} 0 \\ (-1)^{m+1}/(2m+1)! \end{pmatrix} = C_{2m+1},
\end{aligned}$$

and

$$\begin{aligned}
&\begin{pmatrix} 0 & 1/2(m+1) \\ -1/2(m+1) & 0 \end{pmatrix} C_{2m+1} \\
&= \begin{pmatrix} 0 & 1/2(m+1) \\ -1/2(m+1) & 0 \end{pmatrix} \begin{pmatrix} 0 \\ (-1)^{m+1}/(2m+1)! \end{pmatrix} \\
&= \begin{pmatrix} (-1)^{m+1}/(2(m+1))! \\ 0 \end{pmatrix} = C_{2(m+1)}.
\end{aligned}$$

This establishes the induction, with the result that

$$\begin{aligned}
y(x) &= \sum_{k=0}^{\infty} C_k x^k \\
&= \sum_{m=0}^{\infty} C_{2m} x^{2m} + \sum_{m=0}^{\infty} C_{2m+1} x^{2m+1} \\
&= \begin{pmatrix} \displaystyle\sum_{m=0}^{\infty} \frac{(-1)^m}{(2m)!} x^{2m} \\ 0 \end{pmatrix} - \begin{pmatrix} 0 \\ \displaystyle\sum_{m=0}^{\infty} \frac{(-1)^m}{(2m+1)!} x^{2m+1} \end{pmatrix} \\
&= \begin{pmatrix} \cos x \\ -\sin x \end{pmatrix}.
\end{aligned}$$

◇

Having the function `DSolveOrdinarySeries` available certainly takes the drudgery out of finding series solutions about ordinary points. There is an analogous function `GenerateRSPSeries` that finds series solutions in the case that x0 is a regular singular point. This function requires that the system have certain properties that make it relatively simple to obtain a solution. It requires you to manually find the indices and their corresponding characteristic vectors, and then it finds the terms you want from the resulting series.

Exercises 9.2

1. Redo the solution of Example 9.8 using `Initial` $= \{0, 1\}$. Solve the recursion relation and identify the functions represented by the series.

2. Redo the solution of Example 9.8 using `Initial` $= \{\{1, 0\}, \{0, 1\}\}$. Solve the recursion relation. You have already identified the functions represented by the series.

3. Find a fundamental series solution about $x_0 = 0$ of the differential system

$$\frac{d\boldsymbol{y}}{dx} = \begin{pmatrix} 0 & x \\ -x & 0 \end{pmatrix} \boldsymbol{y}.$$

4. Show that the following initial value problem has a series solution in powers of x in which each component is a polynomial.

$$\frac{d\boldsymbol{y}}{dx} = \begin{pmatrix} 0 & 6x^2 & 0 \\ 0 & 0 & 3x^2 \\ 1 & -x^3 & 0 \end{pmatrix} \boldsymbol{y}, \quad \boldsymbol{y}(0) = \begin{pmatrix} 0 \\ 0 \\ 1 \end{pmatrix}.$$

5. Show that the following initial value problem has a series solution in powers of x in which each component is a polynomial.

$$\frac{d\boldsymbol{y}}{dx} = \begin{pmatrix} 0 & 3x & 0 \\ 0 & 0 & 1 \\ 1 & -x^2 & 0 \end{pmatrix} \boldsymbol{y}, \quad \boldsymbol{y}(0) = \begin{pmatrix} 0 \\ 0 \\ 1 \end{pmatrix}.$$

Use the function `DSolveOrdinarySeries` to find at least eight terms of the solution power series about $x_0 = 0$ of these differential systems.

6. $\boldsymbol{Y}'(x) = \begin{pmatrix} 1 & 2 \\ x & 1 \end{pmatrix} \boldsymbol{Y}(x), \quad \boldsymbol{Y}(0) = \boldsymbol{I}_2.$

7. $\boldsymbol{Y}'(x) = \begin{pmatrix} 3 & 2x \\ 1+x & 1-x \end{pmatrix} \boldsymbol{Y}(x), \quad \boldsymbol{Y}(0) = \boldsymbol{I}_2.$

8. $(1+2x)\boldsymbol{Y}'(x) = \begin{pmatrix} 0 & 1 \\ -3(7+10x) & 2(5+8x) \end{pmatrix} \boldsymbol{Y}(x), \quad \boldsymbol{Y}(0) = \boldsymbol{I}_2.$

9. $\boldsymbol{Y}'(x) = \begin{pmatrix} 0 & 3x & 0 \\ 0 & 0 & 1 \\ 1 & -x^2 & 0 \end{pmatrix} \boldsymbol{Y}(x), \quad \boldsymbol{Y}(0) = \boldsymbol{I}_3.$

10. $\boldsymbol{Y}'(x) = \begin{pmatrix} 0 & 6x^2 & 0 \\ 0 & 0 & 3x^2 \\ 1 & -x^3 & 0 \end{pmatrix} \boldsymbol{Y}(x), \quad \boldsymbol{Y}(0) = \boldsymbol{I}_3.$

9.3 Solution About Regular Singular Points

In this section we begin the study of the solution of differential systems in series centered on a point x_0 that is a **regular singular point** of the equation. Any point that is not an **ordinary point** of a differential system is a **singular point** of the system. Some singular points occur in such a way that finding solutions can still be donc in an orderly and somewhat predictable manner.

Definition 9.2 *The point x_0 is a* **regular singular point** *of the differential system*

$$\boldsymbol{y}'(x) = \boldsymbol{P}(x)\boldsymbol{y}(x)$$

provided that the $n \times n$ matrix-valued function $\boldsymbol{P}(x)$ has the form

$$\boldsymbol{P}(x) = \frac{1}{x - x_0}\boldsymbol{Q}(x),$$

where the $n \times n$ matrix-valued function $\boldsymbol{Q}(x)$ is analytic at x_0. This is equivalent to saying that the system can be put in the form

$$(x - x_0)\boldsymbol{y}'(x) = \boldsymbol{Q}(x)\boldsymbol{y}(x),$$

where $\boldsymbol{Q}(x)$ is analytic at x_0.

We can always shift the origin to 0, so we usually say that $x_0 = 0$. The system we are considering thus has the form

$$x\boldsymbol{y}'(x) = \boldsymbol{Q}(x)\boldsymbol{y}(x) \tag{9.8}$$

where $\boldsymbol{Q}(x)$ is analytic at 0.

Because $\boldsymbol{Q}(x)$ is analytic, $\boldsymbol{Q}(x) = \sum_{j=0}^{\infty} \boldsymbol{B}_j x^j$. Therefore, for x near 0, the $n \times n$ matrix-valued function $\boldsymbol{Q}(x)$ is near the $n \times n$ constant matrix $\boldsymbol{Q}(0) = \boldsymbol{B}_0$. This means that for x small, system 9.8 is near the system

$$x\boldsymbol{y}'(x) = \boldsymbol{B}_0\boldsymbol{y}(x). \tag{9.9}$$

This is a Cauchy-Euler system like those discussed in Section 9.1, where we learned that for $x > 0$ every vector solution is a constant vector times a power of x. In Chapter 7, when we first encountered an equation that is near a Cauchy-Euler equation, we learned to assume that the solutions have the form $\boldsymbol{y}(x) = x^r\boldsymbol{v}(x)$, where $\boldsymbol{v}(x)$ is analytic at 0. This is the case here, as this theorem, a variation on one due to Georg Frobenius, explains.

Theorem 9.2 *Consider the system*

$$x\boldsymbol{y}'(x) = \boldsymbol{Q}(x)\boldsymbol{y}(x),$$

where $\boldsymbol{Q}(x) = \sum_{j=0}^{\infty} \boldsymbol{B}_j x^j$ is analytic at $x = 0$ with radius of convergence R.

Then there is at least one (possibly complex) number r such that the function

$$\boldsymbol{y}(x) = x^r \sum_{j=0}^{\infty} \boldsymbol{C}_k x^k$$

is a solution, with the $\boldsymbol{C}_k$ determined by the equations

$$\begin{array}{ll} r\boldsymbol{C}_0 - \boldsymbol{B}_0\boldsymbol{C}_0 = \boldsymbol{0} & (\text{when} \quad k=0) \\ (1+r)\boldsymbol{C}_1 - \boldsymbol{B}_0\boldsymbol{C}_1 = \boldsymbol{B}_1\boldsymbol{C}_0 & (\text{when} \quad k=1) \\ (2+r)\boldsymbol{C}_2 - \boldsymbol{B}_0\boldsymbol{C}_2 = \boldsymbol{B}_1\boldsymbol{C}_1 + \boldsymbol{B}_2\boldsymbol{C}_0 & (\text{when} \quad k=2) \end{array}$$

or, in general,

$$(\boldsymbol{B}_0 - (k+r)\boldsymbol{I}_n)\boldsymbol{C}_k = -\sum_{i=1}^{k} \boldsymbol{B}_i\boldsymbol{C}_{k-i} \quad (\text{any} \quad k).$$

The series portion of this solution has radius of convergence at least R.

The values of r and $\boldsymbol{C}_0$ are determined by the first equation,

$$r\boldsymbol{C}_0 - \boldsymbol{B}_0\boldsymbol{C}_0 = (r\boldsymbol{I}_n - \boldsymbol{B}_0)\boldsymbol{C}_0 = \boldsymbol{0},$$

or

$$(\boldsymbol{B}_0 - r\boldsymbol{I}_n)\boldsymbol{C}_0 = \boldsymbol{0}. \tag{9.10}$$

This is called the **indicial equation** of system 9.8 at the regular singular point $x = 0$. Since every $\boldsymbol{C}_k$ will be determined in terms of $\boldsymbol{C}_0$, it is essential that $\boldsymbol{C}_0 \neq \boldsymbol{0}$, and this means that in order to obtain a solution, we need to be able to solve the **characteristic value problem** 9.10. Once r and $\boldsymbol{C}_0$ have been determined, we can uniquely determine all of the other $\boldsymbol{C}_k$ provided that all of the matrices of the form

$$\boldsymbol{B}_0 - (k+r)\boldsymbol{I}_n$$

are nonsingular. These are the coefficients of $\boldsymbol{C}_k$ for $k > 0$.

The characteristic roots of the indicial equation 9.10 are called the **indices** of system 9.8 at $x_0 = 0$. The distribution of these indices is critical to the solution process. We continue our study with only the simplest case and leave further study to subsequent courses. To continue your study of these systems, review Chapter 7, where properties of Cauchy-Euler equations suggest avenues of approach to new situations. The same is true of systems with variable coefficients. The introduction of natural logarithms is again required to find some solutions.

There is a solution if r is chosen such that $\boldsymbol{B}_0 - (k+r)\boldsymbol{I}_n$ is invertible for all $k \geq 1$. We can then uniquely determine the $\boldsymbol{C}_k$ for all $k \geq 1$ in terms of $\boldsymbol{C}_0$, a characteristic vector corresponding to r. Taking the index r with the largest real part insures that each of the matrices is invertible. Any index that differs from no other index by a positive integer also guarantees a solution. Among those indices for which $\boldsymbol{B}_0 - (k+r)\boldsymbol{I}_n$ is invertible for all $k \geq 1$, any whose **alge-**

braic multiplicity agrees with its **geometric multiplicity** will produce a full complement of solutions. Complex (pairs) of indices that are not repeated roots generate complex solutions that can be converted into pairs of real solutions. Occasionally two indices that differ by an integer will both produce solutions, but usually a natural logarithm is needed to find solutions corresponding to one of the two indices. Except in very simple cases it is not easy to find n linearly independent solutions.

Example 9.10 Find two linearly independent solutions to the singular differential system

$$x\boldsymbol{y}'(x) = \begin{pmatrix} x & -1/2 \\ 1/3 & 5/6 \end{pmatrix} \boldsymbol{y}(x) = \left(\begin{pmatrix} 0 & -1/2 \\ 1/3 & 5/6 \end{pmatrix} + \begin{pmatrix} 1 & 0 \\ 0 & 0 \end{pmatrix} x \right) \boldsymbol{y}(x)$$

for $x > 0$.

Solution. The series portion of our solution will be defined for $x < 0$, but the solution itself may not. At any rate, we use $x_0 = 0$, and assume a solution of the form $\boldsymbol{y}(x) = \sum_{k=0}^{\infty} \boldsymbol{C}_k x^{k+r}$. Set $\boldsymbol{B}_0 = \begin{pmatrix} 0 & -1/2 \\ 1/3 & 5/6 \end{pmatrix}$ and $\boldsymbol{B}_1 = \begin{pmatrix} 1 & 0 \\ 0 & 0 \end{pmatrix}$. Then

$$\sum_{k=0}^{\infty} (k+r)\boldsymbol{C}_k x^{k+r} = \sum_{k=0}^{\infty} \boldsymbol{B}_0 \boldsymbol{C}_k x^{k+r} + \sum_{k=0}^{\infty} \boldsymbol{B}_1 \boldsymbol{C}_k x^{k+r+1}.$$

Adjust the index on the last sum to get

$$\sum_{k=0}^{\infty} (k+r)\boldsymbol{C}_k x^{k+r} = \sum_{k=0}^{\infty} \boldsymbol{B}_0 \boldsymbol{C}_k x^{k+r} + \sum_{k=1}^{\infty} \boldsymbol{B}_1 \boldsymbol{C}_{k-1} x^{k+r}.$$

This is equivalent to

$$(r\boldsymbol{I} - \boldsymbol{B}_0)\boldsymbol{C}_0 x^r + \sum_{k=1}^{\infty} [((k+r)\boldsymbol{I}_2 - \boldsymbol{B}_0)\boldsymbol{C}_k - \boldsymbol{B}_1 \boldsymbol{C}_{k-1}] x^{k+r} = 0.$$

We must have r and $\boldsymbol{C}_0$ chosen to be solutions of the characteristic value problem

$$(\boldsymbol{B}_0 - r\boldsymbol{I})\boldsymbol{C}_0 = \mathbf{0}.$$

Numerically, this is

$$\begin{pmatrix} -r & -1/2 \\ 1/3 & 5/6 - r \end{pmatrix} \boldsymbol{C}_0 = \mathbf{0}.$$

This characteristic value problem has characteristic equation

$$\det \begin{pmatrix} -r & -1/2 \\ 1/3 & 5/6 - r \end{pmatrix} = r^2 - \frac{5}{6} r + \frac{1}{6} = \left(r - \frac{1}{2} \right) \left(r - \frac{1}{3} \right) = 0,$$

which says that $r_1 = 1/2$ and $r_2 = 1/3$. The corresponding characteristic vectors are $\{1, -1\}$ and $\{3, -2\}$, as can be easily verified.

When $r_1 = 1/2$ is used, the other coefficients are found from

$$\boldsymbol{C}_k = \left(\left(k + \frac{1}{2}\right)\boldsymbol{I}_2 - \boldsymbol{B}_0\right)^{-1} \boldsymbol{B}_1 \boldsymbol{C}_{k-1}$$

$$= \begin{pmatrix} -k - (1/2) & -1/2 \\ 1/3 & 1/3 - k \end{pmatrix}^{-1} \begin{pmatrix} 1 & 0 \\ 0 & 0 \end{pmatrix} \boldsymbol{C}_{k-1}, \quad \text{for} \quad k \geq 1.$$

The first several coefficients are

$$\boldsymbol{C}_0 = \begin{pmatrix} 1 \\ -1 \end{pmatrix}, \quad \boldsymbol{C}_1 = \begin{pmatrix} 4/7 \\ 2/7 \end{pmatrix}, \quad \boldsymbol{C}_2 = \begin{pmatrix} 20/91 \\ 4/91 \end{pmatrix},$$

$$\boldsymbol{C}_3 = \begin{pmatrix} 320/5187 \\ 40/5187 \end{pmatrix}, \quad \boldsymbol{C}_4 = \begin{pmatrix} 352/25935 \\ 32/25935 \end{pmatrix}, \quad \boldsymbol{C}_5 = \begin{pmatrix} 1408/574275 \\ 704/4019925 \end{pmatrix}.$$

This gives a vector series solution

$$\boldsymbol{y}_1(x) = \left(\begin{pmatrix} 1 \\ -1 \end{pmatrix} + \begin{pmatrix} 4/7 \\ 2/7 \end{pmatrix} x + \begin{pmatrix} 20/91 \\ 4/91 \end{pmatrix} x^2 + \begin{pmatrix} 320/5187 \\ 40/5187 \end{pmatrix} x^3 \right.$$

$$\left. + \begin{pmatrix} 352/25935 \\ 32/25935 \end{pmatrix} x^4 + \begin{pmatrix} 1408/574275 \\ 704/4019925 \end{pmatrix} x^5 + \cdots \right) x^{1/2}$$

This solution checks. It is not defined for $x < 0$.

When $r_1 = 1/3$ is used, the other coefficients are found from

$$\boldsymbol{C}_k = \left(\left(k + \frac{1}{3}\right)\boldsymbol{I}_2 - \boldsymbol{B}_0\right)^{-1} \boldsymbol{B}_1 \boldsymbol{C}_{k-1}$$

$$= \begin{pmatrix} -k - (1/3) & -1/2 \\ 1/3 & 1/2 - k \end{pmatrix}^{-1} \begin{pmatrix} 1 & 0 \\ 0 & 0 \end{pmatrix} \boldsymbol{C}_{k-1}, \quad \text{for} \quad k \geq 1.$$

The first several coefficients are

$$\boldsymbol{C}_0 = \begin{pmatrix} 3 \\ -2 \end{pmatrix}, \boldsymbol{C}_1 = \begin{pmatrix} 9/5 \\ 6/5 \end{pmatrix}, \boldsymbol{C}_2 = \begin{pmatrix} 81/110 \\ 9/55 \end{pmatrix},$$

$$\boldsymbol{C}_3 = \begin{pmatrix} 81/374 \\ 27/935 \end{pmatrix}, \boldsymbol{C}_4 = \begin{pmatrix} 1701/34408 \\ 81/17204 \end{pmatrix}, \boldsymbol{C}_5 = \begin{pmatrix} 45927/4989160 \\ 1701/2494580 \end{pmatrix}.$$

This gives a vector series solution

$$\boldsymbol{y}_2(x) = \left(\begin{pmatrix} 3 \\ -2 \end{pmatrix} + \begin{pmatrix} 9/5 \\ 6/5 \end{pmatrix} x + \begin{pmatrix} 81/110 \\ 9/55 \end{pmatrix} x^2 + \begin{pmatrix} 81/374 \\ 27/935 \end{pmatrix} x^3 \right.$$

$$\left. + \begin{pmatrix} 1701/34408 \\ 81/17204 \end{pmatrix} x^4 + \begin{pmatrix} 45927/4989160 \\ 1701/2494580 \end{pmatrix} x^5 + \cdots \right) x^{1/3}.$$

This solution also checks. It happens to be defined for $x < 0$, so this second solution is defined on an open interval containing 0.

For positive x near 0 these two solution can be combined into the matrix function $W(x) = (\boldsymbol{y}_1(x) \mid \boldsymbol{y}_2(x))$. This matrix solution is fundamental, since its determinant is near

$$x^{5/6}\det\begin{pmatrix}1 & 3\\ -1 & -2\end{pmatrix} = x^{5/6}$$

which is nonzero (for positive x). ◇

Exercises 9.3

The notebook `GenerateRSPSeries` may be useful for these problems. Modify the necessary cells as you execute them.

1. Solve this regular singular point problem for a solution centered on $x_0 = 0$.

$$t\frac{d\boldsymbol{y}}{dt} = \begin{pmatrix}0 & 2t^2 & t\\ 0 & 1 & 0\\ 0 & 2 & 0\end{pmatrix}\boldsymbol{y}(t), \quad \boldsymbol{y}(1) = \begin{pmatrix}1\\1\\1\end{pmatrix}.$$

Find at least eight terms of the solution in power series about $x_0 = 0$ of these differential systems. You may find the need for $\ln x$ series if there is a problem with the indices.

2. $x\boldsymbol{Y}'(x) = \begin{pmatrix}8+x & -5\\ 10 & -7-x\end{pmatrix}\boldsymbol{Y}(x).$

3. $x\boldsymbol{Y}'(x) = \begin{pmatrix}3 & 0 & x\\ -10 & -2 & 2x\\ 10 & 6 & 1+x\end{pmatrix}\boldsymbol{Y}(x).$

4. $x\boldsymbol{Y}'(x) = \begin{pmatrix}1 & x & 0\\ -4 & -1 & x\\ 8 & 4+2x & 1\end{pmatrix}\boldsymbol{Y}(x).$

5. $x\boldsymbol{Y}'(x) = \begin{pmatrix}-2 & 2 & -8 & -7\\ x & -1 & -8 & -8\\ 2x & -x & -5 & -4\\ 0 & x & 8 & 7\end{pmatrix}\boldsymbol{Y}(x).$

A

Mathematica: General Introduction

Within the Macintosh or Windows environment, **Open** folders and applications (including *Mathematica*) by double-clicking on the appropriate icon. *Mathematica* comes in two parts: a **front end** and a **kernel**. The front end manipulates *Mathematica* windows on your screen, processes your typing, and posts responses from the kernel to your screen. The kernel does all of the calculations which you request. It is not activated until you request that something be done (Press ENTER while in a cell). It takes some time for the kernel to be loaded and configured. During this time you may continue to work in your window. Once the kernel is active it responds to your requests very rapidly. At any time while the kernel is doing something, you may still communicate with the front end in any normal manner.

Things to Note About *Mathematica* Windows

(Most things are standard)

Cell Blocks	]-marks down the right margin of a window; these contain complete *Mathematica* ideas. Nesting is significant.
______________	Horizontal bar across window. Click above or below any cell: this line means a new cell can appear here. Type to make a cell.
Cursor Shape	Indicates what you can do where you are on the screen: select, edit, resize, format/unformat, start cell, hide/show cell contents, etc.
In[n]:=	Input cell n
Out[n]=	Output cell n (same as %n)
Scroll Bars	Standard

Drag Region	Standard
Close Box	Standard
Delete Key	Delete previous character.
COMMAND-X	Cut selection.
COMMAND-C	Copy selection.
COMMAND-V	Paste at cursor position.

General Conventions in *Mathematica*

Names	Names that belong to *Mathematica* always start with a capital. Embedded capitals logically distinguish multiple words within a single name. Objects and expressions need not be named.
`(  )`	Parentheses are used only for grouping expressions. Standard mathematical conventions hold.
`[  ]`	Brackets delimit arguments of functions. They are never used for grouping. Separate arguments by commas.
`{  }`	Braces delimit lists. They may be nested in order to make arrays, etc. Separate entries by commas.
RETURN Key	New line within cell. No action taken.
ENTER Key	Execute cell (or selection of multiple cells). (Same as Shift-RETURN.)

Symbols for Standard Mathematical Operations

(See Appendix C for mathematical functions.)

`+`	`Plus`: $a + b$
`-`	`Minus`: $a - b$; Negative: $-a$
`*` or space	`Times`: $a * b$ or `a b`
`/`	`Divide` a/b
`=`	Substitution (or `Set`): $x = 4$
`==`	`Equal`: `x == y` returns either True if x and y agree or False.
`!`	`Not`: `x != y` returns either True if x and y do not agree or False.
`%`	Last input; %n is input number n, as is In[n].

B

Mathematica Style: Some Considerations

A consistent set of rules for exposition in *Mathematica* have been used throughout the text and in the notebooks. You need not have extensive experience with *Mathematica* prior to studying this book. When you have completed the book you will be able to do simple programming in *Mathematica*. If you choose to study Mathematica deeply, that is good, but is not required for the study of this book. These notes can help you decide how to express ideas in a manner that is consistent with accepted *Mathematica* programming style.

Confusing Alternatives

Return versus Enter

In *Mathematica* these two keys are not equivalent. RETURN serves as the "newline" character within a cell. It causes no action. No cell gets executed when RETURN is pressed. ENTER, when pressed while the insertion point is in a cell or when one or more cells are selected, initiates the execution of the cell or cells.

Set versus SetDelayed ["=" vs. ":="]

Throughout this text and the notebooks in most instances `Set [=]` has been used rather than `SetDelayed [:=]` for one simple reason: `Set` causes the defined expression to be displayed upon execution so that it can be checked for errors. If an expression is defined using `SetDelayed`, it must be manually displayed to check for errors. Here is an example:

```
In[1]:=  z = (x+1)/(x^2+1)

Out[1]=   x + 1
         ------
              2
         1 + x

In[2]:=  z := (x+1)/(x^2+1)
```

Note that nothing was displayed from the last evaluation. To check your definition, ask to see `z`. Its value is displayed for you to check.

```
In[3]:=  z

Out[3]=   x + 1
         ------
              2
         1 + x
```

There are more profound distinctions than the one just illustrated between `Set` and `SetDelayed`. Here is a brief summary:

Use SetDelayed If:

- The expression contains parameters whose values can subsequently change, such as `a`, `b`, and `c` in `p[x_]:= a x^2 + b x + c`;
- Evaluation at the time of definition is inappropriate, as in `g[y_] := D[y, x]`; (`g[y_] = D[y, x]` would cause `g[`y`_] = 0`, because $\partial y/\partial x = 0$.)
- When defining a function that is a `Module` or a `Block`;
- When defining a function that is to be executed conditionally `[/;]`.

Use Set If:

- You want to check your expression at the time it is defined;
- It is appropriate to evaluate the expression at the time of definition, such as `g[x_, y_] = y'[x]`;
- Subsequent calculations such as plotting would be adversely affected. (An expression that contains a delayed derivative would require that the derivative be formally calculated as well as evaluated at each point to be plotted.)

Set versus Equal

`[``='' vs. ``=='']`. Here the actions are completely different. `Set` is the action commonly called "substitution." `Equal` performs a test for equality. It is a boolean function—it normally returns either `True` or `False`. When `Equal` cannot determine whether an express is `True` or `False`, it returns the expression itself.

Simplify versus Expand

It is always appropriate to `Simplify` an expression, but `Simplify` may attempt to use more powerful techniques than are required. As a result when `Expand` does apply, it is much faster than `Simplify`. `Expand` does not work across both sides of an equation, whereas `Simplify` does. Often using `Simplify` before integrating speeds finding the integral.

`x` versus `x_`

When should `x_` be used? The symbol `x_`, pronounced "x-Blank," denotes a "pattern" named "x" that matches any expression; the symbol `x` is a literal, and matches only "x" and nothing else. For example: If `f[x] = x+2`, then `f[3]` and `f[a+b]` will not be evaluated, but `f[x]` will be evaluated to `x+2`. However, if `f[x_]=x+2`, then `f[3]` is 5, `f[a+b]` is `a+b+2`, and `f[x]` is `x+2`.

`y` versus `y[x_]`

When should arguments be used when an expression is named? Basically, the answer is: when it is convenient to do so. For example: If `y = Sin[x+Pi]`, then `Integrate[y, x]` is straight forward, but `Integrate[y, t]` produces `Sin[x+Pi] t`. Furthermore, using `y/.x->Pi/4` to produce `Sin[Pi/4+Pi]` and `D[y, x]/.x->Pi/4` to produce `Cos[Pi/4+Pi]` are not simple.

However, if `y[x_]` = `Sin[x+Pi]`, then `Integrate[y[x], x]` does what you expect, as does `Integrate[y[t], t]`. In addition, `Integrate[y[x],t]` produces `Sin[x+Pi] t`, and `y[Pi/4]` produce `Sin[Pi/4+Pi]`, and `y'[Pi/4]` produces `Cos[Pi/4+Pi]` as they should. These are all simple and easy to understand.

The secret is to **use an expression the way you defined it**. If it is defined with no arguments, refer to it without arguments. If it is defined with arguments use arguments when referring to it.

`x` versus `X`

Mathematica is case sensitive. The variable `x` (lower case) is different from the variable `X` (upper case). Be alert to this difference.

Symbols Used for Grouping

`( ... )` Groups expressions to clarify (or establish) meaning.

`[ ... ]` Used to delimit function arguments. This is the only use of these symbols. The arguments of every function must be delimited by brackets. These symbols are sometimes called "square brackets."

`{ ... }` Used to form lists. This is the only use of these symbols. There are many interpretations for lists in *Mathematica*. You will learn several such interpretations. For example:

`{t, a, b}` denotes a variable t with $a \leq t \leq b$. This usage occurs in definite integrals and to specify ranges for plotting.

`{f1, f2, f3}` can be a vector or a set, depending on whether or not order is important. Be alert to the context. That may help you determine whether or not order is important—the expression itself does not distinguish whether or not order is important.

`{{a, b}, {c, d}}` is a list of lists—a rectangular matrix or array.

Part and [[ ]]

When an expression is made up of several terms, factors, components, or other similar pieces, `Part` can be used to extract any subexpression. Subexpressions are numbered in a definite, but complicated way. The function `FullForm` can be used, along with the function `Position`, to explore the numbering scheme.

We will often need to extract a designated component from a `List` (vector). We do this by using `Part` as the examples below illustrate. `[[ ]]` is an abbreviation for `Part`. `Part[expr,n]` and `expr[[n]]` have the same meaning. Here are some examples of each.

```
In[4]:=  v={a, b, c, d}

Out[4]=  {a, b, c, d}

In[5]:=  Part[v,1]

Out[5]=  a

In[6]:=  v[[1]]

Out[6]=  a

In[7]:=  Part[v,3]

Out[7]=  c

In[8]:=  v[[3]]

Out[8]=  c
```

Observe that it is appropriate to think of `v[[n]]` as v_n.

Cells

Cells appear as] marks along the right margin of a *Mathematica* notebook window. Cells can be nested. They group related *Mathematica* "sentences." Before a cell can be executed, every expression in the cell must be syntactically correct (contain no errors). There is a distinctive marking at the top or bottom of every cell that tells you the attributes of that cell. Only "plain" cells are "active" and hence can be executed.

Suppressing Output

There are times when you do not want to make a calculation but not display the result. This is accomplished by terminating the expression with a semicolon "`;`." For example, `y=x+3;` evaluates `y`, but does not display the result.

One place where you should do this consistently is with `Plot`, `Plot3D`, `ParametricPlot`, `ParamtericPlot3D`, or `ContourPlot`. Each of these functions produces a graphics object as output unless it is suppressed. It is time-consuming to generate these graphics objects (they are not the plot images, but contain information about how the plot was constructed), and they have no value to us. We therefore suppress them.

Another instance where you may want to suppress output is when a large array is to be generated, but there is no compelling reason to display it. You may wish to look at the output for a small sample version to verify that it is correct, and then suppress the output of the larger version. Suppressing output does not affect the calculation—only the display.

Shorthand Notation

Symbol	Meaning	Example
`;`	Suppress output	`Plot[f[x],{x,0,1}];`
`/.`	`ReplaceAll`	`(a+b+c+d)/.p_+q_->p*q`
`//.`	`ReplaceRepeated`	`(a+b+c+d)//.p_+q_->p*q`
`->`	`Rule`	See above
`&`	`Function`	`Function[e,e^2+c]` ≡ `(#^2+c)&`
`#`	`Slot`	See above

Try the two replacement examples to see that they produce different results.

C

Mathematica Functions for Differential Equations

Mathematical Functions

Abs	`Abs[`z`]` gives the absolute value of the real or complex number z.
And	e_1 `&&` e_2 `&&` $\cdots$ is the logical AND function. It evaluates its arguments in order, giving `False` immediately if any of them are `False`, and `True` if they are all `True`.
ArcCos	`ArcCos[`z`]` gives the arc cosine of z.
ArcCosh	`ArcCosh[`z`]` gives the hyperbolic arc cosine of z.
ArcCot	`ArcCot[`z`]` gives the arc cotangent of the complex number z.
ArcCoth	`ArcCoth[`z`]` gives the hyperbolic arc cotangent of the complex number z.
ArcCsc	`ArcCsc[`z`]` gives the arc cosecant of the complex number z.
ArcCsch	`ArcCsch[`z`]` gives the hyperbolic arc cosecant of the complex number z.
ArcSec	`ArcSec[`z`]` gives the arc secant of the complex number z.
ArcSech	`ArcSech[`z`]` gives the hyperbolic arc secant of the complex number z.
ArcSin	`ArcSin[`z`]` gives the arc sine of the complex number z.
ArcSinh	`ArcSinh[`z`]` gives the hyperbolic arc sine of the complex number z.
ArcTan	`ArcTan[`z`]` gives the arc tangent of the complex number z. `ArcTan[x, y]` gives the arc tangent of y/x, taking into account which quadrant the point (x, y) is in.
ArcTanh	`ArcTanh[`z`]` gives the hyperbolic arc tangent of z.
Cos	`Cos[`z`]` gives the cosine of z.
Cosh	`Cosh[`z`]` gives the hyperbolic cosine of z.
Cot	`Cot[`z`]` gives the cotangent of z.
Coth	`Coth[z]` gives the hyperbolic cotangent of z.
Csc	`Csc[`z`]` gives the cosecant of z.
Csch	`Csch[`z`]` gives the hyperbolic cosecant of z.

E	E is the exponential constant e (base of natural logarithms), with numerical value 2.71828....
Exp	`Exp[`z`]` is the exponential function.
FindRoot	`FindRoot[`$lhs == rhs$`,{`x, x_0`}]` gives the numerical solution to the equation `lhs==rhs` which is "near" x_0.
Infinity	`Infinity` is a symbol that represents a positive infinite quantity.
Log	`Log[`z`]` gives the natural logarithm of z (logarithm to base `E`). `Log[b, z]` gives the logarithm to base b.
Max	`Max[`$x_1, x_2, \ldots$`]` yields the numerically largest of the x_i. `Max[{`$x_1, x_2, \ldots$`}, {`$y_1, \ldots$`},...]` yields the largest element of any of the lists.
Min	`Min[`$x_1, x_2, \ldots$`]` yields the numerically smallest of the x_i. `Min[{`$x_1, x_2, \ldots$`}, {`$y_1, \ldots$`},...]` yields the smallest element of any of the lists.
Not	`!expr` is the logical NOT function. It gives False if expr is True, and True if it is False.
Or	$e_1 \parallel e_2 \parallel \ldots$ is the logical OR function. It evaluates its arguments in order, giving `True` immediately if any of them are `True`, and `False` if they are all `False`.
Pi	`Pi` is π, with numerical value 3.14159....
Power	`a^ b` gives a to the power b.
Sec	`Sec[`z`]` gives the secant of z.
Sech	`Sech[`z`]` gives the hyperbolic secant of z.
Sign	`Sign[`x`]` gives -1, 0, or 1 depending on whether the real number x is negative, zero, or positive.
Sin	`Sin[`z`]` gives the sine of z.
Sinh	`Sinh[`z`]` gives the hyperbolic sine of z.
Sqrt	`Sqrt[`z`]` gives the square root of z.
Tan	`Tan[`z`]` gives the tangent of z.
Tanh	`Tanh[`z`]` gives the hyperbolic tangent of z.

Manipulation

Cancel	Remove common factors from a fraction.
Clear	`Clear[`$symbol_1, symbol_2, \ldots$`]` clears values defined for the $symbol_i$. `Clear["`$form_1$`," "`$form_2$`,"...]` clears values for all symbols whose names textually match any of the $form_i$.
Coefficient	`Coefficient[expr, form]` gives the coefficient of form in the polynomial expr. `Coefficient[expr, form, n]` gives the coefficient of $form$^n in *expr*.
CoefficientList	`CoefficientList[poly, var]` gives a list of coefficients of powers of *var* in *poly*, starting with power 0. `CoefficientList [poly, {`$var_1, var_2, \ldots$`}]` gives a matrix of coefficients of the var_i.
ColumnForm	`ColumnForm[matrix]` or `matrix//ColumnForm` prints the elements of matrix as a single column.
Collect	`Collect[expr, x]` collects together terms involving the same power of x. `Collect[expr, {`$x_1, x_2, \ldots$`}]` collects together terms that involve the same powers of $x_1, x_2, \ldots$.

Complement `Complement[`$list, list_1, list_2, \ldots$`]` gives the elements of list that are not in any of the $list_i$.

ComplexExpand `ComplexExpand[expr]` gives `Complex[`$expr_1, expr_2$`]` such that $expr = expr_1 + I expr_2$. Any variables are considered to be real.

Equal `lhs == rhs` tests whether or not lhs and rhs are equal.

Expand `Expand[expr]` expands out products and positive powers in $expr$.

Factor `Factor[poly]` factors a polynomial over the integers. `Factor[poly, Modulus->p]` factors a polynomial modulo a prime p.

Factorial `n!` gives the factorial of n.

FactorList `FactorList[poly]` gives a list of the factors of a polynomial, together with their exponents.

False `False` is the symbol for the Boolean value *false*.

Flatten `Flatten[list]` flattens out nested lists. `Flatten[list, n]` flattens to level n.

FullForm `FullForm[expr]` or `expr//FullForm` prints as the full form of $expr$, with no special syntax.

Function `Function[body]` or `body&` is a pure function.

Intersection `Intersection[`$list_1, list_2, \ldots$`]` gives a sorted list of all the distinct elements that are in common with all of the $list_i$.

Join `Join[`$list_1, list_2, \ldots$`]` concatenates the lists together.

List $\{e_1, e_2, \ldots\}$ is a list of elements.

LogicalExpand `LogicalExpand[expr]` expands out expressions containing connectives such as `&&` and `||`.

MatrixForm `MatrixForm[expr]` or `expr//MatrixForm` causes $expr$ to print out like a matrix.

N `N[expr]` gives the numerical value of $expr$. `N[expr, n]` does computations to n-digit precision.

Null `Null` is a symbol used to indicate the absence of an expression or a result. When it appears as an output expression, no output is printed.

Part `Part[expr,i]` or `expr[[i]]` is the ith element of $expr$.

Partition `Partition[list, n]` partitions list into nonoverlapping sublists of length n.

Product `Product[f, {i, imax}]` evaluates the product of f with i running from 1 to $imax$. `Product[f, {i, imin, imax}]` starts with $i = imin$. `Product[f, {i, imin, imax, di}]` uses steps di. `Product[f, {i, imin, imax}, {j, jmin, jmax}, ...]` evaluates a multiple product.

Random `Random[ ]` gives a uniformly distributed pseudorandom Real in the range 0 to 1. `Random[type, range]` gives a pseudorandom number of the specified type, lying in the specified range. Possible types are: Integer, Real, and Complex. The default range is 0 to 1. You can give the range {min, max} explicitly; a range specification of max is equivalent to {0, max}.

Range `Range[imax]` generates the list $\{1, 2, \ldots, imax\}$. `Range[imin, imax]` generates the list $\{imin, \ldots, imax\}$. `Range[imin, imax, di]` uses step di.

ReplaceAll	`expr /. rules` applies a rule or list of rules to *expr*.
ReplaceRepeated	`expr //. rules` applies a rule or list of rules to *expr* until there are no further changes.
Reverse	`Reverse[expr]` gives the elements of *expr* in reverse order.
Rule	`lhs -> rhs` represents a rule that transforms *lhs* into *rhs*.
RuleDelayed	`lhs :> rhs` represents a rule that transforms *lhs* into *rhs*.
Simplify	`Simplify[expr]` performs a sequence of algebraic and trigonometric transformations on *expr*, and returns the simplest form it finds. `Simplify [expr, Trig->False]` does not perform trigonometric simplification.
True	`True` is the symbol for the Boolean value *true*.
TrueQ	`TrueQ[expr]` yields `True` if expr is `True`, and yields `False` otherwise.
Union	`Union[`$list_1, list_2, \ldots$`]` gives a sorted list of all the distinct elements that appear in any of the $list_i$. `Union[list]` gives a sorted version of a list, in which all duplicated elements have been dropped.
Variables	`Variables[poly]` gives a list of all independent variables in a polynomial.
WorkingPrecision	`WorkingPrecision` is an option for various numerical operations which specifies how many digits of precision should be maintained in internal computations.

Graphics

$DisplayFunction	`$DisplayFunction` gives the default setting for the option `DisplayFunction` in graphics functions.
AspectRatio	`AspectRatio` is an option for `Show` and related functions which specifies the ratio of height to width for a plot.
ContourPlot	`ContourPlot[f[x, y], {x, xmin, xmax}, {y, ymin, ymax}, options]` represents a three-dimensional graphical image as a collection of contours. The number or separation of the contours can be specified.
Graphics	`Graphics[primitives, options]` represents a two-dimensional graphical image.
ListPlot	`ListPlot[{`$y_1, y_2, \ldots$`}]` plots a list of values. The x coordinates for each point are taken to be 1, 2,.... `ListPlot[{{`x_1, y_1`}, {`x_2, y_2`}, ...}]` plots a list of values with specified x and y coordinates.
ParametricPlot	`ParametricPlot[{`f_x, f_y`}, {`$t, tmin, tmax$`}]` produces a parametric plot with x- and y-coordinates f_x and f_y generated as a function of t.
ParametricPlot3D	`ParametricPlot3D[{`f_x, f_y, f_z`}, {t, tmin, tmax}]` produces a three-dimensional space curve parameterized by the variable t which runs from *tmin* to *tmax*. `ParametricPlot3D[{fx, fy, fz}, {t, tmin, tmax}, {u, umin, umax}]` produces a three-dimensional surface parameterized by t and u.

Plot	`Plot[f, {x, xmin, xmax}]` generates a plot of f as a function of x from $xmin$ to $xmax$. `Plot[{`$f_1, f_2, \ldots$`}, {x, xmin, xmax}]` plots several functions f_i.
PlotRange	`PlotRange` is an option for graphics functions that specifies what points to include in a plot.
Show	`Show[graphics, options]` displays two- and three-dimensional graphics, possibly with options changed. `Show[`$g_1, g_2, \ldots$`]` shows several plots combined.

Linear Algebra

Det	`Det[m]` gives the determinant of the square matrix m.
DiagonalMatrix	`DiagonalMatrix[list]` gives a diagonal matrix having as diagonal the entries of the list.
Dimensions	`Dimensions[a]` gives the dimensions of a if a is a matrix. Handles more than two dimensions.
Dot	`a.b` gives the ordinary dot product of a and b. Inner is more general.
Eigensystem	`Eigensystem[m]` gives a list {values, vectors} of the eigenvalues and eigenvectors of the square matrix m.
Eigenvalues	`Eigenvalues[m]` gives a list of the eigenvalues of the square matrix m.
Eigenvectors	`Eigenvectors[m]` gives a list of the eigenvectors of the square matrix m.
Eliminate	`Eliminate[sys, vars]` eliminates $vars$ from the equations sys.
IdentityMatrix	`IdentityMatrix[n]` gives an $n \times n$ identity matrix.
Inverse	`Inverse[matrix]` gives the inverse of the matrix if it exists; an error message is returned otherwise.
LinearSolve	`LinearSolve[m, b]` gives a vector x which solves the matrix equation `m.x==b`.
MatrixExp	`MatrixExp[mat]` gives the matrix exponential of the square matrix mat.
MatrixPower	`MatrixPower[mat,n]` gives the nth power of the square matrix mat.
NullSpace	`NullSpace[m]` gives a list of vectors that forms a basis for the null space of the matrix m.
RowReduce	`RowReduce[m]` gives the row-reduced form of the matrix m.
Solve	`Solve[eqns, vars]` attempts to solve an equation or set of equations for the variables $vars$. `Solve[eqns, vars, elims]` attempts to solve the equations for $vars$, elmininating the variables $elims$.
Transpose	`Transpose[matrix]` gives the transpose of the matrix.

Differential Equations and Calculus

Apart	`Apart[expr]` rewrites a rational expression as a sum of terms with minimal denominators. `Apart[expr, var]` treats all variables other than var as constants.

Apply — `Apply[f, expr]` replaces the head of *expr* by f. Example: `Apply[f, {a, b, c}]` produces `f[a, b, c]`.

D — `D[f, x]` gives the partial derivative of f with respect to x. `D[f, {x, n}]` gives the nth partial derivative with respect to x. `D[f,`$x_1, x_2, \ldots$`]` gives a mixed derivative.

Derivative — `f'` represents the derivative of f. `f''` the second derivative of f. `Derivative[`$n_1, n_2, \ldots, n_k$`][f]` is the general kth partial derivative of f.

DSolve — `DSolve[eqn,y[x],x]` solves differential equations for the function `y[x]` with independent variable x. `DSolve[{`eqn_1, $eqn_2, \ldots$`}, {`$y_1[x], y_2[x], \ldots$`}, {`$x_1, x_2, \ldots$`}]` solves a system of differential equations.

Integrate — `Integrate[`f, x`]` gives the indefinite integral of f with respect to x. `Integrate[f, {x, xmin, xmax}]` gives the definite integral. `Integrate[f, {x, xmin, xmax}, {y, ymin, ymax}]` gives a multiple integral.

Limit — `Limit[expr, x->x`$_0$`]` finds the limiting value of *expr* when x approaches x_0.

NDSolve — `NDSolve[eqn,y,{x,xmin,xmax}]` finds a numerical solution to the differential equations for the function y with independent variable x in the range *xmin* to *xmax*. The result is an InterpolatingFunction object. `NDSolve[eqns, {`$y_1, y_2, \ldots$`}, {x, xmin, xmax}]` finds numerical solutions for the functions y_i.

NIntegrate — `NIntegrate[expr, {`x, x_0, x_1`}]` gives the numerical value of the integral of *expr* over the interval $[x_0, x_1]$. Adaptive quadrature is used.

O — `O[x]^`n represents a term of order x^n. `O[x]^`n is generated to represent omitted higher-order terms in a power series. $O[x, x_0]$^n represents a term of order $(x - x_0)$^n.

Series — `Series[f, {`x, x_0, n`}]` generates a power series expansion for f about the point $x = x_0$ to order $(x - x_0)$^n. `Series[f, {`x, x_0, n_x`}, {`y, y_0, n_y`}]` successively finds series expansions with respect to y, then x.

SeriesData — `SeriesData[`x, x_0`, {`$a_0, a_1, \ldots$`}, `$n_{\min}, n_{\max}, den$`]` represents a power series in the variable x about the point x_0. The a_i are the coefficients in the power series. The powers of $(x - x_0)$ that appear are $n_{\min}/den, (n_{\min+1})/den, \ldots, n_max/den$.

Together — `Together[expr]` puts *expr* over a common denominator and cancels any common factors.

Programming in *Mathematica*

Blank — `Blank[x]` or `x_` stands for an expression to be referenced by x.

Do — `Do[expr, {imax}]` evaluates *expr imax* times. `Do[expr, {i, imax}]` evaluates *expr* with the variable i successively taking on the values 1 through *imax* (in steps of 1). `Do[expr, {i, imin, imax}]` starts with $i = imin$. `Do[expr, {i, imin, imax, di}]` uses steps *di*. `Do[expr, {i, imin, imax}, {j, jmin, jmax}, ...]` evaluates *expr* looping over different values of j, etc., for each i.

Function — `Function[body]` or `body&` is a pure function. The formal parameters are # (or #1), #2, etc. `Function[x, body]` is a pure function with a single formal parameter x. `Function[{`x_1, x_2`, ...}, body]` is a pure function with a list of formal parameters.

Get — `<<filename` reads in a file, evaluating each expression in it, and returning the last one.

If — `If[condition, t, f]` gives t if condition is `True`, and f if it is `False`. `If[condition, t, f, u]` gives u if condition is neither `True` nor `False`.

Map — `Map[f,expr]` applies f to each element of *expr*.

Module — `Module[{`$x_1, x_2, \ldots$`}, expr]` specifies that *expr* is to be evaluated with the symbols x_i taken as local variables. `Module[{`$x_1 = v_1, \ldots$`}, expr]` defines initial values for local variables.

Needs — `Needs[context]` loads a file if the specified context is not already in `$ContextPath`.

NestList — `NestList[`$f, expr, n$`] -> {expr, f[`*expr*`], f[f[`*expr*`]], ...` n `times}`.

Normal — `Normal[expr]` converts *expr* to a normal (not series) expression.

Print — `Print[`$expr_1, expr_2, \ldots$`]` prints the $expr_i$, followed by a newline (line feed).

Return — `Return[expr]` returns the value *expr* from a function. `Return[ ]` returns the value `Null`.

Rule — `lhs -> rhs` represents a rule that transforms *lhs* to *rhs*.

RuleDelayed — `lhs :> rhs` represents a rule that transforms *lhs* to *rhs*, evaluating *rhs* only when the rule is used.

SameQ — `SameQ[`$expr_1, expr_2, \ldots$`]` yields `True` if all the $expr_i$ are identical, and yields `False` otherwise.

Select — `Select[list, criterion]` picks out all elements e_i of the list for which `criterion[`e_i`]` is `True`.

Set — `lhs = rhs` evaluates *rhs* and assigns the result to be the value of *lhs*. *lhs* is then replaced by *rhs* whenever it appears. `{`l_1, l_2`, ...} = {`$r_1, r_2, \ldots$`}` evaluates the r_i, and assigns the results to be the values of the corresponding l_i.

SetDelayed — `lhs := rhs` assigns *rhs* to be the delayed value of *lhs*. *rhs* is maintained in an unevaluated form. When *lhs* appears, it is replaced by *rhs*, evaluated afresh each time.

Sort — `Sort[list]` puts the elements of list in order.

Sum — `Sum[f,{i, imax}]` evaluates the sum of f with i running from 1 to *imax*. `Sum[f, {i, imin, imax}]` starts with $i = imin$. `Sum[f, {i, imin, imax, di}]` uses steps *di*. `Sum[f, {i, imin, imax}, {j, jmin, jmax}, ...]` evaluates a multiple sum.

Table — `Table[expr, {imax}]` generates a list of *imax* copies of *expr*. `Table[expr, {i, imax}]` generates a list of the values of *expr* when i runs from 1 to *imax*. `Table[expr, {i, imin, imax}]` starts with $i = imin$. `Table[expr, {i, imin, imax, di}]` uses steps *di*. `Table[expr, {i, imin, imax},`

	`{j, jmin, jmax}, ...]` gives a nested list. The list associated with i is outermost.
TableForm	`TableForm[list]` prints with the elements of list arranged in tabular form. `TableForm[list, d]` puts elements down to level d in list in tabular form.
Unequal	`lhs != rhs` returns `False` if *lhs* and *rhs* are identical.
While	`While[test, body]` evaluates *test*, then *body*, repetitively, until *test* first fails to give `True`.

Text-Specific Packages in *Mathematica*

Classify.m
DKernel.m
DSolveCauchyEulerSystem.m
DSolveOrdinarySeries.m
GenerateRSPSeries.m
grad.m
LPT.m
BasicLaplaceTransform.m
LaplaceTransform.m

Bibliography

Abramowitz, Milton and I. Stegun. *Handbook of Mathematical Functions.* Dover, New York, 1965.

Asimov, Isaac. *Asimov's Chronology of Science and Discovery.* Harper & Row, New York, 1989.

Ayres, Frank, Jr. *Theory and Problems of Differential Equations.* Schaum's Outline. McGraw-Hill, New York, 1952.

Birkhoff, Garrett and G.-C. Rota. *Ordinary Differential Equations.* Blaisdell, Waltham, MA, 2nd edition, 1969.

Boyce, W. E. and R. C. DiPrima. *Elementary Differential Equations.* Wiley, New York, 4th edition, 1986.

Burton, David M.. *The History of Mathematics: An Introduction.* Allyn & Bacon, Newton, MA, 1984.

Christiansen, Gale E.. *In the Presence of the Creator: Isaac Newton and His Times.* The Free Press, New York, 1984.

Coddington, E. A. and N. Levinson. *Theory of Ordinary Differential Equations.* McGraw-Hill, New York, 1955.

Davenport, J. H., Y. Siret, and E. Tournier. *Computer Algebra: Systems and Algorithms for Algebraic Computation.* Academic Press, New York, 1988.

Fagg, S. V. *Differential Equations.* English Universities Press, London, 1956.

Ford, L. R. *Differential Equations.* McGraw-Hill, New York, 2nd edition, 1955.

Forrester, Jay W. *World Dynamics.* Wright-Allen, Cambridge, MA, 1973.

Fraleigh, J. B. and R. A. Beauregard. *Linear Algebra.* Addison-Wesley, Reading, MA, 2nd edition, 1989.

Friedlander, Gerhart, J. W. Kennedy, E. S. Macias, and J. M. Miller. *Nuclear and Radiochemistry.* Wiley, New York, 3rd edition, 1981.

Geyh, M. A. and H. Schleicher. *Absolute Age Determination: Physical and Chemical Dating Methods and Their Application.* Springer-Verlag, New York, 1990.

Gillispie, C. C., editor. *Dictionary of Scientific Biography.* Scribner's, New York, 1970–1980.

Gleick, J. *Chaos.* Penguin, New York, 1987.

Goldberg, J. L. and A. J. Schwartz. *Systems of Ordinary Differential Equations: An Introduction.* Harper & Row, New York, 1972.

Gray, T. W. and J. Glynn. *Exploring Mathematics with Mathematica.* Addison-Wesley, Reading, MA, 1991.

Hairer, E., C. Lubich, and M. Roche. *The Numerical Solution of Differential-Algebraic Systems by Runge-Kutta Methods.* Lecture Notes in Mathematics. Springer-Verlag, New York, 1989. No. 1409.

Hairer, E., S. P. Nørsett, and G. Wanner. *Solving Ordinary Differential Equations I: Nonstiff Problems.* Springer-Verlag, New York, 1987.

Hethcote, H. W. Qualitative analyses of communicable disease models. *Math. Biosci.*, 28:335–356, 1976.

Hille, E. *Lectures on Ordinary Differential Equations.* Addison-Wesley, Reading, MA, 1969.

Hurewicz, W. *Lectures on Ordinary Differential Equations.* M.I.T. Press, Cambridge, MA, 1958.

Ince, E. L. *Ordinary Differential Equations.* Dover, New York, 1956.

Kamke, E. *Differential Gleichungen: Lösungsmethoden und Lösungen.* Akademische Verlagsgesellsschaft Geest & Portig K.-G., Leipzig, 1956.

Lang, Serge. *Calculus of Several Variables.* Addison-Wesley, Reading, MA, 2nd edition, 1979.

Lazer, A. C. and P. J. McKenna. Large-amplitude periodic oscillations in suspension bridges: Some connections with nonlinear analysis. *SIAM Review*, 32(4):537–578, 1990.

Leigh, E. R. *The Ecological Role of Volterra's Equations: Some Problems in Mathematical Biology.* American Mathematical Society, Providence, RI, 1968.

Leighton, Walter. *An Introduction to the Theory of Differential Equations.* McGraw-Hill, New York, 1952.

Lotka, A. J. *Elements of Mathematical Biology.* Dover, New York, 1956. This is a reissue of a 1925 work.

Maeder, Roman. *Programming in Mathematica.* Addison-Wesley, Reading, MA, 1989.

Martin, W. T. and E. Reissner. *Elementary Differential Equations.* Addison-Wesley, Reading, MA, 1956.

Miller, Richard H. Experimenting with galaxies. *American Scientist*, 80(2):152–163.

Morgan, Frank. Calculus, planets, and general relativity. *SIAM Review*, 34(2):295–299, 1992.

Nitecki, Z. H. and M. M. Guterman. *Differential Equations with Linear Algebra.* Saunders, Philadelphia, 1985.

Osserman, R. *Two-Dimensional Calculus.* Harcourt, 1968. Reprinted by Krieger.

Pennisi, Louis L. *Elements of Ordinary Differential Equations.* Holt, New York, 1972.

Rabenstein, A. L. *Elementary Differential Equations.* Academic, New York, 3rd edition, 1982.

Rainville, E. D. *Intermediate Course in Differential Equations.* Wiley, New York, 1943.

Rainville, E. D. *Elementary Differential Equations.* Macmillan, New York, 1952.

Rapoport, Anatol. Contributions to the mathematical theory of mass behavior: I. the propagation of single acts. *Bull. Math. Biophys.*, 14:159–169, 1952.

Richardson, Lewis F. *Arms and Insecurity.* Boxwood Press, Pittsburgh, 1960.

Richardson, Lewis F. *Statistics of Deadly Quarrels.* Boxwood Press, Pittsburgh, 1960.

Ross, C. C. *Singular Differential Systems.* Ph.D. thesis, University of North Carolina, Chapel Hill, 1964.

Ross, C. C. Why the method of undetermined coefficients works. *American Mathematical Monthly*, 98(8), 1991.

Ross, S. L. *Introduction to Differential Equations.* Wiley, New York, 2nd edition, 1974.

Sánchez, D. A. *Ordinary Differential Equations and Stability Theory: An Introduction.* Freeman, San Francisco, 1968.

Schmidt, Frank and Rodica Simion. Problem 91-1: Even minus odd involutions in the symmetric group. *SIAM Review*, 34(2): 315–317, 1992. April 1992. Solution by Robert B. Israel.

Simmons, G. F. *Differential Equations: With Applications and Historical Notes.* McGraw-Hill, New York, 1972.

Spiegel, Murray R. *Applied Differential Equations.* Prentice-Hall, Englewood Cliffs, NJ, 3rd edition, 1981.

Spiegel, Murray R. *Mathematical Handbook of Formulas and Tables.* Schaum's Outline Series. McGraw-Hill, New York, 1991.

Temple, Blake and C. A. Tracy. From Newton to Einstein. *Amer. Math. Monthly*, 99(6):507–521, 1992.

Thomas, George B., Jr., and Ross L. Finney. *Calculus and Analytic Geometry.* Addison-Wesley, Reading, MA, 8th edition, 1992.

Tudor, David. Modeling the effect of public health campaigns on the spread of aids. *SIAM Review*, 34(2):300–303, 1992.

Turnbull, H. W. *Newton: Correspondence.* Cambridge, England, 1959. Published for the Royal Society at the University, V1–V7, 1959–1977.

Wagon, Stan. *Mathematica in Action.* Freeman, New York, 1991.

Wasow, Wolfgang. *Asymptotic Expansions for Ordinary Differential Equations.* Dover, New York, 1987. (1965, 1976).

Wiggins, Stephen. *Introduction to Applied Nonlinear Dynamical Systems and Chaos.* Texts in Applied Mathematics, 2nd edition, Springer-Verlag, New York, 1990.

Wilcox, L. R. and H. J. Curtis. *Elementary Differential Equations.* International, Scranton, PA, 2nd edition, 1966.

Wolfram, Stephen. *Mathematica: A System for Doing Mathematics by Computer.* Addison-Wesley, Reading, MA, 2nd edition, 1991.

Zill, D. G. *A First Course in Differential Equations with Applications.* PWS-Kent, Boston, MA, 4th edition, 1988.

Zwilliger, Daniel. *Handbook of Differential Equations.* Academic Press, New York, 2nd edition, 1992.

Index of *Mathematica* Commands

Index